The Evolution of Cetacean Societies

The Evolution of Cetacean Societies

Uncovering the Social Complexity of Whales and Dolphins

EDITED BY DARREN P. CROFT,
ANDREW D. FOOTE, ELLEN C. GARLAND,
AND STEPHANIE L. KING

The University of Chicago Press
Chicago and London

Editors are listed in alphabetical order and all contributed equally.

The University of Chicago Press, Chicago 60637
The University of Chicago Press, Ltd., London

Published 2026
Printed in the United States of America

35 34 33 32 31 30 29 28 27 26   1 2 3 4 5

ISBN-13: 978-0-226-84055-0 (cloth)
ISBN-13: 978-0-226-84057-4 (paper)
ISBN-13: 978-0-226-84056-7 (ebook)
DOI: https://doi.org/10.7208/chicago/9780226840567.001.0001

This publication was supported by funding from the Northwest Fisheries Science Center (NOAA Fisheries, USA).

IFAW—International Fund for Animal Welfare

University of St. Andrews, School of Biology, Scottish Oceans Institute

Library of Congress Cataloging-in-Publication Data

Names: Croft, Darren P. editor | Foote, Andrew D. editor | Garland, Ellen C. editor | King, Stephanie (Lecturer in animal communication) editor
Title: The evolution of cetacean societies : uncovering the social complexity of whales and dolphins / edited by Darren P. Croft, Andrew D. Foote, Ellen C. Garland, and Stephanie L. King.
Description: Chicago : The University of Chicago Press, 2026. | Includes bibliographical references and index.
Identifiers: LCCN 2025020912 | ISBN 9780226840550 cloth | ISBN 9780226840574 paperback | ISBN 9780226840567 ebook
Subjects: LCSH: Cetacea—Behavior
Classification: LCC QL737.C4 E96 2026 | DDC 599.5—dc23/eng/20250709
LC record available at https://lccn.loc.gov/2025020912

♾ This paper meets the requirements of ANSI/NISO Z39.48-1992 (Permanence of Paper).

Authorized Representative for EU General Product Safety Regulation (GPSR) queries: **Easy Access System Europe**—Mustamäe tee 50, 10621 Tallinn, Estonia, gpsr.requests@easproject.com
Any other queries: https://press.uchicago.edu/press/contact.html

Contents

Introduction

DARREN P. CROFT, ANDREW D. FOOTE,
ELLEN C. GARLAND, AND
STEPHANIE L. KING

Cetaceans are elusive, difficult to study mammals that live in an alien environment to humans. Due to their aquatic lifestyle, we are only able to catch a small glimpse of a limited portion of their lives. Yet, despite these limitations, cetaceans are profoundly relevant in any comparative understanding given our shared mammalian heritage. Like us, cetaceans are highly social, and social complexity has long been thought of as a potential explanation for brain expansion, with peaks in mammalian brain evolution found in humans and other great apes, elephants, and cetaceans (Connor 2007). Interestingly, these mammalian groups are also characterized by the formation of structured, multilevel societies (hierarchically nested social levels), flexible communication systems, the presence of enduring, affiliative social relationships, and long-term social memory. Indeed, alongside apes and elephants, cetaceans are valuable comparative model systems for understanding human social and cognitive evolution. For example, recent work on intergenerational reproductive conflict in killer whales (Croft et al. 2017) has shed light on the evolution of menopause (found in humans and some cetaceans); studies of specialized dolphin societies (e.g., ecotypes) have found correlated cultural and genetic variation (Foote et al. 2016; Louis et al. 2021) that is also found in humans (Laland et al. 2010); similarities in the transmission of cultural behavior have been found between cetaceans and great apes (Allen et al. 2013; Arnon et al. 2025; Wild et al. 2020); and studies on cooperation among bottlenose dolphins have revealed that the formation of strategic intergroup alliances between unrelated individuals is not unique to humans (Connor et al. 2022). This powerful approach of examining cases of convergence between humans and cetaceans to understand social evolution is not limited to social complexity; there are also striking examples in the vocal domain. From group-level vocal dialects in killer whales and sperm whales that function as symbolic markers of clan identity (Ford 1991; Cantor et al. 2015) akin to ethnic markers in humans (Hersh et al. 2022), to bottlenose dolphins developing individual vocal labels analogous to a human name (Janik et al. 2006; King and Janik 2013; Bruck et al. 2022) that underpin social concepts (King et al. 2021); communication complexity is abundant in cetaceans. These strong parallels between the social lives of whales and dolphins and our own societies have led scientists and the public alike to become fascinated with this enigmatic group of animals.

Our goal in writing *The Evolution of Cetacean Societies* is to showcase the significant advances in our understanding of the formation and maintenance of cetacean societies and the state-of-the-art methods used to unlock these insights. The establishment of new research fields, emergence of new technologies, and the novel techniques used to study cetaceans have allowed researchers to expand the depth and diversity of questions able to be explored. Over 20 years ago *Cetacean Societies* (Mann et al. 2000) presented the first comprehensive synthesis and review of the state of the field. It introduced us to the methodological, comparative, and theoretical context of cetacean social ecology and provided a comprehensive account of field studies on whales and dolphins. *Cetacean Societies* laid the foundations for a generation of young scientists by summarizing the existing knowledge at that time and lit the spark for future research by highlighting what might be possible in studies of these socially complex animals.

To understand where the field is now, it's worth reflecting on the substantial progress made over the last two decades. Photo identification (photo-ID) of uniquely marked individuals has been key to our understanding of the complexity of cetacean societies. Photo-ID was well established as a research method for studying cetaceans when *Cetacean Societies* was first published, yet it is perhaps useful to reflect that this now commonplace approach was a revolutionary advancement when it was first developed (Bigg et al. 1976; Whitehead and Payne 1978; Hammond et al. 1990), as evidenced by a publication in the prestigious journal *Science* (Würsig and Würsig 1977). At the time of the publication of *Cetacean Societies*, most researchers were still using black-and-white negatives and matching negatives to identification reference images by eye. Subsequently, during the early 2000s, a photographic revolution was happening, as digital cameras became more readily available and increasingly capable of producing the high-resolution images needed for individual identification of cetaceans from scars and notches on their bodies. For example, Nikon released the D1 camera the year prior to the publication of *Cetacean Societies*. Over the following years, digital cameras were embraced by marine mammal field scientists, yet this was not without initial resistance and an engaging discussion on the marine mammal (MARMAM) international mailing list in the mid-2000s, in which there was widespread skepticism toward the resolution offered by digital cameras. However, the field was quickly swayed by the arguments for the digital revolution—researchers were no longer restricted by the length of their photography film or the processing time in the darkroom when they returned to the shore. Long-term field studies saw a jump in the number of photos, which has opened the door for more detailed analysis of the social behavior of many whale and dolphin populations. Photographic data from "citizen scientists" also went through a boom due to the digitization of photography, increasing the number of photographs taken per encounter, and the ease of sending large numbers of images to researchers. For example, a photo-ID study on killer whales utilizing primarily opportunistic citizen science photographs reported higher certainty in mark-recapture estimates from 2005 onward, which the authors attributed to the increased use of digital cameras (Beck et al. 2014). Over this same time period, automated matching algorithms have been developed and increasingly implemented to automate photo-ID (Blount et al. 2022). Such approaches show promising results in terms of reducing the time spent matching photographs relative to matches done by eye, and the potential to compare datasets across study areas or across much larger population sizes, as found in some pelagic dolphin species.

As outlined in *Cetacean Societies*, the difficulty of observing cetacean behavior underwater means that studying whale and dolphin sociality in the wild poses some significant methodological and technological challenges. However, in the decades since *Cetacean Societies* was published, there have been dramatic advances in the approaches used to study whale and dolphin societies. Notably, these include technological advancements using unoccupied aerial vehicles (UAVs) and noninvasive sound and movement tags to study cetacean behavior, and computational advancements to analyze large field-based datasets, such as social network analysis. These state-of-the-art technologies have revolutionized the field of marine mammal behavior, providing unparalleled information on the fine-scale behavior of individuals in the wild (Johnson and Tyack 2003; Torres et al. 2018; King and Jensen 2023). UAVs, for example, provide an inexpensive platform that enables cetaceans to be observed for extended periods of time with minimal disturbance. Indeed, the use of UAVs has increased exponentially in recent years (fig. I.1) and is now widely used to address diverse questions, from measuring cetacean body condition to helping understand species resilience to changing ecological conditions (Christiansen et al. 2016, 2021) to collecting data on

FIGURE I.1. The methodological progression of documenting cetacean behavior in the wild. (*Left, top and bottom*) Traditional photo-ID methods used to study cetacean social behavior, where photographs provide information about which individuals spend time in close proximity. (*Middle, top and bottom*) Attempts to record the behavior of cetaceans subsurface led to the use of the first UAV blimps and helikites (blimp-shaped aerostats), which were tethered to the boat and difficult to maneuver. (*Right, top and bottom*) Recent advancements in drone technology mean that UAVs are now inexpensive and easy to deploy and provide high-resolution video or images, enabling cetaceans to be observed for extended periods of time with minimal disturbance.
Source: Photos by Simon J. Allen and Shark Bay Dolphin Research.

direct social interactions subsurface, which are revealing that cetacean social relationships are not being fully characterized by traditional proximity-based measures of association data alone (Weiss et al. 2021). UAVs have allowed us, for the first time, to document group hunting in what were previously thought to be solitary species (Torres Ortiz et al. 2021), food sharing between conspecifics (Ramos et al. 2021), and cetacean responses to sound playback experiments (Barluet de Beauchesne et al. 2022; King et al. 2021), where fine-scale behavioral changes in orientation and movement patterns are recorded even when animals are below the surface. By providing a window into their underwater world and a means to characterize subtle behavioral changes, UAVs have opened the door to researchers wanting to study cetacean social cognition in the wild.

UAVs are not the only technological development that has revolutionized the field of cetacean biology. High-resolution sound and movement biologging tags have allowed us to make some big leaps forward and take our observations below the water's surface, providing new insight into the acoustic and movement behavior of whales that spend the majority of their lives hundreds or thousands of meters below the water's surface (Johnson and Tyack 2003). Attaching tags to marine mammals has a long history, with the first biologging studies being carried out in the 1940s when whales and seals were fitted with depth gauges (Scholander 1940). By the 1990s, tags had developed significantly and were starting to provide information not just on water depth, but on a range of behavioral and environmental variables such as the position, speed, and movement of the animal; the acoustic environment (using tag-based hydrophones); and physiological data (e.g., heart rate). Over the last two decades, the technology for data collection tags on marine mammals has seen a rapid evolution—the tags have become smaller, carry more sensors, and have better transmission capabilities. While these high-resolution sound and movement biologging tags were initially invented to investigate how cetaceans respond to anthropogenic noise, particularly for understanding how deep-diving whales respond to navy sonar, they have also allowed us to investigate cetacean behavior and cognition in the wild, providing unique insight into how cetaceans keep track of each other and coordinate their behavior (King and Jensen 2023). In addition to the development of acoustic tags, advancements in the size and cost of hydrophone arrays have also meant that more researchers can attempt to localize vocalizations to individuals or groups and explore the relationship between observed physical behavior and vocal production, providing new insight into the function of vocal signals and the interaction between communication and social structure.

Just as the technology and tools for collecting data using tags, photos, and videos have developed considerably over the last two decades, so have the tools and software for processing and analyzing data. For example, there have been rapid advancements in machine learning and artificial intelligence, which are making call detection and classification easier and faster (Usman et al. 2020), supporting the analysis of the high-resolution data that we are now able to collect with these exciting new observational technologies. New computational techniques for analyzing social organization have opened up new opportunities to quantify the patterning of social interactions among individuals (Farine 2013; Whitehead 2009; Hart et al. 2023; Ross et al. 2024); and this ability to measure interspecific variation in social complexity is significantly expanding our understanding of the evolution of cetacean societies.

Culture is ubiquitous in human societies, yet it has only recently been recognized and become widely accepted that nonhuman animals have culture (Whiten 2021). Cetaceans have been at the forefront of this revolution in understanding the role of culture in animal societies (Rendell and Whitehead 2001; Whitehead et al. 2004; Whitehead and Rendell 2014). Culture was barely covered in *Cetacean Societies* but has now become a burgeoning field of research. From humpback whale songs (Garland et al. 2011), southern right whale migrations (Carroll et al. 2015), tool use in bottlenose dolphins (Mann et al. 2012; Wild et al. 2019), sophisticated hunting strategies of killer whales

(Srinivasan 2023), symbolic markers in sperm whale codas (Hersh et al. 2022) and many other examples, studies of cetaceans have changed our understanding of the evolution of behavioral variation.

Finally, advancements have also been made in genome sequencing; *Cetacean Societies* was published three years before the first draft of the human genome (International Human Genome Consortium 2004). The human genome project took 13 years to complete using the traditional Sanger method of sequencing small fragments of the genome, one at a time. The subsequent genomics revolution arose from the development of parallelized high-throughput sequencing technologies (initially referred to as next-generation sequencing or NGS), which unlocked the potential for simultaneously generating genomic sequences for multiple individuals in a few days. Scientists studying the ecology and evolution of cetaceans were quick to uptake this new technology (Cammen et al. 2016), being among the first wave of papers to utilize high-throughput sequencing, showcasing the technology by generating complete mitochondrial genomes for an unprecedentedly large global dataset of killer whales (Morin et al. 2010). The advent of the genomic era has given us an increasingly high-resolution reconstruction of the origins of some of the best-studied cetacean societies, including contrasting ecotypes of killer whales and bottlenose dolphins (Foote et al. 2016; Louis et al. 2021). These studies have also reported signatures of adaptation within the genome, providing new insights into the genetic response of cetacean societies to the selective regimes imposed by their preferred ecology and habitat. Thus, we now have the genomic tool kit to explore the interplay between culturally acquired ecological preferences and the associated genetic changes that optimize fitness within a preferred ecological niche.

Historically, studies of cetaceans often used cross-sectional data, sampling individuals at a single point in time. While these data have given many great insights into the life history, social structure, and reproductive biology of numerous species, it has a major shortcoming; it is not possible to track individuals through time, and thus it is not possible to assess the fitness costs and benefits of behavior or reproductive strategies. The importance of individual-based studies, where individually recognizable animals are tracked over time, was clearly recognized in *Cetacean Societies*, and indeed both the research methods for collecting long-term individual-based data and the insights this data had given featured prominently in the book. Thankfully, many of the long-term individual-based studies that featured in *Cetacean Societies* have continued and have celebrated their fortieth anniversary, with some now running for more than half a century. The maturation of these long-term datasets has unlocked our ability to address a wide range of research questions. For example, long-term studies make it possible to look at age effects on behavior and reproductive traits within individuals (Wiszniewski et al. 2012; Gerber et al. 2022). Understanding this variation at the level of the individual is of fundamental importance for assessing how ecological and evolutionary processes operate in natural populations. With considerable numbers of individuals and years of observation added to many of these long-term datasets, it has become possible to link data across different life history stages at the level of the individual. For example, one can ask what the effect of early-life conditions are on survival or behavior in later life (Stanton and Mann 2012; Holmes et al. 2024).

A key benefit of long-term individual-based data is that it allows researchers to study the dynamics of social relationships at the individual, group, community, and population levels. For example, in recent years these long-term data have been used to examine how social structure changes in response to changing ecological conditions (Foster et al. 2012), or how social structure is linked to individual survival and reproductive success (Ellis et al. 2017). The early pioneers who started many of the long-term individual-based studies in cetaceans did so to assess questions at the level of the population, e.g., how big is the population? As these datasets have matured, they have provided a platform for asking questions that require data spanning multiple generations. In the following chapters of this book, we will celebrate many of these key innovations over the last two decades and the discoveries they have led to.

Despite significant progress in our understanding of cetacean societies over the past two decades, many cetacean populations have continued to decline and remain threatened, balancing on a knife-edge of extinction (Beck et al. 2014; Crossman et al. 2024; Jaramillo-Legorreta et al. 2019; Williams et al. 2024). Indeed, since the publication of *Cetacean Societies* we have lost one species of cetacean; in 2006, the Yangtze River dolphin (*Lipotes vexillifer*) was declared functionally extinct (Turvey et al. 2007). There remains a diverse range of threats to cetacean societies, including anthropogenic noise, contaminants, declining prey resources, habitat degradation, ship strikes, inbreeding depression, and mortality through both bycatch and targeted hunting, to name a few. The scale of the conservation challenge is immense, magnified by the large geographic ranges of many cetacean societies. As a result, cetaceans increasingly face a myriad of existential threats, which can act in concert, impacting additively or multiplicatively upon the viability of endangered populations. By shining the spotlight on the strong parallels between human and cetacean societies, we hope to amplify the importance of conserving these enigmatic species with whom we share so many social and cognitive traits.

After examining other societies, it is only right to turn the spotlight on ourselves. Our own societies have become increasingly complex and interconnected over the past decade. When *Cetacean Societies* was being drafted, it was primarily among editors based in North America. In this digital age, as reflected in this volume, we are able to interact with colleagues around the globe on a daily basis. The key innovation underlying this connectivity is probably best introduced in an anecdote shared by Richard Connor, an editor of *Cetacean Societies*. The editors had spent time working on *Cetacean Societies* at the Centre for Advanced Study in the Behavioral Sciences at Stanford University. Richard recalled to us, "When I returned from the center to the University of Michigan for the '95–'96 academic year (or was it the year before?), one of my colleagues said, 'There is this really cool new way to send messages to each other on campus now; it's called *email*.'" The opportunities email and the internet have afforded each of us in our careers make it difficult to imagine how we would have fared without it. We reflect on the potential this increased connectivity has for positive change toward increased inclusivity within our society, that of scientists studying cetaceans.

The original *Cetacean Societies* was edited by three men and one woman and authored by nine men and four women. Only two of the 12 chapters had a female first author. Marine mammal science was dominated by senior, white men from Europe and North America. Today, the field is still skewed toward white men in senior positions, despite a higher proportion of women at the early career researcher (ECR) stage. While this has improved over the last 20 years, there is still a long way to go. One way to change this is to actively encourage diversity in writing teams, which we have strived to do for each chapter in this book. Gender diversity is just one of many challenges marine mammal science faces, similar to other fields (Johri et al. 2021). Racial diversity is rare; glance around many offices and you will see an array of almost exclusively white faces, and our LGBTQ colleagues still face prejudice and exclusion. Supporting, promoting, and recognizing barriers to inclusion in marine mammal science is essential to ensure we grow a diverse community with a variety of life experiences and viewpoints. There also remains a significant underrepresentation of researchers from the Global South; over 50% of papers published between 2013 and 2018 in *Marine Mammal Science* were from North America. Further, most conference locations for the Marine Mammal Society's flagship biennial conference have been in North America (80%), or in Europe (8%). If you live, work, or study in the Southern Hemisphere, the cost of presenting your work at the largest marine mammal conference in the world, and the potential to make new collaborations or networks, is astronomical. However, the world is currently changing post-COVID to provide a more inclusive and affordable conference environment. Another ongoing discussion is around unpaid internships, which are also present in other marine sciences (Osiecka et al. 2021). There are many issues to be addressed in our field; we have highlighted a few in the hopes that, by doing so, each of us reflects on how they can contribute to a better society. The eager ECRs we see entering our field give us hope that change is coming, little by little; it is our job to continue striving forward, breaking glass ceilings as we also reach down to pull up the next generation of scientists.

Structure of the Book

This book is divided into five parts, each representing an area of research. We do, however, want to stress the connection in concepts, theory, and techniques that span multiple parts and have endeavored to highlight these to the reader. Part 1 focuses on the evolutionary history and genetic composition of cetacean societies. The advent of next-generation sequencing in the 2000s has led to a genomic revolution in which high-quality reference genomes have been generated for an increasing number of cetacean species, and population genomic datasets are emerging as the new normal in the field. This technological advance has brought our understanding of the evolutionary histories of cetaceans into a new era, providing new insight into the genetic relationships among individuals, social groups, populations, ecotypes, and species. This section starts by exploring the early evolution of present-day cetaceans and their transition from land to water, the development of specialized feeding and vocal apparatus, and finally brain size and its relationship with the study of cognition. Genomes also provide unprecedented insight into the demographic and evolutionary histories of cetacean societies, which are ultimately shaped at the level of kin structure. Specifically, the role of philopatry, kinship and kin-based interactions within cetacean social groups, and how these roles vary across different ecological contexts and timescales.

Part 2 focuses on the socioecology of cetacean societies. Cetacean social structures vary in both size and complexity, ranging from hierarchically structured family groups to populations with high levels of fission-fusion dynamics. Over the last few decades, a combination of developments in data analysis techniques (particularly in the analysis of social networks), data collection methods (e.g., drones and acoustic tags) and the maturity of long-term individual-based field studies have opened up new insights into the evolution of social behavior in cetaceans. This part, and the chapters therein, explores how social relationships are structured in cetacean societies with a particular emphasis on the interaction between communication and social structure. Cetaceans are well-known for cooperating extensively in the wild, including allomaternal care, feeding strategies that require coordinated action and the formation of long-term cooperative alliances. However, conflicts invariably arise in societies built on cooperation, where individuals balance personal gain with the success of others in the group. Part 3 discusses key advances in our understanding of how cooperation and conflict have shaped cetacean societies, the behavioral processes used to maintain cooperative relationships, particularly among nonkin, and the mechanisms that individuals use to coordinate their cooperative endeavors.

Part 4 focuses on the cultures of cetacean societies. Studies of cetacean culture have emerged over the last two decades and provide unparalleled and unique examples in the field of animal culture. This field of study was just emerging during the initial book, and through the maturation of long-term datasets, and new technologies and analytical methods has become an integral part of the study of cetacean societies. This section explores the multiple facets of culture based on behavioral contexts (e.g., communication and foraging) and the subsequent conservation challenges that emerge from our increasing understanding of cetacean culture. Cetaceans have long held great significance in the lives of humans. From humans and cetaceans engaging in mutually beneficial foraging relationships to humans hunting cetaceans for food and other resources. We start part 5 of the book by reflecting on our own relationships with cetacean societies. We then turn our gaze inward, examining diversity and inclusivity in our own field. While advances in representation over the past decade are to be commended, we advocate for increased efforts to strive toward a society of marine mammal biologists that truly reflects global diversity.

References

Allen, Jenny A., Mason T. Weinrich, William Hoppitt, and Luke Rendell. 2013. "Network-Based Diffusion Analysis Reveals Cultural Transmission of Lobtail Feeding in Humpback Whales." *Science* 340 (6131): 485–88. https://doi.org/10.1126/science.1231976.

Arnon, Inbal, Simon Kirby, Jenny A. Allen, Claire Garrigue, Emma L. Carroll, and Ellen C. Garland. 2025. "Whale Song Shows Language-Like Statistical Structure." *Science* 387 (6734): 649–53. https://doi.org/10.1126/science.adq7055.

Barluet de Beauchesne, Lucie, Mathilde Massenet, Machiel G. Oudejans, Annebelle CM Kok, Fleur Visser, and Charlotte Curé. 2022. "Friend or Foe: Risso's Dolphins Eavesdrop on Conspecific Sounds to Induce or Avoid Intra-Specific Interaction." *Animal Cognition* 25 (2): 287–96. https://doi.org/10.1007/s10071-021-01535-y.

Beck, Suzanne, Andrew D. Foote, Sandra Koetter, et al. 2014. "Using Opportunistic Photo-Identifications to Detect a Population Decline of Killer Whales (*Orcinus orca*) in British and Irish Waters." *Journal of the Marine Biological Association of the United Kingdom* 94 (6): 1327–33. https://doi.org/10.1017/S0025315413001124.

Bigg, Michael Andrew, Ian B. MacAskie, and Graeme M. Ellis. 1976. *Abundance and Movements of Killer Whales off Eastern and Southern Vancouver Island: With Comments on Management*. Arctic Biological Station.

Blount, Drew, Shane Gero, Jon Van Oast, et al. 2022. "Flukebook: An Open-Source AI Platform for Cetacean Photo Identification." *Mammalian Biology* 102 (3): 1005–23. https://doi.org/10.1007/s42991-021-00221-3.

Bruck, Jason N., Sam F. Walmsley, and Vincent M. Janik. 2022. "Cross-Modal Perception of Identity by Sound and Taste in Bottlenose Dolphins." *Science Advances* 8 (20): eabm7684. https://doi.org/10.1126/sciadv.abm7684.

Cammen, Kristina M., Kimberly R. Andrews, Emma L. Carroll, et al. 2016. "Genomic Methods Take the Plunge: Recent Advances in High-Throughput Sequencing of Marine Mammals." *Journal of Heredity* 107 (6): 481–95. https://doi.org/10.1093/jhered/esw044.

Cantor, Mauricio, Lauren G. Shoemaker, Reniel B. Cabral, César O. Flores, Melinda Varga, and Hal Whitehead. 2015. "Multilevel Animal Societies Can Emerge from Cultural Transmission." *Nature Communications* 6 (1): 8091. https://doi.org/10.1038/ncomms9091.

Carroll, Emma Louise, C. Scott Baker, Mandy Watson, et al. 2015. "Cultural Traditions Across a Migratory Network Shape the Genetic Structure of Southern Right Whales Around Australia and New Zealand." *Scientific Reports* 5 (1): 16182. https://doi.org/10.1038/srep16182.

Christiansen, Fredrik, Antoine M. Dujon, Kate R. Sprogis, John P. Y. Arnould, and Lars Bejder. 2016. "Noninvasive Unmanned Aerial Vehicle Provides Estimates of the Energetic Cost of Reproduction in Humpback Whales." *Ecosphere* 7 (10): e01468. https://doi.org/10.1002/ecs2.1468.

Christiansen, Fredrik, Fabian Rodríguez-González, Sergio Martínez-Aguilar, et al. 2021. "Poor Body Condition Associated with an Unusual Mortality Event in Gray Whales." *Marine Ecology Progress Series* 658: 237–52. https://doi.org/10.3354/meps13585.

Connor, Richard C. 2007. "Dolphin Social Intelligence: Complex Alliance Relationships in Bottlenose Dolphins and a Consideration of Selective Environments for Extreme Brain Size Evolution in Mammals." *Philosophical Transactions of the Royal Society B* 362 (1480): 587–602. https://doi.org/10.1098/rstb.2006.1997.

Connor, Richard C., Michael Krützen, Simon J. Allen, William B. Sherwin, and Stephanie L. King. 2022. "Strategic Intergroup Alliances Increase Access to a Contested Resource in Male Bottlenose Dolphins." *Proceedings of the National Academy of Sciences* 119 (36): e2121723119. https://doi.org/10.1073/pnas.2121723119.

Croft, Darren P., Rufus A. Johnstone, Samuel Ellis, et al. 2017. "Reproductive Conflict and the Evolution of Menopause in Killer Whales." *Current Biology* 27 (2): 298–304. https://doi.org/10.1016/j.cub.2016.12.015.

Crossman, Carla A., Philip K. Hamilton, Moira W. Brown, et al. 2024. "Effects of Inbreeding on Reproductive Success in Endangered North Atlantic Right Whales." *Royal Society Open Science* 11 (7): 240490. https://doi.org/10.1098/rsos.240490.

Ellis, Samuel, Daniel W. Franks, Stuart Nattrass, et al. 2017. "Mortality Risk and Social Network Position in Resident Killer Whales: Sex Differences and the Importance of Resource Abundance." *Proceedings of the Royal Society B* 284 (1865): 20171313. https://doi.org/10.1098/rspb.2017.1313.

Farine, Damien R. 2013. "Animal Social Network Inference and Permutations for Ecologists in R Using *asnipe*." *Methods in Ecology and Evolution* 4 (12): 1187–94. https://doi.org/10.1111/2041-210X.12121.

Foote, Andrew D., Nagarjun Vijay, María C. Ávila-Arcos, et al. 2016. "Genome-Culture Coevolution Promotes Rapid Divergence of Killer Whale Ecotypes." *Nature Communications* 7 (1): 11693. https://doi.org/10.1038/ncomms11693.

Ford, John K. B. 1991. "Vocal Traditions Among Resident Killer Whales (*Orcinus orca*) in Coastal Waters of British Columbia." *Canadian Journal of Zoology* 69 (6): 1454–83. https://doi.org/10.1139/z91-206.

Foster, Emma A., Daniel W. Franks, Lesley J. Morrell, et al. 2012. "Social Network Correlates of Food Availability in an Endangered Population of Killer Whales, *Orcinus orca*." *Animal Behaviour* 83 (3): 731–36. https://doi.org/10.1016/j.anbehav.2011.12.021.

Garland, Ellen C., Anne W. Goldizen, Melinda L. Rekdahl, et al. 2011. "Dynamic Horizontal Cultural Transmission of Humpback Whale Song at the Ocean Basin Scale." *Current Biology* 21 (8): 687–91. https://doi.org/10.1016/j.cub.2011.03.019.

Gerber, Livia, Richard C. Connor, Simon J. Allen, et al. 2022. "Social Integration Influences Fitness in Allied Male Dolphins." *Current Biology* 32 (7): 1664–69. https://doi.org/10.1016/j.cub.2022.03.027.

Hammond, Philip S., Sally A. Mizroch, and Gregory P. Donovan. 1990. "Individual Recognition of Cetaceans: Use of Photo-Identification and Other Techniques to Estimate Population Parameters." *Reports of the International Whaling Commission* 12.

Hart, Jordan, Michael N. Weiss, Daniel Franks, and Lauren Brent. 2023. "BISoN: A Bayesian Framework for Inference of Social Networks." *Methods in Ecology and Evolution* 14: 2411–20. https://doi.org/10.1111/2041-210X.14171.

Hersh, Taylor A., Shane Gero, Luke Rendell, et al. 2022. "Evidence from Sperm Whale Clans of Symbolic Marking in Non-Human Cultures." *Proceedings of the National Academy of Sciences* 119 (37): e2201692119. https://doi.org/10.1073/pnas.2201692119.

Holmes, Kathryn G., Michael Krützen, Amanda R. Ridley, et al. 2024. "Juvenile Social Play Predicts Adult Reproductive Success in Male Bottlenose Dolphins." *Proceedings of National Academy of Sciences USA* 121 (25): e2305948121. https://doi.org/10.1073/pnas.2305948121.

International Human Genome Consortium. 2004. "Finishing the Euchromatic Sequence of the Human Genome." *Nature* 431 (7011): 931–45. https://doi.org/10.1038/nature03001.

Janik, Vincent M., Laela S. Sayigh, and Randall S. Wells. 2006. "Signature Whistle Shape Conveys Identity Information to Bottlenose Dolphins." *Proceedings of the National Academy of Sciences* 103 (21): 8293–97. https://doi.org/10.1073/pnas.0509918103.

Jaramillo-Legorreta, Armando M., Gustavo Cardenas-Hinojosa, et al. 2019. "Decline Towards Extinction of Mexico's Vaquita Porpoise (*Phocoena sinus*)." *Royal Society Open Science* 6 (7): 190598. https://doi.org/10.1098/rsos.190598.

Johnson, Mark P., and Peter L. Tyack. 2003. "A Digital Acoustic Recording Tag for Measuring the Response of Wild Marine Mammals to Sound." *IEEE Journal of Oceanic Engineering* 28 (1): 3–12. https://doi.org/10.1109/JOE.2002.808212.

Johri, Shaili, Maria Carnevale, Lindsay Porter, et al. 2021. "Pathways to Justice, Equity, Diversity, and Inclusion in Marine Science and Conservation." *Frontiers in Marine Science* 8: 696180. https://doi.org/10.3389/fmars.2021.696180.

King, Stephanie L., Richard C. Connor, Michael Krützen, and Simon J. Allen. 2021. "Cooperation-Based Concept Formation in Male Bottlenose Dolphins." *Nature Communications* 12 (1): 2373. https://doi.org/10.1038/s41467-021-22668-1.

King, Stephanie L., and Vincent M. Janik. 2013. "Bottlenose Dolphins Can Use Learned Vocal

Labels to Address Each Other." *Proceedings of the National Academy of Sciences* 110 (32): 13216–21. https://doi.org/10.1073/pnas.1304459110.

King, Stephanie L., and Frants H. Jensen. 2023. "Rise of the Machines: Integrating Technology with Playback Experiments to Study Cetacean Social Cognition in the Wild." *Methods in Ecology and Evolution* 14 (8): 1873–86. https://doi.org/10.1111/2041-210x.13935.

Laland, Kevin N., John Odling-Smee, and Sean Myles. 2010. "How Culture Shaped the Human Genome: Bringing Genetics and the Human Sciences Together." *Nature Reviews Genetics* 11 (2): 137–48. https://doi.org/10.1038/nrg2734.

Louis, Marie, Marco Galimberti, Frederick Archer, et al. 2021. "Selection on Ancestral Genetic Variation Fuels Repeated Ecotype Formation in Bottlenose Dolphins." *Science Advances* 7 (44): eabg1245. https://doi.org/10.1126/sciadv.abg1245.

Mann, Janet, Richard C. Connor, Peter L. Tyack, and Hal Whitehead, eds. 2000. *Cetacean Societies: Field Studies of Dolphins and Whales*. University of Chicago Press.

Mann, Janet, Margaret A. Stanton, Eric M. Patterson, Elisa J. Bienenstock, and Lisa O. Singh. 2012. "Social Networks Reveal Cultural Behaviour in Tool-Using Dolphins." *Nature Communications* 3 (1): 980. https://doi.org/10.1038/ncomms1983.

Morin, Phillip A., Frederick I. Archer, Andrew D. Foote, et al. 2010. "Complete Mitochondrial Genome Phylogeographic Analysis of Killer Whales (*Orcinus orca*) Indicates Multiple Species." *Genome Research* 20 (7): 908–16. https://doi.org/10.1101/gr.102954.109.

Osiecka, Anna N., Susanna Quer, Aleksandra Wróbel, and Katarzyna Osiecka-Brzeska. 2021. "Unpaid Work in Marine Science: A Snapshot of the Early-Career Job Market." *Frontiers in Marine Science* 8: 690163. https://doi.org/10.3389/fmars.2021.690163.

Ramos, Eric A., Jeremy J. Kiszka, Victoria Pouey-Santalou, Raul Ramirez Barragan, Andrea Jacqueline Garcia Chavez, and Katherina Audley. 2021. "Food Sharing in Rough-Toothed Dolphins off Southwestern Mexico." *Marine Mammal Science* 37 (1): 352–60. https://doi.org/10.1111/mms.12727.

Rendell, Luke, and Hal Whitehead. 2001. "Culture in Whales and Dolphins." *Behavioral and Brain Sciences* 24 (2): 309–24. https://doi.org/10.1017/S0140525X0100396X.

Ross, Cody T., Richard McElreath, and Daniel Redhead. 2024. "Modelling Animal Network Data in R Using STRAND." *Journal of Animal Ecology* 93 (3): 254–66. https://doi.org/10.1111/1365-2656.14021.

Scholander, Per Fredrik. 1940. "Experimental Investigations on the Respiratory Function in Diving Mammals and Birds." *Hvalradets skrifter* 22: 1.

Srinivasan, Mridula. 2023. "Social Strategies of a Consummate Marine Predator: Mammal-Hunting Killer Whales." In *Social Strategies of Carnivorous Mammalian Predators: Hunting and Surviving as Families*, 229–81. Springer. https://doi.org/10.1007/978-3-031-29803-5_7.

Stanton, Margaret A., and Janet Mann. 2012. "Early Social Networks Predict Survival in Wild Bottlenose Dolphins." *PLOS ONE* 7 (10): e47508. https://doi.org/10.1371/journal.pone.0047508.

Torres, Leigh G., Sharon L. Nieukirk, Leila Lemos, and Todd E. Chandler. 2018. "Drone Up! Quantifying Whale Behavior from a New Perspective Improves Observational Capacity." *Frontiers in Marine Science* 5: 319. https://doi.org/10.3389/fmars.2018.00319.

Torres Ortiz, Sara, Johanna Stedt, Henrik Skov Midtiby, Henrik Dyrberg Egemose, and Magnus Wahlberg. 2021. "Group Hunting in Harbour Porpoises (*Phocoena phocoena*)." *Canadian Journal of Zoology* 99 (6): 511–20. https://doi.org/10.1139/cjz-2020-0289.

Turvey, Samuel T., Robert L. Pitman, Barbara L. Taylor, et al. 2007. "First Human-Caused Extinction of a Cetacean Species?" *Biology Letters* 3 (5): 537–40. https://doi.org/10.1098/rsbl.2007.0292.

Usman, Ayinde M., Olayinka O. Ogundile, and Daniel J. J. Versfeld. 2020. "Review of Automatic Detection and Classification Techniques for Cetacean Vocalization." *IEEE Access* 8: 105181–206. https://doi.org/10.1109/ACCESS.2020.3000477.

Weiss, Michael N., Daniel W. Franks, Deborah A. Giles, et al. 2021. "Age and Sex Influence Social Interactions, but Not Associations, Within a Killer Whale Pod." *Proceedings of the Royal Society B* 288 (1953): 20210617. https://doi.org/10.1098/rspb.2021.0617.

Whitehead, Hal. 2009. "SOCPROG Programs: Analyzing Animal Social Structures." *Behavioral Ecology and Sociobiology* 63: 765–78. https://doi.org/10.1007/s00265-008-0697-y.

Whitehead, Hal, and Roger Payne. 1978. "New Techniques for Assessing Populations of Right Whales Without Killing Them." *Mammals in the Seas* 3: 189.

Whitehead, Hal, and Luke Rendell. 2014. *The Cultural Lives of Whales and Dolphins*. University of Chicago Press.

Whitehead, Hal, Luke Rendell, Richard W. Osborne, and Bernd Würsig. 2004. "Culture and Conservation of Non-Humans with Reference to Whales and Dolphins: Review and New Directions." *Biological Conservation* 120 (3): 427–37. https://doi.org/10.1016/j.biocon.2004.03.017.

Whiten, Andrew. 2021. "The Burgeoning Reach of Animal Culture." *Science* 372 (6537): eabe6514. https://doi.org/10.1126/science.abe6514.

Wild, Sonja, Simon J. Allen, Michael Krützen, Stephanie L. King, Livia Gerber, and William J. E. Hoppitt. 2019. "Multi-Network-Based Diffusion Analysis Reveals Vertical Cultural Transmission of Sponge Tool Use Within Dolphin Matrilines." *Biology Letters* 15 (7): 20190227. https://doi.org/10.1098/rsbl.2019.0227.

Wild, Sonja, William J. E. Hoppitt, Simon J. Allen, and Michael Krützen. 2020. "Integrating Genetic, Environmental, and Social Networks to Reveal Transmission Pathways of a Dolphin Foraging Innovation." *Current Biology* 30 (15): 3024–30. https://doi.org/10.1016/j.cub.2020.05.069.

Williams, Rob, Robert C. Lacy, Erin Ashe, et al. 2024. "Warning Sign of an Accelerating Decline in Critically Endangered Killer Whales (*Orcinus orca*)." *Communications Earth and Environment* 5 (1): 173. https://doi.org/10.1038/s43247-024-01327-5.

Wiszniewski, Joanna, Shannon Corrigan, Luciano B. Beheregaray, and Luciana M. Möller. 2012. "Male Reproductive Success Increases with Alliance Size in Indo-Pacific Bottlenose Dolphins (*Tursiops aduncus*)." *Journal of Animal Ecology* 81 (2): 423–31. https://www.jstor.org/stable/41495986.

Würsig, Bernd, and Melany Würsig. 1977. "The Photographic Determination of Group Size, Composition, and Stability of Coastal Porpoises (*Tursiops truncatus*)." *Science* 198 (4318): 755–56. https://doi.org/10.1126/science.198.4318.755.

PART I

The Evolutionary History and Genetic Composition of Cetacean Societies

1

The Evolutionary History of Cetacean Behavior and Cognition

MICHAEL R. MCGOWEN,
MORGAN L. MCCARTHY, AND
MATTHEW R. MCCURRY

Abstract

Over the last few decades, our knowledge of cetacean evolution and phylogenetic relationships has increased exponentially. Here, we summarize the major themes of cetacean evolution and systematics with a particular focus on behavior and cognition. First, we discuss the early evolution of cetaceans and their transition from land to water including major themes such as the reduction of limbs. Next, we provide a thorough summary of the relationships among the extant families and the important features of each family. Then, we detail the major studies on cetacean brain size and their relationship with the study of cognition in cetaceans that have arisen over the past two decades. Finally, we discuss the evolution of sociality and how phylogeny and ecology have played a role in shaping cetacean societies.

Introduction

Integral to any discussion of behavior is its phylogenetic and evolutionary context (Brooks and McLennan 1991; Harvey and Pagel 1991). Since the publication of the first edition of *Cetacean Societies* over 20 years ago, our understanding of the phylogenetic relationships and evolutionary tree of both extinct and extant cetaceans has improved greatly (Marx et al. 2016; McGowen et al. 2020). Many recent studies have examined cognitive and behavioral evolution in cetaceans and tested earlier hypotheses and suppositions concerning the evolution of brain size and sociality. In this chapter, we outline the major themes in the phylogeny and evolution of cetaceans, particularly as it pertains to the evolution of cognition and sociality. First, we give context to later chapters by discussing the origins of cetaceans and describing how the earliest whales invaded the aquatic environment. Next, we discuss the evolution of modern cetaceans and introduce each extant family. Then, we examine the evolution of cetacean cognitive capacity as measured through brain size and discuss caveats using this approach. Finally, we examine the evolution of cetacean sociality through the lens of phylogeny. The aim of this chapter is to give readers a firm grasp of the phylogenetic relationships of cetaceans and the use of phylogeny in addressing major questions in cetacean cognition, behavioral ecology, and evolution.

The Earliest Whales

Cetaceans belong to the order Artiodactyla, a clade of mammals that mostly includes terrestrial representatives collectively known as even-toed ungulates (i.e., pigs, camels, hippopotamuses, giraffes, deer, antelopes, cows, goats, and sheep). Their closest living relatives among these species are the two extant species of hippopotamus (river and pygmy; Gatesy et al. 2013). (Fig. 1.1 provides an overview of cetacean evolutionary history; subsequent references to groups and species depicted in this figure remain relevant throughout the following discussion.) Together hippos and whales form the clade Cetancodonta within Artiodactyla, which is supported by extensive molecular and morphological evidence (Montgelard et al. 1997; Nikaido et al. 1999; Thewissen et al. 2007; Spaulding et al. 2009; Meredith et al. 2011; O'Leary et al. 2013; Gatesy et al. 2013; Yuan et al. 2021; Foley et al. 2023). Multiple studies revealed hippos and cetaceans likely diverged from their last common ancestor over 54 million years ago (MYA) in the early Eocene (Meredith et al. 2011; Yuan et al. 2021; Foley et al. 2023). Perhaps the most profound morphological evidence that cetaceans are artiodactyls is that the morphology of the pulley-like ankle bone (astragalus) in early cetaceans exactly resembles that of terrestrial artiodactyls (Gingerich et al. 2001; Thewissen et al. 2001, 2007; Marx et al. 2016).

The earliest stem-cetacean in the fossil record is *Himalayacetus subathuensis*, consisting of a partial mandible from India, dated to the early Eocene (~53.5 MYA) and identified as having similar teeth to other early stem-cetaceans (Bajpai and Gingerich 1998). More complete early stem-cetaceans from the early Eocene include the semiaquatic *Indohyus* from India and *Pakicetus* from Pakistan, both of which show evidence of thickened ear bones and $d^{18}O$ isotope values reminiscent of a semiaquatic lifestyle that may have involved bottom walking (Madar 2007; Thewissen et al. 2007; Clementz et al. 2006). Later fossils demonstrated that early whales (or "archaeocetes") acquired further adaptations to the aquatic environment, including *Ambulocetus* with its larger tail and paddle-like limbs and remingtonocetids with their long bodies, short limbs, and increasingly marine lifestyle (Marx et al. 2016).

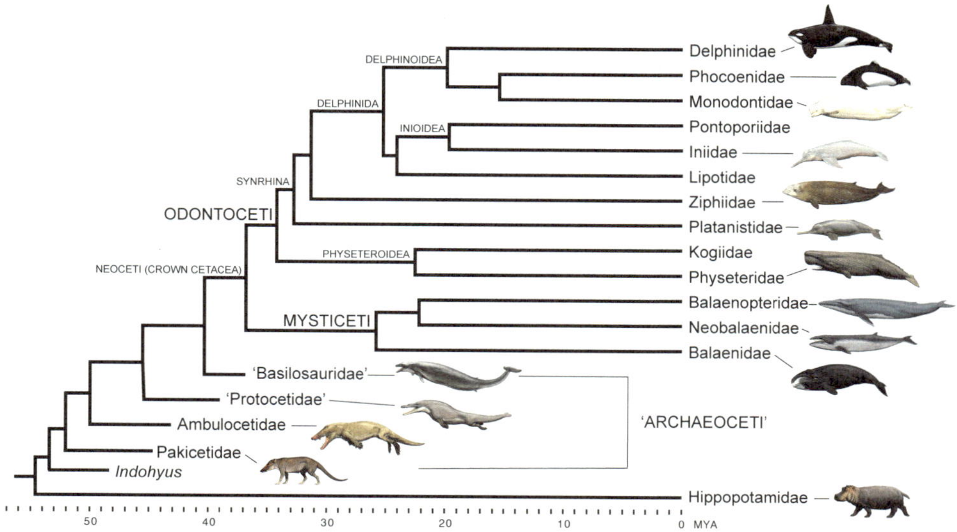

FIGURE 1.1. Phylogenetic tree of living cetacean families with major fossil archaeocete lineages and outgroup hippopotamuses. The tree is derived from McGowen et al. (2020) and Lloyd and Slater (2021) with representative fossil taxa shown. Names in quotes represent paraphyletic groups. Both "Basilosauridae" and "Protocetidae" grades are shown as clades here for ease of representation. Major clades are shown to the left of nodes (e.g., Odontoceti, Synrhina).
Source: Illustrations by Carl Buell; copyright by John Gatesy, used here with approval of John Gatesy.

Around ~48 MYA, we see the appearance of a new group of archaeocetes classified in the paraphyletic family Protocetidae. These early whales were the first to spread out of the ancient Tethys Sea bordering India and Pakistan to the rest of the globe and are all found in coastal marine deposits (Marx et al. 2016). Protocetids could likely still walk on land, but many showed even greater adaptations to a marine lifestyle, such as unfused sacral vertebrae or a reduction in the attachment of the pelvis to the axial skeleton (Uhen 2010). Further adaptation to the aquatic environment occurred in the likely paraphyletic family Basilosauridae, which were the first fully aquatic cetaceans. Basilosaurids had a reduced pelvis with small hind limbs and an ear that was acoustically isolated from the skull for better underwater hearing (Uhen 2010). This group included the eellike genus *Basilosaurus* (~40–34 MYA), which ranged up to 17 m long with a long, slender tail and rudimentary hindlimbs that some have suggested functioned like claspers in sharks (Gingerich et al. 1990). The basilosaurids also included the gigantic *Perucetus colossus* that exhibits robust pachyosteosclerotic vertebrae and, by some estimates, was heavier than an adult blue whale (Bianucci et al. 2023).

All modern cetaceans likely originated from within Basilosauridae or from a common ancestor with Basilosauridae sometime in the late Eocene

(Marx et al. 2016; Geisler 2018). Modern cetaceans (also termed Neoceti or crown Cetacea) are united morphologically in that they possess an immovable elbow joint, extensive telescoping of skull bones by movement of the nares posteriorly, and only one set of teeth (if present) that are not replaced (Geisler 2018). While fossils of early whales have revealed evidence of certain habitats, diets, and swimming behaviors, fossils have given us few clues regarding social behavior. However, the use of phylogenetic comparative analysis has allowed researchers to make some inferences about early whales, which we discuss in the section "Brain-Size Evolution in Cetaceans."

A Taxonomic Tour of Modern Cetaceans

Currently, 94 extant and recently extinct cetacean species are recognized by the Society for Marine Mammalogy, including 15 species of baleen whales (Mysticeti) and 79 species of toothed whales (Odontoceti; Society for Marine Mammalogy Committee on Taxonomy 2025). As many as 12 new species have been officially recognized by the committee since the publication of *Cetacean Societies* (Mann et al. 2000), mainly from splitting existing species with previously unidentified genetic and morphological diversity (Beasley et al. 2005; Braulik et al. 2021; Caballero et al. 2007; Carroll et al. 2021; Costa et al. 2022; Dalebout et al. 2002, 2014; Jefferson and Rosenbaum 2014; Rosel et al. 2021; Wada et al. 2003; Wang et al. 2010; Yamada et al. 2019). These "cryptic species" include both baleen and toothed whales. It is likely that more species will be officially recognized in the next few years from either detailed worldwide studies of existing species using genomics and morphology or quite possibly from completely new sightings of species wholly unknown to science.

Baleen whales (Mysticeti) form a solid monophyletic group known for their lack of teeth as adults and the development of thick, keratinous plates in their palate called baleen, which filter small prey items from the water in bulk (Pivorunas 1979; Gatesy et al. 2013). Mysticetes likely diverged from toothed whales in the late Eocene ~37–39 MYA (Marx and Fordyce 2015; McGowen et al. 2020; Lloyd and Slater 2021; Yuan et al. 2021; Wolf et al. 2023). Extant baleen whales currently are grouped into four families, Balaenidae (four spp., right whales and bowhead), Neobalaenidae or Cetotheriidae (one sp., pygmy right whale), Eschrichtiidae (one sp., gray whale), and Balaenopteridae (nine spp., rorquals; Gatesy and McGowen 2021). Balaenids are the earliest diverging mysticetes (fig. 1.1) and feed primarily via slowly swimming and skimming the upper water column, moving water continually through their baleen filter (Pivorunas 1979; Gatesy et al. 2013). Currently, there is uncertainty in the name of the family Neobalaenidae, which includes the pygmy right whale. It was traditionally known as Neobalaenidae, but many phylogenetic analyses using fossils place the pygmy right whale within the extinct family Cetotheriidae, whose name takes priority over Neobalaenidae (Marx and Fordyce 2015). A recent analysis of rostrum shape showed that the pygmy right whale evolved skim feeding separately from balaenids (Tanaka 2022), and recent whole genome analyses continue to support this (Wolf et al. 2023; Dutoit et al. 2023). Although Eschrichtiidae is still officially recognized as a separate family, molecular data have consistently placed it within Balaenopteridae (Rychel et al. 2004; McGowen et al. 2009, 2020; Marx and Fordyce 2015; Hassanin et al. 2012; Lammers et al. 2019), including recent whole genome analyses (Árnason et al. 2018; Wolf et al. 2023; Yuan et al. 2021). The reticence to include the gray whales within Balaenopteridae is perhaps related to the elaborate engulfment feeding adaptations of balaenopterids, which use their loose mandibular symphysis and large pleated throat pouch to gulp and then filter huge quantities of water for prey such as krill or small fish (Pivorunas 1979; Goldbogen et al. 2017; Gatesy and McGowen 2021). In contrast, gray whales feed on benthic invertebrates on the seafloor through the use of suction. Balaenopterids such as the humpback whale (*Megaptera novaeangliae*) have also evolved elaborate methods of corralling prey using bubble nets and feeding in coordinated groups (Clapham 2018; Dunlop and Parks, this volume; Wild and Rendell, this volume).

Many mysticete species migrate long distances during the year, usually between high-latitude feeding areas and warmer low-latitude breeding and calving grounds (Dunlop and Parks, this volume). Species such as gray whales and humpback whales perform some of the longest migrations among mammals (Bannister 2018). The gray whale is perhaps the most well-known example of long-distance migration; eastern Pacific populations migrate between summer feeding grounds off Alaska to winter birthing lagoons along the coast of Baja California (Bannister 2018). However, the longest migrations are conducted by Southern Hemisphere humpback whales that migrate from Antarctica to Central America, a total distance of about 8,300 km and a record extreme for mammals (Rasmussen et al. 2007). Different species of baleen whales vary in the degree of feeding on calving grounds and the distance they travel. For example, members of the Bryde's whale complex, which solely inhabits subtropical and tropical waters, usually do not migrate long distances (Bannister 2018).

Toothed whales (Odontoceti) form the other major group of living cetaceans and are much more diverse than mysticetes with ten families and at least 79 species total (fig. 1.1). All odontocetes echolocate, a process by which high-frequency sounds are produced in their nasal cavity that then reflect off objects in their surrounding environment (Cranford et al. 1996). These reflected sounds are then received through a thin part of the posterior mandible (the pan bone) and funneled to the inner ear, enabling toothed whales to "see" their surrounding environment through sound (Møhl et al. 1999). Toothed whales also exhibit multiple morphological features associated with echolocation including nasofacial asymmetry and the presence of the melon, a large fatty organ that serves to focus echolocation clicks and gives toothed whales the characteristic bulbous forehead (McKenna et al. 2011; Coombs et al. 2020). The earliest toothed whales (stem odontocetes) are known from the early Oligocene (33.9–27.82 MYA) of the Pacific Northwest and East Coast of North America (Marx et al. 2016). But perhaps the earliest evidence of echolocation comes from the fossil odontocete *Cotylocara macei* from South Carolina (~29 MYA), which shows evidence of the development of the air sacs and broad maxillae associated with echolocation in modern odontocetes (Geisler et al. 2014). Further evidence of the early origins of ultrasonic hearing come from *Echovenator sandersi* (27–24 MYA), also from South Carolina, which shows both primitive and advanced features in the inner ear, suggesting true ultrasonic hearing may have developed from already existing sensitivity to high frequencies in basilosaurid archaeocetes (Churchill et al. 2016).

The monophyly of toothed whales is now well established using genomic data (McGowen et al. 2020; Yuan et al. 2021; Foley et al. 2023), despite some early mitochondrial analyses that placed mysticetes within odontocetes (Milinkovitch et al. 1993). Since the beginning of the Oligocene (~33.7 MYA), odontocetes have radiated into a number of modern forms with a range of adult body size from male sperm whales (70,000 kg, 19.2 m, maximum) to the Commerson's dolphin (25 kg, 1.4 m, maximum; Carwardine 2020). Sperm whales (Physeteroidea) likely split from all other extant odontocetes (Synrhina) ~32–34 MYA and are classified in two families (Physeteridae and Kogiidae), which then later diverged ~22 MYA. (McGowen et al. 2020; fig. 1.1). Both physeterids and kogiids are deep divers, suction feeders, and predominantly feed on cephalopods (Marx et al. 2016). The sperm whale itself is placed in a monotypic family (Physeteridae) and displays the most extreme sexual dimorphism of any cetacean (Whitehead 2003; Parsons et al., this volume). Female sperm whales are highly social and exist in closely knit family units of related individuals (Whitehead 2003). The two other species, the dwarf and pygmy sperm whales, are placed in Kogiidae. These species are much smaller than sperm whales with the shortest rostra of all cetaceans (McAlpine 2014; Coombs et al. 2022). Although there is limited data, these species do not appear to be very social and are encountered in small groups or as solitary individuals (McAlpine 2014). Within Synrhina, the South Asian river dolphins are supported as the sister group of all other extant cetaceans. South Asian river dolphins include two freshwater species (the Ganges and Indus River dolphins) in the family Platanistidae (Braulik et al. 2021; fig. 1.1), a relict group whose relatives used to be mostly marine and much more numerous in the past (Braulik 2014; Marx et al. 2016). Extant platanistids are known for their long beaks, needlelike teeth, bizarre maxillary crests, and greatly reduced eyes (Braulik 2014; Marx et al. 2016).

Beaked whales of the family Ziphiidae split from the remaining odontocetes ~30–31 MYA (McGowen et al. 2020; fig. 1.1). Ziphiidae is the second most speciose family of cetaceans with 24 species of pelagic, deep-diving suction feeders that feed on cephalopods and deep-sea fishes (MacLeod 2014; Carroll et al. 2021). In most ziphiids, males have two large tusks in the lower jaw and no other erupted teeth. Species vary in the size, shape, and position of these tusks, and females generally do not have any erupted teeth. Two genera differ from this pattern: *Berardius* has four teeth in the lower jaw in both males and females and *Tasmacetus* has a full set of teeth in both jaws in adults (MacLeod 2014). Due to sexual dimorphism in most species and scarring in many males, it has been suggested that many beaked whale species engage in male-male competition for mates and/or female choice (Mann et al. 2000; Dalebout et al. 2008; Ellis and Mead 2017). The pelagic nature of ziphiids and the limited time that many spend on the surface make them hard to observe and the life histories of many species remain a mystery. Some species are only known from stranded individuals and have never been observed at sea (Ellis and Mead 2017).

All remaining extant odontocetes are placed in the clade Delphinida, which can be further split into two clades, Inioidea and Delphinoidea, which diverged from one another ~24–26 MYA (Geisler et al. 2011; Gatesy et al. 2013; McGowen et al. 2020; fig. 1.1). Inioidea includes three monotypic families of smaller cetaceans, Iniidae (Amazon River dolphin), Pontoporiidae (franciscana), and the recently extinct Lipotidae (Yangtze River dolphin). Together with *Platanista* these have been termed the "river dolphins" (even though the franciscana lives in coastal and estuarine habitats) and all were classified together in Platanistidae due to their morphological resemblance of long, toothy beaks and reduced orbits (Geisler et al. 2011). Taken together these traits were hypothesized as adaptations for riverine environments, even though they were not found in delphinid freshwater species. Geisler et al. (2011) combined molecular, morphological, and fossil data and found that many of these traits evolved in marine ancestors. In addition, this study also showed that all three riverine lineages (Platanistidae, Lipotidae, and Iniidae) independently evolved from marine ancestors (Geisler et al. 2011).

Delphinoidea includes the three remaining living families: Monodontidae (narwhal and beluga, two species), Phocoenidae (porpoises, seven species) and Delphinidae (true or "oceanic" dolphins, 37 species; fig. 1.1). Monodontids and phocoenids are more closely related to one another than delphinids, having diverged from delphinids in the early Miocene (~18 MYA; McGowen et al. 2020). Belugas and narwhals are adapted for arctic and subarctic environments, while the diminutive porpoises inhabit many environments including pelagic, coastal, and riverine. Delphinids represent the most speciose family of cetaceans and are incredibly diverse in habitat (pelagic, coastal, estuarine, riverine), diet (marine mammals, cephalopods, fishes), size (small *Cephalorhynchus* species [~25–30 kg] to killer whales, *Orcinus orca* [maximum 6,600 kg]) and degree of sociality (Slater et al. 2010; May-Collado et al. 2007; Carwardine 2020; Galatius et al. 2021). Delphinids radiated rapidly ~12 MYA and include some of the most social species of cetaceans (McGowen et al. 2020; May-Collado et al. 2007).

Brain-Size Evolution in Cetaceans

Brain size has been used by many researchers as a rough measure for general cognition in a particular species, and as a result it remains a powerful tool for examining potential trends in evolution regarding intelligence and sociality. Cetaceans boast some of the largest brains known to exist, with two odontocetes, the sperm whale (*Physeter macrocephalus*) and killer whale, sharing the title for the largest absolute brain size of any living thing (Ridgway and Hanson 2014). Although the blue whale is the largest animal, its absolute brain size is not as large as some smaller cetaceans such as the sperm whale. However, to measure brain size accurately, Jerison (1973) argued that body size and the scaling relationship between brain and body size must be taken into account. Some species stand out from general predictions of brain-to-body size ratios and display disproportionately larger relative brain sizes compared to their body size. To define a metric that accounts for this variation, Jerison (1973) developed the encephalization quotient (EQ), which quantifies the extent to which brain size varies from what is expected for a particular species based on its body size. EQs have been given much attention in the study of mammal cognition, including cetaceans, and by many measures delphinid cetaceans have some of the highest encephalization scores documented, surpassing those of nonhuman primates (Marino 1998; Boddy et al.

2012; Montgomery et al. 2013). However, the scaling of brain-to-body size ratio for cetaceans appears to deviate from their terrestrial artiodactyl relatives and it has been proposed that the weightlessness of the aquatic environment, coupled with certain feeding preferences could partly explain why a relatively recent increase in mysticete body size leads to an appearance of under-encephalization, despite displaying characteristics of sociality and complex foraging strategies (McCurry et al. 2021). Indeed, the use of EQs in general has been met with some criticism (Van Schaik et al. 2021; Ridgway et al. 2016; Smaers et al. 2021), and some researchers contend that other indices may more accurately measure cognition such as absolute numbers of neurons or neuron density (Herculano-Houzel 2011).

Fossil finds from early cetaceans have been fundamental in documenting the transition to becoming fully aquatic as well as documenting changes in cetacean brain and body size. The fossil record of archaeocetes provides researchers with a window into the past to estimate when major transitions in EQ across the cetacean lineage occurred. Early whales such as archaeocetes generally had smaller brains than modern groups (but not always) and there is a general trend in increasing relative brain size through time (Montgomery et al. 2013; McCurry et al. 2021). McCurry et al. (2021) notes that some of the larger basilosaurid archaeocetes such as *Dorudon* and *Basilosaurus* also independently developed large relative brain size (fig. 1.2).

Following the emergence of modern cetaceans, notable differences emerged in body and brain-size evolution between mysticetes and odontocetes. Montgomery et al. (2013) reported a directional trend of both increasing brain and body mass in mysticetes. However, body size increases outpaced brain mass, and consequently EQ remained stable or decreased. Notably, enormous increases in body size for the bowhead (*Balaena mysticetus*) and blue (*Balaenoptera musculus*) whales resulted in decreases in EQ. It was proposed that in the case of mysticetes, there may be a decoupling in the evolution of body and brain mass related to the special case of their gigantic size (Montgomery et al. 2013). It remains unclear as to why this decoupling exists; it could be related to the removal of constraints on body size related to the buoyancy of water or that a large amount of the mass of mysticetes whale is made up of blubber, which requires little input from neurological tissues. Research into extremely large body sizes of some mysticetes has suggested that this relatively recent (~3 MYA) growth spurt in baleen whales may be the result of adaptations to increased productivity, patchiness, and prey densities following changes in the Plio-Pleistocene (5.4 MYA–11,700 years BP) ocean dynamics (Slater et al. 2017). Earlier studies suggested that the weightlessness of the aquatic environment promoted large body size in mysticetes and needed to be factored

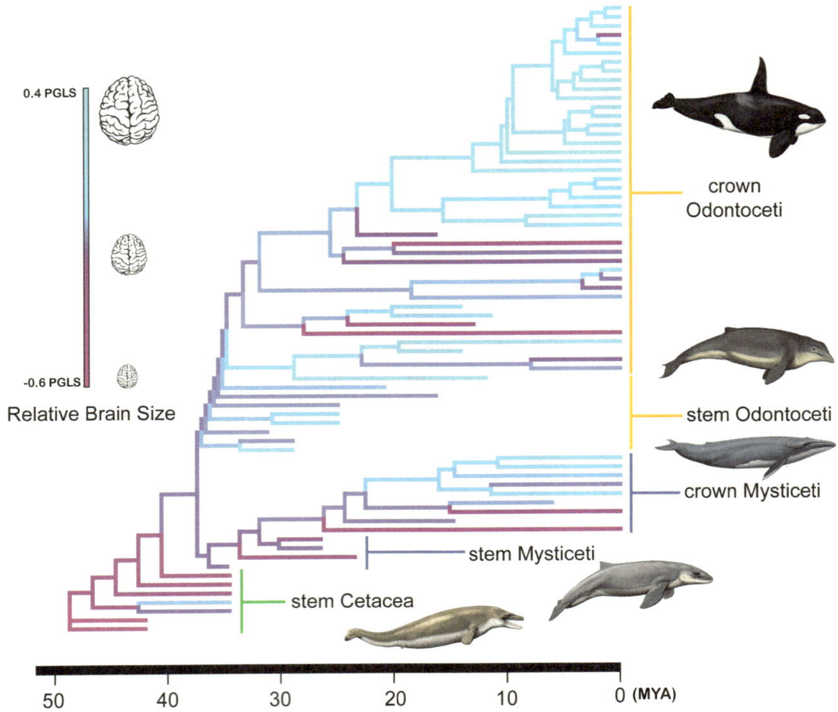

FIGURE 1.2. Relative brain size (encephalization quantified via a phylogenetic least squares [PGLS] approach regressing absolute brain mass on body mass as calculated in McCurry et al. 2021) reconstructed on the phylogeny of fossil and living cetaceans through time from 50 MYA to the present.
Source: McCurry et al. (2021), figure 2. Illustrations by Carl Buell. Reprinted by permission of Oxford University Press.

into calculating EQ (Gingerich 1998), as the rapid increase in body size in mysticetes may be a driver for their relatively low EQ (Montgomery et al. 2013; McCurry et al. 2021). McCurry et al. (2021) added several fossil and modern mysticetes that had been lacking in earlier analyses and found a decrease in rates of brain-size evolution in stem mysticetes but an overall increase in relative brain size in rorquals (fig. 1.2).

In contrast to mysticetes, Montgomery et al. (2013) and Smaers et al. (2021) showed that odontocetes generally decreased in body size with some stasis in brain size over time. It is this decrease in body size that may have driven increases in relative brain size in this group, particularly in Delphinoidea. McCurry et al. (2021) also demonstrated that delphinoids had some of the highest relative brain sizes with the highest rate increases in relative brain size occurring within the Delphinidae (fig. 1.2). Other studies also have indicated that delphinoids or specifically delphinids display a shift to higher EQs (Marino et al. 2004; Waugh and Thewissen 2021; Smaers et al. 2021). There are a few notable exceptions to the general trend of increasing encephalization in odontocetes. The lineages leading to both sperm whales and killer whales had extreme body size increases that outpaced any increase in brain size (Montgomery et al. 2013; McCurry et al. 2021). Goldbogen and Madsen (2018) hypothesized that an increase in body size in sperm whales (and beaked whales) may have resulted from the need to access prey in deeper water habitats following the development of echolocation. Though these examples of decreased encephalization resulted from an outsized increase in body size, the lineage leading to the South Asian river dolphins (*Platanista*) is estimated to have had a decrease in brain size, leading to a decrease in encephalization (Montgomery et al. 2013). McCurry et al. (2021) also noted a similar trend to lower relative brain size in other river dolphin lineages and suggested that some foraging strategies may limit brain size, and that the behavioral snap feeding (perceived as a simpler strategy) of river dolphins, possibly coupled with their less complex social behavior, might result in a constraint on brain size (fig. 1.2).

Many studies have tried to address what might be driving brain-size shifts in cetaceans. Fox et al. (2017) for example, tested the social brain hypothesis in cetaceans, which posits that large brains are associated with rich social environments. They found that EQ is strongly predicted by social complexity (see Weiss et al., this volume) as well as group size and that this may be a strong driver on the evolution of cetacean encephalization. In a separate analysis, Groot et al. (2023) tested whether the cognitive buffer hypothesis (CBH) fit cetacean brain size. This hypothesis predicts that the benefits of having a larger brain relative to body size would provide higher survival rates and better adaptation to the environment (Groot et al. 2023). Since a large brain is metabolically costly and takes longer to develop, the CBH can be extended to expect that the organism should also have a longer lifespan to compensate for the energetic cost of developing and maintaining a larger brain (Groot et al. 2023). The authors, however, found an unexpected negative correlation between brain size and longevity in cetaceans. Cetaceans appear to be outliers in this respect since other large-brained mammals, including primates, have positive correlations between brain size and lifespan (DeCasien et al. 2018; Street et al. 2017). One other hypothesis concerning cetacean encephalization suggests that the cognitive requirements for echolocation were responsible for the relatively large brain in odontocetes (Marino et al. 2007). However, based on the collective work of researchers over the years, it is becoming increasingly clear that the development of echolocation was not the sole driver of encephalization in early cetaceans, as some archaeocetes and mysticetes experienced increases in EQ whereas some odontocetes have experienced decreases (Montgomery et al. 2013; McCurry et al. 2021; fig. 1.2). Taken together, these studies suggest that, as complex structures that serve multiple functions, brains are under a broad array of selective pressures. Future analyses might shift away from testing particular causes to understanding the relative importance of the wide array of different forces impacting brain-size evolution through time. Likely, new data will be needed to undertake this. Some species are particularly underrepresented and difficult to study in the field (e.g., beaked whales; Hooker et al. 2019), and increased behavioral data may help to further pinpoint drivers for large brains in cetaceans (Groot et al. 2023).

Social Evolution in Cetaceans

Since their return to the sea, cetaceans have provided a natural evolutionary experiment in how terrestrial mammal behavior evolves in response to the adaptive pressures of the marine environment. Cetaceans display a remarkable diversity in social and behavioral indices such as group size composition, communication, and vocalizations; foraging strategies; maternal care and social learning; and long-term social bonds. A better understanding of cetacean phylogeny paired with species-specific sociality measures has allowed researchers to better encapsulate overarching patterns across the cetacean tree of life.

Inferring the timing of when sociality became important in cetacean evolution is challenging because fossils provide limited evidence of how social a species may have been, although as previously discussed, EQ may provide clues. Fossils are limited in revealing direct evidence of group sizes or cooperative foraging, among other measures of sociality that are studied in the field (e.g., group foraging and co-parenting). We can, however, look to the closest living relatives of cetaceans to evaluate early trends in sociality. Within the order Artiodactyla, there are highly variable strategies of social organizational structures, though it's thought that the most likely social organization of the ancestors of Artiodactyla was pair living (Jaeggi et al. 2020). The closest living relatives of cetaceans, the two species of hippopotamus, diverge in social composition. The common hippopotamus is generally considered social and lives in groups consisting of adult females, subadults, and calves with a dominant male (Lewison 2011). However, the pygmy hippopotamus seems to be less social and is found alone or in pairs, although much about pygmy hippo behavior is not well-known (Lewison 2011). Recent evidence has found that common hippos produce low-frequency vocalizations to manage intergroup relationships (Thévenet et al. 2022). This leaves us to question whether more complex forms of sociality were already characteristics of the whale-hippo ancestor or whether it evolved independently in both groups.

Although it is unclear whether early whales were social, social communication has evolved across both mysticete and odontocete lineages though the source of vocalizations between suborders varies with mysticete sound production in the larynx and odontocete sound production in the "phonic lips" of the nasal sac system (Cranford et al. 1996; Reidenberg and Laitman 2007). Both mysticetes and odontocetes produce tonal sounds that are used in social communication. Many odontocetes also produce complex high-frequency whistles. Because these were first identified in social delphinids, many researchers assumed that these whistles were directly related to the evolution of complex social systems in delphinids (Podos et al. 2002). However, May-Collado et al. (2007) did not find support for this hypothesis when tested using phylogenetic comparative methods by mapping vocalization and sociality on a tree. May-Collado et al. (2007) found no correlation between the evolution of whistles and the evolution of Delphinidae, but did find that whistles either evolved twice in odontocetes or evolved once and were subsequently lost three times including within social clades. The authors, however, found an association between group size and tonal frequency and proposed that simple tonal sounds were more common among solitary species (e.g., rorquals and river dolphins), and tonal sound complexity increased in the lineage leading to Delphinoidea, which includes the most social odontocete species (May-Collado et al. 2007). Indeed, Madsen et al. (2023) recently showed that odontocetes can make a diverse array of sounds with the phonic lips in multiple vocal registers, producing both echolocation clicks and sounds such as whistles used for communication. This diversity produces the basis for plasticity and vocal learning (Madsen et al. 2023).

Cetacean social dynamics are diverse, and species range from primarily solitary to groups of individuals forming short-term associations and subsequent separation (fission-fusion), to stable groups that spend most of their lives together. There are some broad patterns that are true across the cetacean phylogeny, but the lack of long-term field data limits our ability to make broad generalizations about sociality for some groups (Hooker et al. 2019; Groot et al. 2023; May-Collado et al. 2007). In general, baleen whales are considered less social than odontocetes. They spend considerable time alone, and there is little evidence for maintaining long-term social bonds other than the relationship between mother and offspring (Wade et al. 2012). Some baleen whales form large aggregations around feeding grounds, but it is suspected that this reflects many individuals concentrated around an abundant food source rather than a group with an underlying social structure (Clapham

2000). While this sentiment of relatively unstructured social dynamics is agreed upon in the literature, recent long-term datasets have suggested the possibility that baleen whales may be more social than we think (see Dunlop and Parks, this volume). Humpback whales feeding in a fjord system in the Canadian Pacific were found to have partitioned feeding grounds based on social foraging groups and long-term pair bonds of up to 12 years (Wray et al. 2021). The authors propose that baleen whales may have more complex social lives than our current understanding. Future research in this direction is required to better understand whether these long-term pair bonds are unique to that population, specific to humpback whales, or exist across other mysticetes.

Odontocetes have evolved into multiple lineages with well-documented complex social structure. For instance, stable long-term social structures (usually matrilineal) have evolved at least three times: Delphinidae (e.g., killer whale, pilot whales), Physeteridae (e.g., sperm whales), and Monodontidae (e.g., narwhals; May-Collado et al. 2007). Highly fluid fission-fusion relationships have also evolved in independent lineages including Delphinidae (e.g., bottlenose dolphins; Connor et al. 2000) and Ziphiidae (e.g., Baird's beaked whale; Fedutin et al. 2015). Group living with weak associations has been retained in separate lineages including Phocoenidae (e.g., harbor porpoises), multiple river dolphin lineages, and members of the genus *Cephalorhynchus* in Delphinidae (Gowans et al. 2007; May-Collado et al. 2007). Patterns emerging from these observations show that social evolution in toothed whales seems to be at least partly explained by body size, with small species ranging from low to intermediate levels of group association and matrilineal kinship and larger species having strong group associations and female-based kinship (Rendell et al. 2019). Apart from the varying dynamics of groupings within odontocetes, group size also varies, with strong support for large groups evolving in the lineage leading to Delphinidae (May-Collado et al. 2007).

Advances over the last two decades in our knowledge of the cetacean phylogeny and synthesis of years of field observations studying cetacean sociality has allowed researchers to better understand the evolution of sociality within and among cetacean species. Further work on data deficient and difficult to study species will surely help increase our understanding in the future.

References

Árnason, Úlfur, Fritjof Lammers, Vikas Kumar, Maria A. Nilsson, and Axel Janke. 2018. "Whole-Genome Sequencing of the Blue Whale and Other Rorquals Finds Signatures for

Introgressive Gene Flow." *Science Advances* 4: eaap9873. https://doi.org/10.1126/sciadv.aap9873.

Bajpai, Sunil, and Philip D. Gingerich. 1998. "A New Eocene Archaeocete (Mammalia, Cetacea) from India and the Time of Origin of Whales." *Proceedings of the National Academy of Sciences* 95: 15464–68. https://doi.org/10.1073/pnas.95.26.15464.

Bannister, John L. 2018. "Baleen Whales (Mysticeti)." In *Encyclopedia of Marine Mammals*, 3rd ed., edited by Bernd Würsig, Johannes G. M. Thewissen, and Kit M. Kovacs, 62–69. Academic Press.

Beasley, Isabel, Kelly M. Robertson, and Peter Arnold. 2005. "Description of a New Dolphin, the Australian Snubfin Dolphin *Orcaella heinsohni* sp. n. (Cetacea, Delphinidae)." *Marine Mammal Science* 21: 365–400. https://doi.org/10.1111/j.1748-7692.2005.tb01239.x.

Bianucci, Giovanni, Olivier Lambert, Mario Urbina, et al. 2023. "A Heavyweight Early Whale Pushes the Boundaries of Vertebrate Morphology." *Nature* 620: 824–29. https://doi.org/10.1038/s41586-023-06381-1.

Boddy, Amy M., Michael R. McGowen, Chet C. Sherwood, Lawrence I. Grossman, Morris Goodman, and Derek E. Wildman. 2012. "Comparative Analysis of Encephalization in Mammals Reveals Relaxed Constraints on Anthropoid Primate and Cetacean Brain Scaling." *Journal of Evolutionary Biology* 25: 981–94. https://doi.org/10.1111/j.1420-9101.2012.02491.x.

Braulik, Gill T. 2014. "Family Platanistidae (South Asian River Dolphin)." In *Handbook of the Mammals of the World*, vol. 4, *Sea Mammals*, edited by Don E. Wilson and Russell A. Mittermeier, 358–63. Lynx Edicions.

Braulik, Gill T., Frederick I. Archer, Uzma Khan, et al. 2021. "Taxonomic Revision of the South Asian River Dolphins (*Platanista*): Indus and Ganges River Dolphins Are Separate Species." *Marine Mammal Science* 37: 1022–59. https://doi.org/10.1111/mms.12801.

Brooks, Daniel R., and Deborah A. McLennan. 1991. *Phylogeny, Ecology, and Behavior: A Research Program in Comparative Biology*. University of Chicago Press.

Caballero, Susana, F. Trujillo, J. A. Vianna, et al. 2007. "Taxonomic Status of the Genus *Sotalia*: Species Level Ranking for 'Tucuxi' (*Sotalia fluviatilis*) and 'Costero' (*Sotalia guianensis*) Dolphins." *Marine Mammal Science* 23: 358–86. https://doi.org/10.1111/j.1748-7692.2007.00110.x.

Carroll, Emma L., Michael R. McGowen, Morgan L. McCarthy, et al. 2021. "Speciation in the Deep: Genomics and Morphology Reveal a New Species of Beaked Whale *Mesoplodon eueu*." *Proceedings of the Royal Society B* 288: 20211213. https://doi.org/10.1098/rspb.2021.1213.

Carwardine, Mark. 2020. *Handbook of Whales, Dolphins, and Porpoises*. Bloomsbury.

Churchill, Morgan, Manuel Martínez-Cáceres, Christian de Muizon, Jessica Mnieckowski, Jonathan H. Geisler. 2016. "The Origin of High-Frequency Hearing in Whales." *Current Biology* 26: 2144–49. https://doi.org/10.1016/j.cub.2016.06.004.

Clapham, Phillip J. 2000. "The Humpback Whale: Seasonal Feeding and Breeding in a Baleen Whale." In *Cetacean Societies: Field Studies of Dolphins and Whales*, edited by Janet Mann, Richard C. Connor, Peter L. Tyack, and Hal Whitehead, 173–96. University of Chicago Press.

Clapham, Phillip J. 2018. "Humpback Whale." In *Encyclopedia of Marine Mammals*, 3rd ed., edited by Bernd Würsig, Johannes G. M. Thewissen, and Kit M. Kovacs, 489–92. Academic Press.

Clementz, Mark T., Anjali Goswami, Philip D. Gingerich, and Paul L. Koch. 2006. "Isotopic Records from Early Whales and Sea Cows: Contrasting Patterns of Ecological Transition."

Journal of Vertebrate Paleontology 26: 355–70. https://doi.org/10.1671/0272-4634(2006)26[355:IRFEWA]2.0.CO;2.

Connor, Richard C., Randall S. Wells, Janet Mann, and Andrew J. Read. 2000. "The Bottlenose Dolphin: Social Relationships in a Fission-Fusion Society." In *Cetacean Societies: Field Studies of Dolphins and Whales*, edited by Janet Mann, Richard C. Connor, Peter L. Tyack, and Hal Whitehead, 91–126. University of Chicago Press.

Coombs, Ellen J., Julien Clavel, Travis Park, Morgan Churchill, and Anjali Goswami, A. 2020. "Wonky Whales: The Evolution of Cranial Asymmetry in Cetaceans." *BMC Biology* 18: 86. https://doi.org/10.1186/s12915-020-00805-4.

Coombs, Ellen J., Ryan N. Felice, Julien Clavel, et al. 2022. "The Tempo of Cetacean Cranial Evolution." *Current Biology* 32: 2233–47. https://doi.org/10.1016/j.cub.2022.04.060.

Costa, Ana P. B., Wayne McFee, Lynsey A. Wilcox, Frederick I. Archer, and Patricia E. Rosel. 2022. "The Common Bottlenose Dolphin (*Tursiops truncatus*) Ecotypes of the Western North Atlantic Revisited: An Integrative Taxonomic Investigation Supports the Presence of Distinct Species." *Zoological Journal of the Linnean Society* 196: 1608–36. https://doi.org/10.1093/zoolinnean/zlac025.

Cranford, Ted W., Mats Amundin, and Kenneth S. Norris. 1996. "Functional Morphology and Homology in the Odontocete Nasal Complex: Implications for Sound Generation." *Journal of Morphology* 228: 223–85. https://doi.org/10.1002/(SICI)1097-4687(199606)228:3<223::AID-JMOR1>3.0.CO;2-3.

Dalebout, Merel L., C. Scott Baker, Debbie Steel, et al. 2014. "Resurrection of *Mesoplodon hotaula* Deraniyagala 1963: A New Species of Beaked Whale in the Tropical Indo-Pacific." *Marine Mammal Science* 30: 1081–108. https://doi.org/10.1111/mms.12113.

Dalebout, Merel L., James G. Mead, C. Scott Baker, Alan N. Baker, and Anton L. van Helden. 2002. "A New Species of Beaked Whale *Mesoplodon perrini* sp. n. (Cetacea: Ziphiidae) Discovered Through Phylogenetic Analyses of Mitochondrial DNA Sequences." *Marine Mammal Science* 18: 577–608. https://doi.org/10.1111/j.1748-7692.2002.tb01061.x.

Dalebout, Merel L., Debbie Steel, and C. Scott Baker. 2008. "Phylogeny of the Beaked Whale Genus *Mesoplodon* (Ziphiidae: Cetacea) Revealed by Nuclear Introns: Implications for the Development of Male Tusks." *Systematic Biology* 57: 857–75. https://doi.org/10.1080/10635150802559257.

DeCasien, Alex R., Nicole A. Thompson, Scott A. Williams, and Milena R. Shattuck. 2018. "Encephalization and Longevity Evolved in a Correlated Fashion in Euarchontoglires but Not in Other Mammals." *Evolution* 72: 2617–31. https://doi.org/10.1111/evo.13633.

Dutoit, Ludovic, Kieren J. Mitchell, Nicolas Dussex, et al. 2023. "Convergent Evolution of Skim Feeding in Baleen Whales." *Marine Mammal Science* 39: 1337–43. https://doi.org/10.1111/mms.13047.

Ellis, Richard, and James G. Mead. 2017. *Beaked Whales*. Johns Hopkins University Press.

Fedutin, Ivan D., Olga A. Filatova, Evgeny G. Mamaev, Alexandr M. Burdin, and Erich Hoyt. 2015. "Occurrence and Social Structure of Baird's Beaked Whales, *Berardius bairdii*, in the Commander Islands, Russia." *Marine Mammal Science* 31: 853–65. https://doi.org/10.1111/mms.12204.

Foley, Nicole M., Victor C. Mason, Andrew J. Harris, et al. 2023. "A Genomic Timescale for Placental Mammal Evolution." *Science* 380: eab8189. https://doi.org/10.1126/science.abl8189.

Fox, Kieran C. R., Michael Muthukrishna, and Susanne Shultz. 2017. "The Social and Cultural

Roots of Whale and Dolphin Brains." *Nature Ecology and Evolution* 1 (11): 1699–705. https://doi.org/10.1038/s41559-017-0336-y.

Galatius, Anders, Rachel Racicot, Michael McGowen, and Morten T. Olsen. 2020. "Evolution and Diversification of Delphinid Skull Shapes." *iScience* 23: 101543. https://doi.org/10.1016/j.isci.2020.101543.

Gatesy, John, Jonathan H. Geisler, Joseph Chang, et al. 2013. "A Phylogenetic Blueprint for a Modern Whale." *Molecular Phylogenetics and Evolution* 66: 479–506. https://doi.org/10.1016/j.ympev.2012.10.012.

Gatesy, John, and Michael R. McGowen. 2021. "Higher-Level Phylogeny of Baleen Whales." In *The Bowhead Whale Balaena mysticetus: Biology and Human Interactions*, edited by J. C. George and J. G. M. Thewissen, 3–10. Academic Press. https://doi.org/10.1016/B978-0-12-818969-6.00001-7.

Geisler, Jonathan H. 2018. "Neoceti." In *Encyclopedia of Marine Mammals*, 3rd ed., edited by Bernd Würsig, Johannes G. M. Thewissen, and Kit M. Kovacs, 631–32. Academic Press.

Geisler, Jonathan H., Matthew W. Colbert, and James L. Carew. 2014. "A New Fossil Species Supports an Early Origin for Toothed Whale Echolocation." *Nature* 508: 383–86. https://doi.org/10.1038/nature13086.

Geisler, Jonathan H., Michael R. McGowen, Guang Yang, and John Gatesy. 2011. "A Supermatrix Analysis of Genomic, Morphological, and Paleontological Data from Crown Cetacea." *BMC Evolutionary Biology* 11: 112. https://doi.org/10.1186/1471-2148-11-112.

Gingerich, Philip D. 1998. "Paleobiological Perspectives on Mesonychia, Archaeoceti, and the Origin of Whales." In *The Emergence of Whales. Advances in Vertebrate Paleobiology*, vol. 1, edited by J. G. M. Thewissen, 423–49. Springer. https://doi.org/10.1007/978-1-4899-0159-0_15.

Gingerich, Philip D., B. Holly Smith, and Elwyn L. Simons. 1990. "Hind Limbs of Eocene *Basilosaurus*: Evidence of Feet in Whales." *Science* 249: 154–57. https://doi.org/10.1126/science.249.4965.154.

Gingerich, Philip D., Munir Ul Haq, Iyad S. Zalmout, Intizar Hussain Khan, and M. Sadiq Malkani. 2001. "Origin of Whales from Early Artiodactyls: Hands and Feet of Eocene Protocetidae from Pakistan." *Science* 293: 2239–42. https://doi.org/10.1126/science.1063902.

Goldbogen, Jeremy A., D. E. Cade, J. Calambokidis, et al. 2017. "How Baleen Whales Feed: The Biomechanics of Engulfment and Filtration." *Annual Review of Marine Science* 9: 367–86. https://doi.org/10.1146/annurev-marine-122414-033905.

Goldbogen, Jeremy A., and P. T. Madsen. 2018. "The Evolution of Foraging Capacity and Gigantism in Cetaceans." *Journal of Experimental Biology* 221 (11): jeb166033. https://doi.org/10.1242/jeb.166033.

Gowans, Shannon, Bernd Würsig, and Leszek Karczmarski. 2007. "The Social Structure and Strategies of Delphinids: Predictions Based on an Ecological Framework." *Advances in Marine Biology* 53: 195–294. https://doi.org/10.1016/S0065-2881(07)53003-8.

Groot, Nikita E., Rochelle Constantine, Ellen C. Garland, and Emma L. Carroll. 2023. "Phylogenetically Controlled Life History Trait Meta-Analysis in Cetaceans Reveals Unexpected Negative Brain Size and Longevity Correlation." *Evolution* 77 (2): 534–49. https://doi.org/10.1093/evolut/qpac050.

Harvey, Paul H. and Mark D. Pagel. 1991. *The Comparative Method in Evolutionary Biology*. Oxford University Press.

Hassanin, Alexandre, Frédéric Delsuc, Anne Ropiquet, et al. 2012. "Pattern and Timing of Diversification of Cetartiodactyla (Mammalia, Laurasiatheria), as Revealed by a Comprehensive

Analysis of Mitochondrial Genomes." *Comptes Rendus Biologies* 335: 32–50. https://doi.org/10.1016/j.crvi.2011.11.002.

Herculano-Houzel, Suzana. 2011. "Not All Brains Are Made the Same: New Views on Brain Scaling in Evolution." *Brain Behavior and Evolution* 78: 22–36. https://doi.org/10.1159/000327318.

Hooker, Sascha K., Natacha Aguilar de Soto, Robin W. Baird, et al. 2019. "Future Directions in Research on Beaked Whales." *Frontiers in Marine Science* 5: 514. https://doi.org/10.3389/fmars.2018.00514.

Jaeggi, A. V., M. I. Miles, M. Festa-Bianchet, C. Schradin, and L. D. Hayes. 2020. "Variable Social Organization is Ubiquitous in Artiodactyla and Probably Evolved from Pair-Living Ancestors." *Proceedings of the Royal Society B* 287: 2020035. https://dx.doi.org/10.1098/rspb.2020.0035.

Jefferson, Thomas A. and Howard C. Rosenbaum. 2014. "Taxonomic Revision of the Humpback Dolphins (*Sousa* spp.), and Description of a New Species from Australia." *Marine Mammal Science* 30: 1494–541. https://doi.org/10.1111/mms.12152.

Jerison, Harry J. 1973. *Evolution of the Brain and Intelligence*. Academic Press.

Lammers, Fritjof, Moritz Blumer, Cornelia Rücklé, and Maria A. Nilsson. 2019. "Retrophylogenomics in Rorquals Indicate Large Ancestral Population Sizes and a Rapid Radiation." *Mobile DNA* 10: 5. https://doi.org/10.1186/s13100-018-0143-2.

Lewison, Rebecca L. 2011. "Family Hippopotamidae (Hippopotamuses)." In *Handbook of the Mammals of the World*, vol. 2, *Hoofed Mammals*, edited by Don E. Wilson and Russell A. Mittermeier, 308–19. Lynx Edicions.

Lloyd, Graeme T., and Graham J. Slater. 2021. "A Total-Group Phylogenetic Metatree for Cetaceans and the Importance of Fossil Data in Diversification Analyses." *Systematic Biology* 70: 922–39. https://doi.org/10.1093/sysbio/syab002.

MacLeod, Colin D. 2014. "Family Ziphiidae (Beaked Whales)." In *Handbook of the Mammals of the World*, vol. 4, *Sea Mammals*, edited by Don E. Wilson and Russell A. Mittermeier, 326–57. Lynx Edicions.

Madar, S. I. 2007. "The Postcranial Skeleton of Early Eocene Pakicetid Cetaceans." *Journal of Paleontology* 81: 176–200. https://doi.org/10.1666/0022-3360(2007)81[176:TPSOEE]2.0.CO;2.

Madsen, Peter T., Ursula Siebert, and Coen P. H. Elemans. 2023. Toothed Whales Use Distinct Vocal Registers for Echolocation and Communication. *Science* 379: 928–33. https://doi.org/10.1126/science.adc957.

Mann, Janet. 2000. "Unraveling the Dynamics of Social Life: Long-Term Studies and Observational Methods." In *Cetacean Societies: Field Studies of Dolphins and Whales*, edited by Janet Mann, Richard C. Connor, Peter L. Tyack, and Hal Whitehead, 45–64. University of Chicago Press.

Mann, Janet, Richard C. Connor, Peter L. Tyack, and Hal Whitehead, eds. 2000. *Cetacean Societies: Field Studies of Dolphins and Whales*. University of Chicago Press.

Marino, Lori. 1998. "A Comparison of Encephalization Between Odontocete Cetaceans and Anthropoid Primates." *Brain Behavior and Evolution* 51: 230–38. https://doi.org/10.1159/000006540.

Marino, Lori, Richard C. Connor, R. Ewan Fordyce, et al. 2007. "Cetaceans Have Complex Brains for Complex Cognition." *PLOS Biology* 5 (5): e139. https://doi.org/10.1371/journal.pbio.0050139.

Marino, Lori, Daniel W. McShea, and Mark D. Uhen. 2004. "Origin and Evolution of Large

Brains in Toothed Whales." *Anatomical Record Part A* 281: 1247–55. https://doi.org/10.1002/ar.a.20128.

Marx, Felix G., and R. Ewan Fordyce. 2015. "Baleen Boom and Bust: A Synthesis of Mysticete Phylogeny, Diversity and Disparity." *Royal Society Open Science* 2: 140434. https://dx.doi.org/10.1098/rsos.140434.

Marx, Felix G., Olivier Lambert, and Mark D. Uhen. 2016. *Cetacean Paleobiology*. Chichester, UK: Wiley-Blackwell.

May-Collado, Laura J., Ingi Agnarsson, and Douglas Wartzok. 2007. "Phylogenetic Review of Tonal Sound Production in Whales in Relation to Sociality." *BMC Evolutionary Biology* 7 (1): 136. https://doi.org/10.1186/1471-2148-7-136.

McAlpine, Donald F. 2014. "Family Kogiidae (Pygmy Sperm Whales)." In *Handbook of the Mammals of the World*, vol. 4, *Sea Mammals*, edited by Don E. Wilson and Russell A. Mittermeier, 318–25. Lynx Edicions.

McCurry, Matthew R., Felix G. Marx, Alistair R. Evans, et al. 2021. "Brain Size Evolution in Whales and Dolphins: New Data from Fossil Mysticetes." *Biological Journal of the Linnean Society* 133: 990–98. https://doi.org/10.1093/biolinnean/blab054.

McGowen, Michael R., Michelle Spaulding, and John Gatesy. 2009. "Divergence Date Estimation and a Comprehensive Molecular Tree of Extant Cetaceans." *Molecular Phylogenetics and Evolution* 53: 891–906. https://doi.org/10.1016/j.ympev.2009.08.018.

McGowen, Michael R., Georgia Tsagkogeorga, Sandra Álvarez-Carretero, et al. 2020. "Phylogenomic Resolution of the Cetacean Tree of Life Using Target Sequence Capture." *Systematic Biology* 69 (3): 479–501. https://doi.org/10.1093/sysbio/syz068.

McKenna, Megan F., Ted W. Cranford, Annalisa Berta, and Nicolas D. Pyenson. 2011. "Morphology of the Odontocete Melon and its Implications for Acoustic Function." *Marine Mammal Science* 28: 690–713. https://doi.org/10.1111/j.1748-7692.2011.00526.x.

Meredith, Robert W., Jan E. Janečka, John Gatesy, et al. 2011. "Impacts of the Cretaceous Terrestrial Revolution and KPg Extinction on Mammal Diversification." *Science* 334: 521–24. https://doi.org/10.1126/science.1211028.

Milinkovitch, Michel C., Guillermo Orti, and Axel Meyer. 1993. "Revised Phylogeny of Whales Suggested by Mitochondrial Ribosomal DNA Sequences." *Nature* 361: 346–48. https://doi.org/10.1038/361346a0.

Møhl, Bertel, W. W. L. Au, J. Pawloski, and P. E. Nachtigall. 1999. "Dolphin Hearing: Relative Sensitivity as a Function of Point of Application of a Contact Sound Source in the Jaw and Head Region." *Journal of the Acoustical Society of America* 105: 3421–24. https://doi.org/10.1121/1.426959.

Montgelard, Claudine, François M. Catzeflis, and Emmanuel Douzery. 1997. "Phylogenetic Relationships of Artiodactyls and Cetaceans as Deduced from the Comparison of Cytochrome B and 12S rRNA Mitochondrial Sequences." *Molecular Biology and Evolution* 14: 550–59. https://doi.org/10.1093/oxfordjournals.molbev.a025792.

Montgomery, Stephen H., Jonathan H. Geisler, Michael R. McGowen, Charlotte Fox, Lori Marino, and John Gatesy. 2013. "The Evolutionary History of Cetacean Brain and Body Size." *Evolution* 67: 3339–53. https://doi.org/10.1111/evo.12197.

Nikaido, Masato, Alejandro P. Rooney, Norihiro Okada. 1999. "Phylogenetic Relationships Among Cetartiodactyls Based on Insertions of Short and Long Interspersed Elements: Hippopotamuses are the Closest Extant Relatives of Whales." *Proceedings of the National Academy of Sciences* 96: 10261–66. https://doi.org/10.1073/pnas.96.18.10261.

O'Leary, Maureen A., Jonathan I. Bloch, John J. Flynn, et al. 2013. "The Placental Mammal Ancestor and the Post-K-Pg Radiation of Placentals." *Science* 339: 662–67. https://doi.org/10.1126/science.1229237.

Pivorunas, August. 1979. "The Feeding Mechanisms of Baleen Whales." *American Scientist* 67: 432–40.

Podos, Jeffrey, Vera M. F. da Silva, and Marcos R. Rossi-Santos. 2002. "Vocalizations of Amazon River Dolphins, *Inia geoffrensis*: Insights into the Evolutionary Origins of Delphinid Whistles." *Ethology* 108: 601–12. https://doi.org/10.1046/j.1439-0310.2002.00800.x.

Rasmussen, Kristin, Daniel M. Palacios, John Calambokidis, et al. 2007. "Southern Hemisphere Humpback Whales Wintering off Central America: Insights from Water Temperature into the Longest Mammalian Migration." *Biology Letters* 3 (3): 302–5. https://doi.org/10.1098/rsbl.2007.0067.

Reidenberg, Joy S., and Jeffrey T. Laitman. 2007. "Discovery of a Low Frequency Sound Source in Mysticeti (Baleen Whales): Anatomical Establishment of a Vocal Fold Homolog." *Anatomical Record* 290: 745–59. https://doi.org/10.1002/ar.20544.

Rendell, Luke, Mauricio Cantor, Shane Gero, Hal Whitehead, and Janet Mann. 2019. "Causes and Consequences of Female Centrality in Cetacean Societies." *Philosophical Transactions of the Royal Society B* 374 (1780): 20180066. https://doi.org/10.1098/rstb.2018.0066.

Ridgway, Sam H., Kevin P. Carlin, Kaitlin R. Van Alstyne, Alicia C. Hanson, and Raymond J. Tarpley. 2016. "Comparison of Dolphins' Body and Brain Measurements with Four Other Groups of Cetaceans Reveals Great Diversity." *Brain Behavior and Evolution* 88: 235–57. https://doi.org/10.1159/000454797.

Ridgway, Sam H., and Alicia C. Hanson. 2014. "Sperm Whales and Killer Whales with the Largest Brains of All Toothed Whales Show Extreme Differences in Cerebellum." *Brain Behavior and Evolution* 83: 266–74. https://doi.org/10.1159/000360519.

Rosel, Patricia E., Lynsey A. Wilcox, Tadasu K. Yamada, and Keith D. Mullin. 2021. "A New Species of Baleen Whale (*Balaenoptera*) from the Gulf of Mexico, with a Review of Its Geographic Distribution." *Marine Mammal Science* 37: 577–610. https://doi.org/10.1111/mms.12776.

Rychel, Amanda L., Tod W. Reeder, and Annalisa Berta. 2004. "Phylogeny of Mysticete Whales Based on Mitochondrial and Nuclear Data." *Molecular Phylogenetics and Evolution* 32: 892–901. https://doi.org/10.1016/j.ympev.2004.02.020.

Slater, Graham J., Jeremy A. Goldbogen, and Nicholas D. Pyenson. 2017. "Independent Evolution of Baleen Whale Gigantism Linked to Plio-Pleistocene Ocean Dynamics." *Proceedings of the Royal Society B* 284: 20170546. https://dx.doi.org/10.1098/rspb.2017.0546.

Slater, Graham J., Samantha A. Price, Francesco Santini, Michael E. Alfaro. 2010. "Diversity Versus Disparity and the Radiation of Modern Cetaceans." *Proceedings of the Royal Society B* 277: 3097–104. https://doi.org/10.1098/rspb.2010.0408.

Smaers, Jeroen B., Ryan S. Rothman, Daphne R. Hudson, et al. 2021. "The Evolution of Mammalian Brain Size." *Science Advances* 7: eabe2101. https://doi.org/10.1126/sciadv.abe2101.

Society for Marine Mammalogy Committee on Taxonomy. 2025. "List of Marine Mammal Species and Subspecies." Accessed July 23, 2025. https://marinemammalscience.org/science-and-publications/list-marine-mammal-species-subspecies/.

Spaulding, M., Maureen A. O'Leary, and John Gatesy. 2009. "Relationships of Cetacea (Artiodactyla) Among Mammals: Increased Taxon Sampling Alters Interpretations of Key Fossils and Character Evolution." *PLOS ONE* 4: e7062. https://doi.org/10.1371/journal.pone.0007062.

Street, Sally E., Ana F. Navarrete, Simon M. Reader, and Kevin N. Laland. 2017. "Coevolution of Cultural Intelligence, Extended Life History, Sociality, and Brain Size in Primates." *Proceedings of the National Academy of Sciences* 114: 7908–14. https://doi.org/10.1073/pnas.1620734114.

Tanaka, Yoshihiro. 2022. "Rostrum Morphology and Feeding Strategy of the Baleen Whales Indicate That Right Whales and Pygmy Right Whales Became Skimmers Independently." *Royal Society Open Science* 9: 221353. https://doi.org/10.1098/rsos.221353.

Thévenet, Julie, Nicolas Grimault, Paulo Fonseca, and Nicolas Mathevon. 2022. "Voice-Mediated Interactions in a Megaherbivore." *Current Biology* 32: R55–71. https://doi.org/10.1016/j.cub.2021.12.017.

Thewissen, Johannes G. M., Lisa Noelle Cooper, Mark T. Clementz, Sunil Bajpai, and B. N. Tiwari. 2007. "Whales Originated from the Aquatic Artiodactyls in the Eocene Epoch of India." *Nature* 450: 1190–95. https://doi.org/10.1038/nature06343.

Thewissen, Johannes G. M., E. M. Williams, L. J. Roe, and S. T. Hussain. 2001. "Skeletons of Terrestrial Cetaceans and the Relationships of Whales to Artiodactyls." *Nature* 413: 277–81. https://doi.org/10.1038/35095005.

Uhen, Mark D. 2010. "The Origin(s) of Whales." *Annual Review of Earth and Planetary Sciences* 38: 189–219.

Van Schaik, Carel P., Zegni Triki, Redouan Bshary, and Sandra A. Heldstab. 2021. "A Farewell to the Encephalization Quotient: A New Brain Size Measure for Comparative Primate Cognition." *Brain Behavior and Evolution* 96: 1–12. https://doi.org/10.1159/000517013.

Wada, Shiro, Masayuki Oishi, and Tadasu K. Yamada. 2003. "A Newly Discovered Species of Living Baleen Whale." *Nature* 426: 278–81. https://doi.org/10.1038/nature02103.

Wade, Paul R., Randall R. Reeves, and Sarah L. Mesnick. 2012. "Social and Behavioural Factors in Cetacean Responses to Overexploitation: Are Odontocetes Less 'Resilient' Than Mysticetes?" *Journal of Marine Biology* 15: e567276. https://doi.org/10.1155/2012/567276.

Wang, John Y., Shih Chu Yang, Bi J. Wang, and Lo S. Wang. 2010. "Distinguishing Between Two Species of Finless Porpoises (*Neophocaena phocaenoides* and *N. asiaeorientalis*) in Areas of Sympatry." *Mammalia* 74: 305–10. https://doi.org/10.1515/mamm.2010.029.

Waugh, David A., and Johannes G. M. Thewissen. 2021. "The Pattern of Brain-Size Change in the Early Evolution of Cetaceans." *PLOS ONE* 16: e0257803. https://doi.org/10.1371/journal.pone.0257803.

Whitehead, Hal. 2003. *Sperm Whales: Social Evolution in the Ocean*. University of Chicago Press.

Wolf, Magnus, Konstantin Zapf, Deepak Kumar Gupta, Michael Hiller, Úlfur Árnason, and Axel Janke. 2023. "The Genome of the Pygmy Right Whale Illuminates the Evolution of Rorquals." *BMC Biology* 21: 79. https://doi.org/10.1186/s12915-023-01579-1.

Wray, Janie, Eric M. Keen, and Éadin N. O'Mahony. 2021. "Social Survival: Humpback Whales (*Megaptera novaeangliae*) Use Social Structure to Partition Ecological Niches Within Proposed Critical Habitat." *PLOS ONE* 16 (6): e0245409. https://doi.org/10.1371/journal.pone.0245409.

Yamada, Tadasu K., Shino Kitamura, Syuiti Abe, et al. 2019. "Description of a New Species of Beaked Whale (*Berardius*) Found in the North Pacific." *Scientific Reports* 9: 12723. https://doi.org/10.1038/s41598-019-46703-w.

Yuan, Yuan, Yaolei Zhang, Peijun Zhang, et al. 2021. "Comparative Genomics Provides Insights into the Aquatic Adaptations of Mammals." *Proceedings of the National Academy of Sciences* 118: e2106080118. https://doi.org/10.1073/pnas.2106080118.

2

Genomic Insights into Demographic and Evolutionary Histories

ANDREW D. FOOTE, MARIE LOUIS, ALANA ALEXANDER, EMMA L. CARROLL, AND MICHAEL C. FONTAINE

Abstract

The formation and shaping of cetacean societies (broadly defined here as genetically and socially interconnected individuals) is often associated with genetic revolutions—founder effects, range expansions, and genetic replacement of populations through migration—all of which can dramatically change the genetic composition of a population. Newly formed societies may also encounter novel selection pressures that act upon the phenotype but also shape genomic variation within a society. In species capable of adaptation through innovation and social learning, cultural and genetic evolution can react to the same selection pressures and act synergistically. Cultural and genetic variation can also be shaped by the same demographic processes, so that they become correlated without either mode of evolution driving variation in the other. Both types of interaction have been categorized as gene-culture coevolution. In this chapter, we highlight how the advent of the genomics era has provided unprecedented resolution of the ancestral processes that shaped present-day cetacean societies. We focus on the cetacean species for which there is comparative genomic and sociality data. Throughout we draw comparisons with the evolution of human societies, which share much in common with cetacean societies. Rather than providing an exhaustive account of all the insights genomics can provide, we select a subset of themes we believe are fundamental to understanding the formation of cetacean societies.

Introduction

Cetacean Societies (Mann et al. 2000) preceded publication of the first human genome (Venter et al. 2001). This landmark achievement took more than a decade to complete, due to the use of traditional Sanger sequencing approaches to painstakingly construct the genome one small piece at a time. The same laborious approach was also used on a subset of diverse taxa for comparative genomic analyses (see Lindblad-Toh et al. 2011) and included the first cetacean genome, Turtru1.0, published in 2008 from a bottlenose dolphin, *Tursiops truncatus*. Thankfully, the advent of high-throughput sequencing in the late 2000s changed the field of genomics for non-model organisms, allowing rapid, parallel sequencing of many individuals for a reasonable cost (Kircher and Kelso 2010; Morin et al. 2010; Davey et al. 2011). Through subsequent iterations of sequencing platforms, capacity increased and costs dropped, ushering in the era of population genomics (Pennisi 2010).

Genomes as Records of Past Population Dynamics

The biparentally inherited portion of the genome (e.g., the autosomes) of diploid organisms consists of blocks of ancestry from each parent, randomly shuffled by recombination during meiosis. These blocks contain ancestry from four grandparents, eight great-grandparents, and so on (fig. 2.1). By sequencing a single genome, we potentially sample the genetic contributions of up to 1,024 genealogical ancestors from just the past 10 generations (approximately the past 250 years for humans and killer whales [*Orcinus orca*], assuming a generation time of 25 years). This estimate is based upon the theoretical number of genealogical ancestors from each past generation. However, when we consider evolutionary timescales, for example, the past 25,000 years (approximately 1,000 human and killer whale generations), the number of theoretical genealogical ancestors from the thousandth generation in the past would be an astronomical 2^{1000}, i.e., more than the number of humans or killer whales alive at that time or that have ever existed (Haub 1995). We know this could not possibly be true, but how do we reconcile the theoretical expectation of the number of ancestors with reality? Genealogies collapse and intersect back in time, as ancestors from one generation may share common ancestors from prior generations (Derrida et al. 2000; Coop 2020; fig. 2.1). In addition to genealogical collapse, meiotic recombination stochastically reshuffles genetic blocks from each parent, resulting in only a subset of genealogical ancestors contributing their genetic variation to the present-day generation (Coop 2020). The more generations back in time we consider, the less genetic variation each genealogical ancestor contributes to the present-day ancestry (Coop 2020, 2022 fig. 2.1). Thus, the exponential relationship between the number of genealogical ancestors from each prior generation quickly breaks down with each past generation when we consider effective genetic ancestors (those that contributed ancestry to the current population).

FIGURE 2.1. These schematic genealogies show the genetic contribution of ancestors from the previous four generations to the genome of a single individual. (a) The number of theoretical genealogical ancestors doubles each generation back in time, but the genetic contribution varies stochastically (indicated by shading) due to the random processes of recombination and transmission. As we see in panel (b), genealogies collapse, reducing the actual number of genetic ancestors. In this example, the collapse is due to two grandparents being sired by the same father from the previous generation. This single collapse results in a reduction of 128 and 131,072 potential ancestors from the tenth and twentieth generations back in time, respectively. Such collapses will be increasingly common each generation back, as the number of genealogical ancestors increases. The rate of such collapses at each past generation provides insights into the past demography of populations.

When genealogies collapse due to sharing a common ancestor, we say that the genetic variation coalesces in this common ancestor. These coalescent events within individual genomes and among the genomes of individuals are the basis of coalescent theory (Wakeley 2009; Rosenberg and Nordborg 2002), and the associated statistical framework upon which much of modern population genetics is based. For example, patterns of coalescence can inform us of changes in population size through time. An intuitive way to visualize this is to compare the chances of both your parents sharing a recent common ancestor under two scenarios. The first scenario is that they live in a bustling cosmopolitan metropolis like New York City, where inhabitants from all over the world have migrated in the last 10 generations. The second scenario is that they live on a remote isolated island with no further migration since it was first colonized. Clearly, under the second scenario, it would be more likely that your parents share common ancestors and that their genomes would share many genealogies. Consequently, your genome would comprise a large proportion of identical DNA sequences from each parent (known as runs of homozygosity, or ROH). We can therefore use the rate of coalescence and the distribution of lengths of identical-by-descent tracts across the genome to reconstruct effective population size through time (e.g., Li and Durbin 2011; Kirin et al. 2010). Our genomes (and the genomes of cetaceans) are thus a record of our demographic and evolutionary history: a composite of the changes in population size through time, migration between populations, and signatures of strong selection on specific regions of the genome (Ceballos et al. 2018).

Genomic Revolutions and the Formation of Cetacean Societies

The term "genomic revolution" refers to the dramatic change in genetic composition at the wave front of a population expansion as a species colonizes new territory and expands from a small founder group to occupy the available habitat (Barton and Charlesworth 1984; Excoffier and Ray 2008; fig. 2.2). There are some parallels with the "cultural revolutions" in humpback whales (*Megaptera novaeangliae*), whereby song characteristics sweep across an ocean basin, replacing previous song variants (Garland et al. 2011; Hersh et al., this volume). Ancestral migrations of human populations out of Africa and across Eurasia (Deshpande et al. 2009; Henn et al. 2012, 2016), postglacial range expansions of terrestrial flora and fauna (Hewitt 2000), and even experiments of bacterial and yeast colonies on a petri dish (Hallatschek et al. 2007) leave these same genetic patterns at the range margins. The small population size at the leading edge of an expanding population results in increased genetic drift and thus higher fixation rates of alleles, a phenomenon known as "allele surfing" (Excoffier and Ray 2008; Excoffier et al. 2009). This in turn reduces heterozygosity, which can result in an increase in deleterious recessive mutations being expressed in homozygote genotypes (e.g., Lohmueller et al. 2008) where they had been previously harbored neutrally in heterozygous genotypes (Peischl et al. 2013; Peischl and Excoffier 2015).

Patterns of genetic variation resulting from genomic revolutions are also dependent upon the social relationships among individuals and, in particular, who mates with whom (see Parsons et al., this volume). For example, the insular nature of the resident killer whale populations of the North Pacific results in genetic variation within each resident population coalescing in a relatively recent common ancestor (Barrett-Lennard 2000; Hoelzel et al. 1998a; Parsons et al. 2013; Filatova et al. 2018; Morin et al. 2015; Foote et al. 2016). The matrilineal social structure of most killer whale populations results in increased coalescence rates through time for the maternally inherited mitochondrial DNA (mtDNA), and mtDNA haplotypes are often fixed

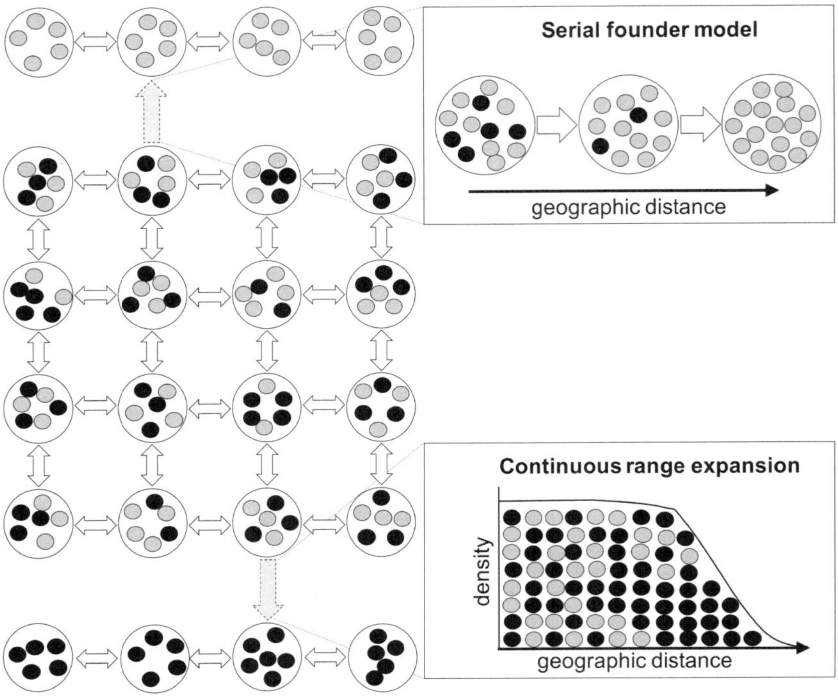

FIGURE 2.2. A simple schematic model showing allele frequencies (for two alleles, black and gray) at a polymorphic site in the genome and which are at equilibrium due to low levels of episodic admixture between demes (large circles) in the core range, while alleles drift to extinction or fixation in demes at the range margins due to serial founder effects or continuous range expansion (adapted from Foote et al. 2021b). The founder population can then expand to fill the emerging habitat at the range margins. Post expansion variation will coalesce in the founder population. Patterns consistent with these processes are found in bottlenose dolphins, killer whales, and sperm whales.

within each resident population (Barrett-Lennard 2000; Hoelzel et al. 1998a; Parsons et al. 2013; Filatova et al. 2018; Morin et al. 2010, 2015). The simplest visualization of this pattern of coalescence is arguably in a phylogenetic tree of the maternally inherited mtDNA (see Morin et al. 2015), in which all branches with a given ecotype at their tip stem from a shared ancestral node in the phylogenetic tree. For example, we see that the mitochondrial DNA sequenced from 110 resident killer whales from populations across much of the ecotype's known range, all share a recent (<20,000 years BP) common ancestor (Morin et al. 2015; Filatova et al. 2018). This is not just true of the resident ecotype, but also other high-latitude populations (Morin et al. 2015).

Coalescent rates within the nuclear genomes of these high-latitude killer whale populations tell the same story as the mitochondrial genomes (Foote et al. 2016, 2019, 2021b). The length of ROH and estimated changes in effective population size through time all suggest recent expansion from a late Pleistocene–early Holocene bottleneck for the ancestors of these high-latitude populations (Foote et al. 2021b). In each example, it was a shared ancestral population that underwent the greatest genetic changes, presumably when founding emerging high-latitude habitats. As a result, a portion of the nuclear genetic variation coalesces in a common ancestor shared with all other living descendants. The independent replication of loss of genetic diversity (relative to lower-latitude populations) in high-latitude killer whale societies that currently inhabit regions inaccessible to killer whales during the last glacial maximum (LGM) is compelling evidence of the impact of parallel range expansions and shifts.

Killer whales are not the only species that harbor profound signatures of past expansions associated with climate. The genetic impact of the founding of new societies (or populations) is found in other social odontocetes, including bottlenose dolphins (Nykänen et al. 2019), bottlenose whales *Hyperoodon ampullatus* (De Greef et al. 2022), and sperm whales (Morin et al. 2018). Common bottlenose dolphins have repeatedly transited from pelagic to coastal waters throughout their range as inferred from mitochondrial DNA and microsatellites studies (Costa et al. 2021; Hoelzel et al. 1998b; Louis et al. 2014; Nykänen et al. 2019), and more recently whole genome sequences (Louis et al. 2021). In each of those studies, coastal populations have been inferred to have reduced genetic diversity relative to the source pelagic populations. For example, in the eastern North Atlantic, Nykänen et al. (2019) inferred that coastal waters were colonized after the LGM by a small number of individuals, comprising as few as two maternal lineages, from the pelagic population.

Sperm whales show genomic signatures consistent with an expansion from a global refugia, potentially located in the Pacific, coincident with the Eemian interglacial approximately 125,000 years ago (Morin et al. 2018). Because of the shared ancestry tracing back to this refugia, genetic diversity and divergence among extant sperm whale populations is low, particularly in the mtDNA, where further coalescences have likely occurred due to the partial matrilineality of sperm whale societies (Morin et al. 2018; Alexander et al. 2016; Sarano et al. 2021; Konrad et al. 2018; see Parsons et al., this volume). These patterns are particularly pronounced in regions of the Atlantic, where dispersal limitations imposed by geographic features such as relatively (for a sperm whale!) shallow habitat (e.g., Gulf of Mexico) or narrow straits (e.g., Mediterranean Sea) further restrict the number of potential founding ancestors of a population (Violi et al. 2023; Alexander et al. 2016; Morin et al. 2018).

A counterexample of an "unsocial" odontocete is provided by the harbor porpoise (*Phocoena phocoena*) in box 2.1. The social structure of this small odontocete is much less pronounced (or even absent to the best of our knowledge) compared to the killer whale, bottlenose dolphin, and other highly social cetaceans. This has a tremendous impact on genetic diversity, geographic patterns of population structure, and population genetic processes.

Box 2.1. A Genetic Perspective on the Harbor Porpoise: The "Unsocial" Odontocete

Yacine Ben Chehida and Michael C. Fontaine

Harbor porpoises recolonized the Nordic habitats (i.e., North Sea, Baltic Sea, and other Nordic coastal waters further north) after they were freed from ice at the end of the last ice age. However, in contrast to killer whales and bottlenose dolphins, no latitudinal decrease in genetic diversity (leading-edge effect) was observed in harbor porpoises (Ben Chehida et al. 2021). Harbor porpoises have a loose and fluid social structure compared to the killer whale, bottlenose dolphins, and other highly social cetaceans (Read 1999; Teilmann and Sveegaard 2019). This small cetacean displays a rapid and demanding reproductive strategy of one calf every two years (Read 1999; Read and Hohn 1995; Lockyer 2007). The only clear and strong social bond identified so far is between mother and calf, which lasts until the next calf is born (Read 1999, 2018; Teilmann et al. 2007). Females are known to display site fidelity and philopatric behavior, returning to their native site to reproduce during the breeding season (Teilmann and Sveegaard 2019; Ben Chehida et al. 2021). Males, on the other hand, are spermatic competitors, trying to mate with as many females as possible during the breeding season (Fontaine and Barrette 1997; Keener et al. 2018). Such life history traits and fluid social structure greatly impact harbor porpoise genetic diversity and population structure. Population genetic structure is primarily driven by dispersal constraints, historical and present-day habitats, and resource variation, as well as by large oceanic barriers to dispersal with little-known influence of social and cultural structure (Ben Chehida et al. 2021; Fontaine 2016; Fontaine et al. 2007; Fontaine et al. 2014). Genetic structure has been characterized by a genetic isolation-by-distance pattern where geographically close individuals are genetically more similar to each other than more distant ones are (Ben Chehida et al. 2021; Fontaine et al. 2007). Detection of isolation-by-distance typically suggests that an equilibrium between individual dispersal and genetic drift has been met (a.k.a. migration-drift equilibrium; Leblois et al. 2004; Rousset 2004; Hutchison and Templeton 1999). So, in other words, individual intergenerational dispersal related to the fluid social structure since the last ice age was strong enough to attenuate any leading-edge effect left by the postglacial recolonization (Ben Chehida et al. 2021). No doubt fluid social structure facilitated the establishment of the migration-drift equilibrium, removing any evidence of a leading-edge effect, in contrast to other more social species like the killer whales and the bottlenose dolphins, where such leading-edge effects are prominent.

In baleen whales, climatic processes are also thought to influence both population sizes and connectivity. In both Antarctic blue (*Balaenoptera musculus musculus*) and bowhead (*Balaena mysticetus*) whales, decreases in effective population sizes are associated with the decline of their sea ice habitat after the LGM (Attard et al. 2015; Phillips et al. 2013), whereas an increase in population connectivity is seen between Indo-Pacific and South Atlantic southern right whales after the LGM (Carroll et al. 2019). Crossman et al. (2024) use whole genome sequences to disentangle the impacts of ancient bottlenecks associated with the founding of different right whale populations/species from the more recent impacts of historical whaling. The contiguity of whole genome sequences means that the timing of inbreeding can be inferred through the length of ROH (Foote et al. 2021b; Crossman et al. 2024). Whole genome sequences have also differentiated long-term small effective population size from the impacts of whaling for two fin whale populations, in the eastern North Pacific and the Gulf of California (Nigenda-Morales et al. 2023).

Adaptation to Novel Selection Pressures

The founding of new populations (or societies) following the colonization of new habitats often entails encountering novel selection pressures, including the newly acquired expansion load. Natural selection acts upon the genetic underpinnings of phenotypes and can cause small and large changes in genetic variation within and among populations (Jones et al. 2012; Almén et al. 2016). Selection can cause changes in the frequency of genetic variation underpinning beneficial phenotypes that increase fitness (positive selection), and phenotypes that result in reduced fitness (negative selection), and selection can also act to retain alternative genetic variants within a population (balancing selection). Furthermore, selection can target an individual locus or a few loci, particularly if they have a large effect on a particular (oligogenic) trait, or many loci across the genome, each having a small additive effect on a (polygenic) trait. The action of selection impacts not only the specific sites in the genome that have an effect on fitness, but also neighboring sites, sometimes stretching for thousands of base pairs adjacent to the target of selection. The impact on linked loci of positive selection (commonly called hitchhiking; Maynard Smith and Haigh 1974) and negative selection (known as background selection; Charlesworth 1994) are increasingly thought to have a genome-wide influence on genetic diversity (Hahn 2008; Pouyet et al. 2018). Linked selection therefore has an impact on estimates of effective population size and heterozygosity, which we typically estimate using methods that assume evolution under neutrality (Charlesworth and Jensen 2021). This is likely to have exacerbated the downward trajectory of effective population size through time in studies of the vaquita (Robinson et al. 2022), the type D killer whale (Foote et al. 2023), and likely others.

Positive selection has arguably attracted the most attention during the first decade of cetacean population genomics studies. Genome scans identify loci that differ in allele frequencies between populations. These methods typically use a metric related to F_{ST} to quantify genetic differentiation per site or per window of (e.g.) 10 thousand sites between two populations or ecotypes, and then set a cutoff (e.g., based on simulating neutrally evolving allele frequencies; Beaumont and Nichols 1996; or simply by selecting the top 0.01% of loci or windows) to identify genetic variation putatively evolving under positive selection (Cammen et al. 2015; Moura et al. 2014; Pratt et al. 2022; Foote et al. 2016; Louis et al. 2021; Batley et al. 2019, 2021; Zhou et al. 2018; Celemín et al. 2023). However, F_{ST} outlier peaks can also arise from other processes such as increased background selection in low recombination regions of the genome (Bank et al. 2014; Foote et al. 2016).

Robust inference that genetic changes were caused by natural selection due to ecological variation can come from studying examples of parallel adaptation (e.g., Jones et al. 2012; Schlötterer 2023). Comparison of the genomes of three coastal and pelagic bottlenose dolphin ecotype pairs found repeated adaptation to coastal waters (Louis et al. 2021). Pelagic and coastal ecotype pairs in the Atlantic and the Pacific have evolved independently, while those in the Northeast and Northwest Atlantic have a partially shared demographic and evolutionary history (Louis et al. 2021). To identify alleles at linked sites that were repeatedly associated with coastal habitat, Louis et al. (2021) used a genome scan method (Galimberti et al. 2020) that identifies loci evolving under divergent selection between ecotypes, but that also results in the sharing of alleles among coastal ecotypes. The term "linked" is an important caveat here, as single nucleotide polymorphisms (SNPs) identified by any genome scan method can be the focal target of selection, or loci physically linked on the genome to this target can change in frequency through hitchhiking (Charlesworth 1994; Maynard Smith and Haigh 1974). Habitat-associated genomic regions under repeated adaptation were older than the rest of the genome, and coastal-associated alleles were also present at low to intermediate frequency in the pelagic populations (Louis et al. 2021). Ancestral standing genetic variation present in the pelagic populations is likely the key to the bottlenose dolphins' ability to repeatedly adapt to coastal waters during interglacial periods. Genes within genomic regions inferred to be under repeated adaptation to coastal habitat were linked to cognitive abilities and feeding, suggesting coevolution with social behavior facilitating the ability of bottlenose dolphins to adapt to novel conditions. Similar patterns of parallel evolution have been documented in humans and have emphasized the interaction between genetic and cultural variation (Balentine and Bolnick 2022).

Gene-Culture Coevolution

Like humans, cetacean societies evolve through genetic inheritance, and through a second inheritance system of social learning, i.e., cultural inheritance (Whitehead and Rendell 2014; Wild and Rendell, this volume; Whiten et al. 2017). Patterns of cultural and genetic inheritance can be correlated, or even interact, so that culturally mediated traits expose successive generations to novel selection pressures that then drive changes in adaptive genetic variation. The quantitative study of gene-culture coevolution was pioneered by Cavalli-Sforza and Feldman (1973, 1981), who highlighted the relative complexity of cultural inheritance compared to genetic inheritance. For example, the modes of transmission of cultural traits include vertical transmission from parent to offspring as is the case of genetic inheritance, but also horizontal transmission among peers within the same generation, and intergenerational oblique transmission between nonparent-offspring pairs (Cavalli-Sforza and Feldman 1981).

The socially learned vocal repertoires of northeastern Pacific resident killer whales have a mostly matrilineal transmission, which results in a correlation between mtDNA and vocal repertoire (Yurk et al. 2002; Filatova et al. 2015; Hersh et al., this volume). Arguably the best-known cultural trait in cetaceans, humpback whale song (Payne and McVay 1971), has a more complex inheritance pattern. Song is produced by males on the breeding grounds, with conformity driven by horizontal cultural transmission on each breeding ground within a season (Payne and Payne 1985; Hersh et al., this volume). There is also horizontal transmission between breeding grounds; sporadically this takes the form of "song revolutions": rapid and dramatic song change across ocean basins (Garland et al. 2011; Hersh et al., this volume). Despite the decoupling of maternally inherited mtDNA variation and horizontal transmission of song among males, Garland et al. (2015) found that South Pacific humpback whale song characteristics and mtDNA haplotype sharing are broadly correlated, which they attributed to maternal site fidelity to breeding grounds (Baker et al. 1986; Jackson et al. 2015; Larsen et al. 1996).

The current era of population genomics provides unprecedented opportunities to explore the interaction of culturally and genetically heritable phenotypes (Laland et al. 2010; Whitehead et al. 2019). The primarily vertical transmission pathways of both culture and genetic variation in killer whale societies make them an attractive model system to investigate this dynamic (Foote et al. 2016). Findings include signatures of selection in two mammal-eating ecotypes on different genes within the pathway that converts dietary protein into the essential amino acid methionine; and selection on genes associated with adaptation to the cold in Antarctic killer whales (Foote et al. 2016). However, as highlighted in the section "Adaptation to Novel Selection Pressures," inferring that genetic variation is evolving under natural selection acting due to some specific selective ecological variable(s) remains challenging. Similarly, social learning of behaviors and traits is often indirectly inferred through patterns of behavioral variation, without direct observation of teaching, or through experimental manipulation (Thornton and Raihani 2010; but see Wild and Rendell, this volume, for an overview of advances over the last two decades from the ethnographic method of exclusion to more quantitative approaches to identifying the diffusion of putative cultural traits). We therefore acknowledge the challenges of robustly inferring the interaction between these two evolutionary processes primarily through correlation (Brakes et al. 2021).

Let us consider the best-studied example of gene-culture coevolution in humans: selection on the lactase gene that promotes the continued expression of lactase (lactase persistence or LP) and therefore milk tolerance into adulthood in populations with a long history of pastoral farming (Feldman and Laland 1996; Laland et al. 2010). Genomics studies identified LP as the strongest selected trait in human populations during the past 10,000 years (Sabeti et al. 2006; Enattah et al. 2008). Milk consumption was widespread throughout Europe during the Neolithic (past 10,000 years) as evidenced by fat residues on pottery and other archaeological artifacts. Yet recent ancient

DNA studies have found that the LP allele was at low frequency in central and northern Europe until approximately 3,000 years ago (Burger et al. 2020; Mathieson et al. 2015; Olalde et al. 2018). Temporal changes in LP frequency are therefore not well correlated with changes in milk exploitation (Evershed et al. 2022). Furthermore, analyses of ancient genomes have found selection on the region of the genome containing the LP allele predating the emergence of the LP allele by several thousand years, indicating other sites were the initial target of selection (Irving-Pease et al. 2024). Analyses of UK Biobank genomic data for 500,000 contemporary Europeans also found no indication of health or fitness benefits associated with LP, and only a weak association with milk consumption (Evershed et al. 2022).

While an association of the LP allele and dairy farming (culture) is not contested, it appears milk tolerance is not a strong enough selection pressure on its own, and additional selection pressures (e.g., famine conditions or increased pathogen exposure) would have been required to drive selection for the LP allele to increase it to its present high frequency in European populations (Evershed et al. 2022). Similar observations from ancient human genomes have been made in respect to other functional genetic variation and the timing of agricultural practices (Mathieson and Mathieson 2018).

Difficulties identifying the causation between the onset of a culture and changes in the frequency of associated genetic variants, and the spatiotemporal context of that genetic change, even in this exemplar of gene-culture coevolution are informative of the challenges involved. In this case, human material culture and a densely sampled time series of ancient genomes have been key to providing a more nuanced understanding of the interaction of culture, natural selection, and genetic variation. We rarely have such temporal genomic data (see section "Paleogenomics"), and no material culture, to study the spatiotemporal context of gene-culture coevolution in cetaceans. While it is often not possible to directly corroborate cultural cause and genetic effect, or vice versa, additional support can be gained through parallel studies in multiple systems, or using forward-in-time simulations that can combine cultural traits and genetic outcomes (Haller and Messer 2019; see Guez et al. 2023).

Cultural Hitchhiking

Genomic data have provided new insight into the cultural hitchhiking hypothesis (Whitehead 1998). The premise of this innovative and intuitive hypothesis is that in matrilineal cetacean societies, neutral mtDNA haplotypes can "hitchhike" on a beneficial cultural variant that increases individual and group fitness. Selection acting on the cultural variant, and indirectly on the mitochondrial haplotype, causes both to rise to high frequency in the population (analogous to the hitchhiking of neutral genetic variation linked to a selected genetic variant; Maynard Smith and Haigh 1974). The cultural hitchhiking model was extended to patriarchal human societies to explain low Y-chromosome variation (Whitehead et al. 2002) and was suggested to be a partial fit to simulated patterns in Pleistocene hominins (Premo et al. 2009). The Holocene bottleneck in Y-chromosome diversity in modern humans has also been hypothesized to be a result of the coinciding spread of Neolithic cultures, under the condition that male reproductive success was culturally mediated (Karmin et al. 2015; Zeng et al. 2018).

Subsequent empirical studies on global mtDNA datasets for matrilineal cetacean species have argued that demographic patterns associated with the glacial cycles provide the most parsimonious explanation for global variation in mtDNA diversity in killer whales and sperm whales (Hoelzel et al. 2002; Morin et al. 2015, 2018; see section on genomic revolutions). In killer whales coalescent-based estimates of the rate of birth and death of new mitochondrial lineages balance each other out through time, consistent with a stable global effective matrilineal population size (Morin et al. 2015). As noted in "Genomic Revolutions," the local changes in coalescent rates in both the mitochondrial and nuclear genomes and consequential reduced genetic diversity in populations inhabiting high-latitude waters that were inaccessible during the LGM best-fit demographic patterns associated with founder effects and subsequent expansion (Morin et al. 2015, 2018; Hoelzel et al. 2007; Foote et al. 2016, 2021b). A correlation between culture and genetic variation may also be part of this process. For example, a behavioral trait that is advantageous upon colonizing a new area or habitat might be culturally (vertically) transmitted down generations within a matriline so that there is a correlation between culture and mitochondrial variation (such correlations are found in killer whales). Additionally, demographic bottlenecks associated with range expansions could also result in the stochastic loss and fixation of cultural variants just as they do genetic variants.

While the role of cultural hitchhiking as the principal driver of global patterns of genetic diversity in matrilineal whale species is not well supported by genomic data, there is evidence that this mechanism can reduce local genetic diversity and increase genetic structuring among neighboring, culturally divergent groups. A series of detailed studies have found such patterns of genetic variation associated with a remarkable example of tool use in cetaceans. A subset of the bottlenose dolphins inhabiting Shark Bay carry sponges on

FIGURE 2.3. Gene-culture coevolution in the sponging bottlenose dolphins of Shark Bay. (*Top*) "Dodger" carrying a sponge on her rostrum, with her offspring "Bluejay" just below her. (*Bottom*) Studies on cultural and genetic variation in Shark Bay dolphins have found that sponging is primarily passed on from mother to female offspring. The transmission of sponging therefore correlates with the transmission of maternally inherited mitochondrial DNA sequence. In the schematic we see the transmission pathway (gray shading) for both the mitochondrial haplotype and sponging behavior, and that it represents a single matrilineal genealogy. The rigorous testing and rejection of alternative hypotheses make this an exemplar of correlative gene-culture coevolution in cetacean societies.
Source: (*Top*) Photo by Stephanie King, Shark Bay Dolphin Research.

their rostrum when foraging (Smolker et al. 1997). All but one of these dolphins share the same mtDNA haplotype (Krützen et al. 2005). Cultural transmission of "sponging" is primarily from mother to female offspring (Kopps et al. 2014; Wild et al. 2019), resulting in correlated mtDNA and foraging behavior (fig. 2.3). Bacher et al. (2010) systematically excluded the remote possibility that advantageous mitochondrial variation either determined this behavioral trait or was under selection for metabolic performance. Thus, the relationship between mtDNA haplotype and cultural variant appears to represent an example of cultural hitchhiking (Kopps et al. 2014) in which the matrilineal mode of cultural transmission shapes local matrilineally inherited genetic diversity and structuring.

Paleogenomics: Sampling the Ancestors of Today's Cetacean Societies

In the first sections of this chapter, we have outlined how coalescent patterns can be used to infer demographic and evolutionary changes in the past. The advances in sequencing ancient genomes mean that we now have the potential to directly sample the genomes of individuals from ancestral populations. The first studies of cetacean "paleogenomes" have utilized single individuals, or small (less than five individuals) datasets (Foote et al. 2021a; Louis et al. 2023). Despite this limitation, these studies have provided novel insights into evolutionary histories during the formation of cetacean societies. It may not surprise the reader to learn that the two species used in these studies are two of the focal species of chapters throughout this book: the killer whale and the bottlenose dolphin. In both studies the ancient genomes provided a dated calibration point with which to estimate the timing and tempo of the action of natural selection (fig. 2.4).

In the common bottlenose dolphin, Louis et al. (2023) harnessed the power of paleogenomics to infer the chronology, mode, and tempo of parallel adaptation to coastal habitat. They sequenced the genomes of four bottlenose dolphins using DNA extracted from partially mineralized bones, known as subfossils, which originated from the southern North Sea at the time of formation of local coastal habitats (8,610–5,626 years BP) following postglacial

FIGURE 2.4. (A) Ancient DNA lab work to extract DNA from an 8,600-year-old dolphin mandible. Any DNA from such an old specimen that has been buried on the seabed for millennia and stored in nonsterile conditions in museum collections will be a "metagenomic" sample of dolphin, microbial, and human DNA. Dr. Marie Louis demonstrates the stringent measures taken to avoid further contaminating ancient DNA samples with modern DNA in the clean lab, which include wearing full bodysuits, double-layered gloves, and face mask. (B) Schematic of the action of purifying selection on the genomes of Norwegian killer whales, time calibrated using the genome of a 7,500-year-old subfossil killer whale from Denmark. Changes in the proportion of nonsynonymous to synonymous mutations along branches of a tree are represented by shading dark to light, reflecting high to low ratio of nonsynonymous mutations (i.e., mutations that change the amino acid being expressed). The genome is approximately ancestral to the Norwegian killer whale genome, and it shows evidence of experiencing some, but not all, of the purging of nonsynonymous mutations by purifying selection found in the Norwegian genome. From this pattern, we deduce most of the purging took place in the subsequent 7,500 years. (C) Schematic depiction of the action of positive selection on coastal-associated alleles in bottlenose dolphins. Parallel changes in the frequency of coastal-associated alleles occur along the branches to the ancient sample and to the present-day coastal population. The ancient sample provides temporal context, showing that coastal alleles were already at detectable frequencies at the time of the formation of new coastal habitat. (D) The ancient killer whale genome shares genetic variation with the present-day Norwegian population but is not a direct ancestor. By sequencing this ancient genome, we sample genetic variation not passed on to the present-day population.

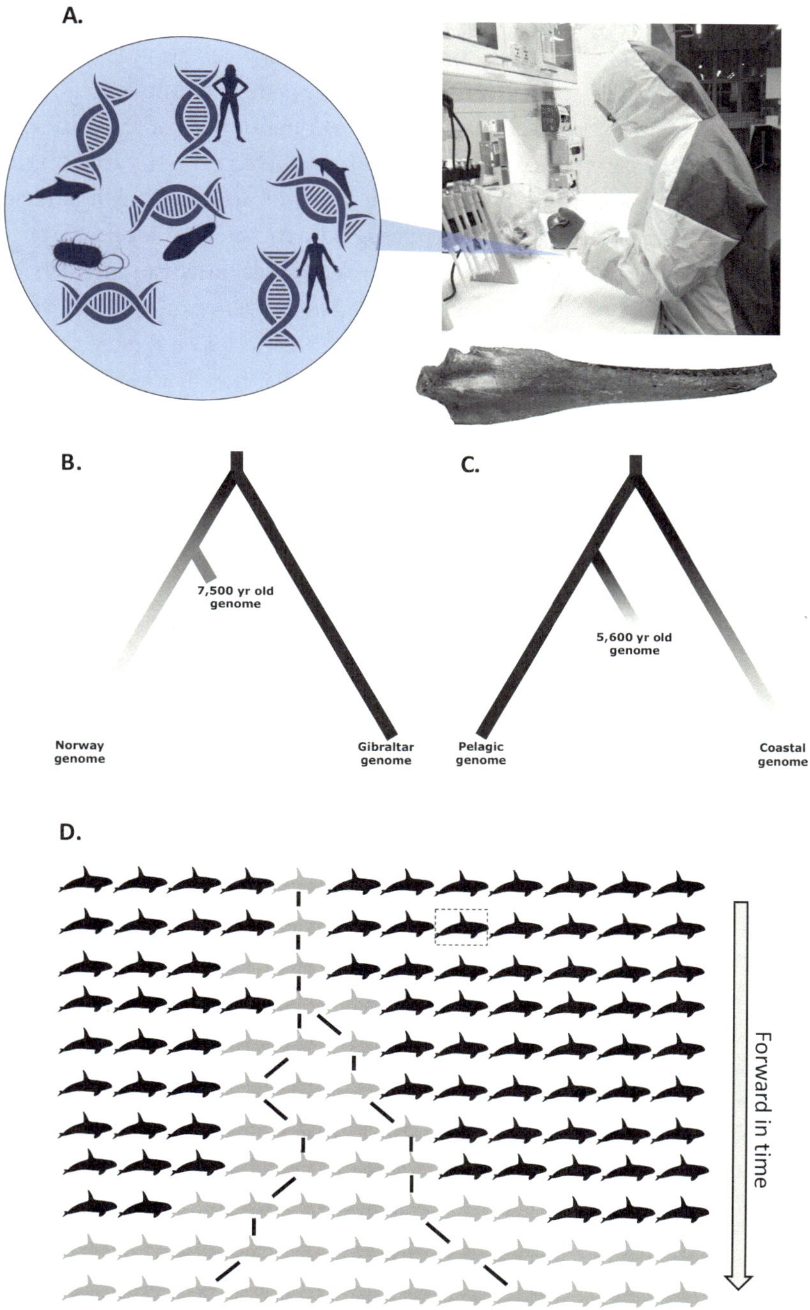
A.
B.
7,500 yr old genome
Norway genome
Gibraltar genome
C.
5,600 yr old genome
Pelagic genome
Coastal genome
D.
Forward in time

sea-level rise (Shennan et al. 2000). The oldest subfossil sample (8,610 years BP) genetically clusters with the genomes of present-day pelagic dolphins, while the three younger subfossil samples show more genetic affinity with present-day North Atlantic coastal dolphins. The genome of the youngest subfossil dolphin (5,626 years BP) contained coastal-associated genetic variation that, as noted earlier, was inferred to be under selection to coastal habitat (Louis et al. 2023). This observation sheds new light on the speed of local adaptation and suggests that parallel adaptation occurred rapidly after the emergence of new coastal habitats, likely from standing genetic variation retained at low frequency in the pelagic population. Coastal-associated ancestry could be reintroduced into pelagic populations by past gene flow from coastal populations into pelagic populations, making adaptive alleles available for selection when pelagic populations found new coastal habitats in the future (i.e., Schluter and Conte's [2009] "transporter mechanism").

A genome sequence recovered from a 7,500-year-old killer whale subfossil from Denmark was found to be approximately ancestral to the present-day metapopulation of killer whales that spans from Norway to West Greenland (Foote et al. 2021a). By comparing the ratio of synonymous (presumed neutral) to nonsynonymous (presumed weakly deleterious) mutations among a global dataset of genomes sampled from present-day killer whales, Foote et al. (2021a) found that the genomes of the Norwegian and Icelandic killer whales showed evidence of purging deleterious mutations. The ancient genome also showed this signature of purging, but less so than in the contemporary Norwegian and Icelandic genomes. Therefore, by incorporating an approximately ancestral genome to the contemporary genomes, Foote et al. (2021a) were able to calibrate the timing of the action of natural selection as concurrent to the timing of the colonization of the high-latitude habitat by killer whales in the Atlantic after the last glacial period (Aaris-Sørensen et al. 2010). As noted, such range expansions can result in reduced heterozygosity and a higher recessive mutation load, but as recessive mutations are increasingly exposed to selection when in homozygous genotypes, this affords the opportunity for purging of deleterious variation by purifying selection (Robinson et al. 2022, 2023).

Summary

In this chapter we have highlighted how the resolution of contiguity, linkage, and marker density afforded by whole genome sequences have provided an unprecedented perspective into the deep evolutionary past of individuals, populations, and species (see also Taylor et al. 2021). Even though most genomic studies to date have sequenced only tens of genomes at most, we have highlighted how these can be used to give new insights into demographic and evolutionary histories and processes by sampling ancestry from thousands of genetic ancestors. Emerging studies are harnessing the genome sequences of close kin to estimate key variables such as mutation and recombination rates and provide fresh insights into inbreeding and past population size (Bergeron et al. 2023; Kardos et al. 2023; Suárez-Menéndez et al. 2023). The tools are now available to sample the genealogies of millions of individual genomes and to incorporate the tree sequences into simulation tools (Harris 2019). We anticipate that whole genomes will become the marker of choice over the coming decade and that studies will routinely incorporate hundreds of individuals in their datasets. The application of such genomic datasets to more social-based research questions is enticing. We hope this chapter has inspired you, the reader, to believe that even a single genome is an incredible record of the evolutionary history of that individual and the forces (selection, drift, migration, and mutation) that have acted upon the genetic variation within. Thus, genomic datasets have the power to provide insights extending millennia before even the longest longitudinal field study, providing a complementary perspective to individual identification data in understanding the cultural and genetic foundations of cetacean societies.

References

Aaris-Sørensen, Kim, Kaare Lund Rasmussen, Carl Kinze, and Kaj Strand Petersen. 2010. "Late Pleistocene and Holocene Whale Remains (Cetacea) from Denmark and Adjacent Countries: Species, Distribution, Chronology, and Trace Element Concentrations." *Marine Mammal Science* 26 (2): 253–81. https://doi.org/10.1111/j.1748-7692.2009.00356.x.

Alexander, Alana, Debbie Steel, Kendra Hoekzema, et al. 2016. "What Influences the Worldwide Genetic Structure of Sperm Whales (*Physeter macrocephalus*)?" *Molecular Ecology* 25 (12): 2754–72. https://doi.org/10.1111/mec.13638.

Almén, Markus Sällman, Sangeet Lamichhaney, Jonas Berglund, et al. 2016. "Adaptive Radiation of Darwin's Finches Revisited Using Whole Genome Sequencing." *BioEssays* 38 (1): 14–20. https://doi.org/10.1002/bies.201500079.

Attard, Catherine R. M., Luciano B. Beheregaray, K. Curt S. Jenner, et al. 2015. "Low Genetic Diversity in Pygmy Blue Whales Is Due to Climate-Induced Diversification Rather Than Anthropogenic Impacts." *Biology Letters* 11 (5): 20141037. https://dx.doi.org/10.1098/rsbl.2014.1037.

Bacher, Kathrin, Simon Allen, Anna K. Lindholm, Lars Bejder, and Michael Krützen. 2010. "Genes or Culture: Are Mitochondrial Genes Associated with Tool Use in Bottlenose Dolphins (*Tursiops* sp.)?" *Behavior Genetics* 40: 706–14. https://doi.org/10.1007/s10519-010-9375-8.

Baker, C. Scott, Louis M. Herman, Anjanette Perry, et al. 1986. "Migratory Movement and Population Structure of Humpback Whales (*Megaptera novaeangliae*) in the Central and Eastern

North Pacific." *Marine Ecology Progress Series* 31 (2): 105–19. https://www.jstor.org/stable/24817567.

Balentine, Christina M., and Deborah A. Bolnick. 2022. "Parallel Evolution in Human Populations: A Biocultural Perspective." *Evolutionary Anthropology: Issues, News, and Reviews* 31 (6): 302–16. https://doi.org/10.1002/evan.21956.

Bank, Claudia, Gregory B. Ewing, Anna Ferrer-Admettla, Matthieu Foll, and Jeffrey D. Jensen. 2014. "Thinking Too Positive? Revisiting Current Methods of Population Genetic Selection Inference." *Trends in Genetics* 30 (12): 540–46. https://doi.org/10.1016/j.tig.2014.09.010.

Barrett-Lennard, Lance Godfrey. 2000. "Population Structure and Mating Patterns of Killer Whales (*Orcinus orca*) as Revealed by DNA Analysis." PhD diss., University of British Columbia.

Barton, Nicholas H., and Brian Charlesworth. 1984. "Genetic Revolutions, Founder Effects, and Speciation." *Annual Review of Ecology and Systematics* 15 (1): 133–64. https://www.jstor.org/stable/2096945.

Batley, Kimberley C., Jonathan Sandoval-Castillo, Catherine M. Kemper, et al. 2019. "Genome-Wide Association Study of an Unusual Dolphin Mortality Event Reveals Candidate Genes for Susceptibility and Resistance to Cetacean Morbillivirus." *Evolutionary Applications* 12 (4): 718–32.

Batley, Kimberley C., Jonathan Sandoval-Castillo, Catherine M. Kemper, et al. 2021. "Whole Genomes Reveal Multiple Candidate Genes and Pathways Involved in the Immune Response of Dolphins to a Highly Infectious Virus." *Molecular Ecology* 30 (23): 6434–48. https://doi.org/10.1111/eva.12747.

Beaumont, Mark A., and Richard A. Nichols. 1996. "Evaluating Loci for Use in the Genetic Analysis of Population Structure." *Proceedings of the Royal Society B* 263 (1377): 1619–26. https://doi.org/10.1098/rspb.1996.0237.

Ben Chehida, Yacine, Roisin Loughnane, Julie Thumloup, et al. 2021. "No Leading-Edge Effect in North Atlantic Harbor Porpoises: Evolutionary and Conservation Implications." *Evolutionary Applications* 14 (6): 1588–611. https://doi.org/10.1111/eva.13227.

Bergeron, Lucie A., Søren Besenbacher, Jiao Zheng, et al. 2023. "Evolution of the Germline Mutation Rate Across Vertebrates." *Nature* 615 (7951): 285–91. https://doi.org/10.1038/s41586-023-05752-y.

Brakes, Philippa, Emma L. Carroll, Sasha R. X. Dall, et al. 2021. "A Deepening Understanding of Animal Culture Suggests Lessons for Conservation." *Proceedings of the Royal Society B* 288 (1949): 20202718. https://doi.org/10.1098/rspb.2020.2718.

Burger, Joachim, Vivian Link, Jens Blöcher, et al. 2020. "Low Prevalence of Lactase Persistence in Bronze Age Europe Indicates Ongoing Strong Selection over the Last 3,000 Years." *Current Biology* 30 (21): 4307–15. https://doi.org/10.1016/j.cub.2020.08.033.

Cammen, Kristina M., Thomas F. Schultz, Patricia E. Rosel, Randall S. Wells, and Andrew J. Read. 2015. "Genomewide Investigation of Adaptation to Harmful Algal Blooms in Common Bottlenose Dolphins (*Tursiops truncatus*)." *Molecular Ecology* 24 (18): 4697–710. https://doi.org/10.1111/mec.13350.

Carroll, Emma L., R. Alderman, J. L. Bannister, et al. 2019. "Incorporating Non-Equilibrium Dynamics into Demographic History Inferences of a Migratory Marine Species." *Heredity* 122 (1): 53–68. https://doi.org/10.1038/s41437-018-0077-y.

Cavalli-Sforza, Luigi L., and Marcus W. Feldman. 1973. "Cultural Versus Biological Inheritance:

Phenotypic Transmission from Parents to Children (A Theory of the Effect of Parental Phenotypes on Children's Phenotypes)." *American Journal of Human Genetics* 25 (6): 618.

Cavalli-Sforza, Luigi L., and Marcus W. Feldman. 1981. *Cultural Transmission and Evolution: A Quantitative Approach.* No. 16. Princeton University Press.

Ceballos, Francisco C., Peter K. Joshi, David W. Clark, Michèle Ramsay, and James F. Wilson. 2018. "Runs of Homozygosity: Windows into Population History and Trait Architecture." *Nature Reviews Genetics* 19 (4): 220–34. https://doi.org/10.1038/nrg.2017.109.

Celemín, Enrique, Marijke Autenrieth, Anna Roos, et al. 2023. "Evolutionary History and Seascape Genomics of Harbour Porpoises (*Phocoena phocoena*) Across Environmental Gradients in the North Atlantic and Adjacent Waters." *Molecular Ecology Resources*, https://doi.org/10.1111/1755-0998.13860.

Charlesworth, Brian. 1994. "The Effect of Background Selection Against Deleterious Mutations on Weakly Selected, Linked Variants." *Genetics Research* 63 (3): 213–27.

Charlesworth, Brian, and Jeffrey D. Jensen. 2021. "Effects of Selection at Linked Sites on Patterns of Genetic Variability." *Annual Review of Ecology, Evolution, and Systematics* 52: 177–97. https://doi.org/10.1017/S0016672300032365.

Coop, Graham. 2020. Population and Quantitative Genetics, version c0a7708. https://github.com/cooplab/popgen-notes/blob/master/popgen_notes.pdf.

Coop, Graham. 2022. "Genetic Similarity and Genetic Ancestry Groups." *arXiv*, January. https://doi.org/10.48550/arxiv.2207.11595.

Costa, Ana P. B., Pedro F. Fruet, Eduardo R. Secchi, et al. 2021. "Ecological Divergence and Speciation in Common Bottlenose Dolphins in the Western South Atlantic." *Journal of Evolutionary Biology* 34 (1): 16–32. https://doi.org/10.1111/jeb.13575.

Crossman, Carla A., Michael C. Fontaine, and Timothy R. Frasier. 2024. "A Comparison of Genomic Diversity and Demographic History of the North Atlantic and Southwest Atlantic Southern Right Whales." *Molecular Ecology* 33 (20): e17099. https://doi.org/10.1111/mec.17099.

Davey, John W., Paul A. Hohenlohe, Paul D. Etter, Jason Q. Boone, Julian M. Catchen, and Mark L. Blaxter. 2011. "Genome-Wide Genetic Marker Discovery and Genotyping Using Next-Generation Sequencing." *Nature Reviews Genetics* 12 (7): 499–510. https://doi.org/10.1038/nrg3012.

De Greef, Evelien, Anthony L. Einfeldt, Patrick J. O. Miller, et al. 2022. "Genomics Reveal Population Structure, Evolutionary History, and Signatures of Selection in the Northern Bottlenose Whale, *Hyperoodon ampullatus*." *Molecular Ecology* 31 (19): 4919–31. https://doi.org/10.1111/mec.16643.

Derrida, Bernard, Susanna C. Manrubia, and Damián H. Zanette. 2000. "On the Genealogy of a Population of Biparental Individuals." *Journal of Theoretical Biology* 203 (3): 303–15. https://doi.org/10.1006/jtbi.2000.1095.

Deshpande, Omkar, Serafim Batzoglou, Marcus W. Feldman, and L. Luca Cavalli-Sforza. 2009. "A Serial Founder Effect Model for Human Settlement Out of Africa." *Proceedings of the Royal Society B* 276 (1655): 291–300. https://doi.org/10.1098/rspb.2008.0750.

Enattah, Nabil Sabri, Tine G. K. Jensen, Mette Nielsen, et al. 2008. "Independent Introduction of Two Lactase-Persistence Alleles into Human Populations Reflects Different History of Adaptation to Milk Culture." *American Journal of Human Genetics* 82 (1): 57–72. https://doi.org/10.1016/j.ajhg.2007.09.012.

Evershed, Richard P., George Davey Smith, Mélanie Roffet-Salque, et al. 2022. "Dairying, Diseases and the Evolution of Lactase Persistence in Europe." *Nature* 608 (7922): 336–45. https://doi.org/10.1038/s41586-022-05010-7.

Excoffier, Laurent, Matthieu Foll, and Rémy J. Petit. 2009. "Genetic Consequences of Range Expansions." *Annual Review of Ecology, Evolution, and Systematics* 40: 481–501. https://doi.org/10.1146/annurev.ecolsys.39.110707.173414.

Excoffier, Laurent, and Nicolas Ray. 2008. "Surfing During Population Expansions Promotes Genetic Revolutions and Structuration." *Trends in Ecology and Evolution* 23 (7): 347–51. https://doi.org/10.1016/j.tree.2008.04.004.

Feldman, Marc W., and Kevin N. Laland. 1996. "Gene-Culture Coevolutionary Theory." *Trends in Ecology and Evolution* 11 (11): 453–57.

Filatova, Olga A., Ekaterina A. Borisova, Ilya G. Meschersky, et al. 2018. "Colonizing the Wild West: Low Diversity of Complete Mitochondrial Genomes in Western North Pacific Killer Whales Suggests a Founder Effect." *Journal of Heredity* 109 (7): 735–43. https://doi.org/10.1093/jhered/esy037.

Filatova, Olga A., Filipa I. P. Samarra, Volker B. Deecke, John K. B. Ford, Patrick J. O. Miller, and Harald Yurk. 2015. "Cultural Evolution of Killer Whale Calls: Background, Mechanisms and Consequences." *Behaviour* 152 (15): 2001–38. https://doi.org/10.1163/1568539X-00003317.

Fontaine, Michael C. 2016. "Harbour Porpoises, *Phocoena phocoena*, in the Mediterranean Sea and Adjacent Regions: Biogeographic Relicts of the Last Glacial Period." *Advances in Marine Biology* 75: 333–58. https://doi.org/10.1016/bs.amb.2016.08.006.

Fontaine, Michael C., Stuart J. E. Baird, Sylvain Piry, et al. 2007. "Rise of Oceanographic Barriers in Continuous Populations of a Cetacean: The Genetic Structure of Harbour Porpoises in Old World Waters." *BMC Biology* 5: 1–16. https://doi.org/10.1186/1741-7007-5-30.

Fontaine, Michael C., Kathleen Roland, Isabelle Calves, et al. 2014. "Postglacial Climate Changes and Rise of Three Ecotypes of Harbour Porpoises, *Phocoena phocoena*, in Western Palearctic Waters." *Molecular Ecology* 23 (13): 3306–21. https://doi.org/10.1111/mec.12817.

Fontaine, P. M., and C. Barrette. 1997. "Megatestes: Anatomical Evidence for Sperm Competition in the Harbor Porpoise." *Mammalia* 61 (1): 65–72. https://doi.org/10.1515/mamm.1997.61.1.65.

Foote, Andrew D., Alana Alexander, Lisa T. Ballance, et al. 2023. "'Type D' Killer Whale Genomes Reveal Long-Term Small Population Size and Low Genetic Diversity." *Journal of Heredity* 114 (2): 94–109. https://doi.org/10.1093/jhered/esac070.

Foote, Andrew D., M. Thomas P. Gilbert, Shyam Gopalakrishnan, et al. 2021a. "Evidence of Long-Term Purging of Mutation Load in Killer Whale Genomes." Preprint. *bioRxiv*. https://doi.org/10.1101/2021.08.21.457186.

Foote, Andrew D., Rebecca Hooper, Alana Alexander, et al. 2021b. "Runs of Homozygosity in Killer Whale Genomes Provide a Global Record of Demographic Histories." *Molecular Ecology* 30 (23): 6162–77. https://doi.org/10.1111/mec.16137.

Foote, Andrew D., Michael D. Martin, Marie Louis, et al. 2019. "Killer Whale Genomes Reveal a Complex History of Recurrent Admixture and Vicariance." *Molecular Ecology* 28 (14): 3427–44. https://doi.org/10.1111/mec.15099.

Foote, Andrew D., Nagarjun Vijay, María C. Ávila-Arcos, et al. 2016. "Genome-Culture Coevolution Promotes Rapid Divergence of Killer Whale Ecotypes." *Nature Communications* 7 (1): 11693. https://doi.org/10.1038/ncomms11693.

Galimberti, Marco, Christoph Leuenberger, Beat Wolf, Sándor Miklós Szilágyi, Matthieu Foll, and Daniel Wegmann. 2020. "Detecting Selection from Linked Sites Using an F-Model." *Genetics* 216 (4): 1205–15. https://doi.org/10.1534/genetics.120.303780.

Garland, Ellen C., Anne W. Goldizen, Matthew S. Lilley, et al. 2015. "Population Structure of Humpback Whales in the Western and Central South Pacific Ocean as Determined by Vocal Exchange Among Populations." *Conservation Biology* 29 (4): 1198–207. https://doi.org/10.1111/cobi.12492.

Garland, Ellen C., Anne W. Goldizen, Melinda L. Rekdahl, et al. 2011. "Dynamic Horizontal Cultural Transmission of Humpback Whale Song at the Ocean Basin Scale." *Current Biology* 21 (8): 687–91. https://doi.org/10.1016/j.cub.2011.03.019.

Guez, Jérémy, Guillaume Achaz, François Bienvenu, et al. 2023. "Cultural Transmission of Reproductive Success Impacts Genomic Diversity, Coalescent Tree Topologies, and Demographic Inferences." *Genetics* 223 (4): iyad007. https://doi.org/10.1093/genetics/iyad007.

Hahn, Matthew W. 2008. "Toward a Selection Theory of Molecular Evolution." *Evolution* 62 (2): 255–65. https://doi.org/10.1111/j.1558-5646.2007.00308.x.

Hallatschek, Oskar, Pascal Hersen, Sharad Ramanathan, and David R. Nelson. 2007. "Genetic Drift at Expanding Frontiers Promotes Gene Segregation." *Proceedings of the National Academy of Sciences* 104 (50): 19926–30. https://doi.org/10.1073/pnas.0710150104.

Haller, Benjamin C., and Philipp W. Messer. 2019. "SLiM 3: Forward Genetic Simulations Beyond the Wright-Fisher Model." *Molecular Biology and Evolution* 36 (3): 632–37. https://doi.org/10.1093/molbev/msy228.

Harris, Kelley. 2019. "From a Database of Genomes to a Forest of Evolutionary Trees." *Nature Genetics* 51 (9): 1306–7. https://doi.org/10.1038/s41588-019-0492-x.

Haub, Carl. 1995. "How Many People Have Ever Lived on Earth?" *Population Today* 23 (2): 4–5.

Henn, Brenna M., Laura R. Botigué, Stephan Peischl, et al. 2016. "Distance from Sub-Saharan Africa Predicts Mutational Load in Diverse Human Genomes." *Proceedings of the National Academy of Sciences* 113 (4): E440–49. https://doi.org/10.1073/pnas.1510805112.

Henn, Brenna M., L. Luca Cavalli-Sforza, and Marcus W. Feldman. 2012. "The Great Human Expansion." *Proceedings of the National Academy of Sciences* 109 (44): 17758–64. https://doi.org/10.1073/pnas.1212380109.

Hewitt, Godfrey. 2000. "The Genetic Legacy of the Quaternary Ice Ages." *Nature* 405 (6789): 907–13. https://doi.org/10.1038/35016000.

Hoelzel, A. Rus, M. Dahlheim, and S. J. Stern. 1998a. "Low Genetic Variation Among Killer Whales (*Orcinus orca*) in the Eastern North Pacific and Genetic Differentiation Between Foraging Specialists." *Journal of Heredity* 89 (2): 121–28. https://doi.org/10.1093/jhered/89.2.121.

Hoelzel, A. Rus, Jody Hey, Marilyn E. Dahlheim, Colin Nicholson, Vladimir Burkanov, and Nancy Black. 2007. "Evolution of Population Structure in a Highly Social Top Predator, the Killer Whale." *Molecular Biology and Evolution* 24 (6): 1407–15. https://doi.org/10.1093/molbev/msm063.

Hoelzel, A. Rus, Ada Natoli, Marilyn E. Dahlheim, Carlos Olavarria, Robin W. Baird, and Nancy A. Black. 2002. "Low Worldwide Genetic Diversity in the Killer Whale (*Orcinus orca*): Implications for Demographic History." *Proceedings of the Royal Society B* 269 (1499): 1467–73. https://doi.org/10.1098/rspb.2002.2033.

Hoelzel, A. Rus, Charles W. Potter, and Peter B. Best. 1998b. "Genetic Differentiation Between Parapatric 'Nearshore' and 'Offshore' Populations of the Bottlenose Dolphin." *Proceedings of the Royal Society B* 265 (1402): 1177–83. https://doi.org/10.1098/rspb.1998.0416.

Hutchison, Delbert W., and Alan R. Templeton. 1999. "Correlation of Pairwise Genetic and Geographic Distance Measures: Inferring the Relative Influences of Gene Flow and Drift on the Distribution of Genetic Variability." *Evolution* 53 (6): 1898–914. https://doi.org/10.1111/j.1558-5646.1999.tb04571.x.

Irving-Pease, Evan K., Alba Refoyo-Martínez, William Barrie, et al. 2024. "The Selection Landscape and Genetic Legacy of Ancient Eurasians." *Nature* 625 (7994): 312–20. https://doi.org/10.1038/s41586-023-06705-1.

Jackson, Jennifer A., Andrea Ross-Gillespie, Doug Butterworth, et al. 2015. "Southern Hemisphere Humpback Whale Comprehensive Assessment: A Synthesis and Summary: 2005–2015." Report No. SC/66a/SH3 to the Scientific Committee of the International Whaling Commission, 1–38.

Jones, Felicity C., Manfred G. Grabherr, Yingguang Frank Chan, et al. 2012. "The Genomic Basis of Adaptive Evolution in Threespine Sticklebacks." *Nature* 484 (7392): 55–61. https://doi.org/10.1038/nature10944.

Kardos, Marty, Yaolei Zhang, Kim M. Parsons, et al. 2023. "Inbreeding Depression Explains Killer Whale Population Dynamics." *Nature Ecology and Evolution* 7 (5): 675–86. https://doi.org/10.1038/s41559-023-01995-0.

Karmin, Monika, Lauri Saag, Mário Vicente, et al. 2015. "A Recent Bottleneck of Y Chromosome Diversity Coincides with a Global Change in Culture." *Genome Research* 25 (4): 459–66. https://doi.org/10.1101/gr.186684.114.

Keener, William, Marc A. Webber, Isidore D. Szczepaniak, Tim M. Markowitz, and Dara N. Orbach. 2018. "The Sex Life of Harbor Porpoises (*Phocoena phocoena*): Lateralized and Aerial Behavior." *Aquatic Mammals* 44 (6): 620–32. https://doi.org/10.1578/am.44.6.2018.620.

Kircher, Martin, and Janet Kelso. 2010. "High-Throughput DNA Sequencing—Concepts and Limitations." *BioEssays* 32 (6): 524–36. https://doi.org/10.1002/bies.200900181.

Kirin, Mirna, Ruth McQuillan, Christopher S. Franklin, Harry Campbell, Paul M. McKeigue, and James F. Wilson. 2010. "Genomic Runs of Homozygosity Record Population History and Consanguinity." *PLOS ONE* 5 (11): e13996. https://doi.org/10.1371/journal.pone.0013996.

Konrad, Christine M., Shane Gero, Timothy R. Frasier, and Hal Whitehead. 2018. "Kinship Influences Sperm Whale Social Organization Within, but Generally Not Among, Social Units." *Royal Society Open Science* 5 (8): 180914. https://doi.org/10.1098/rsos.180914.

Kopps, Anna M., Corinne Y. Ackermann, William B. Sherwin, Simon J. Allen, Lars Bejder, and Michael Krützen. 2014. "Cultural Transmission of Tool Use Combined with Habitat Specializations Leads to Fine-Scale Genetic Structure in Bottlenose Dolphins." *Proceedings of the Royal Society B* 281 (1782): 20133245. https://doi.org/10.1098/rspb.2013.3245.

Krützen, Michael, Janet Mann, Michael R. Heithaus, Richard C. Connor, Lars Bejder, and William B. Sherwin. 2005. "Cultural Transmission of Tool Use in Bottlenose Dolphins." *Proceedings of the National Academy of Sciences* 102 (25): 8939–43. https://doi.org/10.1073/pnas.0500232102.

Laland, Kevin N., John Odling-Smee, and Sean Myles. 2010. "How Culture Shaped the Human Genome: Bringing Genetics and the Human Sciences Together." *Nature Reviews Genetics* 11 (2): 137–48. https://doi.org/10.1038/nrg2734.

Larsen, Anja Holm, Jóhann Sigurjónsson, Nils Øien, Gísli Víkingsson, and Per Palsbøll. 1996. "Populations Genetic Analysis of Nuclear and Mitochondrial Loci in Skin Biopsies Collected from Central and Northeastern North Atlantic Humpback Whales (*Megaptera novaeangliae*):

Population Identity and Migratory Destinations." *Proceedings of the Royal Society B* 263 (1376): 1611–18. https://doi.org/10.1098/rspb.1996.0236.

Leblois, Raphael, François Rousset, and Arnaud Estoup. 2004. "Influence of Spatial and Temporal Heterogeneities on the Estimation of Demographic Parameters in a Continuous Population Using Individual Microsatellite Data." *Genetics* 166 (2): 1081–92. https://doi.org/10.1093/genetics/166.2.1081.

Li, Heng, and Richard Durbin. 2011. "Inference of Human Population History from Individual Whole-Genome Sequences." *Nature* 475 (7357): 493–96. https://doi.org/10.1038/nature10231.

Lindblad-Toh, Kerstin, Manuel Garber, Or Zuk, et al. 2011. "A High-Resolution Map of Human Evolutionary Constraint Using 29 Mammals." *Nature* 478 (7370): 476–82. https://doi.org/10.1038/nature10530.

Lockyer, Christina. 2007. "All Creatures Great and Smaller: A Study in Cetacean Life History Energetics." *Journal of the Marine Biological Association of the United Kingdom* 87 (4): 1035–45. https://doi.org/10.1017/S0025315407054720.

Lohmueller, Kirk E., Amit R. Indap, Steffen Schmidt, et al. 2008. "Proportionally More Deleterious Genetic Variation in European Than in African Populations." *Nature* 451 (7181): 994–97. https://doi.org/10.1038/nature06611.

Louis, Marie, Michael C. Fontaine, Jérôme Spitz, et al. 2014. "Ecological Opportunities and Specializations Shaped Genetic Divergence in a Highly Mobile Marine Top Predator." *Proceedings of the Royal Society B* 281 (1795): 20141558. https://doi.org/10.1098/rspb.2014.1558.

Louis, Marie, Marco Galimberti, Frederick Archer, et al. 2021. "Selection on Ancestral Genetic Variation Fuels Repeated Ecotype Formation in Bottlenose Dolphins." *Science Advances* 7 (44): eabg1245. https://doi.org/10.1126/sciadv.abg1245.

Louis, Marie, Petra Korlević, Milaja Nykänen, et al. 2023. "Ancient Dolphin Genomes Reveal Rapid Repeated Adaptation to Coastal Waters." *Nature Communications* 14 (1): 4020. https://doi.org/10.1038/s41467-023-39532-z.

Mann, Janet, Richard C. Connor, Peter L. Tyack, and Hal Whitehead, eds. 2000. *Cetacean Societies: Field Studies of Dolphins and Whales*. University of Chicago Press.

Mathieson, Iain, Iosif Lazaridis, Nadin Rohland, et al. 2015. "Genome-Wide Patterns of Selection in 230 Ancient Eurasians." *Nature* 528 (7583): 499–503. https://doi.org/10.1038/nature16152.

Mathieson, Sara, and Iain Mathieson. 2018. "FADS1 and the Timing of Human Adaptation to Agriculture." *Molecular Biology and Evolution* 35 (12): 2957–70. https://doi.org/10.1093/molbev/msy180.

Maynard Smith, John, and John Haigh. 1974. "The Hitch-Hiking Effect of a Favourable Gene." *Genetics Research* 23 (1): 23–35. https://doi.org/10.1017/S0016672300014634.

Morin, Phillip A., Frederick I. Archer, Andrew D. Foote, et al. 2010. "Complete Mitochondrial Genome Phylogeographic Analysis of Killer Whales (*Orcinus orca*) Indicates Multiple Species." *Genome Research* 20 (7): 908–16. https://doi.org/10.1101/gr.102954.109.

Morin, Phillip A., Andrew D. Foote, Charles Scott Baker, et al. 2018. "Demography or Selection on Linked Cultural Traits or Genes? Investigating the Driver of Low mtDNA Diversity in the Sperm Whale Using Complementary Mitochondrial and Nuclear Genome Analyses." *Molecular Ecology* 27 (11): 2604–19. https://doi.org/10.1111/mec.14698.

Morin, Phillip A., Kim M. Parsons, Frederick I. Archer, et al. 2015. "Geographic and Temporal Dynamics of a Global Radiation and Diversification in the Killer Whale." *Molecular Ecology* 24 (15): 3964–79. https://doi.org/10.1111/mec.13284.

Moura, Andre E., John G. Kenny, Roy Chaudhuri, et al. 2014. "Population Genomics of the Killer Whale Indicates Ecotype Evolution in Sympatry Involving Both Selection and Drift." *Molecular Ecology* 23 (21): 5179–92. https://doi.org/10.1111/mec.12929.

Nigenda-Morales, Sergio F., Meixi Lin, et al. 2023. "The Genomic Footprint of Whaling and Isolation in Fin Whale Populations." *Nature Communications* 14 (1): 5465. https://doi.org/10.1038/s41467-023-40052-z.

Nykänen, Milaja, Kristin Kaschner, Willy Dabin, 2019. "Postglacial Colonization of Northern Coastal Habitat by Bottlenose Dolphins: A Marine Leading-Edge Expansion?" *Journal of Heredity* 110 (6): 662–74. https://doi.org/10.1093/jhered/esz039.

Olalde, Iñigo, Selina Brace, Morten E. Allentoft, et al. 2018. "The Beaker Phenomenon and the Genomic Transformation of Northwest Europe." *Nature* 555 (7695): 190–96. https://doi.org/10.1038/nature25738.

Parsons, Kim M., John W. Durban, Alexander M. Burdin, et al. 2013. "Geographic Patterns of Genetic Differentiation Among Killer Whales in the Northern North Pacific." *Journal of Heredity* 104 (6): 737–54. https://doi.org/10.1093/jhered/est037.

Payne, Katharine, and Roger Payne. 1985. "Large Scale Changes over 19 Years in Songs of Humpback Whales in Bermuda." *Zeitschrift für Tierpsychologie* 68 (2): 89–114. https://doi.org/10.1111/j.1439-0310.1985.tb00118.x.

Payne, Roger S., and Scott McVay. 1971. "Songs of Humpback Whales: Humpbacks Emit Sounds in Long, Predictable Patterns Ranging over Frequencies Audible to Humans." *Science* 173 (3997): 585–97. https://doi.org/10.1126/science.173.3997.585.

Peischl, Stephan, Isabelle Dupanloup, Mark Kirkpatrick, and Laurent Excoffier. 2013. "On the Accumulation of Deleterious Mutations During Range Expansions." *Molecular Ecology* 22 (24): 5972–82. https://doi.org/10.1111/mec.12524.

Peischl, Stephan, and Laurent Excoffier. 2015. "Expansion Load: Recessive Mutations and the Role of Standing Genetic Variation." *Molecular Ecology* 24 (9): 2084–94. https://doi.org/10.1111/mec.13154.

Pennisi, Elizabeth. 2010. "1000 Genomes Project Gives New Map of Genetic Diversity." *Science* (2010): 574–75. https://doi.org/10.1126/science.330.6004.574.

Phillips, C. D., J. I. Hoffman, J. C. George, et al. 2013. "Molecular Insights into the Historic Demography of Bowhead Whales: Understanding the Evolutionary Basis of Contemporary Management Practices." *Ecology and Evolution* 3 (1): 18–37. https://doi.org/10.1002/ece3.374.

Pouyet, Fanny, Simon Aeschbacher, Alexandre Thiéry, and Laurent Excoffier. 2018. "Background Selection and Biased Gene Conversion Affect More Than 95% of the Human Genome and Bias Demographic Inferences." *eLife* 7: e36317. https://doi.org/10.7554/eLife.36317.

Pratt, Eleanor A. L., Luciano B. Beheregaray, Kerstin Bilgmann, et al. 2022. "Seascape Genomics of Coastal Bottlenose Dolphins Along Strong Gradients of Temperature and Salinity." *Molecular Ecology* 31 (8): 2223–41. https://doi.org/10.1111/mec.16389.

Premo, Luke S., and Jean-Jacques Hublin. 2009. "Culture, Population Structure, and Low Genetic Diversity in Pleistocene Hominins." *Proceedings of the National Academy of Sciences* 106 (1): 33–37. https://doi.org/10.1073/pnas.0809194105.

Read, Andrew J. 1999. "Harbour Porpoise *Phocoena phocoena* (Linnaeus, 1758)." In *Handbook of Marine Mammals: The Second Book of Dolphins and the Porpoises*, vol. 6, edited by S. Ridgway and R. Harrison, 323–50. Academic Press. https://doi.org/10.1016/B978-0-12-373553-9.00209-1.

Read, Andrew J. 2018. "Porpoise, Overview." In *Encyclopedia of Marine Mammals*, 3rd ed., edited by Bernd Würsig, Johannes G. M. Thewissen, and Kit M. Kovacs. Academic Press.

Read, Andrew J., and Aleta A. Hohn. 1995. "Life in the Fast Lane: The Life History of Harbor Porpoises from the Gulf of Maine." *Marine Mammal Science* 11 (4): 423–40. https://doi.org/10.1111/j.1748-7692.1995.tb00667.x.

Robinson, Jacqueline A., Christopher C. Kyriazis, Sergio F. Nigenda-Morales, et al. 2022. "The Critically Endangered Vaquita Is Not Doomed to Extinction by Inbreeding Depression." *Science* 376 (6593): 635–39. https://doi.org/10.1126/science.abm1742.

Robinson, Jacqueline, Christopher C. Kyriazis, Stella C. Yuan, and Kirk E. Lohmueller. 2023. "Deleterious Variation in Natural Populations and Implications for Conservation Genetics." *Annual Review of Animal Biosciences* 11: 93–114. https://doi.org/10.1146/annurev-animal-080522-093311.

Rosenberg, Noah A., and Magnus Nordborg. 2002. "Genealogical Trees, Coalescent Theory and the Analysis of Genetic Polymorphisms." *Nature Reviews Genetics* 3 (5): 380–90. https://doi.org/10.1038/nrg795.

Rousset, François. 2004. *Genetic Structure and Selection in Subdivided Populations*. Vol. 40. Princeton University Press.

Sabeti, Pardis C., Stephen F. Schaffner, Ben Fry, et al. 2006. "Positive Natural Selection in the Human Lineage." *Science* 312 (5780): 1614–20. https://doi.org/10.1126/science.1124309.

Sarano, François, Justine Girardet, Véronique Sarano, et al. 2021. "Kin Relationships in Cultural Species of the Marine Realm: Case Study of a Matrilineal Social Group of Sperm Whales off Mauritius Island, Indian Ocean." *Royal Society Open Science* 8 (2): 201794. https://doi.org/10.1098/rsos.201794.

Schlötterer, Christian. 2023. "How Predictable Is Adaptation from Standing Genetic Variation? Experimental Evolution in Drosophila Highlights the Central Role of Redundancy and Linkage Disequilibrium." *Philosophical Transactions of the Royal Society B* 378 (1877): 20220046. https://doi.org/10.1098/rstb.2022.0046.

Schluter, Dolph, and Gina L. Conte. 2009. "Genetics and Ecological Speciation." *Proceedings of the National Academy of Sciences* 106 (supplement 1): 9955–62. https://doi.org/10.1073/pnas.0901264106.

Shennan, Ian, K. Lambeck, R. Flather, et al. 2000. "Modelling Western North Sea Palaeogeographies and Tidal Changes During the Holocene." *Geological Society London Special Publications* 166 (1): 299–319. https://doi.org/10.1144/GSL.SP.2000.166.01.15.

Smolker, Rachel, Andrew Richards, Richard Connor, Janet Mann, and Per Berggren. 1997. "Sponge Carrying by Dolphins (Delphinidae, *Tursiops* sp.): A Foraging Specialization Involving Tool Use?" *Ethology* 103 (6): 454–65. https://doi.org/10.1111/j.1439-0310.1997.tb00160.x.

Suárez-Menéndez, Marcos, Martine Bérubé, Fabrício Furni, et al. 2023. "Wild Pedigrees Inform Mutation Rates and Historic Abundance in Baleen Whales." *Science* 381 (6661): 990–95. https://doi.org/10.1126/science.adf2160.

Taylor, Rebecca S., Evelyn L. Jensen, David W. Coltman, Andrew D. Foote, and Sangeet Lamichhaney. 2021. "Seeing the Whole Picture: What Molecular Ecology is Gaining from Whole Genomes." *Molecular Ecology* 30 (23): 5917–22. https://doi.org/10.1111/mec.16282.

Teilmann, Jonas, Finn Larsen, and Genevieve Desportes. 2007. "Time Allocation and Diving Behaviour of Harbour Porpoises (*Phocoena phocoena*) in Danish and Adjacent Waters." *Journal of Cetacean Research and Management* 9: 201–10.

Teilmann, Jonas, and Signe Sveegaard. 2019. "Porpoises the World Over: Diversity in Behavior and Ecology." *Ethology and Behavioral Ecology of Odontocetes*: 449–64. https://doi.org/10.1007/978-3-030-16663-2_21.

Thornton, Alex, and Nichola J. Raihani. 2010. "Identifying Teaching in Wild Animals." *Learning and Behavior* 38 (3): 297–309. https://doi.org/10.3758/LB.38.3.297.

Venter, J. Craig, Mark D. Adams, Eugene W. Myers, et al. 2001. "The Sequence of the Human Genome." *Science* 291 (5507): 1304–51. https://doi.org/10.1126/science.1058040.

Violi, Biagio, Menno J. de Jong, Alexandros Frantzis, et al. 2023. "Genomics Reveals the Role of Admixture in the Evolution of Structure Among Sperm Whale Populations Within the Mediterranean Sea." *Molecular Ecology* 32: 2715–31. https://doi.org/10.1111/mec.16898.

Wakeley, John H. 2009. *Coalescent Theory: An Introduction*. Roberts and Company Publishers.

Whitehead, Hal. 1998. "Cultural Selection and Genetic Diversity in Matrilineal Whales." *Science* 282 (5394): 1708–11. https://doi.org/10.1126/science.282.5394.1708.

Whitehead, Hal, Kevin N. Laland, Luke Rendell, Rose Thorogood, and Andrew Whiten. 2019. "The Reach of Gene-Culture Coevolution in Animals." *Nature Communications* 10 (1): 2405. https://doi.org/10.1038/s41467-019-10293-y.

Whitehead, Hal, and Luke Rendell. 2014. *The Cultural Lives of Whales and Dolphins*. University of Chicago Press.

Whitehead, Hal, Peter J. Richerson, and Robert Boyd. 2002. "Cultural Selection and Genetic Diversity in Humans." *Selection* 3 (1): 115–25. https://doi.org/10.1556/select.3.2002.1.9.

Whiten, Andrew, Francisco J. Ayala, Marcus W. Feldman, and Kevin N. Laland. 2017. "The Extension of Biology Through Culture." *Proceedings of the National Academy of Sciences* 114 (30): 7775–81. https://doi.org/10.1073/pnas.1707630114.

Wild, Sonja, Simon J. Allen, Michael Krützen, Stephanie L. King, Livia Gerber, and William J. E. Hoppitt. 2019. "Multi-Network-Based Diffusion Analysis Reveals Vertical Cultural Transmission of Sponge Tool Use Within Dolphin Matrilines." *Biology Letters* 15 (7): 20190227. https://doi.org/10.1098/rsbl.2019.0227.

Yurk, Harald, Lance Barrett-Lennard, J. K. B. Ford, and C. O. Matkin. 2002. "Cultural Transmission Within Maternal Lineages: Vocal Clans in Resident Killer Whales in Southern Alaska." *Animal Behaviour* 63 (6): 1103–19. https://doi.org/10.1006/anbe.2002.3012.

Zeng, Tian Chen, Alan J. Aw, and Marcus W. Feldman. 2018. "Cultural Hitchhiking and Competition Between Patrilineal Kin Groups Explain the Post-Neolithic Y-Chromosome Bottleneck." *Nature Communications* 9 (1): 2077. https://doi.org/10.1038/s41467-018-04375-6.

Zhou, Xuming, Xuanmin Guang, Di Sun, et al. 2018. "Population Genomics of Finless Porpoises Reveal an Incipient Cetacean Species Adapted to Freshwater." *Nature Communications* 9 (1) 1276. https://doi.org/10.1038/s41467-018-03722-x.

3

The Role of Kinship in Shaping Cetacean Societies

KIM M. PARSONS, MICHAEL KRÜTZEN, LIVIA GERBER, EVE JOURDAIN, CHRISTINE M. KONRAD CLARKE, KAREN K. MARTIEN, AND ANDREW D. FOOTE

Abstract

Social relationships among kin can influence individual fitness, thereby shaping sociality through the action of selection. Since the publication of *Cetacean Societies*, the number of long-term field studies has increased and matured, providing greater comparative and longitudinal perspectives on the role of kinship in structuring cetacean societies. In this chapter, we consider the role of philopatry, kinship, and kin-based interactions within social groups, and how these roles vary across different ecological contexts and timescales. We focus on four well-studied odontocete taxa for which there are both long-term social and genetic data: bottlenose dolphins (*Tursiops* sp.), killer whales (*Orcinus orca*), sperm whales (*Physeter macrocephalus*), and false killer whales (*Pseudorca crassidens*). By considering both fitness benefits (in terms of passing genes to the next generation), and fitness costs of reproductive conflict between kin, we provide a synthesis of studies comparing social and genetic data and how these costs and benefits have shaped cetacean sociality.

Introduction

Kin selection provides a powerful framework for understanding the evolution of social traits (Hamilton 1964). Social interactions among related individuals can provide "inclusive" fitness benefits by supporting the survival and reproduction of relatives, thereby increasing the transmission success of an individual's alleles. In his seminal paper, Hamilton (1964) suggested that the degree of relatedness necessary to produce indirect benefits may arise through social viscosity (where kin cluster together socially) and "kin discrimination," i.e., individuals preferentially interacting with close relatives. The benefits of helping kin, however, need to be traded off against the costs of interacting with kin, which can include the costs of competition with kin, which may harm the fitness of relatives (Taylor 1992), and the costs of inbreeding (Charlesworth and Willis 2009).

The costs and benefits of interacting with kin will differ depending on the social and ecological conditions leading to a diversity of kin structure across animal societies (Hatchwell 2010). In the following sections, we discuss patterns of kinship structure in four well-studied cetacean species—killer whales (*Orcinus orca*), false killer whales (*Pseudorca crassidens*), sperm whales (*Physeter macrocephalus*), and bottlenose dolphins (*Tursiops* sp.). We focus on these species due to the extent of existing data regarding both kinship and cooperation. However, we note that their societies are not representative of all cetaceans. Three of the species (killer whales, false killer whales, and sperm whales) exhibit highly matrilineal social organization, a pattern that is likely uncommon among cetaceans as a whole. Bottlenose dolphins, in contrast, exhibit a fission-fusion pattern of social organization that may be more typical of cetacean species. We review approaches to estimating kinship between associating individuals; demonstrate variation in patterns of association and kinship among cooperating individuals, populations, and species; and consider the fitness costs and benefits associated with kin-based societies in cetaceans.

Genetic Approaches for Examining the Role of Kinship in Social Groups

Kinship among individuals, referred to as "relatedness," can be inferred by genetic analyses. Each parent passes half of its genome to their offspring; thus, parent-offspring pairs share half of their genetic material, while grandparental, half sibling, and aunt/uncle/niece/nephew relationships, for example, are expected, on average, to share a quarter (see Foote et al., this volume). We use the terms r_0, r_1, and r_2 to denote the probability of zero, one, or two alleles being shared between two individuals due to recent common ancestry, referred to as identical-by-descent. The probabilities can be estimated per site or averaged across the genome. For parent-offspring pairs $r_1 = 1$, as each parent passes on one allele from each site of their genome to their offspring. Full siblings share one allele across approximately half their genome ($r_1 = 0.5$), and zero or two alleles across a quarter of the genome each ($r_0 = 0.25$, $r_2 = 0.25$). Values of r_1 and r_2 decrease and values of r_0 increase with decreasing kinship (see fig. 3.1a). There are nine potential combinations of allelic inheritance that make up the possible r_0, r_1, and r_2 outcomes (see fig. 3.1; Jacquard 1972), and different combinations can provide distinct insights into the patterns of

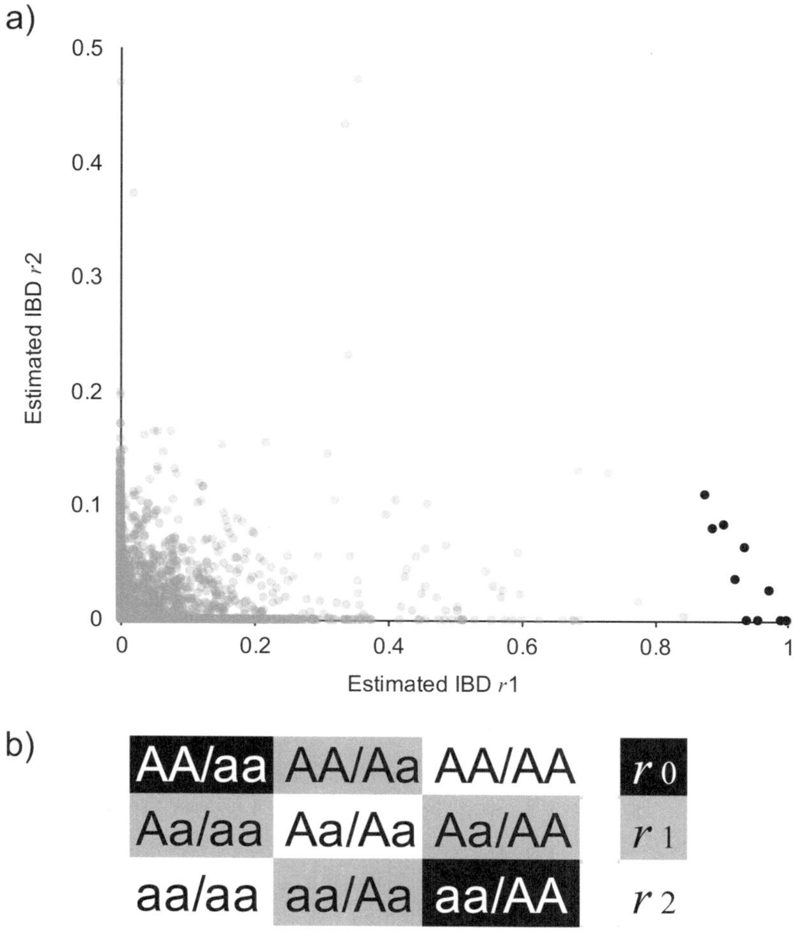

FIGURE 3.1. (a) Estimated coefficient of kinship in a dataset of 106 Norwegian killer whales genotyped at 1,346 SNPs (Jourdain et al. 2024). Relatedness estimates (r_x) due to identity-by-descent from a recent common ancestor are shown for all possible combinations of two individuals. Data points inferred to reflect parent-offspring relationships are shaded black. Parent-offspring relationships are characterized by a probability of approximately 1 of sharing a single allele (r_1) at each polymorphic site of the genome. There appear to be few or no full sibling relations in this dataset, based on expectations of a probability of 0.25 of both alleles (r_2) at polymorphic sites being shared between full siblings. With the exception of the putative parent-offspring relationships, the relatively continuous distribution of data points reflects the difficulty of establishing precise relationships between individuals from genetic data alone. (b) The nine possible allele combinations at a given site in a comparison of two individuals. The background shading highlights whether a combination results in the two individuals sharing zero, one, or two alleles.

identity-by-descent between two individuals (see Hanghøj et al. 2019). Many cetacean societies are descended from small founder groups, or from more recently bottlenecked populations (Crossman et al. 2024; Foote et al. 2021; Foote et al., this volume; Nigenda-Morales et al. 2023; Suárez-Menéndez et al. 2023). An important consideration is to estimate relatedness using appropriate statistics that account for the level of background inbreeding within the population (Hedrick and Lacy 2015).

Molecular genetic approaches can be used to estimate unilineal (matrilineal or patrilineal) genetic distance using uniparentally inherited markers such as mitochondrial DNA (mtDNA) to infer matrilineal relationships (Konrad et al. 2018), or Y-chromosome markers to infer patrilineal relationships (Nash 2004). Kinship or relatedness among pairs of individuals can be estimated from multilocus genotypes and then used to infer close "relationships" (e.g., parent-offspring; see fig. 3.1). Nuclear markers such as microsatellites (short tandem repeat motifs) and single nucleotide polymorphisms (SNPs) can be used to construct individual genotypes based on many loci throughout the genome. A relatively small number of microsatellites (10 to 20) can provide highly reliable paternity information (Harrison et al. 2013). However, a much higher number is needed to resolve more distant genetic relationships, such as half siblings or cousins (Attard et al. 2018; Städele and Vigilant 2016; Taylor 2015). With the introduction of high-throughput sequencing (HTS) technologies, sequencing complete mitochondrial and nuclear genomes and genotyping at thousands of SNPs from multiplexed pools of individuals has become increasingly accessible and cost-effective (Kircher and Kelso 2010; Morin et al. 2010; Davey et al. 2011). Kinship estimates of cetaceans have been made from whole genome sequences (e.g., Robinson et al. 2022; Foote et al. 2023), data generated from reduced representation library methods such as restriction site–associated DNA sequencing (RAD-seq) and genotyping by sequencing (GBS; e.g., Attard et al. 2018; Gerber et al. 2021), and data from target enrichment capture of SNPs (Jourdain et al. 2024; Baumgartner et al. 2025). The democratization of HTS technologies has greatly increased opportunities for expanding studies that were previously limited by both scope and scale, increasing the power to explore questions related to the role of kinship in the evolution and maintenance of associations among cetaceans.

The inference of shared genetic material allows biologists to unveil hidden pedigrees in wild populations. Age estimates are often crucial for identifying the direction of these relationships, that is, determining which individual is the offspring and which is the parent. The most detailed genetic-derived pedigrees are therefore often those associated with long-term studies of identifiable individuals. For example, the southern resident killer whale (SRKW) population has been the subject of half a century of data collection, providing an exemplary record of social composition through time (Parsons et al. 2009). While the photo-ID data revealed mother-offspring relationships for calves born during the study period (Bigg et al. 1990; Ford et al. 1994), paternities were unknown until relatively recent genetic studies on the population (Barrett-Lennard 2000; Ford et al. 2011, 2018). By genotyping 105 identified individuals, Ford et al. (2018) found that the two oldest adult males in the population had sired over half of the calves born since 1990, with the breeding success of males J1 and L42 increasing with age (Ford et al. 2018). This skewed breeding success is one of the causes of inbreeding in the SRKW population (Ford et al. 2018; Kardos et al. 2023; see section "Costs of Kin-Based Groups"). It also increases relatedness among siblings, which will increase selection on cooperative kin-directed behaviors.

Associations, Cooperation, and Interactions Between Kin

Social complexity is partially determined by the types of social interaction among associates (Weiss et al., this volume). In the following sections we highlight patterns of kinship underpinning these associations and interactions.

Males are mostly limited in their reproductive success by access to females, and female choice, and are thus expected, and often observed, to compete with each other to increase their access to mates (Mann et al. 2000). The reproductive success of females, in contrast, is limited by access to food and safety (Mann et al. 2000). It is therefore not surprising that in most group-living mammals, females form persisting cooperative relationships while males engage in one-off coalitions. Such coalitions often serve to attain higher ranking males in order to increase access to females (e.g., savanna baboons, *Papio anubis*; Noë and Sluijter 1995) or to prevent territorial takeovers by other males (e.g., hyenas, *Hyaena hyaena*; Caro 1994; Wagner et al. 2008). Yet in some species, primarily restricted to mammals and birds, males form long-lasting cooperative social bonds that are linked to adaptive benefits. Groups of two or more males that repeatedly cooperate to outcompete other males over access to females are referred to as alliances (Harcourt and de Waal 1992). Alliances have been described in humans (Chagnon 1988), nonhuman primates (savanna baboons, genus: *Papio*; Noë 1992; chimpanzees, *Pan troglodytes*; Mitani et al. 2002), lions (*Panthera leo*; Packer et al. 1991), Camargue horses (*Equus caballus*; Feh 1999), cheetahs (*Acinonyx jubatus*; Caro 1994), lekking birds (long-tailed manakins, *Chiroxiphia linearis*; McDonald and Potts 1994), and bottlenose dolphins (*Tursiops* spp.; Connor et al. 1992; see Connor and King, this volume). Because allied males cooperate to gain access to females, but only a single male succeeds to sire a female's offspring, kin selection has long been considered the most parsimonious explanation of male alliances (Díaz-Muñoz et al. 2014).

However, studies on various bottlenose dolphin populations have revealed that the role of kinship in the formation of male alliances is not broadly generalizable. In the Bahamas (*T. truncatus*; Parsons et al. 2003) and South Australia (*T.* cf. *australis*; Diaz-Aguirre et al. 2018), alliance partners were more closely related than expected by chance. This was not the case in Sarasota (*T. truncatus*; Wells 2014), southeastern Australia (*T. aduncus*; Möller et al. 2001), and Shark Bay (*T. aduncus*; Krützen et al. 2003; Gerber et al. 2020; Gerber et al. 2021), highlighting that the formation of kin-based alliances is neither confined to a species nor a specific region. It is important to note that apart from one study in Shark Bay (Gerber et al. 2021), all of the aforementioned studies assessed the potential influence of kinship on alliance formation based on estimates of average relatedness within alliances relative to a population mean. The calculation of average relatedness within alliances may omit critical information on the dyadic level where alliances consist of males with a diverse genetic background (i.e., non-siblings). Dyadic relatedness may be impacted by limited sampling of the larger population, and the inability to estimate background relatedness for calibrating pairwise relatedness (see Hedrick and Lacy 2015). Future efforts to compare relatedness within and among male alliances across *Tursiops* spp. populations globally within the context of greater population-level coverage are still needed to shed light on the enigmatic phenomenon and the evolutionary advantages of male alliance formation in dolphins.

In other social odontocetes, kinship may have little to no role in determining male-male social interactions. For example, male sperm whales (*Physeter macrocephalus*) become increasingly solitary as they mature and disperse from their natal social units before reaching maturity (Best 1979) and making their way to higher latitude foraging grounds (Kobayashi et al. 2020; Whitehead 2003). The role of kinship in associations among male sperm whales has not been examined, but circumstances suggest it is unlikely to play a key role. While two or more young male relatives may disperse from their social unit together, given the relatively small size of social units (typically 5–12 members [Gero et al. 2014 Whitehead et al. 2012] though a social unit as large as 28 individuals has been described [Sarano et al. 2021]), this circumstance is expected to be uncommon. The long interbirth interval for sperm whales (around five years; Best et al. 1984) also makes it unlikely for half siblings to disperse together, limiting the degree of relatedness between potential co-dispersers. This pattern of male dispersal followed by weaker and more transitory social associations among males than females is shared between sperm whales and African elephants (*Loxodonta africana*; Weilgart et al. 1996; Chiyo et al. 2011). However, in African elephants, these associations appear to be weakly correlated with relatedness (Chiyo et al. 2011).

Associations among female cetaceans are often influenced by kinship (but see Louis et al. 2018). Within sperm whale social units, females preferentially associate with close kin rather than more distant relatives (Konrad et al. 2018). Similarly, although female humpback whales are predominantly solitary, when in association they are more likely to be found with female kin (Weinrich et al. 2006). Strong(er) bonds among maternal relatives are also observed in open and more dynamic fission-fusion societies of bottlenose dolphins (*Tursiops* sp.), which may be composed of multiple matrilines (Möller et al. 2006; Frère et al. 2010b; Diaz-Aguirre et al. 2018). In Shark Bay, Western Australia, the social networks of female dolphins are shaped by the social networks of their mothers (Frère et al. 2010a). Given that calving success and offspring survival is linked to social bonds with other females in this population, the presence of female relatives impacts reproductive success and thus, both direct and indirect fitness. This contrasts to many male cetaceans, where cooperation among nonkin benefits only direct fitness (e.g., Gerber et al. 2022).

Matrilineal cetacean societies are examples of some of the most enduring female kin-based social bonds in nature (see Foroughirad et al., this volume). Matrilineal descent groups were defined by anthropologist Schneider (1961) as those in which culture is inherited by offspring from the mother, rather than from the father. Matriarchal kin groups were further distinguished as those which are female led (Schneider and Gough 1961). They framed unilineal descent groups, such as matrilines, as cultural solutions to ecological and social challenges, arguing that the form of the unilineal inheritance would reflect the combination of these challenges and the cultural resources available to tackle them. Finally, they proposed that the persistence of matrilineal descent will depend on the persistence of the ecological and social challenges, and the extent to which matrilineal inheritance remains a functionally adaptive solution.

Matrilineal kin-based social structure is found within several of the larger social odontocetes, including killer whales, false killer whales, and sperm whales. In these species, natal philopatry of offspring provides the basis for the matrilineal transmission of culture. Such matrilineal cetacean societies can be female led (i.e., matriarchal as defined by Schneider and Gough 1961; e.g., Brent et al. 2015), and are often multigenerational and multitiered, in which the extent of kinship varies within and among tiers, and through time (Bigg et al. 1990; Ivkovich et al. 2010; Whitehead et al. 2012; Tavares et al.

2017; Konrad et al. 2018). Matrilineal social structure and cultural descent provides the opportunity for mothers and grandmothers to contribute toward inclusive fitness (see Ellis et al., this volume). However, among species and even among populations within a species, there is variation in the extent to which cultural descent and philopatry within kin groups is strictly matrilineal. We therefore discuss the evolutionary advantages of this variation in unilineal descent relative to the ecological and social challenges facing different species and populations.

Among social odontocetes, killer whales are perhaps the most often cited species exhibiting lifelong matrilineal philopatry (Bigg et al. 1990; Ford et al. 1994; Rendell et al. 2019). The photo-identification study of eastern North Pacific resident killer whales, which has become foundational to research on the species, was the first to describe this matrilineal social structure and gradual fission as social units grow in size and become more structured, commonly when there is no longer a living common maternal ancestor (Bigg et al. 1990; Ford et al. 1994; Stredulinsky et al. 2021). The matrilineal social groups of the North Pacific resident ecotype are among the most cohesive and stable matrilineal groups of killer whales studied to date worldwide. However, some degree of matrilineal philopatry has been described for several other killer whale populations (Baird and Whitehead 2000; Beck et al. 2012; Esteban et al. 2016; Nielsen et al. 2023), but with greater fission-fusion dynamics in some populations (Ivkovich et al. 2010; Reisinger et al. 2017; Denkinger et al. 2020), or contexts (e.g., Tavares et al. 2017; Busson et al. 2019; Jordaan et al. 2021; see sections "Temporal Variation in Kin-Based Social Bonds" and "Ecological and Geographic Variation in Kin-Based Social Bonds").

Most of what we know about false killer whale sociality comes from a long-term study around the main Hawaiian Islands (MHI), where a small ($N < 200$; Bradford et al. 2018) island-associated resident population has been identified (Baird et al. 2008, 2010; Martien et al. 2014). Extensive photo-identification data, including opportunistic photographs dating back to 1986 (Baird et al. 2008; Mahaffy et al. 2023) revealed the existence of at least four "social clusters" within the population. The clusters are analogous to killer whale pods, with stable membership spanning decades (Mahaffy et al. 2023; Martien et al. 2019). Relatedness within social clusters is higher than among social clusters, suggesting that kinship is a major driver of social cluster membership, though statistical significance was limited to the two social clusters with the largest sample size (Martien et al. 2019). The distribution of mitochondrial DNA haplotypes among social clusters suggests strong natal fidelity to social clusters by both sexes (Martien et al. 2019; Mahaffy et al. 2023). Similarly, genetic parentage analysis identified 13 mother-offspring pairs in which both members of the pair were photographically identified, including many offspring whose sighting histories extend well into adulthood (Martien et al. 2019). In all cases, the mother and offspring belonged to the same social cluster, regardless of the offspring sex, again indicating both sexes exhibit strong fidelity to natal social clusters. In contrast, two out of five father-offspring pairs in which both individuals had known social cluster affiliation involved individuals from different social clusters (Martien et al. 2019), suggesting that mating occurs equally both within and between social clusters.

Sperm whales also live in matrilineally based groups, called "social units," with stable membership across years. However, in contrast to what has been described in many killer whale societies and false killer whales, males leave their natal social units after reaching maturity (Whitehead 2003). In this regard, as well as in relation to patterns of kinship among females in social units, sperm whales bear much similarity to African elephants (Weilgart et al. 1996; Konrad et al. 2018). Sperm whale social units are typically composed of maternal relatives but are not strictly matrilineal, in that they may not include a living common ancestor and can sometimes include unrelated individuals (Mesnick 2001; Konrad et al. 2018; Sarano et al. 2021). These matrilineally based social units are the foundational building blocks of sperm whales' multilevel social structure, within which cooperation thrives and culture is learned (Gero et al. 2016; Konrad et al. 2019). Maternal kin relationships likely also contribute, indirectly, to higher levels of their social structure. Associations between sperm whale social units are strongly structured by higher-level cultural groups, known as clans. Clans comprise hundreds to thousands of individuals who share vocal dialects (Rendell and Whitehead 2003) and other behavioral commonalities (Vachon et al. 2022; Whitehead and Rendell 2004; Hersh et al., this volume). Social units exclusively associate with other social units from within the same clan, who are not necessarily close maternal relatives; in the eastern Caribbean, associations between social units were generally not correlated with kinship (Konrad et al. 2018). Clans are not strictly kin-based, but assessment of mtDNA is consistent with the hypothesis that clans arise through vertical maternal cultural transmission and oblique transmission within social units (Rendell et al. 2012). Thus, even at this higher level of social structure, shared matrilineal ancestry appears to be the foundation from which cultural groups emerge.

Kin Selection and Fitness Benefits of Social Philopatry

Sperm whale social units are highly cooperative, and kin selection appears to play an important role in maintaining cooperation as a successful strategy.

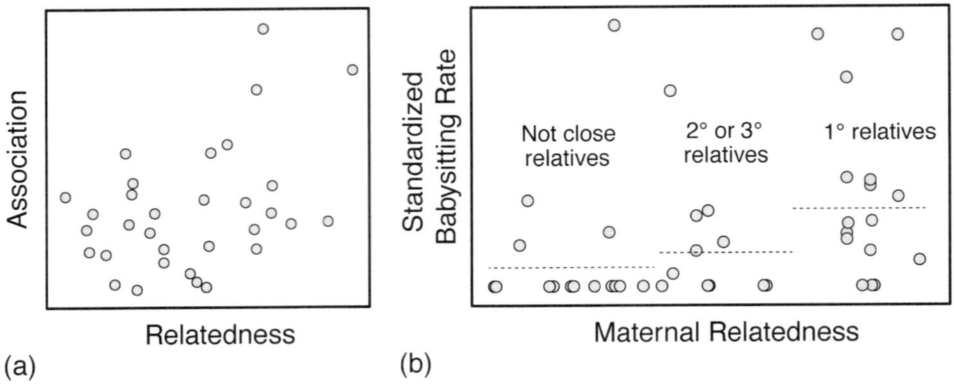

FIGURE 3.2. Among eastern Caribbean sperm whales, (a) genetic relatedness correlates with social association within social units (Unit F; matrix correlation = 0.4; $p = 0.002$; Konrad et al. 2018) and (b) maternal genetic relatedness between calves and non-calf unit members correlates with babysitting rate scaled by the calf's association rate with its mother (Spearman's rank correlation = 0.40; $p = 0.02$, 50,000 permutations; Konrad et al. 2019). On average (dashed lines), likely first-degree relatives of the mother (half siblings or grandmothers of the calf) babysat twice as much as second- or third-degree relatives of the mother (e.g., half aunts of the calf), who babysat twice as much as individuals that were not close relatives of the mother.

In addition to the maternal social units within which cooperation occurs, finer-scale correlations between cooperation and kinship have been described (Konrad et al. 2018, 2019; fig. 3.2). Sperm whale social structure is hypothesized to be driven by the need for cooperative care of young (Gero et al. 2013; Whitehead 1996; Best 1979). Babysitting and cooperative defense against predators are forms of cooperative care that have been documented in sperm whales (Weller et al. 1996; Whitehead 1996). Female sperm whales have also been observed nursing the offspring of other social unit members (allonursing; Gero et al. 2009; Konrad et al. 2019; Sarano et al. 2023). In the eastern Caribbean, maternal kinship was positively correlated with rates of babysitting and the likelihood of allonursing (Konrad et al. 2019; fig. 3.2). In addition, the greater the apparent cost of an alloparental behavior, the more likely it was only performed by closer maternal relatives (Konrad et al. 2019).

Shared paternity within sperm whale social units is not uncommon (Pinela et al. 2009; Konrad et al. 2018; Girardet et al. 2022), and this can increase kinship within social units and amplify inclusive fitness benefits of cooperative behaviors. However, kinship is not the only factor driving cooperative care of young. Alloparents are sometimes not close kin of the calves they nurse or babysit (Konrad et al. 2019; Sarano et al. 2021; Sarano et al. 2023), such as in one eastern Caribbean social unit, within which mothers did not have close kin (fig. 3.3). In such cases, alloparents may derive direct benefits from gained maternal experience, group augmentation, reciprocal altruism, or some combination thereof (Konrad et al. 2019).

False killer whales are also known to engage in social activities among kin, including prey sharing and cooperative hunting (Baird et al. 2008; Baird 2016). Both behaviors can increase inclusive fitness when association is correlated with kinship and may also serve to reinforce social cohesion within clusters (Wright et al. 2016). Cooperative hunting may require a high level of coordination between individuals (see King et al., this volume) and could therefore represent a barrier to new individuals joining a social cluster, especially if social clusters have different hunting strategies. This could be the case for the main Hawaiian Island insular false killer whales. Both photo-identification and satellite tag data show that spatial use within the population's range differs between social clusters (Baird et al. 2012; Mahaffy et al. 2023; Bradford et al. 2018), which may result in contrasting hunting strategies. The near absence of gene flow between the insular population and the surrounding pelagic population could also result from reduced fitness of potential migrants that are unfamiliar with the population's hunting strategy or are socially excluded from cooperative hunting (Martien et al. 2014), as has been suggested for some killer whale populations (Baird and Whitehead 2000).

Philopatry in matrilineal killer whale populations is thought to be driven by the fitness benefits of kin-directed food sharing (Wright et al. 2016), and transmission of ecological and social knowledge through female leadership (Brent et al. 2015). Prey sharing among eastern North Pacific killer whale

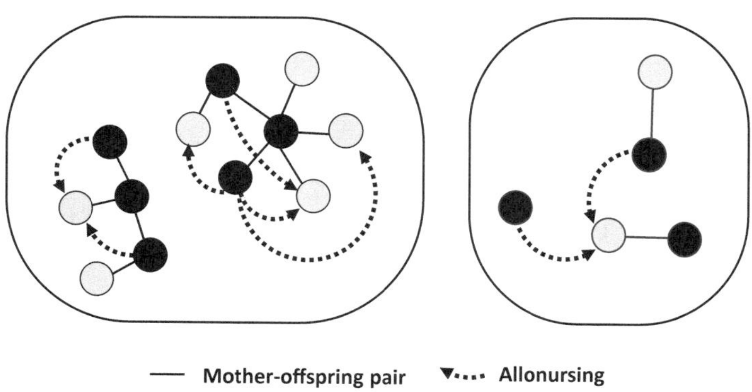

FIGURE 3.3. Patterns of allonursing differ among eastern Caribbean sperm whale social units. Nodes represent individual whales from social units (indicated by oval outlines), with black shading indicating mature or juvenile females and gray shading indicating calves. Mother-offspring relationships are indicated by solid lines between individuals. Dashed arrows point from a mature or juvenile female to the calf observed attempting to nurse from her.
Source: Data from Konrad et al. 2019.

kin has been proposed as a mechanism that reinforces kin-based associations and promotes philopatry (Wright et al. 2016). Data on shared foraging for resident killer whales indicated that prey sharing was more frequent among close maternal relatives. The inclusive fitness benefits associated with provisioning maternal kin may also reduce the potential benefits to be gained from dispersal.

Costs of Kin-Based Groups

The strong natal fidelity to social clusters and evidence of mating within social clusters raises concern regarding the effects of mating among closely related individuals, or inbreeding. Common behavioral strategies for avoiding inbreeding include mating between individuals from different social groups, a pattern that has been observed among long-finned pilot whales (Amos et al. 1991), mating between females and unrelated males regardless of the social group to which they belong, as has been suggested to occur in northern resident killer whales (e.g., Barrett-Lennard 2000), or the dispersal of males from their natal social units altogether, as in sperm whales (Whitehead 2003). However, the endangered SRKW population, where high levels of inbreeding have been characterized based on targeted SNP analyses and whole genome sequence data (Ford et al. 2018; Kardos et al. 2023), does not appear to effectively employ any of these strategies. Main Hawaiian Island insular false killer whales also exhibit high levels of mating within social groups (Martien et al. 2019); however, it is not known whether intragroup matings are between individuals that are more, or less, related than average. The apparent lack of inbreeding avoidance mechanisms within the SRKW and MHI false killer whale populations may be a direct result of their recent precipitous declines in abundance (Reeves et al. 2009; Bigg and Wolman 1975; Wiles 2004), which may have reduced opportunities for intergroup mating (Ford et al. 2018; Martien et al. 2019). Alternatively, if social groups were much larger prior to the declines, there may have been little selective pressure to favor mating outside of social groups.

Kin-based associations can also come at a cost in terms of fitness consequences due to competition for resources. Competition is thought to be a driver of dispersal in some species or populations (e.g., Baird and Dill 1996). A specific form of competition incurred in philopatric species and populations is reproductive conflict between females within a kin-based group (Cant and Johnstone 2008; Johnstone and Cant 2010). Given that philopatry is a trait associated with some of the best-studied social cetacean species, they are expected to have evolved strategies to counter this selective pressure. One such

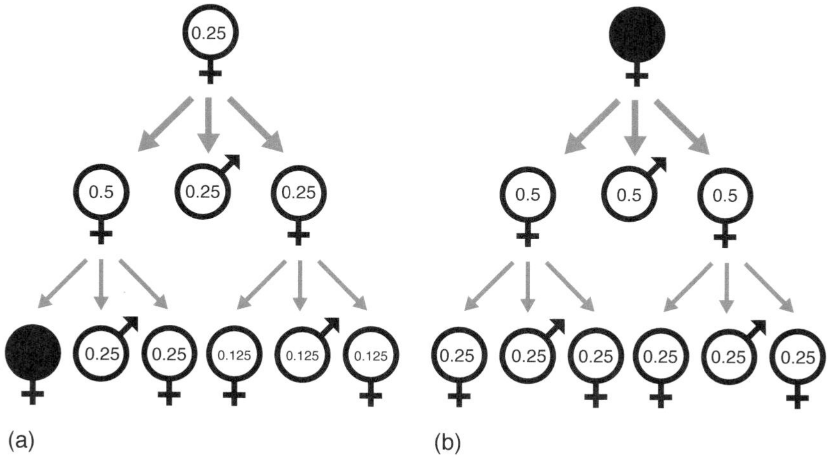

FIGURE 3.4. Schematic figure of increasing local relatedness with age of a female in a three-generation matrilineal pedigree. In panel (a) the focal female (filled black symbol) is a juvenile. In panel (b) the focal female is now the oldest female in the matriline (i.e., the matriarch), and all other group members are her offspring or grandoffspring. Pairwise relatedness between the focal female and each individual is shown. All siblings are assumed to be half siblings. Although the demographic structure of the matriline does not change, the average local relatedness of the focal female to others in the group changes from 0.236 in panel (a) to 0.333 in panel (b). This change in local relatedness theoretically increases the cost of reproductive conflict and increases selection for a post-reproductive life span in females (see Ellis et al., this volume).

strategy is the evolution of a multiyear female menopause, which has been reported in several species with matrilineal kin-based social units, including killer whales, false killer whales, and short-finned pilot whales (see Ellis et al., this volume). The presence of a long post-reproductive lifespan has been shown to reduce intergenerational conflict in resident killer whales (Croft et al. 2017). The increasing average local relatedness with age of a female to kin within her matriline (see fig. 3.4) selects against competition with her daughter for resources by continuing to reproduce, and instead confers gains in inclusive fitness through investing in her existing kin (Croft et al. 2017). The presence of a long post-reproductive female lifespan in short-finned pilot whales, and its absence in long-finned pilot whales, has also been attributed to the presence of reproductive conflict in the former and its absence in the latter (McCormack et al. 2023).

Temporal Variation in Kin-Based Social Bonds

Fluidity in the social structure among closely related individuals plays an important role in creating opportunities for cooperative behaviors, mate selection, reciprocity, and cultural transmission of foraging traditions, affecting not only social behaviors, but also competition for resources (e.g., Cohas et al. 2009; Crook et al. 1976; Gerber et al. 2021; Whitehead et al. 2004; Wild et al. 2020). Relatedness among individuals within groups can also change over time (i.e., kinship dynamics), and in turn affect social behavior and the interactions among associates (Croft et al. 2021). Temporal changes in the strength of associations within the larger matrifocal groups or pods suggest that interannual variability in the number of social clusters in the SRKW population is largely driven by kin-based associations among small groups of individuals (Parsons et al. 2009). SRKW pods were originally defined as individuals seen together more often than not (association index >0.5), based on observations spanning multiple decades (Bigg et al. 1990). While the average rate of association among pod mates was lowest in the largest of the three "pods," associations were significantly greater among maternally related whales, suggesting that smaller matrilineal kin-based units increase in significance as the size of the larger social "pod" grows. Associations among more distantly related members may become too costly, with costs of resource procurement offsetting the indirect fitness benefits of kin-based social units (Parsons et al. 2009). SRKW prey exclusively on fish species, and primarily on Chinook salmon (*Oncorhynchus tshawytscha*) during the summer months (Ford et al. 2016; Ford and Ellis 2006; Hanson et al. 2021). In years with higher Chinook salmon abundance, the SRKW social network was more interconnected than in years when this primary summer prey species was scarcer, over a 23-year period (Foster et al. 2012). This relationship between social network stability and food availability demonstrates a population-level response to environmental variables, with implications for demographic and evolutionary processes (Blumstein et al. 2023).

In both killer whales and sperm whales, fission or fusion of social groups appears to occur in such a way that maximizes maternal kinships within groups (Konrad et al. 2018; Stredulinsky et al. 2021). Within northern resident killer whales, fission occurred such that animals dispersed in parallel with their closest maternal kin, and larger groups were more likely to split into smaller matrilineal units (Stredulinsky et al. 2021). Similarly, in a well-studied sperm whale population in the eastern Caribbean, both group fission and fusion occurred along maternal kin lines, seemingly in response to changes in group size (Konrad et al. 2018). One large social unit appeared to be in the process of fission, along matrilineal lines. In contrast, two smaller social units, the two most closely related social units in the study, slowly merged into a single social unit as the group size of each dwindled (Konrad et al. 2018). A similar fission of a large sperm whale social unit, splitting in a way that kept all first-degree maternal relatives together, has also been reported in the Indian Ocean (Sarano et al. 2023). A parallel influence of kinship on group fission and fusion has also been described in African elephants (Archie et al. 2006). These processes appear to reflect the trade-offs in inclusive fitness gains and other costs (e.g., increased foraging competition and reproductive conflict) and benefits (e.g., in sperm whales, enhanced cooperative defense against predators) associated with increasing group size.

Ecological and Geographic Variation in Kin-Based Social Bonds

All the taxa discussed in this chapter are found across a large range of diverse marine habitats (Leatherwood and Reeves 1989; Whitehead 2003), with some also exhibiting diverse ecological roles. This variation in ecological and geographic environments can be reflected in variation in patterns of kinship in social bonds. For example, among bottlenose dolphins (*Tursiops* spp.), variation in formation of alliances among kin versus nonkin (Parsons et al. 2003; Diaz-Aguirre et al. 2018; Wells 2014; Möller et al. 2001; Krützen et al. 2003; Gerber et al. 2020; Gerber et al. 2021), might reflect adaptations to their respective ecological (e.g., water depth, prey type, predation pressure) and social environments. Across these populations, behavioral studies have revealed that patterns of dispersal, site fidelity, life history parameters, and population size and density vary. Each of these variables may affect whether and among whom alliances are formed.

Until recently, the MHI insular population was the only known small, resident population of false killer whales in the world. However, resident populations have now been identified in the Northwest Hawaiian Islands (Baird et al. 2013), off the northern coast of Australia (Palmer et al. 2023), and ongoing studies suggest the existence of resident populations around New Zealand (Zaeschmar et al. 2014; E. Carroll, pers. comm.) and the Pacific coast of Mexico (Lazcano-Pacheco et al. 2023; Douglas et al. 2023). Though little is known about patterns of kinship within these other populations, genetic data suggest that the Northwest Hawaiian Islands population experiences gene flow with the surrounding pelagic population at a much higher rate than does the MHI population. Martien et al. (2014) suggest that the differences in genetic isolation between the two insular populations may be due to oceanographic influences that result in greater resource specialization in the MHI population, and such differences in ecological pressures may be reflected in contrasting patterns of association and the role of kinship among close associates. Further research is needed to examine patterns of association and kinship in other coastal and insular false killer whale populations. What little is known about association patterns in pelagic false killer whales comes from mass strandings and pelagic encounters in which tissue samples were taken from multiple individuals. Despite the much higher haplotypic diversity in pelagic false killer whales compared to the insular populations, most groups encountered at sea contain only one or two haplotypes (Martien et al. 2019; Crofts et al. 2019), indicating that pelagic groups may also exhibit a strong correlation between association and matrilineal kinship.

Sperm whales also exhibit geographic variation in the patterns of kinship within their social structure. Social units in the Pacific Ocean tend to be larger and more often include unrelated individuals (demonstrated by different mtDNA haplotypes) than social units in the Atlantic Ocean (Whitehead et al. 2012; Konrad et al. 2018). Additionally, social units in the Pacific are more likely to form groups with other social units, spending time with individuals who are presumably not close relatives (Whitehead et al. 2012). Potential drivers of these differences include greater risk of predation by killer whales in the Pacific Ocean, favoring larger group sizes, and heavy modern whaling in the Pacific Ocean perhaps disturbing typical patterns of matrilineally based social units (Whitehead et al. 2012). Further studies in other areas (and ongoing studies in the Pacific as populations recover from whaling) will provide data with which to better investigate these possible drivers. For example, a recent study in the Indian Ocean described a social unit that was large (28 individuals) compared to typical social unit sizes described elsewhere and contained maternal kin as well as one individual with a different mtDNA haplotype than the rest of the social unit; (Sarano et al. 2021). Also, in contrast to observations in the Atlantic Ocean (Konrad et al. 2019), an examination of allonursing in the Indian Ocean did not find maternal kinship to be well correlated with this behavior (Sarano et al. 2023). This suggests sperm whales in this ocean basin may face different ecological pressures than those in the Pacific and Atlantic Oceans. Evidence for shared paternity within social units, however, is consistent in the Atlantic (Pinela et al. 2009; Konrad et al. 2018) and Indian (Girardet et al. 2022) Oceans, indicating this may be a common supplementary force promoting kin-directed cooperation in this species.

In the North Pacific, sympatric, divergent killer whale lineages with contrasting patterns of dispersal, prey preferences, acoustic dialects, and behaviors have been documented in detail through long-term individual-based observations, providing unique insights into dispersal (or lack thereof) and social affiliations (Ford et al. 1998; Ford and Ellis 2014; Saulitis et al. 2000; Saulitis et al. 2005; Riesch and Deecke 2011). These lineages, or "ecotypes," include a fish-eating specialist (the so-called resident killer whales) and a mammal-eating specialist (the so-called transient or Bigg's killer whales). The behavioral ecology and sociobiology of the two ecotypes is shaped by their different prey preferences. For example, resident killer whales can hunt in large groups and coordinate their movement through culturally transmitted vocal dialects (Ford 1989, 1991); in contrast, Bigg's killer whales hunt in small silent groups so as not to alert their acoustically sensitive marine mammal prey (Deecke et al. 2005). The resident killer whales exhibit an absence of natal dispersal, and social bonds among maternally related kin fluctuate interannually but often persist over a lifetime (Bigg et al. 1990; Parsons et al. 2009; Pilot et al. 2010). There is also a high degree of male philopatry in Bigg's killer whale (Baird and Whitehead 2000; Nielsen et al. 2023). Over 40 years of demographic data have shown the probability of a "constant companion" relationship between mother and first male offspring remained high (mean of approximately 0.75 or higher) throughout a male's lifespan (Nielsen et al. 2023). In contrast, there is gradual dispersal of female offspring, typically after a daughter has her first offspring (Baird and Whitehead 2000; Nielsen et al. 2023).

In contrast to the population-level ecological specializations described across the North Pacific, killer whales in the northeastern Atlantic exhibit foraging variation among individuals within populations (Foote et al. 2009; Remili et al. 2023; Samarra and Foote 2015; Samarra et al. 2017; Vongraven and Bisther 2014; Jourdain et al. 2017, 2020). Off Norway, for example, killer whales feed on Atlantic herring (*Clupea harengus*) in winter when it is fattest and most abundant (Similä et al. 1996; Jourdain et al. 2020). As the herring becomes less profitable later in the year, some killer whale groups switch to feeding on alternative energy-rich prey resources such as (spawning) lumpfish (*Cyclopterus lumpus*) or pinnipeds (Vongraven and Bisther 2014; Jourdain et al. 2017, 2019). A variety of overlapping diets consisting exclusively of fish or a mix of fish and mammals, albeit in various group-specific proportions, were identified (Jourdain et al. 2020).

Consistent with North Pacific populations, matrilineal social structure and long-term, kin-based social philopatry also appear to be the building blocks of killer whale social structure in Norwegian waters. Long-term individual photo identification combined with genetic data confirmed matrifocal groups as the core unit of social organization, with mother-offspring bonds lasting long into adulthood, regardless of diet (Jourdain et al. 2024). These social bonds become increasingly flexible with decreasing relatedness, leading to frequent nonkin associations, including between piscivorous individuals and individuals with a more generalist diet (i.e., whales that feed on both fish and mammals). Where North Pacific killer whale ecotypes maintain their

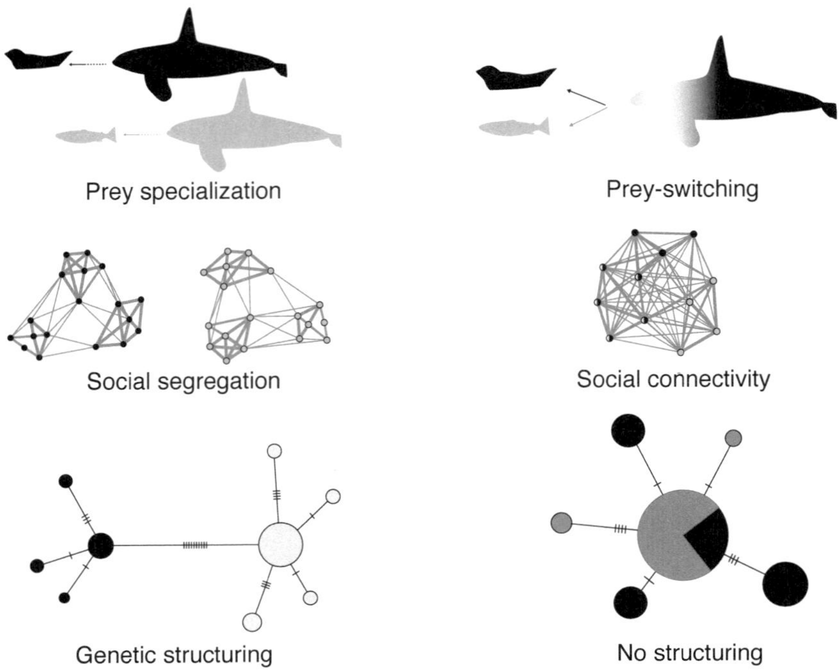

FIGURE 3.5. When kin-based social groups are foraging specialists, cultural transmission of dietary specializations among matriline members can lead to behaviorally assorted groups. Cultural barriers, whereby individuals preferentially associate and mate with others that adopt similar prey preferences, may further constrain social interactions among prey specialists only and lead to reproductive isolation. In contrast, prey-switching killer whales may have a greater range of possible associates—as promoted by shared habitat and common feeding techniques owing to overlapping diets and increasing social connectivity.

divergent prey preferences throughout the year, killer whale groups around Norway can target a common prey species seasonally (e.g., herring feeding in winter, Jourdain et al. 2020), and switch to different but geographically overlapping food sources. This geographic overlap likely promotes associations between groups despite the different degrees of variation in their diet. Given that killer whales in Norway are capable of prey switching (Vongraven and Bisther 2014), they should theoretically be able to disperse from their natal group to join others with different, if overlapping, diets without experiencing reduced fitness (fig. 3.5).

Within Norwegian killer whales, the existence of behaviorally diverse kin groups, and fluid relationships between them, promotes a highly connected social landscape that may be key to maintaining ecological diversity at the population level (Jourdain et al. 2024). Highly stable kin associations at the core of primary social units provide a foundation for cultural transfer, and thus persistence, of feeding behaviors through kin (Brent et al. 2015). These behaviors can then be socially learned and transferred to nonkin through dynamic social relationships between groups (Wild et al. 2020).

Increasing numbers of field studies in diverse ecological contexts suggest killer whale sociality to be more varied than previously thought. Kinship appears as a common driver of social affiliation regardless of ecological context, albeit to various degrees. However, more fluid relationships, less strictly defined by prey species may be more common among killer whales in many geographic regions (e.g., Tavares et al. 2017, 2018; Reisinger et al. 2017; Jordaan et al. 2021; Beck et al. 2012; Jourdain et al. 2024), than the strict matrilineal philopatry reported in the North Pacific.

Through the lens of conservation, the variable patterns of social structure and behavioral diversity that characterize all four of our focal species highlight the importance of conserving genetic, social, and ecological diversity for enhanced adaptive potential in the face of changing marine habitats (Brakes et al. 2019; Brakes and Whitehead, this volume).

Summary

In this chapter, we have reviewed the patterns of kinship and sociality found in four of the best-studied social odontocete species. Such data-rich examples highlight the fitness costs and benefits of associating and interacting with close kin, and their evolutionary outcomes. However, these focal species may not be representative of all, or even the majority of cetacean species. For example, temporally varying social bonds and cooperative behavior have been described among humpback whales (Ramp et al. 2010; Wray et al. 2021), but often in the absence of close kinship bonds (Valsecchi et al. 2002; Pomilla and Rosenbaum 2006; Weinrich et al. 2006). Looking forward, we anticipate that during the coming decades, the wealth of ongoing field studies covering a wide range of cetacean taxa, in diverse ecological contexts, will continue to uncover the diversity of kinship and social interactions with cetacean societies.

References

Amos, Bill, John Barrett, and Gabriel A. Dover. 1991. "Breeding Behaviour of Pilot Whales Revealed by DNA Fingerprinting." *Heredity* 67 (1): 49–55. https://doi.org/10.1038/hdy.1991.64.

Archie, Elizabeth A., Cynthia J. Moss, and Susan C. Alberts. 2006. "The Ties That Bind: Genetic Relatedness Predicts the Fission and Fusion of Social Groups in Wild African Elephants." *Proceedings of the Royal Society B* 273 (1586): 513–22. https://doi.org/10.1098/rspb.2005.3361.

Attard, Catherine R. M., Luciano B. Beheregaray, and Luciana M. Möller. 2018. "Genotyping-by-Sequencing for Estimating Relatedness in Nonmodel Organisms: Avoiding the Trap of Precise Bias." *Molecular Ecology Resources* 18 (3): 381–90. https://doi.org/10.1111/1755-0998.12739.

Baird, Robin W. 2016. *The Lives of Hawai'i's Dolphins and Whales: Natural History and Conservation*. University of Hawai'i Press.

Baird, Robin W., and Lawrence M. Dill. 1996. "Ecological and Social Determinants of Group Size in Transient Killer Whales." *Behavioral Ecology* 7 (4): 408–16. https://doi.org/10.1093/beheco/7.4.408.

Baird, Robin W., Antoinette M. Gorgone, Daniel J. McSweeney, et al. 2008. "False Killer Whales (*Pseudorca crassidens*) Around the Main Hawaiian Islands: Long-Term Site Fidelity, Inter-Island Movements, and Association Patterns." *Marine Mammal Science* 24 (3): 591–612. https://doi.org/10.1111/j.1748-7692.2008.00200.x.

Baird, Robin W., M. Bradley Hanson, Gregory S. Schorr, et al. 2012. "Range and Primary Habitats of Hawaiian Insular False Killer Whales: Informing Determination of Critical Habitat." *Endangered Species Research* 18 (1): 47–61. https://doi.org/10.3354/esr00435.

Baird, Robin W., Erin M. Oleson, Jay Barlow, Allan D. Ligon, Antoinette M. Gorgone, and Sabre D. Mahaffy. 2013. "Evidence of an Island-Associated Population of False Killer Whales (*Pseudorca crassidens*) in the Northwestern Hawaiian Islands." *Pacific Science* 67 (4): 513–21. https://doi.org/10.2984/67.4.2.

Baird, Robin W., Gregory S. Schorr, Daniel L. Webster, Daniel J. McSweeney, M. Bradley Hanson, and Russel D. Andrews. 2010. "Movements and Habitat Use of Satellite-Tagged False Killer Whales Around the Main Hawaiian Islands." *Endangered Species Research* 10: 107–21. https://doi.org/10.3354/esr00258.

Baird, Robin W., and Hal Whitehead. 2000. "Social Organization of Mammal-Eating Killer Whales: Group Stability and Dispersal Patterns." *Canadian Journal of Zoology* 78 (12): 2096–105. https://doi.org/10.1139/z00-155.

Barrett-Lennard, Lance Godfrey. 2000. "Population Structure and Mating Patterns of Killer Whales (*Orcinus orca*) as Revealed by DNA Analysis." PhD diss., University of British Columbia.

Baumgartner, Chérine D., Eve Jourdain, Sebastian Bonhoeffer, et al. 2025. "Kinship Clustering Within an Ecologically Diverse Killer Whale Metapopulation." *Heredity* 134 (1): 109–19. https://doi.org/10.1038/s41437-024-00740-y.

Beck, Suzanne, Sanna Kuningas, Ruth Esteban, and Andrew D. Foote. 2012. "The Influence of Ecology on Sociality in the Killer Whale (*Orcinus orca*)." *Behavioral Ecology* 23 (2): 246–53. https://doi.org/10.1093/beheco/arr151.

Best, Peter B. 1979. "Social Organization in Sperm Whales, *Physeter macrocephalus*." In *Behavior of Marine Animals*, edited by Howard E. Winn and Bori L. Olla, 227–89. Springer.

Best, Peter B., P. A. S. Canham, and N. Macleod. 1984. "Patterns of Reproduction in Sperm Whales, *Physeter macrocephalus*." *Report of the International Whaling Commission* 6: 51–79.

Bigg, Michael A., Peter F. Olesiuk, Graeme M. Ellis, John K. B. Ford, and Kenneth C. Balcomb. 1990. "Social Organization and Genealogy of Resident Killer Whales (*Orcinus orca*) in the Coastal Waters of British Columbia and Washington State." *Report of the International Whaling Commission* 12: 383–405.

Bigg, Michael A., and Allen A. Wolman. 1975. "Live-Capture Killer Whale (*Orcinus orca*) Fishery, British Columbia and Washington, 1962–73." *Journal of the Fisheries Board of Canada* 32 (7): 1213–21. https://doi.org/10.1139/f75-140.

Blumstein, Daniel T., Loren D. Hayes, and Noa Pinter-Wollman. 2023. "Social Consequences of Rapid Environmental Change." *Trends in Ecology and Evolution* 38 (4): 337–45. https://doi.org/10.1016/j.tree.2022.11.005.

Bradford, Amanda L., Robin W. Baird, Sabre D. Mahaffy, et al. 2018. "Abundance Estimates for Management of Endangered False Killer Whales in the Main Hawaiian Islands." *Endangered Species Research* 36: 297–313. https://doi.org/10.3354/esr00903.

Brakes, Philippa, Sasha R. X. Dall, Lucy M. Aplin, et al. 2019. "Animal Cultures Matter for Conservation." *Science* 363 (6431): 1032–34. https://doi.org/10.1126/science.aaw3557.

Brent, Lauren J. N., Daniel W. Franks, Emma A. Foster, Kenneth C. Balcomb, Michael A. Cant, and Darren P. Croft. 2015. "Ecological Knowledge, Leadership, and the Evolution of Menopause in Killer Whales." *Current Biology* 25 (6): 746–50. https://doi.org/10.1016/j.cub.2015.01.037.

Busson, Marine, Matthieu Authier, Christophe Barbraud, et al. 2019. "Role of Sociality in the Response of Killer Whales to an Additive Mortality Event." *Proceedings of the National Academy of Sciences* 116 (24): 11812–17. https://doi.org/10.1073/pnas.1817174116.

Cant, Michael A., and Rufus A. Johnstone. 2008. "Reproductive Conflict and the Separation of Reproductive Generations in Humans." *Proceedings of the National Academy of Sciences* 105 (14): 5332–36. https://doi.org/10.1073/pnas.0711911105.

Caro, Timothy M. 1994. *Cheetahs of the Serengeti Plains: Group Living in an Asocial Species.* University of Chicago Press.

Chagnon, Napoleon A. 1988. "Life Histories, Blood Revenge, and Warfare in a Tribal Population." *Science* 239 (4843): 985–92. https://doi.org/10.1126/science.239.4843.985.

Charlesworth, Deborah, and John H. Willis. 2009. "The Genetics of Inbreeding Depression." *Nature Reviews Genetics* 10 (11): 783–96. https://doi.org/10.1038/nrg2664.

Chiyo, Patrick I., Elizabeth A. Archie, Julie A. Hollister-Smith, et al. 2011. "Association Patterns of African Elephants in All-Male Groups: The Role of Age and Genetic Relatedness." *Animal Behaviour* 81 (6): 1093–99. https://doi.org/10.1016/j.anbehav.2011.02.013.

Cohas, Aurélie, and Dominique Allainé. 2009. "Social Structure Influences Extra-Pair Paternity in Socially Monogamous Mammals." *Biology Letters* 5 (3): 313–16. https://doi.org/10.1098/rsbl.2008.0760.

Connor, Richard C., Rachel A. Smolker, and Andrew F. Richards. 1992. "Two Levels of Alliance Formation Among Male Bottlenose Dolphins (*Tursiops* sp.)." *Proceedings of the National Academy of Sciences* 89 (3): 987–90. https://doi.org/10.1073/pnas.89.3.987.

Croft, Darren P., Rufus A. Johnstone, Samuel Ellis, et al. 2017. "Reproductive Conflict and the Evolution of Menopause in Killer Whales." *Current Biology* 27 (2): 298–304. https://doi.org/10.1016/j.cub.2016.12.015.

Croft, Darren P., Michael N. Weiss, Mia L. K. Nielsen, et al. 2021. "Kinship Dynamics: Patterns and Consequences of Changes in Local Relatedness." *Proceedings of the Royal Society B* 288 (1957): 20211129. https://doi.org/10.1098/rspb.2021.1129.

Crofts, Sarah, Karen K. Martien, Kelly M. Robertson, Andrew Stanworth, Steve Massam, and Caroline R. Weir. 2019. "First Record of False Killer Whales (*Pseudorca crassidens*) in the Falkland Islands (Malvinas)." *Polar Biology* 42 (10): 1923–29. https://doi.org/10.1007/s00300-019-02554-9.

Crook, John H., J. E. Ellis, and J. D. Goss-Custard. 1976. "Mammalian Social Systems: Structure and Function." *Animal Behaviour* 24 (2): 261–74. https://doi.org/10.1016/S0003-3472(76)80035-8.

Crossman, Carla A., Michael C. Fontaine, and Timothy R. Frasier. 2024. "A Comparison of Genomic Diversity and Demographic History of the North Atlantic and Southwest Atlantic Southern Right Whales." *Molecular Ecology* 33 (20): e17099. https://doi.org/10.1111/mec.17099.

Davey, John W., Paul A. Hohenlohe, Paul D. Etter, et al. 2011. "Genome-Wide Genetic Marker Discovery and Genotyping Using Next-Generation Sequencing." *Nature Reviews Genetics* 12 (7): 499–510. https://doi.org/10.1038/nrg3012.

Deecke, Volker B., John K. B. Ford, and Peter J. B. Slater. 2005. "The Vocal Behaviour of Mammal-Eating Killer Whales: Communicating with Costly Calls." *Animal Behaviour* 69 (2): 395–405. https://doi.org/10.1016/j.anbehav.2004.04.014.

Denkinger, Judith, Daniela Alarcon, Bitinia Espinosa, et al. 2020. "Social Structure of Killer Whales (*Orcinus orca*) in a Variable Low-Latitude Environment, the Galápagos Archipelago." *Marine Mammal Science* 36 (3): 774–85. https://doi.org/10.1111/mms.12672.

Diaz-Aguirre, Fernando, Guido J. Parra, Cecilia Passadore, and Luciana Möller. 2018. "Kinship Influences Social Bonds Among Male Southern Australian Bottlenose Dolphins (*Tursiops* cf. *australis*)." *Behavioral Ecology and Sociobiology* 72: 1–13. https://doi.org/10.1007/s00265-018-2621-4.

Díaz-Muñoz, Samuel L., Emily H. DuVal, Alan H. Krakauer, and Eileen A. Lacey. 2014. "Cooperating to Compete: Altruism, Sexual Selection and Causes of Male Reproductive Cooperation." *Animal Behaviour* 88: 67–78. https://doi.org/10.1016/j.anbehav.2013.11.008.

Douglas, Annie B., Frank Garita Alpízar, Alejandro Acevedo-Gutiérrez, et al. 2023. "False Killer Whales (*Pseudorca crassidens* Cetacea: Delphinidae) Along the Pacific Coast of Central America and Mexico: Long-Term Movements, Association Patterns and Assessment of Fishery Interactions." *Revista de Biología Tropical* 71 (S4): e57189. https://doi.org/10.15517/rev.biol.trop..v71iS4.57189.

Esteban, Ruth, Philippe Verborgh, Pauline Gauffier, Joan Giménez, Christophe Guinet, and Renaud de Stephanis. 2016. "Dynamics of Killer Whale, Bluefin Tuna and Human Fisheries in the Strait of Gibraltar." *Biological Conservation* 194: 31–38. https://doi.org/10.1016/j.biocon.2015.11.031.

Feh, Claudia. 1999. "Alliances and Reproductive Success in Camargue Stallions." *Animal Behaviour* 57 (3): 705–13. https://doi.org/10.1006/anbe.1998.1009.

Foote, Andrew D., Alana Alexander, Lisa T. Ballance, et al. 2023. "'Type D' Killer Whale Genomes Reveal Long-Term Small Population Size and Low Genetic Diversity." *Journal of Heredity* 114 (2): 94–109. https://doi.org/10.1093/jhered/esac070.

Foote, Andrew D., Rebecca Hooper, Alana Alexander, et al. 2021. "Runs of Homozygosity in Killer Whale Genomes Provide a Global Record of Demographic Histories." *Molecular Ecology* 30 (23): 6162–77. https://doi.org/10.1111/mec.16137.

Foote, Andrew D., Jason Newton, Stuart B. Piertney, Eske Willerslev, and M. Thomas P. Gilbert. 2009. "Ecological, Morphological and Genetic Divergence of Sympatric North Atlantic Killer Whale Populations." *Molecular Ecology* 18 (24): 5207–17. https://doi.org/10.1111/j.1365-294X.2009.04407.x.

Ford, John K. B. 1989. "Acoustic Behaviour of Resident Killer Whales (*Orcinus orca*) off Vancouver Island, British Columbia." *Canadian Journal of Zoology* 67 (3): 727–45. https://doi.org/10.1139/z89-105.

Ford, John K. B. 1991. "Vocal Traditions Among Resident Killer Whales (*Orcinus orca*) in Coastal

Waters of British Columbia." *Canadian Journal of Zoology* 69 (6): 1454–83. https://doi.org/10.1139/z91-206.

Ford, John K. B., and Graeme M. Ellis. 2006. "Selective Foraging by Fish-Eating Killer Whales *Orcinus orca* in British Columbia." *Marine Ecology Progress Series* 316: 185–99. https://doi.org/10.3354/meps316185.

Ford, John K. B., and G. M. Ellis. 2014. "You Are What You Eat: Ecological Specializations and Their Influence on the Social Organization and Behaviour of Killer Whales." In *Primates and Cetaceans: Field Research and Conservation of Complex Mammalian Societies*, edited by J. Yamagiwa and L. Karczmarski, 75–98. Springer. https://doi.org/10.1007/978-4-431-54523-1_4.

Ford, John K. B., Graeme M. Ellis, and Kenneth C. Balcomb. 1994. *Killer Whales*. UBC Press.

Ford, John K. B., Graeme M. Ellis, Lance G. Barrett-Lennard, Alexandra B. Morton, Rod S. Palm, and Kenneth C. Balcomb III. 1998. "Dietary Specialization in Two Sympatric Populations of Killer Whales (*Orcinus orca*) in Coastal British Columbia and Adjacent Waters." *Canadian Journal of Zoology* 76 (8): 1456–71. https://doi.org/10.1139/z98-089.

Ford, Michael J., M. Bradley Hanson, Jennifer A. Hempelmann, et al. 2011. "Inferred Paternity and Male Reproductive Success in a Killer Whale (*Orcinus orca*) Population." *Journal of Heredity* 102 (5): 537–53. https://doi.org/10.1093/jhered/esr067.

Ford, Michael J., Jennifer Hempelmann, M. Bradley Hanson, et al. 2016. "Estimation of a Killer Whale (*Orcinus orca*) Population's Diet Using Sequencing Analysis of DNA from Feces." *PLOS ONE* 11 (1): e0144956. https://doi.org/10.1371/journal.pone.0144956.

Ford, Michael J., K. M. Parsons, E. J. Ward, et al. 2018. "Inbreeding in an Endangered Killer Whale Population." *Animal Conservation* 21 (5): 423–32. https://doi.org/10.1111/acv.12413.

Foster, Emma A., Daniel W. Franks, Sonia Mazzi, et al. 2012. "Adaptive Prolonged Postreproductive Life Span in Killer Whales." *Science* 337 (6100): 1313. https://doi.org/10.1126/science.1224198.

Frère, Céline H., Michael Krützen, Janet Mann, et al. 2010a. "Home Range Overlap, Matrilineal and Biparental Kinship Drive Female Associations in Bottlenose Dolphins." *Animal Behaviour* 80 (3): 481–86. https://doi.org/10.1016/j.anbehav.2010.06.007.

Frère, Céline H., Michael Krützen, Janet Mann, Richard C. Connor, Lars Bejder, and William B. Sherwin. 2010b. "Social and Genetic Interactions Drive Fitness Variation in a Free-Living Dolphin Population." *Proceedings of the National Academy of Sciences* 107 (46): 19949–54. https://doi.org/10.1073/pnas.1007997107.

Gerber, Livia, Richard C. Connor, Simon J. Allen, et al. 2022. "Social Integration Influences Fitness in Allied Male Dolphins." *Current Biology* 32 (7): 1664–69. https://doi.org/10.1016/j.cub.2022.03.027.

Gerber, Livia, Richard C. Connor, Stephanie L. King, et al. 2020. "Affiliation History and Age Similarity Predict Alliance Formation in Adult Male Bottlenose Dolphins." *Behavioral Ecology* 31 (2): 361–70. https://doi.org/10.1093/beheco/arz195.

Gerber, Livia, Samuel Wittwer, Simon J. Allen, et al. 2021. "Cooperative Partner Choice in Multi-Level Male Dolphin Alliances." *Scientific Reports* 11 (1): 1–10. https://doi.org/10.1038/s41598-021-85583-x.

Gero, Shane, Dan Engelhaupt, Luke Rendell, and Hal Whitehead. 2009. "Who Cares? Between-Group Variation in Alloparental Caregiving in Sperm Whales." *Behavioral Ecology* 20 (4): 838–43. https://doi.org/10.1093/beheco/arp068.

Gero, Shane, Jonathan Gordon, and Hal Whitehead. 2013. "Calves as Social Hubs: Dynamics of the Social Network Within Sperm Whale Units." *Proceedings of the Royal Society B* 280 (1763): 20131113. https://doi.org/10.1098/rspb.2013.1113.

Gero, Shane, Marina Milligan, Caroline Rinaldi, et al. 2014. "Behavior and Social Structure of the Sperm Whales of Dominica, West Indies." *Marine Mammal Science* 30 (3): 905–22. https://doi.org/10.1111/mms.12086.

Gero, Shane, Hal Whitehead, and Luke Rendell. 2016. "Individual, Unit and Vocal Clan Level Identity Cues in Sperm Whale Codas." *Royal Society Open Science* 3 (1): 150372. https://doi.org/10.1098/rsos.150372.

Girardet, Justine, François Sarano, G. Richard, et al. 2022. "Long Distance Runners in the Marine Realm: New Insights into Genetic Diversity, Kin Relationships and Social Fidelity of Indian Ocean Male Sperm Whales." *Frontiers in Marine Science* 9: 815684. https://doi.org/10.3389/fmars.2022.815684.

Hamilton, William D. 1964. "The Genetical Evolution of Social Behaviour. I–II." *Journal of Theoretical Biology* 7 (1): 1–52. https://doi.org/10.1016/0022-5193(64)90039-6.

Hanghøj, Kristian, Ida Moltke, Philip Alstrup Andersen, Andrea Manica, and Thorfinn Sand Korneliussen. 2019. "Fast and Accurate Relatedness Estimation from High-Throughput Sequencing Data in the Presence of Inbreeding." *GigaScience* 8 (5): giz034. https://doi.org/10.1093/gigascience/giz034.

Hanson, M. Bradley, Candice K. Emmons, Michael J. Ford, et al. 2021. "Endangered Predators and Endangered Prey: Seasonal Diet of Southern Resident Killer Whales." *PLOS ONE* 16 (3): e0247031. https://doi.org/10.1371/journal.pone.0247031.

Harcourt, Alexander H., and Frans de Waal, eds. 1992. *Coalitions and Alliances in Humans and Other Animals*. Oxford University Press.

Harrison, Hugo B., Pablo Saenz-Agudelo, Serge Planes, Geoffrey P. Jones, and Michael L. Berumen. 2013. "Relative Accuracy of Three Common Methods of Parentage Analysis in Natural Populations." *Molecular Ecology* 22 (4): 1158–70. https://doi.org/10.1111/mec.12138.

Hatchwell, Ben J. 2010. "Cryptic Kin Selection: Kin Structure in Vertebrate Populations and Opportunities for Kin-Directed Cooperation." *Ethology* 116 (3): 203–16. https://doi.org/10.1111/j.1439-0310.2009.01732.x.

Hedrick, Philip W., and Robert C. Lacy. 2015. "Measuring Relatedness Between Inbred Individuals." *Journal of Heredity* 106 (1): 20–25. https://doi.org/10.1093/jhered/esu072.

Ivkovich, Tatiana, Olga A. Filatova, Alexandr M. Burdin, Hal Sato, and Erich Hoyt. 2010. "The Social Organization of Resident-Type Killer Whales (*Orcinus orca*) in Avacha Gulf, Northwest Pacific, as Revealed Through Association Patterns and Acoustic Similarity." *Mammalian Biology* 75: 198–210. https://doi.org/10.1016/j.mambio.2009.03.006.

Jacquard, Albert. 1972. "Genetic Information Given by a Relative." *Biometrics* 28 (4): 1101–14. https://doi.org/10.2307/2528643.

Johnstone, Rufus A., and Michael A. Cant. 2010. "The Evolution of Menopause in Cetaceans and Humans: The Role of Demography." *Proceedings of the Royal Society B* 277 (1701): 3765–71. https://doi.org/10.1098/rspb.2010.0988.

Jordaan, Rowan K., Ryan R. Reisinger, W. Chris Oosthuizen, and P. J. Nico de Bruyn. 2021. "Seasonal Fission and Fusion of Killer Whale, *Orcinus orca*, Social Structure at Sub-Antarctic Marion Island." *Animal Behaviour* 177: 223–30. https://doi.org/10.1016/j.anbehav.2021.05.007.

Jourdain, Eve, Clare Andvik, Richard Karoliussen, Anders Ruus, Dag Vongraven, and Katrine Borgå. 2020. "Isotopic Niche Differs Between Seal and Fish-Eating Killer Whales (*Orcinus orca*) in Northern Norway." *Ecology and Evolution* 10 (9): 4115–27. https://doi.org/10.1002/ece3.6182.

Jourdain, Eve, Richard Karoliussen, Jacques de Vos, Stanislav E. Zakharov, and Christelle Tougard. 2019. "Killer Whales (*Orcinus orca*) Feeding on Lumpfish (*Cyclopterus lumpus*) in Northern Norway." *Marine Mammal Science* 36 (1): 89–102. https://doi.org/10.1111/mms.12618.

Jourdain, Eve, Richard Karoliussen, Sarah Fordyce Martin, et al. 2024. "Social and Genetic Connectivity Despite Ecological Variation in a Killer Whale Network." *Proceedings of the Royal Society B* 291 (2021): 20240524. https://doi.org/10.1098/rspb.2024.0524.

Jourdain, Eve, Dag Vongraven, Anna Bisther, and Richard Karoliussen. 2017. "First Longitudinal Study of Seal-Feeding Killer Whales (*Orcinus orca*) in Norwegian Coastal Waters." *PLOS ONE* 12 (6): e0180099. https://doi.org/10.1371/journal.pone.0180099.

Kardos, Marty, Yaolei Zhang, Kim M. Parsons, et al. 2023. "Inbreeding Depression Explains Killer Whale Population Dynamics." *Nature Ecology and Evolution* 7 (5): 675–86. https://doi.org/10.1038/s41559-023-01995-0.

Kircher, Martin, and Janet Kelso. 2010. "High-Throughput DNA Sequencing—Concepts and Limitations." *BioEssays* 32 (6): 524–36. https://doi.org/10.1002/bies.200900181.

Kobayashi, H., Hal Whitehead, and M. Amano. 2020. "Long-Term Associations Among Male Sperm Whales (*Physeter macrocephalus*)." *PLOS ONE* 15 (12): e0244204. https://doi.org/10.1371/journal.pone.0244204.

Konrad, Christine M., Timothy R. Frasier, Hal Whitehead, and Shane Gero. 2019. "Kin Selection and Allocare in Sperm Whales." *Behavioral Ecology* 30 (1): 194–201. https://doi.org/10.1093/beheco/ary143.

Konrad, Christine M., Shane Gero, Timothy R. Frasier, and Hal Whitehead. 2018. "Kinship Influences Sperm Whale Social Organization Within, but Generally Not Among, Social Units." *Royal Society Open Science* 5 (8): 180914. https://doi.org/10.1098/rsos.180914.

Krützen, Michael, William B. Sherwin, Richard C. Connor, et al. 2003. "Contrasting Relatedness Patterns in Bottlenose Dolphins (*Tursiops* sp.) with Different Alliance Strategies." *Proceedings of the Royal Society B* 270 (1514): 497–502. https://doi.org/10.1098/rspb.2002.2229.

Lazcano-Pacheco, Camila, Acilegna J. Castillo-Sánchez, Christian D. Ortega-Ortiz, et al. 2023. "Ecological Aspects of False Killer Whales (*Pseudorca crassidens*) from Mexican Pacific and Southern California Waters." *Marine Mammal Science* 39 (4): 1344–55. https://doi.org/10.1111/mms.13056.

Leatherwood, Stephen, and Randall R. Reeves, eds. 1989. *The Bottlenose Dolphin*. Academic Press.

Louis, Marie, Benoit Simon-Bouhet, Amélia Viricel, et al. 2018. "Evaluating the Influence of Ecology, Sex and Kinship on the Social Structure of Resident Coastal Bottlenose Dolphins." *Marine Biology* 165: 1–12. https://doi.org/10.1007/s00227-018-3341-z.

Mahaffy, Sabre D., Robin W. Baird, Annette E. Harnish, et al. 2023. "Identifying Social Clusters of Endangered Main Hawaiian Islands False Killer Whales." *Endangered Species Research* 51: 249–68. https://doi.org/10.3354/esr01258.

Mann, Janet, Richard C. Connor, Peter L. Tyack, and Hal Whitehead, eds. 2000. *Cetacean Societies: Field Studies of Dolphins and Whales*. University of Chicago Press.

Martien, Karen K., S. J. Chivers, R. W. Baird, et al. 2014. "Nuclear and Mitochondrial Patterns of Population Structure in North Pacific False Killer Whales (*Pseudorca crassidens*)." *Journal of Heredity* 105: 611–26. https://doi.org/10.1093/jhered/esu029.

Martien, Karen K., B. L. Taylor, S. J. Chivers, S. D. Mahaffy, A. M. Gorgone, and R. W. Baird. 2019. "Fidelity to Natal Social Groups and Mating Within and Between Social Groups in an Endangered False Killer Whale Population." *Endangered Species Research* 40: 219–30. https://doi.org/10.3354/esr00995.

McCormack, Jack L., Kevin Arbuckle, Karen Fullard, William Amos, and Hazel J. Nichols. 2023. "Lack of Intergenerational Reproductive Conflict, Rather Than Lack of Inclusive Fitness Benefits, Explains Absence of Post-Reproductive Lifespan in Long-Finned Pilot Whales." *Behavioral Ecology* 34 (6): 950–59. https://doi.org/10.1093/beheco/arad062.

McDonald, David B., and Wayne K. Potts. 1994. "Cooperative Display and Relatedness Among Males in a Lek-Mating Bird." *Science* 266 (5187): 1030–32. https://doi.org/10.1126/science.7973654.

Mesnick, Sarah L. 2001. "Genetic Relatedness in Sperm Whales: Evidence and Cultural Implications." *Behavioral and Brain Sciences* 24 (2): 346–47. https://doi.org/10.1017/S0140525X01463965.

Mitani, John C., David P. Watts, and Martin N. Muller. 2002. "Recent Developments in the Study of Wild Chimpanzee Behavior." *Evolutionary Anthropology: Issues, News, and Reviews* 11 (1): 9–25. https://doi.org/10.1002/evan.10008.

Möller, Luciana M., Luciano B. Beheregaray, Simon J. Allen, and Robert G. Harcourt. 2006. "Association Patterns and Kinship in Female Indo-Pacific Bottlenose Dolphins (*Tursiops aduncus*) of Southeastern Australia." *Behavioral Ecology and Sociobiology* 61: 109–17. https://doi.org/10.1007/s00265-006-0241-x.

Möller, Luciana M., Luciano B. Beheregaray, Robert G. Harcourt, and Michael Krützen. 2001. "Alliance Membership and Kinship in Wild Male Bottlenose Dolphins (*Tursiops aduncus*) of Southeastern Australia." *Proceedings of the Royal Society B* 268 (1479): 1941–47. https://doi.org/10.1098/rspb.2001.1756.

Morin, Phillip A., Frederick I. Archer, Andrew D. Foote, et al. 2010. "Complete Mitochondrial Genome Phylogeographic Analysis of Killer Whales (*Orcinus orca*) Indicates Multiple Species." *Genome Research* 20 (7): 908–16. https://doi.org/10.1101/gr.102954.109.

Nash, Catherine. 2004. "Genetic Kinship." *Cultural Studies* 18 (1): 1–33. https://doi.org/10.1080/0950238042000181593.

Nielsen, Mia L. K., Samuel Ellis, Michael N. Weiss, et al. 2023. "Temporal Dynamics of Mother-Offspring Relationships in Bigg's Killer Whales: Opportunities for Kin-Directed Help by Post-Reproductive Females." *Proceedings of the Royal Society B* 290 (2000): 20230139. https://doi.org/10.1098/rspb.2023.0139.

Nigenda-Morales, Sergio F., Meixi Lin, Paulina G. Nuñez-Valencia, et al. 2023. "The Genomic Footprint of Whaling and Isolation in Fin Whale Populations." *Nature Communications* 14 (1): 5465. https://doi.org/10.1038/s41467-023-40052-z.

Noë, Ronald. 1992. "Alliance Formation Among Male Baboons: Shopping for Profitable Partners." In *Coalitions and Alliances in Humans and Other Animals*, edited by Alexander H. Harcourt and Frans de Waal, 285–321. Oxford University Press.

Noë, Ronald, and Albertha A. Sluijter. 1995. "Which Adult Male Savanna Baboons Form Coalitions?" *International Journal of Primatology* 16: 77–105. https://doi.org/10.1007/bf02700154.

Packer, Craig, Dennis A. Gilbert, A. E. Pusey, and S. J. O'Brieni. 1991. "A Molecular Genetic

Analysis of Kinship and Cooperation in African Lions." *Nature* 351 (6327): 562–65. https://doi.org/10.1038/351562a0.

Palmer, C., K. K. Martien, H. Raudino, et al. 2023. "Evidence of Resident Coastal Population(s) of False Killer Whales (*Pseudorca crassidens*) in Northern Australian Waters." *Frontiers in Marine Science* 9: 1067660. https://doi.org/10.3389/fmars.2022.1067660.

Parsons, Kim M., Kenneth C. Balcomb III, John K. B. Ford, and John W. Durban. 2009. "The Social Dynamics of Southern Resident Killer Whales and Conservation Implications for This Endangered Population." *Animal Behaviour* 77 (4): 963–71. https://doi.org/10.1016/j.anbehav.2009.01.018.

Parsons, Kim M., John W. Durban, Diane E. Claridge, Ken C. Balcomb, Les R. Noble, and Paul M. Thompson. 2003. "Kinship as a Basis for Alliance Formation Between Male Bottlenose Dolphins, *Tursiops truncatus*, in the Bahamas." *Animal Behaviour* 66 (1): 185–94. https://doi.org/10.1006/anbe.2003.2186.

Pilot, Malgorzata, Marilyn E. Dahlheim, and A. Rus Hoelzel. 2010. "Social Cohesion Among Kin, Gene Flow Without Dispersal and the Evolution of Population Genetic Structure in the Killer Whale (*Orcinus orca*)." *Journal of Evolutionary Biology* 23 (1): 20–31. https://doi.org/10.1111/j.1420-9101.2009.01887.x.

Pinela, A. M., S. Quérouil, S. Magalhães, et al. 2009. "Population Genetics and Social Organization of the Sperm Whale (*Physeter macrocephalus*) in the Azores Inferred by Microsatellite Analyses." *Canadian Journal of Zoology* 87 (9): 802–13. https://doi.org/10.1139/Z09-066.

Pomilla, Cristina, and Howard C. Rosenbaum. 2006. "Estimates of Relatedness in Groups of Humpback Whales (*Megaptera novaeangliae*) on Two Wintering Grounds of the Southern Hemisphere." *Molecular Ecology* 15 (9): 2541–55. https://doi.org/10.1111/j.1365-294X.2006.02943.x.

Ramp, Christian, Wilhelm Hagen, Per Palsbøll, Martine Bérubé, and Richard Sears. 2010. "Age-Related Multi-Year Associations in Female Humpback Whales (*Megaptera novaeangliae*)." *Behavioral Ecology and Sociobiology* 64 (10): 1563–76. https://doi.org/10.1007/s00265-010-0970-8.

Reeves, Randall R., S. Leatherwood, and R. W. Baird. 2009. "Evidence of a Possible Decline Since 1989 in False Killer Whales (*Pseudorca crassidens*) Around the Main Hawaiian Islands." *Pacific Science* 63: 253–61. https://doi.org/10.2984/049.063.0207.

Reisinger, Ryan R., Charlene Beukes, A. Rus Hoelzel, and P. J. Nico de Bruyn. 2017. "Kinship and Association in a Highly Social Apex Predator Population, Killer Whales at Marion Island." *Behavioral Ecology* 28 (3): 750–59. https://doi.org/10.1093/beheco/arx034.

Remili, Anaïs, Rune Dietz, Christian Sonne, et al. 2023. "Quantitative Fatty Acid Signature Analysis Reveals a High Level of Dietary Specialization in Killer Whales Across the North Atlantic." *Journal of Animal Ecology* 92 (6): 1216–29. https://doi.org/10.1111/1365-2656.13920.

Rendell, Luke, Mauricio Cantor, Shane Gero, Hal Whitehead, and Janet Mann. 2019. "Causes and Consequences of Female Centrality in Cetacean Societies." *Philosophical Transactions of the Royal Society B* 374 (1780): 20180066. https://doi.org/10.1098/rstb.2018.0066.

Rendell, Luke, Sarah L. Mesnick, Merel L. Dalebout, Jessica Burtenshaw, and Hal Whitehead. 2012. "Can Genetic Differences Explain Vocal Dialect Variation in Sperm Whales, *Physeter macrocephalus*?" *Behavior Genetics* 42: 332–43. https://doi.org/10.1007/s10519-011-9513-y.

Rendell, Luke E., and Hal Whitehead. 2003. "Vocal Clans in Sperm Whales (*Physeter macrocephalus*)." *Proceedings of the Royal Society B* 270 (1512): 225–31. https://doi.org/10.1098/rspb.2002.2239.

Riesch, Rüdiger, and Volker B. Deecke. 2011. "Whistle Communication in Mammal-Eating Killer Whales (*Orcinus orca*): Further Evidence for Acoustic Divergence Between Ecotypes." *Behavioral Ecology and Sociobiology* 65: 1377–87. https://doi.org/10.1007/s00265-011-1148-8.

Robinson, Jacqueline A., Christopher C. Kyriazis, Sergio F. Nigenda-Morales, et al. 2022. "The Critically Endangered Vaquita Is Not Doomed to Extinction by Inbreeding Depression." *Science* 376 (6593): 635–39. https://doi.org/10.1126/science.abm1742.

Samarra, Filipa I. P., and A. D. Foote. 2015. "Seasonal Movements of Killer Whales Between Iceland and Scotland." *Aquatic Biology* 24 (1): 75–79. https://doi.org/10.3354/ab00637.

Samarra, Filipa I. P., Sara B. Tavares, Julie Béesau, et al. 2017. "Movements and Site Fidelity of Killer Whales (*Orcinus orca*) Relative to Seasonal and Long-Term Shifts in Herring (*Clupea harengus*) Distribution." *Marine Biology* 164: 1–15. https://doi.org/10.1007/s00227-017-3187-9.

Sarano, François, Justine Girardet, Véronique Sarano, et al. 2021. "Kin Relationships in Cultural Species of the Marine Realm: Case Study of a Matrilineal Social Group of Sperm Whales off Mauritius Island, Indian Ocean." *Royal Society Open Science* 8 (2): 201794. https://doi.org/10.1098/rsos.201794.

Sarano, François, Véronique Sarano, Modan-Lou Tonietto, et al. 2023. "Nursing Behavior in Sperm Whales (*Physeter macrocephalus*)." *Animal Behavior and Cognition* 10 (2): 105–31. https://doi.org/10.26451/abc.10.02.02.2023.

Saulitis, Eva, Craig Matkin, Lance Barrett-Lennard, Kathy Heise, and Graeme Ellis. 2000. "Foraging Strategies of Sympatric Killer Whale (*Orcinus orca*) Populations in Prince William Sound, Alaska." *Marine Mammal Science* 16 (1): 94–109. https://doi.org/10.1111/j.1748-7692.2000.tb00906.x.

Saulitis, Eva L., Craig O. Matkin, and Francis H. Fay. 2005. "Vocal Repertoire and Acoustic Behavior of the Isolated AT1 Killer Whale Subpopulation in Southern Alaska." *Canadian Journal of Zoology* 83 (8): 1015–29. https://doi.org/10.1139/z05-089.

Schneider, David M. 1961. "The Distinctive Features of Matrilineal Descent Groups." In *Matrilineal Kinship*, edited by David M. Schneider and Kathleen Gough, 1–29. University of California Press.

Schneider, David M., and Kathleen Gough, eds. 1961. *Matrilineal Kinship*. University of California Press.

Similä, Tiu, Jens Christian Holst, and Ivar Christensen. 1996. "Occurrence and Diet of Killer Whales in Northern Norway: Seasonal Patterns Relative to the Distribution and Abundance of Norwegian Spring-Spawning Herring." *Canadian Journal of Fisheries and Aquatic Sciences* 53 (4): 769–79. https://doi.org/10.1139/f95-253.

Städele, Veronika, and Linda Vigilant. 2016. "Strategies for Determining Kinship in Wild Populations Using Genetic Data." *Ecology and Evolution* 6 (17): 6107–20. https://doi.org/10.1002/ece3.2346.

Stredulinsky, Eva H., Chris T. Darimont, Lance Barrett-Lennard, Graeme M. Ellis, and John K. B. Ford. 2021. "Family Feud: Permanent Group Splitting in a Highly Philopatric Mammal, the Killer Whale (*Orcinus orca*)." *Behavioral Ecology and Sociobiology* 75: 1–17. https://doi.org/10.1007/s00265-021-02992-8.

Suárez-Menéndez, Marcos, Martine Bérubé, Fabrício Furni, et al. 2023. "Wild Pedigrees Inform Mutation Rates and Historic Abundance in Baleen Whales." *Science* 381 (6661): 990–95. https://doi.org/10.1126/science.adf2160.

Tavares, Sara B., Filipa I. P. Samarra, and Patrick J. O. Miller. 2017. "A Multilevel Society of

Herring-Eating Killer Whales Indicates Adaptation to Prey Characteristics." *Behavioral Ecology* 28 (2): 500–514. https://doi.org/10.1093/beheco/arw179.

Tavares, Sara B., Filipa I. P. Samarra, Sonia Pascoal, Jeff A. Graves, and Patrick J. O. Miller. 2018. "Killer Whales (*Orcinus orca*) in Iceland Show Weak Genetic Structure Among Diverse Isotopic Signatures and Observed Movement Patterns." *Ecology and Evolution* 8 (23): 11900–913. https://doi.org/10.1002/ece3.4646.

Taylor, Helen R. 2015. "The Use and Abuse of Genetic Marker-Based Estimates of Relatedness and Inbreeding." *Ecology and Evolution* 5 (15): 3140–50. https://doi.org/10.1002/ece3.1541.

Taylor, Peter D. 1992. "Inclusive Fitness in a Homogeneous Environment." *Proceedings of the Royal Society B* 249 (1326): 299–302. https://doi.org/10.1098/rspb.1992.0118.

Vachon, Felicia, Ana Eguiguren, Luke Rendell, Shane Gero, and Hal Whitehead. 2022. "Distinctive, Fine-Scale Distribution of Eastern Caribbean Sperm Whale Vocal Clans Reflects Island Fidelity Rather Than Environmental Variables." *Ecology and Evolution* 12 (11): e9449. https://doi.org/10.1002/ece3.9449.

Valsecchi, Elena, Peter Hale, Peter Corkeron, and William Amos. 2002. "Social Structure in Migrating Humpback Whales (*Megaptera novaeangliae*)." *Molecular Ecology* 11 (3): 507–18. https://doi.org/10.1046/j.0962-1083.2001.01459.x.

Vongraven, Dag, and Anna Bisther. 2014. "Prey Switching by Killer Whales in the North-East Atlantic: Observational Evidence and Experimental Insights." *Journal of the Marine Biological Association of the United Kingdom* 94 (6): 1357–65. https://doi.org/10.1017/S0025315413001707.

Wagner, Aaron P., Laurence G. Frank, and Scott Creel. 2008. "Spatial Grouping in Behaviourally Solitary Striped Hyaenas, *Hyaena hyaena*." *Animal Behaviour* 75 (3): 1131–42. https://doi.org/10.1016/j.anbehav.2007.08.025.

Weilgart, Linda, Hal Whitehead, and Katharine Payne. 1996. "A Colossal Convergence." *American Scientist* 84 (3): 278–87. https://www.jstor.org/stable/29775673.

Weinrich, Mason T., Howard Rosenbaum, C. Scott Baker, Alexis L. Blackmer, and Hal Whitehead. 2006. "The Influence of Maternal Lineages on Social Affiliations Among Humpback Whales (*Megaptera novaeangliae*) on Their Feeding Grounds in the Southern Gulf of Maine." *Journal of Heredity* 97 (3): 226–34. https://doi.org/10.1093/jhered/esj018.

Weller, David W., Bernd Würsig, Hal Whitehead, et al. 1996. "Observations of an Interaction Between Sperm Whales and Short-Finned Pilot Whales in the Gulf of Mexico." *Marine Mammal Science* 12 (4): 588–94. https://doi.org/10.1111/j.1748-7692.1996.tb00071.x.

Wells, Randall S. 2014. "Social Structure and Life History of Bottlenose Dolphins Near Sarasota Bay, Florida: Insights from Four Decades and Five Generations." In *Primates and Cetaceans: Field Research and Conservation of Complex Mammalian Societies*, edited by J. Yamagiwa and L. Karczmarski, 149–72. Springer. https://doi.org/10.1007/978-4-431-54523-1_8.

Whitehead, Hal. 1996. "Babysitting, Dive Synchrony, and Indications of Alloparental Care in Sperm Whales." *Behavioral Ecology and Sociobiology* 38 (4): 237–44. https://doi.org/10.1007/s002650050238.

Whitehead, Hal. 2003. *Sperm Whales: Social Evolution in the Ocean*. University of Chicago Press.

Whitehead, Hal, R. Antunes, Shane Gero, S. N. Wong, D. Engelhaupt, and Luke Rendell. 2012. "Multilevel Societies of Female Sperm Whales (*Physeter macrocephalus*) in the Atlantic and Pacific: Why Are They So Different?" *International Journal of Primatology* 33 (5): 1142–64. https://doi.org/10.1007/s10764-012-9598-z.

Whitehead, Hal, and Luke Rendell. 2004. "Movements, Habitat Use and Feeding Success of

Cultural Clans of South Pacific Sperm Whales." *Journal of Animal Ecology* 73 (1): 190–96. https://doi.org/10.1111/j.1365-2656.2004.00798.x.

Whitehead, Hal, Luke Rendell, Richard W. Osborne, and Bernd Würsig. 2004. "Culture and Conservation of Non-Humans with Reference to Whales and Dolphins: Review and New Directions." *Biological Conservation* 120 (3): 427–37. https://doi.org/10.1016/j.biocon.2004.03.017.

Wild, Sonja, William J. E. Hoppitt, Simon J. Allen, and Michael Krützen. 2020. "Integrating Genetic, Environmental, and Social Networks to Reveal Transmission Pathways of a Dolphin Foraging Innovation." *Current Biology* 30 (15): 3024–30. https://doi.org/10.1016/j.cub.2020.05.069.

Wiles, Gary J. 2004. *Washington State Status Report for the Killer Whale.* Washington Department of Fish and Wildlife.

Wray, Janie, Eric M. Keen, and Éadin N. O'Mahony. 2021. "Social Survival: Humpback Whales (*Megaptera novaeangliae*) Use Social Structure to Partition Ecological Niches Within Proposed Critical Habitat." *PLOS ONE* 16 (6): e0245409. https://doi.org/10.1371/journal.pone.0245409.

Wright, Brianna M., Eva H. Stredulinsky, Graeme M. Ellis, and John K. B. Ford. 2016. "Kin-Directed Food Sharing Promotes Lifetime Natal Philopatry of Both Sexes in a Population of Fish-Eating Killer Whales, *Orcinus orca*." *Animal Behaviour* 115 (May): 81–95. https://doi.org/10.1016/j.anbehav.2016.02.025.

Zaeschmar, Jochen R., Ingrid N. Visser, Dagmar Fertl, et al. 2014. "Occurrence of False Killer Whales (*Pseudorca crassidens*) and Their Association with Common Bottlenose Dolphins (*Tursiops truncatus*) off Northeastern New Zealand." *Marine Mammal Science* 30 (2): 594–608. https://doi.org/10.1111/mms.12065.

PART II

The Socioecology of Cetacean Societies

4

Quantifying Social Complexity in Cetaceans

MICHAEL N. WEISS, INGERID H. HELGESTAD, AND DARREN P. CROFT

Abstract

The complexity of cetacean societies has fascinated researchers for decades, motivating attempts to understand the causes and consequences of social complexity in this taxon. Such analyses require rigorous definitions of social complexity, and methods to quantify it. This is challenging in any animal society but becomes particularly difficult when studying highly mobile marine species such as cetaceans. This chapter presents a definition of social complexity based on the uncertainty individuals face in their social environment, along with a statistical method to quantify this complexity using the kinds of data commonly collected from wild cetaceans. A comparative analysis of social complexity across Cetacea is then presented, demonstrating the great diversity of social structure in this taxonomic group. The chapter closes with suggestions for future analyses of social complexity in cetaceans and other taxa.

Introduction

If there's one fact that scientists who study cetacean sociality can agree on, it might be this: Cetacean societies are complex. Cetaceans live in social worlds that are both dynamic and highly individualized. Most populations exhibit very dense social networks, with every individual in the population interacting (or at least associating) with many partners. These interactions are dynamic: partners change over days, weeks, months, and years. Individuals must track and manage their relationships with cooperative partners, potential mates, and competitors across various contexts, with fission-fusion dynamics ensuring that the individuals in one's social environment regularly change. Some populations appear to exhibit social structures of truly remarkable complexity, with multilevel cooperative alliances, culturally defined vocal clans, and large-scale social transmission of novel behavior (Bigg et al. 1990; Cantor et al. 2015; Connor and Krützen 2015; Garland et al. 2011). There is a great deal of academic interest in understanding, quantifying, and comparing social structure and complexity across cetaceans. However, given the difficulties inherent in studying the societies of marine animals, the relevant behavioral measures needed to conduct such analyses are often sorely lacking (Groot et al. 2023).

Why study the diversity of social complexity in cetaceans (or any taxa)? Comparative studies of social complexity are often motivated by the "social brain hypothesis" (Dunbar 1998). Stated briefly, this hypothesis suggests that the cognitive challenges associated with mapping and managing complex social environments have driven cognitive evolution, leading to enlarged brains. Originally developed as an explanation for the extreme brain sizes and cognitive abilities of humans and other primates, this hypothesis has also been invoked to explain the evolution of cetacean brain size, particularly the extremely large brains of dolphins and sperm whales (Connor 2007; Fox et al. 2017; Marino et al. 2007). Further interest in social complexity is driven by hypotheses linking the social environment to the evolution of intricate communication systems (Freeberg et al. 2012; Sewall 2015), a hypothesis also highly relevant to whales and dolphins, which often exhibit complex vocal behavior. Furthermore, understanding which evolutionary pressures lead to social complexity is fundamentally interesting in and of itself. Theoretical and empirical work has investigated how longevity (Lucas and Keller 2020; Silk and Hodgson 2021), ecology (He et al. 2019), and kinship (Lukas and Clutton-Brock 2018) influence social complexity. As a taxon with immense variation in ecology, life history, and social organization, cetaceans represent a unique opportunity to examine and test these hypotheses.

To answer questions relating to drivers and consequences of cetacean social complexity, we need to develop well-motivated, quantitative methods for measuring social complexity, and then apply these methods to numerous species of cetaceans. In this chapter, we present a new approach to describe and understand the diversity of social relationship complexity in cetaceans. Our method and definition focus on understanding the uncertainty, or unpredictability, of individuals' social environments. We first discuss what social complexity is and provide our definition of a particular kind of social complexity, related to the uncertainty or unpredictability that individuals face in their social environment. We then describe a method to quantify this kind of complexity. Finally, we summarize our measure of social complexity across multiple species of cetaceans and, building on these initial results, suggest priority areas for future research.

What Do We Mean When We Say "Social Complexity"?

Let's take a step back, and start with a simple question: what do biologists mean when they say a society is "complex"? How can we classify or quantify the degree of complexity in a particular animal group or population?

We'll start with a brief consideration of the term "complexity." There is a vast body of literature on the philosophical and mathematical nature of complexity, much of it beyond the scope of what we can cover here. Some authors have argued that "complexity" is just the opposite of "simplicity," but we find this definition somewhat unhelpful in this context since it doesn't get us any closer to something that we can measure. One widespread concept is that complexity lies somewhere between complete order and complete randomness; there must be structure in complex systems, but not so much structure that the behavior of the system is entirely predictable (Page 2010). Slightly more concretely, complex systems have been described as those that have many elements of many different types, which interact in diverse ways (McShea and Brandon 2010). For a more in-depth discussion and review, see Freeberg et al. (2012).

How can we apply these concepts to societies? Working off the framework introduced by Hinde (1976), we can consider individuals in a society to represent the "elements" of the system that interact socially. Individuals of different types (e.g., age, sex, and reproductive state), from the perspective of sociality, may occupy different "social roles" (such as helpers and breeders). Moreover, interactions between individuals (elements) scale up to various forms of social relationships (Hinde 1976), e.g., cooperative, aggressive, or competitive. So, using this framework, we can define a complex society as one in which individuals have many different social roles and diverse social relationships. These two features of complex societies were formalized by Lukas and Clutton-Brock (2018) into the concepts of "organizational" and "relational" complexity. Societies with high "organizational" complexity exhibit a wide array of well-defined social roles, often with strong division of labor and cooperative breeding with distinct breeders and helpers. In contrast, "relational" complexity is found in societies with diverse social interactions, such as coalition formation, dominance hierarchies, and aggression. These two types of complexity do not, it seems, go hand in hand in animal societies. Indeed, they seem to emerge from different evolutionary pressures, particularly with respect to kinship structure (Lukas and Clutton-Brock 2018).

Here, we focus on "relational" complexity. Our reasons are twofold, both motivated more by practicality than scientific interest. First, methods for measuring continuous variation in relational complexity are arguably more developed than methods for measuring organizational complexity, as measuring the latter largely relies on reports of the presence or absence of particular features of social organization. Second, we generally have more information about the structure of social relationships in cetacean societies (Möller 2012; Rendell et al. 2019; Weiss et al. 2021) than we do about social roles. The diversity of social roles in cetacean societies likely warrants further attention, as the studies conducted to date have produced intriguing results. In killer whales (*Orcinus orca*) for example, there appear to be age-sex-based roles in social units, with older females being more likely to lead group movement (Brent et al. 2015) and different age-sex classes performing different behaviors during group foraging (Baird and Dill 1995). Similarly, in bottlenose dolphins, there is consistent differentiation between leaders and followers in group movement (Lewis et al. 2011; Lusseau 2007) and evidence of role specialization during foraging (Gazda et al. 2005), although the underlying factors determining social roles are less clear in this species. The presence of alloparental care in several cetacean species is also indicative of organizational complexity within social groups (Augusto et al. 2017; Gero et al. 2009). These findings, however, are typically limited to well-studied species and often describe social roles within specific contexts, making it difficult to conduct broad comparative studies of organizational complexity in cetaceans.

In addition to the concept of diverse social interactions, we would like to add one more key feature of social complexity, motivated by evolutionary hypotheses of social complexity as a driver of cognition and communication: complex societies should pose a challenge to the individuals within them. We do not go as far as Bergman and Beehner (2015) to suggest that measures of social complexity must reflect a *cognitive* challenge, as we wish to avoid circular reasoning in potential tests of the social intelligence hypothesis; there may be many ways for animals to deal with the challenges of complex societies, but we propose that the challenge must be present. We concede, however, that cognition seems the most probable toolkit for coping with unpredictability in an animals' environment generally, including the social environment.

Combining the concept of diversity in interactions and relationships with the idea of social challenge, we propose a broad working definition of relational social complexity: We suggest relational complexity can be operationalized as the *uncertainty* or *unpredictability* that individuals face in their social environment (fig. 4.1). To illustrate this idea, let's consider three extreme hypothetical cases. First, imagine an animal population where every individual has exactly one social partner, and categorically does not interact or associate with any other individuals. These individuals would have extremely predictable social environments, even if there were cognitively complex fac-

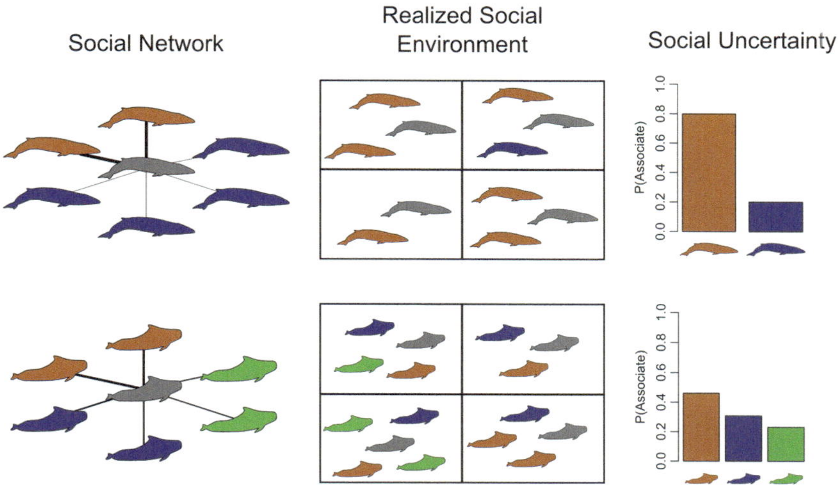

FIGURE 4.1. Social uncertainty for two hypothetical social structures. In the top row, the focal individual (gray) maintains two types of social relationship in their social network, indicated by color. These relationship types are distinguishable by the association rate (indicated by edge thickness), with one class being very strong relationships and the other being weak. This social network results in a social environment where the focal individual's associates are almost always of one relationship type, resulting in low social uncertainty. In contrast, in the second row, the focal individual maintains three types of relationship that are less divergent in their rate of association. This results in a much more varied social environment, leading to higher social uncertainty.
Source: Cetacean outlines by Chris Huh are used, edited, and distributed under a CC BY-SA 3.0 DEED license.

tors at play (such as the recognition of their social partner). Second, consider a population where individuals associate with others at random, but all individuals are functionally identical in terms of their social interactions. While each animal may have a wide array of social associates, their social environment is functionally constant; they are best served by behaving consistently in all their social interactions, just like the first animal with a single constant associate. Finally, consider a population with a large, open social network, but where every single relationship is unique and therefore consists of different types of social interactions. Animals in such a population would have very little baseline certainty in the types of interactions they were going to have in any given social group. Under our definition, the first two societies would both have very low social complexity, despite being completely divergent social systems, while the third example would be an extremely complex society.

This definition captures the idea that the highest complexity lies between a completely ordered (i.e., the first example with single, exclusive social relationships) and random (i.e., the second example with random undifferentiated relationships) system. It also has similarities to Bergman and Beehner's

(2015) idea of complexity being "the number of differentiated relationships individuals have" and Fischer et al.'s (2017) quantitative measure of individuals' diversity in relationship types. The key addition is the conceptual underpinning of social challenge, as represented by the uncertainty in an individual's social environment.

Other Concepts and Measures of Social Complexity in Cetaceans

Having established what we consider to be a useful definition of "social complexity" for comparative studies, we now turn to a brief review of how previous work, particularly in cetaceans, has operationalized the concept of social complexity, with an eye toward how these measures relate to social uncertainty.

Perhaps the most well-known method for quantifying social complexity is to measure group size (e.g., Dunbar 1992; Dunbar and Bever 1998; Pérez-Barbería et al. 2007). This measure has most prominently been used in studies of primates (Dunbar 1992; Dunbar and Bever 1998; Dunbar and Shultz 2021). Note that in primates, the term "group" often refers to bonded social units, while behavioral ecologists working with cetaceans (and indeed any other taxa that are less likely to form bonded units) are more likely to use the term to describe temporary sets of associated individuals. Throughout this chapter, we'll use "group" in this latter sense.

Group size may be a reasonable measure for social complexity in primates. Primate groups are typically composed of one or more social units, and relationships are highly individualized. It makes sense, then, to assume that each individual in the group likely represents a distinct relationship that a focal individual would need to manage. Group size has, however, also been used as a measure of social complexity in other mammalian taxa, where the link with the diversity of social relationships becomes less clear (Dunbar and Bever 1998; Pérez-Barbería and Gordon 2005). This is a problem for studies of cetaceans. Many cetacean species form extremely large temporary aggregations, and it is unknown if individuals in these groups maintain individualized relationships of any kind with all the others. When group size has been used in comparative studies of cetacean sociality, it has often been suggested that the most socially complex societies exhibit "midsized" groups, in contrast to solitary species or huge "mega-pods" (e.g., Fox et al. 2017). Thus, studies of cetacean sociality should not, and typically do not, assume that animals found in larger groups inherently have more complex social lives.

A related measure of social complexity is the number of social relationships individuals maintain (e.g., Dunbar 1992; Kudo and Dunbar 2001; Lehmann and Dunbar 2009). This measure attempts to provide a better indication of the size of individuals' relevant social environment. For animals that form well-defined social units, such as many gregarious primates, the size of bonded units can serve as a useful proxy for this quantity (Dunbar 1992). Things are a bit more difficult in cetaceans, as most species do not appear to form social units. We do not know of any comparative study using the number of social relationships per individual as a proxy for social complexity in cetaceans, however this concept is often invoked to emphasize the challenges inherent to navigating the open societies maintained by some populations of bottlenose dolphin (*Tursiops aduncus*, e.g., Connor 2007). One thing that's currently unclear is whether having many social relationships is universally challenging. As with group size, if all partners that individuals maintain relationships with are functionally the same in terms of social interactions, it would perhaps not be particularly challenging to navigate a large social network. On the other hand, if social relationships are individualized (as appears to be the case in most, if not all, cetaceans), the number of relationships would seem to be a strong predictor of social complexity.

Another way to conceive of social complexity is to use more general descriptions of social systems. Such frameworks may involve describing the population average strength or stability of social associations, establishing the presence of stable social units, summarizing the size and composition of social units, and determining whether units are situated within a multilevel society (e.g., Prox and Farine 2020; Shultz and Dunbar 2007). Because these measures are typically categorical, rather than continuous, they are somewhat less vulnerable to a lack of precision in quantitative data, and have therefore been extensively used in studies of terrestrial mammals (Pérez-Barbería et al. 2007; Pérez-Barbería and Gordon 2005; Shultz and Dunbar 2007) and birds (Hardie and Cooney 2023; Shultz and Dunbar 2010). A similar categorical framework was employed by May-Collado et al. (2007) to examine the coevolution of tonal sounds and social complexity in cetaceans, where species were classified as solitary, group living with weak associations, group living with some long-lasting associations, or group living with stable social units. One potential pitfall here is that it's not self-evident which social systems are the most "complex." It's often been suggested that multilevel societies, where social units are nested within larger social entities, are particularly complex and potentially cognitively challenging; however, this claim has not been carefully examined (Grueter et al. 2020). Indeed, as Rendell et al. (2019) point out, cetacean multilevel societies are typically composed of maternal kin and thus exhibit higher within-group relatedness compared to other cetacean societies (see Parsons et al., this volume), which is generally predicted to reduce relational social complexity (Lukas and Clutton-Brock 2018). Additionally, there is some evidence that individuals in multilevel societies may not always cognitively attend to the different levels of social structure they participate in (Maciej et al. 2013; see King et al., this volume, for discussion of social cognition in cetaceans), further highlighting a lack of clear hierarchy of complexity between different social systems.

Some studies of cetacean social complexity have used the presence of particular social behaviors, such as group hunting and alloparental care, as indicators of social complexity. This method was utilized by Fox et al. (2017) in their analysis of the relationship between sociality and brain size in Cetacea. Importantly, as the authors point out, this is not a measure of the complexity of social structure but rather measures the related idea of complexity in social behavior. This has some relation to our concept of social uncertainty as a measure of social complexity, as a wider repertoire of behaviors may indicate less predictable social interactions. While Fox et al. (2017) did find a correlation between social group size and their "social repertoire score," it is not clear whether more complex social structures necessarily correspond to societies that exhibit a wide array of social behaviors. It's also important to note that some of the features often included in these assessments, such as alloparental care, are more representative of "organizational," rather than "relational," social complexity (Lukas and Clutton-Brock 2018).

Measuring Relational Social Complexity

As none of the measures previously used in comparative studies adequately capture our concept of "social uncertainty," we propose that we need to develop and apply other methods for quantifying social complexity in cetaceans. A few years ago, some of the authors of this chapter published a paper (Weiss et al. 2019) presenting a method for estimating relational social complexity. This method uses data on "association indices" between identified individuals. These take the form of a ratio between the number of occasions each pair was observed together and the number of occasions on which either individual in the pair was observed (together or apart). This method therefore requires data on the co-occurrence of identified individuals at spatial and temporal scales that are biologically relevant. In practice, cetacean researchers typically define groups of individuals during photo-identification surveys and say all individuals in the group are associated.

These association data are then inputted into a statistical model to estimate social complexity. We won't go into the technical details of our proposed method here, but the basic idea is that we use a statistical algorithm, called a "binomial mixture model," to estimate the number of relationship types present in a population, with the types being defined by their average association rate. We then estimate two parameters for each type of relationship: the average association rate, which we'll call μ, and the portion of dyads that belong to that type, which we'll call α. We can then estimate the probability that any given pair of associated animals have a relationship of type k:

$$P\text{ (associates have relationship } k) = q_k = \frac{\mu_k \alpha_k}{\Sigma_k^K \mu_k \alpha_k}.$$

How does this get us to a measure of social complexity? We can borrow a measure from information theory, Shannon's entropy index, to quantify the uncertainty in the relationship that any two associated animals have. Shannon's entropy is the natural measure for this task, as it is specifically developed to quantify the level of uncertainty in the outcome of a random variable (in this case, the type of relationship a pair of associates have). As the possible outcomes become more numerous (i.e., more types of relationships), and the likelihood of these outcomes become more similar (i.e., less certainty in which of the types an individual will encounter), the entropy of the variable increases. We'll call this measure S (following notation in information theory), and it can be calculated as:

$$S = -\Sigma_k^K q_k \log(q_k).$$

In our original paper, we showed the characteristics of this measure on simulated data and then estimated the complexity of sperm whale (*Physeter macrocephalus*) and northern bottlenose whale (*Hyperoodon ampullatus*) social networks, finding that sperm whales seemed to have greater social complexity than bottlenose whales. When discussing this measure in the remainder of this chapter, we'll call it "relational social complexity," but we emphasize that this is just one measure of social complexity.

This quantity is not to be confused with the commonly used measure of "social differentiation," despite both being abbreviated with S and being proposed as measures of social complexity (Whitehead 2008). Social differentiation is a model-based estimate of the coefficient of variation (the standard deviation divided by the mean) of association probabilities, assuming the underlying probabilities can be approximated with a beta distribution. To see how this measure diverges from our proposed measure of relational complexity, consider a population with two classes of association, one with a near-zero rate of association but very high prevalence and another with high association but low prevalence. Our measure of social complexity would assign this society a relatively low complexity score, as associations are dominated by one class of associate. This society would, however, have very high social differentiation, due to the low mean and high variance of the underlying distribution.

There are a few technical considerations to bear in mind here. The process for estimating *S* can be data-intensive; in simulations, depending on the specific method used, it can take between 20 and 100 sampling periods per dyad (not per individual) to get good estimates of *S*; lower sampling will result in downward biases in the estimation of number of components and social complexity. We have accounted for this by using the average denominator of the association index as a covariate in our analysis. Furthermore, the structure of the society itself may present challenges; the greater the number of relationship types, the more data are needed to accurately estimate *S*.

In the following sections, we'll fit this model to datasets from numerous cetacean species. Our goal here is twofold. First, we hope to provide a general taxonomic overview of relational social complexity in cetaceans to promote further thinking about the causes and consequences of social complexity in this taxon. Second, we provide a comparison of our measure of complexity to some of the measures that have previously been utilized as proxies of social complexity in cetaceans.

DATASET

The bulk of our datasets were found during a systematic review carried out in 2021 (Weiss et al. 2021). Of the 181 papers included in that review, 15 included the data necessary to calculate our measure of social complexity (a full list of references can be found in table 4.1). In order to expand our overview to baleen whales, we have included data on humpback whale (*Megaptera novaeangliae*) associations published by Wray et al. (2021) and data on fin whale (*Balaenoptera physalus*) associations published by Keen et al. (2021). We have also included data from our own field studies on associations between southern resident killer whales. Finally, we include the social complexity estimates for bottlenose whales and sperm whales originally published in Weiss et al. (2019). All told, we obtained relational social complexity estimates from 22 datasets across 11 species. Our sample is strongly taxonomically biased toward the delphinids, and particularly common bottlenose dolphins (*Tursiops truncatus*), which are represented in nine different datasets. The only other species for which we found multiple datasets was killer whales, with four datasets including the southern resident killer whale data.

TABLE 4.1. Summary of our comparative dataset and data sources

Species	*Common name*	*Location*	*N*[a]	*Type of social structure*	*References*
Balaenoptera physalus	Fin whale	Kitimat Fjord System, British Columbia CA	94	Solitary	Keen et al. 2021
Berardius bairdii	Baird's beaked whale	Commander Islands, Russia	28	Fission-fusion	Fedutin et al. 2015
Cephalorhynchus hectori	Hector's dolphin	Porpoise Bay, New Zealand	18	Fission-fusion	Bejder et al. 1998
Globicephala macrorhynchus	Short-finned pilot whale	Madeira, Portugal	44	Multilevel	Alves et al. 2013
Hyperoodon ampullatus	Northern bottlenose whale	The Gully, Nova Scotia, CA	30	Fission-fusion	Gowans et al. 2001[b]
Megaptera novaeangliae	Humpback whale	Kitimat Fjord System, British Columbia, CA	77	Solitary	Wray et al. 2021
Orcinus orca	Killer whale	Marion Island, South Africa	54	Multilevel	Reisinger et al. 2017
		Crozet Islands, Indian Ocean	147 54	Multilevel	Busson et al. 2019
		Salish Sea, Washington, USA	75	Multilevel	Center for Whale Research 2023[b]
Physeter macrocephalus	Sperm whale	Dominica, Caribbean Sea	77	Multilevel	Gero et al. 2014[c]
Sousa sahulensis	Australian humpback dolphin	Northwest Cape, Western Australia	84	Fission-fusion	Hunt et al. 2019
Tursiops aduncus	Indo-Pacific bottlenose dolphin	Shark Bay, Western Australia	176	Fission-fusion / multilevel	Foroughirad et al. 2019
Tursiops truncatus	Common bottlenose dolphin	Sado Estuary, Portugal	27	Fission-fusion	Augusto et al. 2011
		Sardina, Italy	124	Fission-fusion	Díaz López 2019
		Patos Lagoon, Brazil	122	Fission-fusion	Genoves et al. 2018
		Savannah, Georgia, USA	346	Fission-fusion	Kovacs et al. 2017
		Laguna, Brazil	41	Fission-fusion	Machado et al. 2019
		Ria de Arousa, Spain	171	Fission-fusion	Methion and Díaz López 2019
		Golfo Dulce, Costa Rica	79	Fission-fusion	Moreno and Acevedo-Gutiérrez 2016
		Western Ireland	49 39	Fission-fusion	Nykänen et al. 2018

a. Number of individuals in dataset. Multiple figures indicate more than one dataset from the study.
b. Unpublished data.
c. Estimates from these datasets were first presented in Weiss et al. (2019).

ANALYSIS

For all datasets included, we extracted the numerator and denominator of the simple ratio indices of association between all identified individuals. We then used the flexmix R package (Leisch 2004) to run a binomial mixture model with a maximum of five components on all datasets. We allowed the model to drop empty components during the model fitting process to arrive at a final number of components (relationship classes), along with parameter estimates for the association probability and frequency of occurrence for each class. From these estimates, we then calculated our estimate of relational social complexity as described in the "Dataset" section.

To examine correlations between our measure of social complexity and possible covariates, we used models with the form:

$$S_i \sim N(\mu_i, \sigma_{\text{residual}}),$$
$$\mu_i = \beta_0 + \beta_1 \log(D_i) + \cdots + \delta_{\text{species}_i} + \gamma_{\text{species}_i},$$
$$\gamma \sim \text{MVN}(0, \Sigma),$$
$$\Sigma = A\sigma_{\text{phylo}},$$
$$\delta \sim N(0, \sigma_{\text{species}}).$$

Here, A is a phylogenetic correlation matrix derived from the phylogeny presented in McGowen et al. (2019), D is the average denominator of the association index, indicating sampling intensity, the σ are estimated standard deviations, and the β are estimated fixed effects. The ellipsis indicates additional, analysis-specific terms. The term MVN(0, Σ) represents a multivariate normal distribution with mean 0 and covariance matrix Σ. We set $N(0,1)$ priors for fixed effects, and broad half-t priors for estimated standard deviations. Given the small size of our dataset, we added only a single predictor variable at a time to this model for any given analysis. We fit these models using the brms R package (Bürkner 2017).

Social Uncertainty Across Cetacean Species

Across our datasets, the number of identified social relationship types ranged from two to four, with resulting estimates of relational social complexity ranging from 0.12 to 1.33 (fig. 4.2). The lowest estimated values of relational social complexity were from the two baleen whale datasets, with $S = 0.12$ for fin whales and $S = 0.42$ for humpback whales. Relational complexity was slightly higher in the beaked whale species where we estimated $S = 0.61$ and $S = 0.69$ for Baird's beaked whales and northern bottlenose whales, respectively.

Among the delphinids, relational social complexity was lowest in Hec-

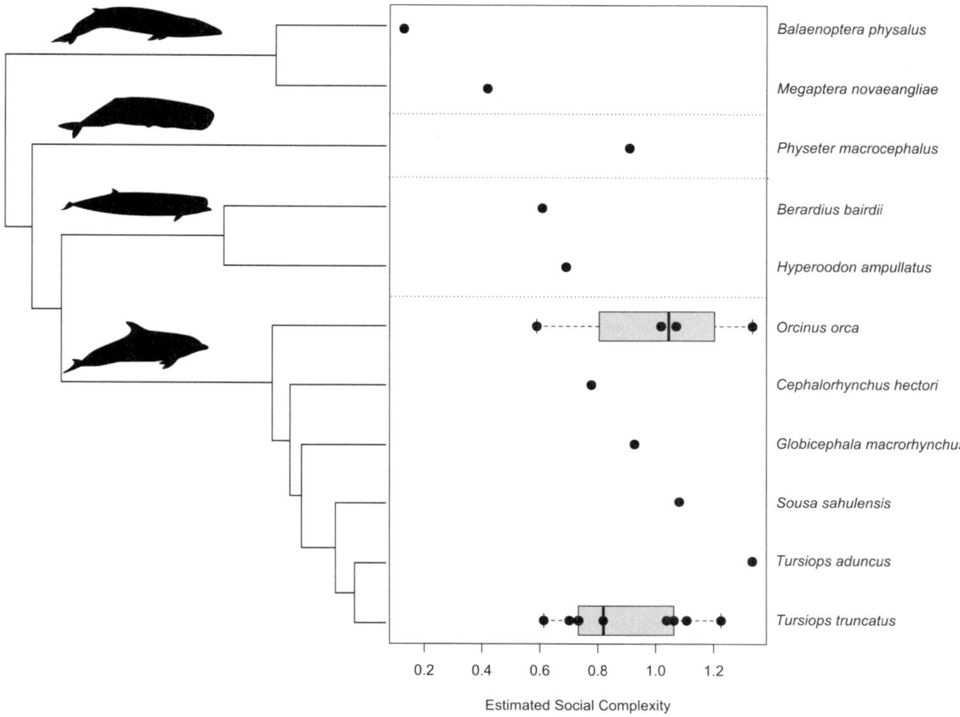

FIGURE 4.2. Estimated social complexity in cetaceans. Each point represents an estimate from a single dataset. Where multiple estimates were made for a species, box plots are also drawn. The phylogeny on the left is based on that presented by McGowen et al. (2019). Silhouettes and dotted lines demarcate families (from top to bottom: Balaenopteridae, Physeteridae, Ziphiidae, and Delphinidae).
Source: Cetacean outlines by Chris Huh are used, edited, and distributed under a CC BY-SA 3.0 DEED license.

tor's dolphin (*Cephalorhynchus hectori*, $S = 0.77$), and peaked in Indo-Pacific bottlenose dolphins and southern resident killer whales ($S = 1.33$). Relational social complexity in sperm whales fits well within the range of complexity estimates for delphinids, most closely resembling that of short-finned pilot whales (sperm whales = 0.91, pilot whales = 0.92). A particularly interesting estimate of social complexity comes from Australian humpback dolphins (*Sousa sahulensis*), with relational social complexity similar to sperm whales, pilot whales, and killer whales. Importantly, in the two species where we had multiple datasets to analyze, we found a significant degree of variability between datasets. In common bottlenose dolphins, estimates of S ranged from 0.61 to 1.23, while in killer whales, estimates ranged from 0.59 to 1.33.

Examining our estimates of social complexity (and the estimated number of "relationship types") across species, our statistical model clearly underestimates social complexity. This is particularly obvious in populations that have been studied in detail for many years. For example, in the Shark Bay bottlenose dolphins, our model estimates four types of relationships, and a resulting social complexity of 1.33. However, from long-term studies, it is known that among males in this population, there are first-, second-, and third-order alliances, and relationships between unallied individuals (see Connor and King, this volume). Additionally, there are weak and strong female-female relationships (see Foroughirad et al., this volume). Finally, there are the relationships between males and females, which clearly consist of very different social interactions than intrasexual relationships (see Connor and Krützen [2015] for a detailed overview of this society). Thus, we know there are (at least) seven different types of social relationships in this population. This discrepancy highlights the degree to which measures of social structure from association, while instructive, inherently underestimate social complexity. There are also clear signs that sampling effort influences our measures, with a strong positive correlation between estimated complexity and average association index denominator ($r_s = 0.66$).

By fitting our model (see "Analysis"), we can investigate whether there is evidence for a phylogenetic signal in our dataset by quantifying the portion of between-species variance (after accounting for sampling intensity) accounted for by the phylogeny as opposed to uncorrelated species-level differences, which we'll refer to as λ. There is evidence of phylogenetic signal in this dataset, although with such a small dataset there is a good deal of uncertainty ($\lambda = 0.86$, 95% CI = [0.1, 1]). More broadly, it seems that the main divergence in social complexity occurred between the toothed and baleen whales (although we have few baleen whale data in this analysis). This makes intuitive sense, as large baleen whales are thought to be primarily solitary or live in highly fluid societies (but see Dunlop and Parks, this volume), while toothed whales are generally group living to some extent. Greater attention to and study of the social relationships of baleen whales will help us expand and refine this picture.

How Does Social Uncertainty in Cetaceans Relate to Other Measures of Social Complexity?

We described three general measures that have been used as proxies of social complexity in previous studies: number of social relationships, social structure classification, and social repertoire. While none of these directly measure (or indeed attempt to directly measure) social uncertainty, it may be helpful to determine whether these features correlate with our estimates of social complexity across cetaceans (fig. 4.3).

First, we can examine how the number of social relationships per indi-

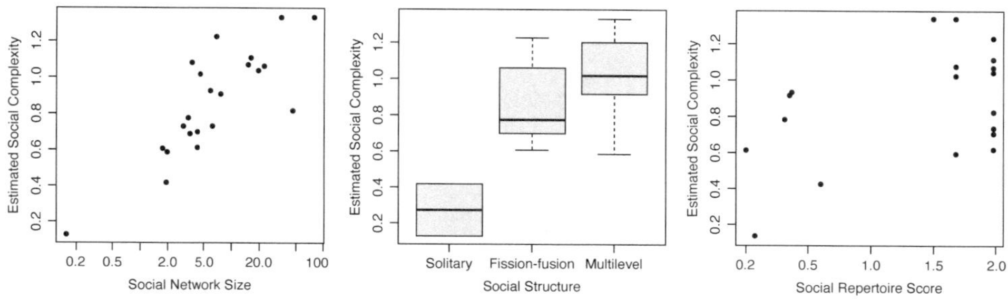

FIGURE 4.3. Comparison of our measure of social complexity with other proxies of social complexity in cetaceans.

vidual correlates with social relationship complexity. There are two ways to go about estimating the number of social relationships in our cetacean association networks. We could consider all nonzero association indices as relationships and therefore use the average number of nonzero indices per individual as our measure. However, there may be many individuals who associate in aggregations but don't maintain a social relationship. An alternative way to approach this question is to use the output of our mixture models to estimate the size of the social network that individuals experience. In our analysis, we naturally segment dyads into those with "no relationship" (i.e., the relationship class with the lowest association rate, typically very close to zero) and those with some stronger relationship (all other classes). Therefore, we can measure the number of social relationships as the average number of these stronger relationships that individuals in the society have. We view this measure as more comparable to the measure of social group size used in studies of primates, as it is an estimate of the number of partners individuals maintain some kind of social bond with. In our dataset, this measure ranged from individuals maintaining fewer than one relationship on average (in fin whales) to individuals maintaining over 40 relationships on average (as in southern resident killer whales). We find strong evidence for a correlation between the logarithm of the number of social relationships per individual and social relationship complexity ($\beta \pm \text{SE} = 0.12 \pm 0.06$, posterior $P[\beta > 0] = 0.98$). We note, however, that these two measures are derived from the same dataset and model fitting process, so some caution should be exercised in interpreting this positive correlation.

What about other features of social structure? Are, for example, multilevel societies truly more complex than other social structures? While more fine-scale analysis would be ideal, we roughly categorized our species as either being primarily solitary (typical group size <2), group living with individualized fission-fusion (typical group size ≥2, no stable social units) or living in multilevel societies (stable social units). Again fitting our model, we can see clear evidence that both fission-fusion and multilevel societies have higher estimated complexities than the primarily solitary societies (multilevel versus solitary: $\beta \pm SE = 0.52 \pm 0.27$, posterior $P[\beta > 0] = 0.97$; fission-fusion versus solitary: $\beta \pm SE = 0.45 \pm 0.26$, posterior $P[\beta > 0] = 0.95$). However, we find very little evidence for a difference between fission-fusion and multilevel societies ($\beta \pm SE = 0.07 \pm 0.15$, posterior $P[\beta > 0] = 0.71$). The Indo-Pacific bottlenose dolphins of Shark Bay represent a special case, as this population exhibits multilevel alliances nested within a generally fission-fusion social system. Altering our classifications to consider this population "fission-fusion" rather than "multilevel," or excluding them from the analysis, did not qualitatively change these results.

Finally, we examine whether the "social repertoire score" developed by Fox et al. (2017) correlates with our measure of social complexity. As this was not designed to be a measure of social structure, but rather social behavior, there is no structural reason why these two measures should be correlated, but there are perhaps evolutionary hypotheses that would propose such a relationship. In our sample, we found very little evidence that this measure correlates with our measure of social complexity ($\beta \pm SE = 0.03 \pm 0.15$, posterior $P[\beta > 0] = 0.6$).

In summary, it appears that the presence of particular social behaviors is not a useful proxy for the diversity or uncertainty of social relationships. Furthermore, while the contrast between group living and primarily solitary species provides some information about social uncertainty, it does not appear that different types of social structure necessarily correspond to different levels of social complexity as defined here. Rather than multilevel sociality being a form of particularly complex social structure, it may instead be just one way in which a society can be complex.

In contrast, the number of social relationships individuals maintain appears to be a useful correlate of social uncertainty. While the precise reason why this is the case in our dataset is unclear, it seems reasonable to suggest that maintaining many social relationships is a necessary prerequisite for having many different types of social relationships and, ultimately, a strongly varied social environment. This also relates to our finding that solitary baleen whales score lower on this form of social complexity than toothed whales. While these baleen whales may encounter many different individuals at various points in time, they appear (in the datasets we examined) to maintain very few strong, consistent relationships, resulting in lower estimates of social complexity (for a more detailed discussion of social complexity in baleen whales see Dunlop and Parks, this volume).

Conclusions and Future Directions

Our goal here is to make progress in operationalizing relational social complexity and provide an initial, quantitative overview of social complexity in cetaceans. Our initial survey of this data demonstrates significant variation in social complexity across cetaceans with an apparently strong phylogenetic signal. Interestingly, this variation can't be fully explained by variation in basic social structure and does not strongly correlate with behavioral repertoires. The number of relationships that individuals maintain, while correlated with relational complexity, also cannot fully explain variation in this measure. An obvious next step is to determine what factors, particularly those related to life history and ecology, predict complex social structure. We also strongly suspect that analyses addressing the social brain and communicative complexity hypotheses using social uncertainty as a predictor may reveal patterns and relationships that have not been clear thus far.

Of particular interest is the within-species variation in killer whales and common bottlenose dolphins. It's worth considering what this variation might mean for hypotheses about the causes and consequences of social complexity. It seems unlikely that species-level traits, such as morphology and brain size, could be predictably driven by the "social environment" when the social environment itself is so variable between populations of the same species. In contrast, this may point to an important role for habitat and ecology in shaping social systems and their resulting complexity. Well-reasoned socio-ecological frameworks for delphinids have been proposed (Gowans et al. 2007; Möller 2012); however, quantitative comparative tests of these frameworks are rarely conducted.

We suggest that future studies expand the taxonomic breadth of the datasets included. Missing from our analysis are any river dolphins, porpoises, or non-rorqual baleen whales. This could be addressed by directly reaching out to field researchers to collate datasets from the numerous individual-level studies currently being conducted on a wide variety of species, or by more raw data being made available along with peer-reviewed publications.

Our method only considers association rates when clustering relationships into types, as this is the primary form of data available to cetacean researchers through photographic identification studies. However, as novel research technologies, especially unoccupied aerial systems and animal-borne tags, make it easier to monitor social interactions (King and Jensen 2023), it may become possible to incorporate other measures of social relationships into this method. The mixture model framework would then have to be expanded to be multivariate, considering multiple forms of association and interaction. Furthermore, we are ultimately interested in the uncertainty of social interactions themselves, with relationship types serving as a rough proxy. Observations of direct social interactions and their unpredictability could provide a much richer picture of the complexity of cetaceans' social worlds.

Another crucial avenue may be to more directly incorporate aspects of communication into our understanding of social complexity in cetaceans. Acoustic playback experiments may help us test whether the different "relationship types" we identify statistically are cognitively relevant to the individuals within the society (King et al., this volume). Furthermore, incorporating acoustic communication may reveal levels of social connectivity not apparent from typical methods for analyzing social structure, particularly in baleen whale species that may interact acoustically without being in close physical proximity (Dunlop and Parks, this volume).

While our focus has been firmly on understanding cetacean sociality, our methods and results may help inform studies of social complexity more generally. We've highlighted here that the occurrence of particular social behaviors and types of social structure cannot be considered indicative of "social complexity" a priori. Instead, we strongly suggest that studies of social complexity focus on well-motivated hypotheses about how sociality may present a challenge to the individuals and measure features of social structure that correspond to these hypotheses. There are encouraging signs in primatology that studies of social complexity may move away from simple analyses based on group size and toward quantitative measures of social structure focused on the diversity of relationships individuals maintain (Fischer et al. 2017; Morrison et al. 2020; Ramos-Fernandez et al. 2018). In other taxa, there is a tendency to use very general aspects of social structure, such as the presence and structure of social groups, as measures of "social complexity." Studies of cetaceans, and our proposed measure of social complexity, emphasize the importance of studying social systems at the level of relationships and individuals, building a bottom-up picture of social complexity. We recommend that future comparative studies of social complexity across taxa make greater attempts to incorporate individual-based studies of social structure; more generally, many taxa would benefit from greater research into their social structure on an individual level.

Understanding the evolutionary causes and consequences of social complexity requires us to operationalize, measure, and compare complexity between societies. We hope that this chapter provides a useful early step toward this goal in cetaceans, a taxon that is inarguably invaluable for studies of social complexity in the wild.

Acknowledgments

M. W. and D. P. C. gratefully acknowledge funding from the NERC Standard Grant (No. NE/S010327/1). The network database was first constructed by I. H. H. during their MSc project supervised by Dr. Sam Ellis, who has continued to support the development of the database for this chapter. The authors thank Dr. Sam Ellis and the members of the Centre for Research in Animal Behaviour at the University of Exeter for their valuable support, insights, and discussions during the development of this chapter.

References

Alves, Filipe, Sophie Quérouil, Ana Dinis, et al. 2013. "Population Structure of Short-Finned Pilot Whales in the Oceanic Archipelago of Madeira Based on Photo-Identification and Genetic Analyses: Implications for Conservation." *Aquatic Conservation: Marine and Freshwater Ecosystems* 23 (5): 758–76. https://doi.org/10.1002/aqc.2332.

Augusto, Joana, Timothy R. Frasier, and Hal Whitehead. 2017. "Characterizing Alloparental Care in the Pilot Whale (*Globicephala melas*) Population That Summers off Cape Breton, Nova Scotia, Canada." *Marine Mammal Science* 33 (2): 440–56. https://doi.org/10.1111/mms.12377.

Augusto, Joana, Patrícia Rachinas-Lopes, and Manuel Eduardo Dos Santos. 2011. "Social Structure of the Declining Resident Community of Common Bottlenose Dolphins in the Sado Estuary, Portugal." *Journal of the Marine Biological Association of the United Kingdom* 92 (8): 1773–82. https://doi.org/10.1017/s0025315411000889.

Baird, Robin W., and Lawrence M. Dill. 1995. "Occurrence and Behaviour of Transient Killer Whales: Seasonal and Pod-Specific Variability, Foraging Behaviour, and Prey Handling." *Canadian Journal of Zoology* 73 (7): 1300–1311. https://doi.org/10.1139/z95-154.

Bejder, Lars, David Fletcher, and Stefan Bräger. 1998. "A Method for Testing Association Patterns of Social Animals." *Animal Behaviour* 56 (3): 719–25. https://doi.org/10.1006/anbe.1998.0802.

Bergman, Thore J., and Jacinta C. Beehner. 2015. "Measuring Social Complexity." *Animal Behaviour* 103 (May): 203–9. https://doi.org/10.1016/j.anbehav.2015.02.018.

Bigg, Michael A., Peter F. Olesiuk, Graeme M. Ellis, John K. B. Ford, and Kenneth C. Balcomb. 1990. "Social Organization and Genealogy of Resident Killer Whales (*Orcinus orca*) in the Coastal Waters of British Columbia and Washington State." *Report of the International Whaling Commission* 12: 383–405.

Brent, Lauren J. N., Daniel W. Franks, Emma A. Foster, Kenneth C. Balcomb, Michael A. Cant, and Darren P. Croft. 2015. "Ecological Knowledge, Leadership, and the Evolution of Menopause in Killer Whales." *Current Biology* 25 (6): 746–50. https://doi.org/10.1016/j.cub.2015.01.037.

Bürkner, Paul-Christian. 2017. "brms: An R Package for Bayesian Multilevel Models Using Stan." *Journal of Statistical Software* 80: 1–28.

Busson, Marine, Matthieu Authier, Christophe Barbraud, et al. 2019. "Role of Sociality in the Response of Killer Whales to an Additive Mortality Event." *Proceedings of the National Academy of Sciences* 116 (24): 11812–17. https://doi.org/10.1073/pnas.1817174116.

Cantor, Mauricio, Lauren G. Shoemaker, Reniel B. Cabral, César O. Flores, Melinda Varga, and Hal Whitehead. 2015. "Multilevel Animal Societies Can Emerge from Cultural Transmission." *Nature Communications* 6 (1): 8091. https://doi.org/10.1038/ncomms9091.

Connor, Richard C. 2007. "Dolphin Social Intelligence: Complex Alliance Relationships in Bottlenose Dolphins and a Consideration of Selective Environments for Extreme Brain Size Evolution in Mammals." *Philosophical Transactions of the Royal Society B* 362 (1480): 587–602. https://doi.org/10.1098/rstb.2006.1997.

Connor, Richard C., and Michael Krützen. 2015. "Male Dolphin Alliances in Shark Bay: Changing Perspectives in a 30-Year Study." *Animal Behaviour* 103 (May): 223–35. https://doi.org/10.1016/j.anbehav.2015.02.019.

Díaz López, Bruno. 2019. "'Hot Deals at Sea': Responses of a Top Predator (Bottlenose Dolphin, *Tursiops truncatus*) to Human-Induced Changes in the Coastal Ecosystem." *Behavioral Ecology* 30 (2): 291–300. https://doi.org/10.1093/beheco/ary162.

Dunbar, Robin. 1992. "Neocortex Size as a Constraint on Group Size in Primates." *Journal of Human Evolution* 22 (6): 469–93. https://doi.org/10.1016/0047-2484(92)90081-j.

Dunbar, Robin. 1998. "The Social Brain Hypothesis." *Evolutionary Anthropology: Issues, News, and Reviews* 6: 178–90. https://doi.org/10.1002/(SICI)1520-6505(1998)6:5<178::AID-EVAN5>3.0.CO;2-8.

Dunbar, Robin, and Julian Bever. 1998. "Neocortex Size Predicts Group Size in Carnivores and Some Insectivores." *Ethology* 104 (8): 695–708. https://doi.org/10.1111/j.1439-0310.1998.tb00103.x.

Dunbar, Robin, and Susanne Shultz. 2021. "Social Complexity and the Fractal Structure of Group Size in Primate Social Evolution." *Biological Reviews* 96 (5): 1889–906. https://doi.org/10.1111/brv.12730.

Fedutin, Ivan D., Olga A. Filatova, E. G. Mamaev, Alexandr M. Burdin, and Erich Hoyt. 2015. "Occurrence and Social Structure of Baird's Beaked Whales, *Berardius bairdii*, in the Commander Islands, Russia." *Marine Mammal Science* 31 (3): 853–65. https://doi.org/10.1111/mms.12204.

Fischer, Julia, Max S. Farnworth, Holger Sennhenn-Reulen, and Kurt Hammerschmidt. 2017. "Quantifying Social Complexity." *Animal Behaviour* 130 (August): 57–66. https://doi.org/10.1016/j.anbehav.2017.06.003.

Foroughirad, Vivienne, Alexis L. Levengood, Janet Mann, and Céline H. Frère. 2019. "Quality and Quantity of Genetic Relatedness Data Affect the Analysis of Social Structure." *Molecular Ecology Resources* 19 (5): 1181–94. https://doi.org/10.1111/1755-0998.13028.

Fox, Kieran C. R., Michael Muthukrishna, and Susanne Shultz. 2017. "The Social and Cultural Roots of Whale and Dolphin Brains." *Nature Ecology and Evolution* 1 (11): 1699–705. https://doi.org/10.1038/s41559-017-0336-y.

Freeberg, Todd M., Robin Dunbar, and Terry J. Ord. 2012. "Social Complexity as a Proximate and Ultimate Factor in Communicative Complexity." *Philosophical Transactions of the Royal Society B* 367 (1597): 1785–801. https://doi.org/10.1098/rstb.2011.0213.

Garland, Ellen C., Anne W. Goldizen, Melinda L. Rekdahl, et al. 2011. "Dynamic Horizontal Cultural Transmission of Humpback Whale Song at the Ocean Basin Scale." *Current Biology* 21 (8): 687–91. https://doi.org/10.1016/j.cub.2011.03.019.

Gazda, Stefanie K., Richard C. Connor, Robert K. Edgar, and Frank Cox. 2005. "A Division of Labour with Role Specialization in Group-Hunting Bottlenose Dolphins (*Tursiops truncatus*) off Cedar Key, Florida." *Proceedings of the Royal Society B* 272 (1559): 135–40. https://doi.org/10.1098/rspb.2004.2937.

Genoves, Rodrigo Cézar, Pedro Fruet, Juliana C. Di Tullio, Luciana M. Möller, and Eduardo R. Secchi. 2018. "Spatiotemporal Use Predicts Social Partitioning of Bottlenose Dolphins with Strong Home Range Overlap." *Ecology and Evolution* 8 (24): 12597–614. https://doi.org/10.1002/ece3.4681.

Gero, Shane, Dan Engelhaupt, Luke Rendell, and Hal Whitehead. 2009. "Who Cares? Between-Group Variation in Alloparental Caregiving in Sperm Whales." *Behavioral Ecology* 20 (4): 838–43. https://doi.org/10.1093/beheco/arp068.

Gero, Shane, Marina Milligan, Caroline Rinaldi, et al. 2014. "Behavior and Social Structure of the Sperm Whales of Dominica, West Indies." *Marine Mammal Science* 30 (3): 905–22. https://doi.org/10.1111/mms.12086.

Gowans, Shannon, Hal Whitehead, and Sascha K. Hooker. 2001. "Social Organization in Northern Bottlenose Whales, *Hyperoodon ampullatus*: Not Driven by Deep-Water Foraging?" *Animal Behaviour* 62 (2): 369–77. https://doi.org/10.1006/anbe.2001.1756.

Gowans, Shannon, Bernd Würsig, and Leszek Karczmarski. 2007. "The Social Structure and Strategies of Delphinids: Predictions Based on an Ecological Framework." *Advances in Marine Biology* 53: 195–294. https://doi.org/10.1016/s0065-2881(07)53003-8.

Groot, Nikita E., Rochelle Constantine, Ellen C. Garland, and Emma L. Carroll. 2023. "Phylogenetically Controlled Life History Trait Meta-Analysis in Cetaceans Reveals Unexpected Negative Brain Size and Longevity Correlation." *Evolution* 77 (2): 534–49. https://doi.org/10.1093/evolut/qpac050.

Grueter, Cyril C., Xiao-Guang Qi, Dietmar Zinner, et al. 2020. "Multilevel Organisation of Animal Sociality." *Trends in Ecology and Evolution* 35 (9): 834–47. https://doi.org/10.1016/j.tree.2020.05.003.

Hardie, Jasmine L., and Christopher R. Cooney. 2023. "Sociality, Ecology and Developmental Constraints Predict Variation in Brain Size Across Birds." *Journal of Evolutionary Biology* 36 (1): 144–55. https://doi.org/10.1111/jeb.14117.

He, Peng, Adriana Alexandra Maldonado-Chaparro, and Damien R. Farine. 2019. "The Role of Habitat Configuration in Shaping Social Structure: A Gap in Studies of Animal Social Complexity." *Behavioral Ecology and Sociobiology* 73 (1). https://doi.org/10.1007/s00265-018-2602-7.

Hinde, Robert A. 1976. "Interactions, Relationships and Social Structure." *Man* 11 (1): 1. https://doi.org/10.2307/2800384.

Hunt, Tim, S. J. Allen, Lars Bejder, and Guido J. Parra. 2019. "Assortative Interactions Revealed in a Fission-Fusion Society of Australian Humpback Dolphins." *Behavioral Ecology* 30 (4): 914–27. https://doi.org/10.1093/beheco/arz029.

Keen, Eric M., James F. Pilkington, Éadin O'Mahony, 2021. "Fin Whales of the Great Bear Rainforest: *Balaenoptera physalus velifera* in a Canadian Pacific Fjord System." *PLOS ONE* 16 (9): e0256815. https://doi.org/10.1371/journal.pone.0256815.

King, Stephanie L., and Frants H. Jensen. 2023. "Rise of the Machines: Integrating Technology with Playback Experiments to Study Cetacean Social Cognition in the Wild." *Methods in Ecology and Evolution* 14 (8): 1873–86. https://doi.org/10.1111/2041-210x.13935.

Kovacs, Carolyn J., Robin M. Perrtree, and Tara Cox. 2017. "Social Differentiation in Common Bottlenose Dolphins (*Tursiops truncatus*) That Engage in Human-Related Foraging Behaviors." *PLOS ONE* 12 (2): e0170151. https://doi.org/10.1371/journal.pone.0170151.

Kudo, Hiroko, and Robin Dunbar. 2001. "Neocortex Size and Social Network Size in Primates." *Animal Behaviour* 62 (4): 711–22. https://doi.org/10.1006/anbe.2001.1808.

Lehmann, Julia, and Robin Dunbar. 2009. "Network Cohesion, Group Size and Neocortex Size in Female-Bonded Old World Primates." *Proceedings of the Royal Society B* 276 (1677): 4417–22. https://doi.org/10.1098/rspb.2009.1409.

Leisch, Friedrich. 2004. "flexmix: A General Framework for Finite Mixture Models and Latent Class Regression in R." *Journal of Statistical Software* 11 (8). https://doi.org/10.18637/jss.v011.i08.

Lewis, Jennifer, Douglas Wartzok, and Michael R. Heithaus. 2011. "Highly Dynamic Fission-Fusion Species Can Exhibit Leadership when Traveling." *Behavioral Ecology and Sociobiology* 65 (5): 1061–69. https://doi.org/10.1007/s00265-010-1113-y.

Lucas, Eric R., and Laurent Keller. 2020. "The Co-Evolution of Longevity and Social Life." *Functional Ecology* 34 (1): 76–87. https://doi.org/10.1111/1365-2435.13445.

Lukas, Dieter, and Tim H. Clutton-Brock. 2018. "Social Complexity and Kinship in Animal Societies." *Ecology Letters* 21 (8): 1129–34. https://doi.org/10.1111/ele.13079.

Lusseau, David. 2007. "Evidence for Social Role in a Dolphin Social Network." *Evolutionary Ecology* 21 (3): 357–66. https://doi.org/10.1007/s10682-006-9105-0.

Machado, Alexandre M. S., Mauricio Cantor, Ana P. B. Costa, et al. 2019. "Homophily Around Specialized Foraging Underlies Dolphin Social Preferences." *Biology Letters* 15 (4): 20180909. https://doi.org/10.1098/rsbl.2018.0909.

Maciej, Peter, Annika Patzelt, I. Malick Ndao, Kurt Hammerschmidt, and Julia Fischer. 2013. "Social Monitoring in a Multilevel Society: A Playback Study with Male Guinea Baboons." *Behavioral Ecology and Sociobiology* 67 (1): 61–68. https://doi.org/10.1007/s00265-012-1425-1.

Marino, Lori, Richard C. Connor, R. Ewan Fordyce, et al. 2007. "Cetaceans Have Complex Brains for Complex Cognition." *PLOS Biology* 5 (5): e139. https://doi.org/10.1371/journal.pbio.0050139.

May-Collado, Laura J., Ingi Agnarsson, and Douglas Wartzok. 2007. "Phylogenetic Review of Tonal Sound Production in Whales in Relation to Sociality." *BMC Evolutionary Biology* 7 (1): 136. https://doi.org/10.1186/1471-2148-7-136.

McGowen, Michael R., Georgia Tsagkogeorga, Sandra Álvarez-Carretero, et al. 2019. "Phylogenomic Resolution of the Cetacean Tree of Life Using Target Sequence Capture." *Systematic Biology* 69 (3): 479–501. https://doi.org/10.1093/sysbio/syz068.

McShea, Daniel W., and Robert N. Brandon. 2010. *Biology's First Law: The Tendency for Diversity and Complexity to Increase in Evolutionary Systems*. University of Chicago Press.

Methion, Séverine, and Bruno Díaz López. 2019. "Individual Foraging Variation Drives Social Organization in Bottlenose Dolphins." *Behavioral Ecology*, September. https://doi.org/10.1093/beheco/arz160.

Möller, Luciana M. 2012. "Sociogenetic Structure, Kin Associations and Bonding in Delphinids." *Molecular Ecology* 21 (3): 745–64. https://doi.org/10.1111/j.1365-294x.2011.05405.x.

Moreno, Kelsey R., and Alejandro Acevedo-Gutiérrez. 2016. "The Social Structure of Golfo Dulce Bottlenose Dolphins (*Tursiops truncatus*) and the Influence of Behavioural State." *Royal Society Open Science* 3 (8): 160010. https://doi.org/10.1098/rsos.160010.

Morrison, Robin E., Winnie Eckardt, Tara S. Stoinski, and Lauren J. N. Brent. 2020. "Comparing Measures of Social Complexity: Larger Mountain Gorilla Groups Do Not Have a Greater Diversity of Relationships." *Proceedings of the Royal Society B* 287 (1931): 20201026. https://doi.org/10.1098/rspb.2020.1026.

Nykänen, Milaja, Eileen Dillane, Anneli Englund, et al. 2018. "Quantifying Dispersal Between Marine Protected Areas by a Highly Mobile Species, the Bottlenose Dolphin, *Tursiops truncatus*." *Ecology and Evolution* 8 (18): 9241–58. https://doi.org/10.1002/ece3.4343.

Page, Scott E. 2010. *Diversity and Complexity*. Princeton University Press.

Pérez-Barbería, F. Javier, and Iain J. Gordon. 2005. "Gregariousness Increases Brain Size in Ungulates." *Oecologia* 145 (1): 41–52. https://doi.org/10.1007/s00442-005-0067-7.

Pérez-Barbería, F. Javier, Susanne Shultz, and Robin Dunbar. 2007. "Evidence for Coevolution of Sociality and Relative Brain Size in Three Orders of Mammals." *Evolution* 61 (12): 2811–21. https://doi.org/10.1111/j.1558-5646.2007.00229.x.

Prox, Lea, and Damien R. Farine. 2020. "A Framework for Conceptualizing Dimensions of Social Organization in Mammals." *Ecology and Evolution* 10 (2): 791–807. https://doi.org/10.1002/ece3.5936.

Ramos-Fernandez, Gabriel, Andrew J. King, Jacinta C. Beehner, et al. 2018. "Quantifying Uncertainty Due to Fission-Fusion Dynamics as a Component of Social Complexity." *Proceedings of the Royal Society B* 285 (1879): 20180532. https://doi.org/10.1098/rspb.2018.0532.

Reisinger, Ryan R., Charlene Beukes, A. Rus Hoelzel, and P. J. Nico De Bruyn. 2017. "Kinship and Association in a Highly Social Apex Predator Population, Killer Whales at Marion Island." *Behavioral Ecology* 28 (3): 750–59. https://doi.org/10.1093/beheco/arx034.

Rendell, Luke, Mauricio Cantor, Shane Gero, Hal Whitehead, and Janet Mann. 2019. "Causes and Consequences of Female Centrality in Cetacean Societies." *Philosophical Transactions of the Royal Society B* 374 (1780): 20180066. https://doi.org/10.1098/rstb.2018.0066.

Sewall, Kendra B. 2015. "Social Complexity as a Driver of Communication and Cognition." *Integrative and Comparative Biology* 55 (3): 384–95. https://doi.org/10.1093/icb/icv064.

Shultz, Susanne, and Robin Dunbar. 2007. "The Evolution of the Social Brain: Anthropoid Primates Contrast with Other Vertebrates." *Proceedings of the Royal Society B* 274 (1624): 2429–36. https://doi.org/10.1098/rspb.2007.0693.

Shultz, Susanne, and Robin Dunbar. 2010. "Social Bonds in Birds Are Associated with Brain Size and Contingent on the Correlated Evolution of Life-History and Increased Parental Investment." *Biological Journal of the Linnean Society* 100 (1): 111–23. https://doi.org/10.1111/j.1095-8312.2010.01427.x.

Silk, Matthew J., and David J. Hodgson. 2021. "Differentiated Social Relationships and the Pace-of-Life-History." *Trends in Ecology and Evolution* 36 (6): 498–506. https://doi.org/10.1016/j.tree.2021.02.007.

Weiss, Michael N., Samuel Ellis, and Darren P. Croft. 2021. "Diversity and Consequences of Social Network Structure in Toothed Whales." *Frontiers in Marine Science* 8: 688842. https://doi.org/10.3389/fmars.2021.688842.

Weiss, Michael N., Daniel W. Franks, Darren P. Croft, and Hal Whitehead. 2019. "Measuring the Complexity of Social Associations Using Mixture Models." *Behavioral Ecology and Sociobiology* 73 (1): 8. https://doi.org/10.1007/s00265-018-2603-6.

Whitehead, Hal. 2008. "Precision and Power in the Analysis of Social Structure Using Associations." *Animal Behaviour* 75 (3): 1093–99. https://doi.org/10.1016/j.anbehav.2007.08.022.

Wray, Janie, Eric M. Keen, and Éadin N. O'Mahony. 2021. "Social Survival: Humpback Whales (*Megaptera novaeangliae*) Use Social Structure to Partition Ecological Niches Within Proposed Critical Habitat." *PLOS ONE* 16 (6): e0245409. https://doi.org/10.1371/journal.pone.0245409.

5

Causes and Consequences of Female-Biased Relationships

VIVIENNE FOROUGHIRAD, MOLLY H. F. MCENTEE, AND JANET MANN

Abstract

Cetaceans have markedly diverse social structures driven by trade-offs between resource competition and the benefits of group living. Female ecology—energetic requirements, foraging strategies, and environmental conditions—and kinship are foundational in forming these social structures. Movement ecology and sex-biased dispersal patterns interact to shape female kin availability, with the majority of species retaining access to female kin through fidelity to maternally inherited ranges or migration routes. Females invest heavily in each offspring; baleen whales focus on fast growth and rapid milk transfer and toothed whales focus more on slow growth, extensive maternal care, and social learning. These differences set the stage for disparate female social strategies ranging from avoidance at breeding sites between mysticete mother-calf pairs, to cooperative care among female kin in some toothed whale species. Using a comparative framework, the socioecological model, we examine how trade-offs for females between social bonds, resource acquisition, and calf care operate in varying environments. We evaluate maternal investment, cooperative foraging, and avoidance of predation and male harassment as the underlying drivers of female social bonds.

Introduction

Females have often been considered the ecological sex, referring to the idea that their grouping and social behavior is primarily determined by trade-offs that maximize foraging efficiency and minimize predation on offspring, while male sociality then develops in response to female socioecology. Consequently, females play a foundational role in shaping mammalian social structure and resulting behaviors. The complexity of female social bonding is expected to be driven by competitive interactions for resources, reproduction, and opportunities for inclusive fitness benefits through within-group female relatedness patterns (Lukas and Clutton-Brock 2018). Much research on intrasexual social bonding in cetaceans has focused on males (Ermak et al. 2017; King et al. 2022); however, new technologies have expanded our understanding of female bonds across a broader range of species and behaviors, contributing to a more comprehensive understanding of cetacean social dynamics.

Within-group relatedness is shaped by dispersal patterns and immigration, which in turn are shaped by resource distribution and the demands of reproduction. Cetacean females invest heavily in each offspring, with virtually all mothers giving birth to a single offspring and providing sole parental care. Like most mammals, female-biased philopatry and male dispersal is characteristic of many toothed whales with some notable exceptions of bisexual philopatry (e.g., killer whales, *Orcinus orca*, Pilot et al. 2010). Toothed whales are income breeders with long periods of calf dependency, focusing on slow growth, extensive maternal care, and social learning (Mann 2019). Such intensive and extensive maternal investment requires additional resources that promote fidelity to familiar habitats to increase foraging efficiency (Lawson Handley and Perrin 2007; Clutton-Brock and Lukas 2012). Allegiance to such habitats facilitates close kin associations and information sharing, including strong vertical or matrilineal cultures (Möller 2012). Less is known about wide-ranging pelagic species of toothed whales, though some may still show fidelity to oceanographic features such as seamounts or canyons, or rely on traveling together in large super aggregations to retain access to social and familial connections across vast spaces (Mesnick et al. 2019).

In contrast, baleen whales are capital breeders, focusing on fast growth and rapid milk transfer while fasting. Among baleen whales, female kin bonds are consistently weaker than in toothed whales, though maternal kinship can still play a significant role in shaping associations through structuring migration routes and feeding or breeding site fidelity (Barendse et al. 2013). Although offspring learn migration routes from their mother, information sharing, including foraging tactics, tends to spread laterally in baleen whales (Allen et al. 2013; Wild and Rendell, this volume). Thus, for both toothed and baleen whales, movement and foraging ecology interact with life history to pattern spatial proximity to kin. This shapes the evolutionary structure and function of female sociality, with consequences for reproduction, calf care, and survival.

Differences such as those between toothed and baleen whales set the stage for disparate female social strategies ranging from avoidance at breeding sites between mysticete mother-calf pairs, to cooperative care in some toothed whale species. We follow a comparative framework, the socioecological model originally developed for primates (Wrangham 1980; Isbell and Young 2002) and later formally extended to other taxa (Gowans et al. 2007; Moscovice et al. 2020), with additional insights from developments in kinship, dispersal, and genetic relatedness (Möller 2012; see Parsons et al., this volume). The model predicts that differences in the extent and complexity of social bonding can be explained by differences in local ecology, especially the predictability and consistency of food resources, which in turn shape the degree of resource competition and movement patterns, impacting local relatedness. Through this lens we examine how scientific advances over the past decades have informed our understanding of the trade-offs between social bonds, resource acquisition, mate choice, and calf care from the female perspective.

Spatial Ecology as a Driver of Social Strategy

Spatial ecology and movement drive opportunities to form and maintain social relationships with kin and nonkin. Unlike terrestrial vertebrates, female cetaceans cannot cache their offspring and are in constant motion, unrestricted by the need to return to a nest or den. While this complicates tracking and knowledge of movement patterns, data obtained through biotelemetry, genomics, and longitudinal data across the lifespans of individuals have been critical in informing how the landscape of potential associates translates into a realized network of social bonds and differentiated relationships.

Generally, cetacean ranging patterns should reflect energetic needs and resource availability and predictability, which in turn shape social strategies (Gowans et al. 2007). For almost all species of cetaceans, female movement ecology allows individuals to retain some ability to continue associating with kin post-weaning, either through maternal inheritance of migration routes, or high natal site or group fidelity.

Among the toothed whales, virtually all patterns of migration and dispersal (or lack thereof) can be found, ranging from complete seasonal migration in belugas (Colbeck et al. 2013), to sex-biased dispersal in sperm whales (Lyrholm et al. 1999; Engelhaupt et al. 2009), to complete bisexual natal residency in bottlenose dolphins (Tsai and Mann 2013), and nomadism in short-finned pilot whales (Foley 2018). Links between spatial ecology and social structure are evident; species with continuous or irregular movement patterns, often termed nomadism, tend to travel in highly cohesive family groups (e.g., short-finned pilot whales, Alves et al. 2013), while more residential species tend to have more flexible social structures with varying degrees of fission-fusion dynamics. Migration may happen with or without the family group.

The vast majority of baleen whales migrate from high-latitude feeding grounds to low-latitude breeding grounds, with several well-studied populations showing fidelity to their maternal migration routes (Carroll et al. 2016; Valenzuela et al. 2009; Garland and Carroll 2022; Dunlop and Parks, this volume). Sex-specific selection pressures have been invoked in shaping these routes, as females must make a critical trade-off between accumulating the necessary energetic stores for capital breeding in cold water habitats (Irvine et al. 2017) while alternately seeking warm water refuges in which to give birth. Two predominant non-mutually exclusive explanations for calving in warm waters include calf thermoregulation (e.g., Rasmussen et al. 2007) and avoiding calf predation by killer whales (Corkeron and Connor 1999; Reeves et al. 2006). Though there may not be a single explanation, calf survival is likely a central driver. Despite the lack of persistent groups, migration routes can be maternally inherited (e.g., southern right whales; Elwen and Best 2004; humpback whales; Barendse et al. 2013; Baker et al. 1986), which creates opportunities for encounters between matrilineal kin.

Structure of Female Social Bonds—Strength, Distribution, and Duration

Where cetacean societies fall in terms of social complexity is a compelling question, given these animals' long lives, large brains, and fission-fusion dynamics. Most toothed whale species appear to live primarily in groups (Groot et al. 2023), with the exception of those in the genus *Kogia* (Dunphy-Daly et al. 2008) and possibly Commerson's dolphin (Coscarella et al. 2010). However, while social complexity requires group living, group living does not necessarily result in social complexity. Furthermore, the concept of social complexity is difficult to operationalize (Kappeler et al. 2019; Weiss et al., this volume). Generally, complexity is thought to scale with the presence of differentiated social relationships or social bonding, also known as "relational complexity." Group size and social bonding have been proposed to have a curvilinear relationship (Schweitzer et al. 2011), with group living therefore a necessary but not sufficient prerequisite to complexity. Under a relational complexity framework (as defined by Lukas and Clutton-Brock 2018; see also Weiss et al., this volume), mammalian societies are typically deemed complex if the variation in social relationships influences access to resources, including food and reproductive opportunities. In contrast, organizational complexity is a type of social complexity focused on division of labor and defined social roles (e.g., in eusocial insects). Differentiated relationships are common to almost all well-studied toothed whales, as well as some populations of humpback whales (Ramp et al. 2010), but whether and how they influence resource access is largely unknown. Here we review the evidence for female social bonds, which we consider as persistent individually specific affiliative relationships, through examining social differentiation (Bergman and Beehner 2015), modularity, and relationship duration; we then consider plausible scenarios under which social bonds might influence resource access and fitness.

Social differentiation can be measured simply as the coefficient of variation among rates of association, *S* (Whitehead 2008), though more complex metrics including those based on mixture models can be used (see Weiss et al. 2019; Weiss et al., this volume). Modularity, Q, represents the proportion of associations that are concentrated within rather than between social units providing a measure of clustering (Girvan and Newman 2002). Differentiation and modularity are typically positively correlated with one another across empirical networks (Weiss et al. 2021) though they measure different network features. Because sex-specific data on cetaceans are rare, the metrics presented here include both sexes (see, e.g., fig. 5.2 below).

Clustering patterns in cetaceans range from tight lifetime social relationships between kin in matrilineal sperm whales (Sarano et al. 2021) and killer whales (Parsons et al. 2009; Parsons et al., this volume) to looser networks characterized by high fission-fusion dynamics in smaller delphinids (Möller 2012). It has been proposed that cetacean societies characterized by high modularity among females always center around lineages of close female relatives (Rendell et al. 2019). For these matrilineal species, individual matrilines are socially limited by group size, experiencing fissions when group size presumably becomes too large for efficient foraging, helping avoid competition with kin for resources (Baird and Dill 1996). These matrilineal groups then continue to associate at lower levels post-fission, creating a hierarchical, multilevel social structure (Parsons et al. 2009; Konrad et al. 2018). Significant post-reproductive lifespans in some species of matrilineal odontocetes effectively divide adult female group members into breeders and nonbreeders, with older females taking on unique leadership roles in foraging as repositories of ecological knowledge (Brent et al. 2015), and these species may approach the definition of organizational social complexity typically reserved for eusocial insects (McAuliffe and Whitehead 2005; Lukas and Clutton-Brock 2018). However, not all matrilineal species have post-reproductive lifespans (Ellis et al. 2017; Ellis et al., this volume). In addition to group-level structure, long-term differentiated bonds have been well-documented for these matrilineal species (sperm whales, *Physeter macrocepha-*

lus [Gero et al. 2015], and resident killer whales, *Orcinus orca* [Baird 2000]) even when individuals spend most of their lifetime in close spatial proximity, providing evidence for both high organizational and high relational complexity in these species.

For the less modular species that live in open, contiguous, fission-fusion communities with low organizational complexity, relational social complexity should be evidenced through competitive interactions, linear dominance hierarchies, and cooperation and alliance formation. Some of these aspects are well-documented for males (e.g., Ermak et al. 2017), but evidence for females is scant. Dominance hierarchies among wild female cetaceans have not been demonstrated, and aggressive interactions between females are rarely observed, even among intensively studied populations (Scott et al. 2005). Female cooperation has been observed, and likely drives social tolerance and affiliation, while female competition likely results in social avoidance behaviors rather than direct conflict (Strickland et al. 2017; Levengood et al. 2022).

Within high fission-fusion communities, total network size corresponds to variation in modularity, with small networks/populations generally having relatively less rigid community structure (e.g., Lusseau et al. 2003) and larger networks exhibiting higher modularity (Weiss et al. 2021). This is likely a function of larger populations exhibiting a more fractal structure (Hill et al. 2008), in which larger populations are made up of smaller communities separated by distance, while small populations effectively form a single community. Notably, even within large populations low modularity can be a consequence of biased subsampling of only the most frequently sighted social groups (e.g., Guiana dolphins; Beirão-Campos et al. 2016), and can vary substantially, depending on whether one or both sexes are included. Within a population, sex-specific selection pressures and sex segregation can create very different structures between males and females, with males for example exhibiting higher modularity due to alliance formation (e.g., Risso's dolphins; Hartman et al. 2008), and female subcommunities organized around less modular extended kinship-based structures (e.g., bottlenose dolphins; Zanardo et al. 2016).

For bottlenose dolphins, female network structure is often referred to as "looser" than that of males (Smolker et al. 1992), in the sense that females typically lack the consistent, structured levels of relationships comprising male mating alliances (Connor and Krützen 2015). However, females do retain variably sized networks of preferred partners (Frère et al. 2010a; Wiszniewski et al. 2010), and these relationships can persist over decades (Miketa 2018), often ending only with the death of a partner. Intriguingly, years-long relationships between individual females have also been documented among species with such disparate ecologies as deep-diving northern bottlenose whales (O'Brien 2013) and Blainville's beaked whales (Badenas et al. 2022). Likewise, though typically found offshore in pelagic habitats, common dolphins in southeastern Australia form measurably strong bonds (Mason et al. 2021) while displaying high site fidelity to a shallow embayment, similar to other coastal delphinids.

Proximate Mechanisms Underlying Female Social Bonding

For the majority of species in which females form differentiated bonds, kinship plays a strong role in their development, especially among delphinids (Möller 2012). Matrilineal kinship dominates the larger-bodied toothed whale societies and plays a considerable role in most midsized cetaceans as well. Even humpback whales, though they lack strong social associations that are biased toward kin, still have nonrandom associations within matrilineal lineages based on mtDNA haplotype data (Weinrich et al. 2006), possibly due to strong maternal fidelity to specific feeding grounds. When females primarily bond socially with their kin, it is assumed they are reaping inclusive fitness benefits by, for example, providing care or protection to the offspring of their female relatives.

Lifelong mother-daughter social bonds are regularly present in at least six species; killer whales, sperm whales, both long- and short-finned pilot whales, false killer whales and pygmy killer whales (Baird et al. 2008; McSweeney et al. 2009). Indo-Pacific bottlenose dolphin mother-daughter pairs also retain lifelong social relationships, though with some variation among dyads (Tsai and Mann 2013; Miketa 2018). At least occasional lifelong social bonds may be present in many delphinid species, subject only to availability of data on known individuals across the lifespan (Mann and Karniski 2017). Interestingly, for Indo-Pacific bottlenose dolphins in Shark Bay, Australia, adult maternal sisters were slightly more likely to be socially bonded than adult mother-daughter dyads, perhaps because of the added benefit of age similarity (Foroughirad et al. 2023).

Social affiliation between females can be based on both paternal and matrilineal relatedness (Frère et al. 2010a; Foroughirad et al. 2023), but mechanisms for kin recognition in cetaceans remain largely unknown. Acoustic communication is thought to be a primary route in individual dolphin identification through signature whistles, but unlike other species such as rhesus macaques (Rendall et al. 1996), dolphins don't show attention to individually specific vocal cues (Sayigh et al. 2017) or use similar whistles among kin (Janik et al. 2006). However, in other species, acoustic cues can constitute

FIGURE 5.1. A female Indo-Pacific bottlenose dolphin sponge forages with her three-year-old calf in infant position.
Source: Photo by Ewa Krzyszczyk, Shark Bay Dolphin Project.

unique dialects that can be used to differentiate matrilineal groups (Filatova et al. 2015), even if they don't correspond to the degree of relatedness, which can reinforce modular social bonds among kin.

Aside from kinship, homophily in other dimensions can also predict social bond formation between females. For Indo-Pacific bottlenose dolphins, male-male associations tend to be structured by cohort, while female-female associations are impervious to age differences and instead are more shaped by relatedness (Foroughirad et al. 2019). In humpback whales, age-related female associations have been documented, but their function is unclear (Ramp et al. 2010). Shared reproductive status also influences female social relationships (Möller and Harcourt 2008; Frère et al. 2010b; Nishita et al. 2017; Diaz-Aguirre et al. 2020), possibly due to the benefit of shared predator vigilance, deterrence of male harassment for females with young calves, or social opportunities afforded to calves.

Shared foraging ecology can similarly predict social bonds. Female dolphins in Shark Bay that use marine sponges as tools are more cliquish and have stronger bonds with each other than with dolphins that do not use tools, even when controlling for relatedness and home range overlap (Mann et al. 2012; fig. 5.1). Distinct vocalizations and social behaviors are also present in female social groups in species such as killer whales (Filatova et al. 2015) and sperm whales (Cantor and Whitehead 2015), but these shared behaviors are likely a consequence of maternally inherited behavioral traits that are reinforced within kin groups rather than a driver of affiliation.

Evolutionary Drivers of Female Social Bonding

Proposed evolutionary drivers for variation in social structure within and between cetacean species can be loosely organized into three interrelated sets of selection pressures, each corresponding to different fitness benefits that female-female social relationships may provide. These include the degree of investment in offspring (through maternal or allomaternal care and predator defense), the distribution and accessibility of foraging resources, and the threat of male harassment.

EXTENDED MATERNAL CARE

The strength and duration of the mother-calf relationship has been proposed as a driving force leading to more generalized female social bonding (Bräger 1999), with mother-daughter (Tsai and Mann 2013) and mother-son (Nielsen et al. 2023) social relationships lasting well beyond the nursing period. Therefore, we might expect the minimum duration of the mother-calf bond (the nursing period), to predict the extent of social bonding or social complexity present in the population. With regard to lactation length, studies using long-term data to observe cessation of nursing of individually marked surviving juveniles (Karniski et al. 2018), as well as those using new stable isotope technology to investigate nutritional transitions (e.g., Newsome et al. 2009; Evacitas et al. 2017; Feyrer et al. 2020), have revealed lactation periods much longer than those reported in earlier literature based on strandings or stomach content analysis (Mann and Karniski 2017). Indeed, most toothed whales have lactation periods that last longer than a single year. Multiyear lactation periods are a phenomenon only exhibited by a small minority of other mammalian species (e.g., elephants; Lee and Moss 1986; walrus; Schulz and Bowen 2005; great apes; van Noordwijk et al. 2013). Intriguing evidence from orangutans suggests that milk intake may fluctuate seasonally among these multiyear periods, varying to compensate for reduced foraging access during times of low fruit availability (Smith et al. 2017). Extended lactation periods in odontocetes (4–8 years; Newsome et al. 2009; Whitehead and Gero 2015; Karniski et al. 2018), relative to early ability of calves to catch their own fish (by 4–6 months; Sargeant and Mann 2009) may thus serve a buffering purpose during times of food scarcity. Under this assumption, wider-

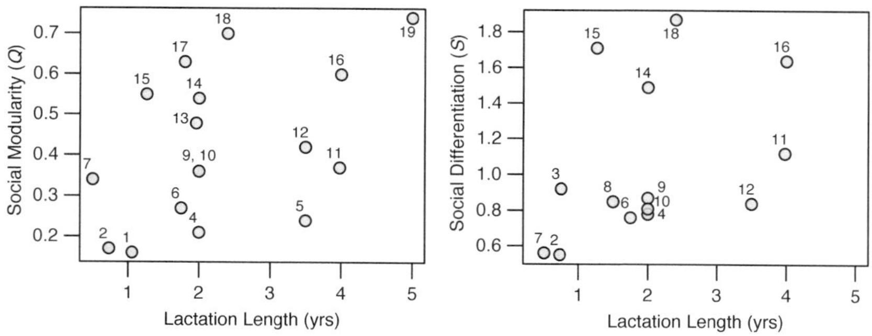

FIGURE 5.2. Visualization of the relationships between social modularity and social differentiation and length of lactation for odontocete species for which information was available. (1) *Delphinus delphis*, (2) *Sotalia guianensis*, (3) *Cephalorhynchus commersonii*, (4) *Stenella frontalis*, (5) *Hyperoodon ampullatus*, (6) *Tursiops truncatus*, (7) *Berardius bairdii*, (8) *Lagenorhynchus obscurus*, (9) *Orcaella heinsohni*, (10) *Sousa plumbea*, (11) *Tursiops aduncus*, (12) *Sousa chinensis*, (13) *Grampus griseus*, (14) *Sousa sahulensis*, (15) *Globicephala melas*, (16) *Orcinus orca*, (17) *Pseudorca crassidens*, (18) *Globicephala macrorhynchus*, (19) *Physeter macrocephalus*.

Sources: Social data obtained from Weiss et al. (2021), Mason et al. (2021), and Santacesaria et al. (2022). Lactation-length data obtained from Subramanian et al. (1988), Herzing (1997), Würsig et al. (2007), Newsome et al. (2009), Möller (2012), Riccialdelli et al. (2013), Braulik et al. (2015), Whitehead and Gero (2015), Chang et al. (2016), Evacitas et al. (2017), Smith (2018), Karniski et al. (2018), Feyrer et al. (2020), Allen (2021), and Palmer et al. (2022).

ranging or seasonally migrating species for which resource availability is inconsistent should have longer lactation durations, and ultimately, enhanced female social bonding. However, species with long lactation durations have a wide variety of movement patterns, suggesting instead that species may exhibit either increased lactation duration (some toothed whales) or migration (baleen whales) as alternative strategies to overcome limited foraging access during unfavorable seasons. As improved lactation-length data are available for a substantial number of species, we can compare their relationship to published metrics of social differentiation and modularity (fig. 5.2).

ALLOMATERNAL CARE

Allomaternal care, defined as care of offspring by a non-mother that provides benefits for the mother and calf, has been proposed as a driver of female social bonds (Rosenbaum and Gettler 2018). However, the link between allomaternal behavior and fitness benefits to mother or calf is difficult to demonstrate. The best-documented case of allomaternal care occurs in sperm whales, where calves remain at or near the surface with social associates while mothers engage in deep foraging dives (Whitehead 1996; Konrad et al. 2019; see Parsons et al., this volume). Allonursing, or calf suckling from a non-maternal individual, has also been documented in sperm whales, although the lactation status of the allonurses was not known (Konrad et al. 2019). While these behaviors seem integral to sperm whale social lives, definitive fitness benefits of allocare to the mother or calf have not been demonstrated.

Allonursing has occasionally been observed in other species in captivity (e.g., belugas, Leung et al. 2010; common bottlenose dolphins, Dudzinski et al. 2022), but would be hard to observe in the wild. In baleen whales, some cases of allonursing are clearly milk theft rather than allomaternal care (Best et al. 2015). For long-finned pilot whales, close association between calves and nonmaternal adults has also been advanced as allomaternal care, though whether the presence of the companion benefits the calf or reduces the energetic output of the mother is unknown (Augusto et al. 2017). In these cases, the primary benefit to the carer is thought to be through inclusive fitness. In Indo-Pacific bottlenose dolphins, calves also sometimes swim in infant position with unrelated juvenile females, though allonursing from a lactating female has never been observed. This behavior does not appear to benefit the mother, as mothers maintain the same foraging activity budgets whether their calves are with them or temporarily engaged with another associate (Mann and Smuts 1998). Newborn calves appear to be especially attractive to juvenile or reproductively inexperienced females who frequently join mother-calf groups and occasionally "steal" or induce neonates into swimming alongside them and away from the mother. In these cases the "learning-to-parent hypothesis," which suggests that inexperienced females benefit by associating with young calves seems the most likely explanation for conspecific interest in calves (Mann and Smuts 1998).

PREDATOR DEFENSE

Many cetacean calves are subject to predation risk to the point that predation was once considered to be the primary force shaping group living in cetaceans (Norris and Dohl 1980; fig. 5.3). Pack-hunting killer whales are the primary predator for baleen species, sperm whales, beaked whales, and some smaller odontocetes. Large sharks prey on delphinids in many coastal populations (Mann and Watson-Capps 2005), and belugas and narwhals are also subject to predation by polar bears in the Arctic (Weller 2018). Most predators target young animals, which may also put mothers at risk if they intervene to protect their calves (Mann and Barnett 1999). In Indo-Pacific bottlenose dolphins, females with young calves have a significantly elevated risk of tiger shark injury compared to females with older calves or no calves (Knotek

FIGURE 5.3. A neonate short-finned pilot whale with obvious injuries possibly attributable to a predation attempt swims alongside its mother.
Source: Photo courtesy of University of North Carolina Wilmington/Ryan J. McAlarney. NMFS Permit #16185.

et al. unpublished data), and adult females and mother-calf pairs are less likely to use risky habitats despite higher food availability (Heithaus and Dill 2002).

Group living can provide protection from predation through increased vigilance and detection probability, and the benefit received from grouping can be a driver of social complexity as well if, for example, coordinated predator defense tactics are needed (Groenewoud et al. 2016). Large group sizes are especially characteristic of bottlenose dolphin mothers with newborn infants (Mann et al. 2000; Dwyer et al. 2014), which may be an important antipredator strategy.

Group defense of calves has been described in several species. Sperm whales have been observed forming defensive formations around calves (e.g., marguerite position, Pitman et al. 2001). The best evidence for social cooperation in defense may come from humpback whales, which have been documented mobbing killer whale groups, and responding to attacks by killer whales by grouping tightly together and mounting a vigorous physical defense, including slashing tail flukes, rolling and thrashing, and head shaking (Ford and Reeves 2008; Pitman et al. 2017). Southern right whales have also been observed forming a "rosette" around vulnerable calves, and this has been observed once for northern right whales as well (Ford and Reeves 2008). There is a notable distinction in responses between the faster non-humpback rorquals, which typically engage in flight strategies, and the slower but more robust baleen species, which may be more likely to fight and therefore benefit from coordinated calf defense (Ford and Reeves 2008; see Dunlop and Parks, this volume). Indo-Pacific bottlenose dolphin females have been observed jointly chasing, mobbing, and attacking tiger sharks (Mann and Barnett 1999; Mann and Watson-Capps 2005).

COOPERATIVE FORAGING

Enhanced resource acquisition may also be a driving force of bonding with individuals that adopt similar feeding behaviors (Cantor and Farine 2018) or divide labor during apparent complex cooperative tactics or exchanges (Similä and Ugarte 1993; Fox and Young 2012; Samuni et al. 2018). In baleen whales, the best evidence for female social bonds is correlated with the unique cooperative foraging strategy of bubble-net feeding (Wray et al. 2021; Mastick et al. 2022). Unlike most baleen whales, individuals that engage in bubble-net feeding show stable associations (Weinrich et al. 1991). Recent data on blue whales also suggests a role for social communication in exploiting large prey patches (Cade et al. 2021). While most bottlenose dolphins do not forage cooperatively, a subset of bottlenose dolphins in Florida have been observed engaging in the cooperative foraging strategy of driver-barrier feeding, although the sex of participants is not known (Gazda 2016). Social bonds may also facilitate access to food resources through providing ecological knowledge (Brent et al. 2015), or through prey sharing, such as through kin-directed food sharing (Wright et al. 2016). For killer whales, there is good evidence that these social associations produce tangible health and fitness benefits (Ellis et al. 2018; Nielsen et al. 2023). In terrestrial mammals, female social bonds and cooperation function to exclude other individuals from discrete food patches (Wrangham 1980). However, with few defensible food resources available to cetaceans, female coalitions against other females appear to be absent.

In many pelagic delphinids that forage in large groups, coordinated foraging on schooling fish need not entail the formation of differentiated social bonds among females, but can instead be a form of by-product mutualism (see King et al., this volume). That is, it does not matter whether the individuals recognize each other or have a unique relationship in order to benefit from association. For example, spinner dolphins engage in coordinated foraging in groups of 16–28 individuals, which significantly increases prey density relative to solitary foraging (Benoit-Bird and Au 2009). Common dolphins (Ruiz-Cooley et al. 2021) and harbor porpoises have also been observed cooperatively foraging in small groups, with evidence for task specialization in harbor porpoise foraging (Torres Ortiz et al. 2021). Thus, cooperative foraging occurs even though there is no evidence for lasting social bonds in these species. This incentive for association differs from other social species like bottlenose dolphins, where large groups are more likely to be resting or socializing, and foraging is a solitary affair (Galezo et al. 2018).

MALE HARASSMENT

A further proposed role for the development of female social bonding is to provide support and defense against male harassment, sexual coercion, and infanticide (Gowans et al. 2007; Möller 2012; Palombit 2015). Males in many large oceanic delphinids display sexual dimorphism in size over females (Dines et al. 2015), while in monomorphic species males may instead form alliances to consort and mate guard estrous females (Connor et al. 1996). These alliances engage in sexual coercion based on evidence of aggression in the form of tooth rakes on cycling females (Scott et al. 2005), as well as observations of direct aggression (Connor et al. 1996), and disruption of normal female ranging and behavior (Wallen et al. 2016). As a result, female avoidance of adult males (Galezo et al. 2018) and sexual social segregation have been observed in a variety of cetacean species.

Female social bonds may buffer against male harassment through providing coalitionary support against aggression and through stress reduction (Wittig et al. 2008; Young et al. 2014; Wittig et al. 2016). One piece of evidence for the role of female social bonding as a mechanism to cope with male harassment comes from bottlenose dolphins in Shark Bay, Australia, which display a sex-specific affiliative contact behavior known as "bonding" that occurs most often between pairs of females in male-biased groups (Connor et al. 2006). Bonding involves pectoral fin to body contact, presumed to be affiliative in other populations as well (Dudzinski and Ribic 2017; fig. 5.4).

Female social bonds might also deter infanticide. In several descriptions of attacks, mothers were assisted in calf defense by female associates or female kin (Robinson 2014; Towers et al. 2018; da Silva et al. 2021). Groups of females with calves might deter infanticidal attacks or inflict injury on the perpetrators (McEntee et al. 2023).

FIGURE 5.4. Two juvenile Indo-Pacific bottlenose dolphins engage in a bout of petting, an affiliative behavior analogous to grooming where they use the pectoral fins or flukes to stroke another individual's body. *Source*: Photo by Ewa Krzyszczyk, Shark Bay Dolphin Project.

Conclusion

Long lifespans, prey dynamics, and prolonged calf care interact to drive the diversity of long-term social bonds between female kin and nonkin. Generally speaking, indicators of social bonding, especially among female kin (Briga et al. 2012), tend to vary positively with degree of maternal investment as defined by lactation length (fig. 5.2). This mirrors the evolutionary explanations for male-female pair bonding as a consequence of the benefits of offspring care (Mock and Fujioka 1990). The presence of helper females socially bonded to the mother should increase her reproductive output (Heldstab et al. 2019) and benefit related helpers via inclusive fitness, though this has yet to be demonstrated for cetaceans. When females form more temporary groups or groups with unrelated females, they tend to be for the purpose of coordinated foraging.

Nevertheless, females, their offspring, and their social relationships are the foundation of cetacean societies. Strong female-biased kin bonding has a suite of sociogenetic consequences, the effects of which are still being disentangled. For example, close matrilineal bonds and maternally inherited cultural differences in diet and foraging strategies may lead to severely reduced mitochondrial diversity through cultural hitchhiking (Whitehead 1998; Whitehead et al. 2017). However, other species with mixed kin and nonkin bonding maintain mitochondrial diversity alongside cultural transmission (Kopps et al. 2014; Vachon et al. 2018). Understanding the relationship between female social relationships and calf survival is pivotal to our understanding of population dynamics (Wade et al. 2012; Manlik et al. 2016), especially resilience and recovery in face of anthropogenic impacts.

There is widespread evidence for highly differentiated long-term social bonds between females in numerous cetacean species. However, their fitness benefits, if any, have yet to be determined, though recently benefits of social relationships between sexes or between animals of unknown sex have been demonstrated (Busson et al. 2019). In other species, female social bonds predict longevity or offspring survival, for example in baboons (Silk et al. 2010), house mice (Weidt et al. 2008), and feral horses (Cameron et al. 2009). In bottlenose dolphins, intriguing evidence shows that reproductively successful females preferentially associate (Frère et al. 2010b). However, there has not been evidence of fitness costs to females that choose to remain largely solitary, as observed in most baleen whales, or even specific individuals within otherwise highly social populations (Mann et al. 2008). Evidence of fitness benefits of social bonding in males has been well established (Stanton and Mann 2012; Gerber et al. 2022; Rankin et al. 2022), and continued research effort on the more variable and diffuse social bonds of females is critical to understanding cetacean social lives.

References

Allen, Jenny A., Mason T. Weinrich, William Hoppitt, and Luke Rendell. 2013. "Network-Based Diffusion Analysis Reveals Cultural Transmission of Lobtail Feeding in Humpback Whales." *Science* 340 (6131): 485–88. https://doi.org/10.1126/science.1231976.

Alves, Filipe, Sophie Quérouil, Ana Dinis, et al. 2013. "Population Structure of Short-Finned Pilot Whales in the Oceanic Archipelago of Madeira Based on Photo-Identification and Genetic Analyses: Implications for Conservation." *Aquatic Conservation: Marine and Freshwater Ecosystems* 23 (5): 758–76. https://doi.org/10.1002/aqc.2332.

Augusto, Joana, Timothy R. Frasier, and Hal Whitehead. 2017. "Characterizing Alloparental Care in the Pilot Whale (*Globicephala melas*) Population That Summers off Cape Breton, Nova Scotia, Canada." *Marine Mammal Science* 33 (2): 440–56. https://doi.org/10.1111/mms.12377.

Badenas, Anja, Ana Dinis, Rita Ferreira, et al. 2022. "Behavioural Ecology Traits of Elusive Deep-Diver Whales Unravel a Complex Social Structure Influenced by Female Philopatry and Defence Polygyny." *Frontiers in Marine Science* 9: 809902. https://www.frontiersin.org/articles/10.3389/fmars.2022.809902.

Baird, Robin W. 2000. "The Killer Whale." In *Cetacean Societies: Field Studies of Dolphins and Whales*, edited by Janet Mann, Richard C. Connor, Peter L. Tyack, and Hal Whitehead, 127–53. University of Chicago Press.

Baird, Robin W., and Lawrence M. Dill. 1996. "Ecological and Social Determinants of Group Size in Transient Killer Whales." *Behavioral Ecology* 7 (4): 408–16. https://doi.org/10.1093/beheco/7.4.408.

Baird, Robin W., Antoinette M. Gorgone, Daniel J. McSweeney, et al. 2008. "False Killer Whales (*Pseudorca crassidens*) Around the Main Hawaiian Islands: Long-Term Site Fidelity, Inter-Island Movements, and Association Patterns." *Marine Mammal Science* 24 (3): 591–612. https://doi.org/10.1111/j.1748-7692.2008.00200.x.

Baker, C. Scott, Louis M. Herman, Anjanette Perry, et al. 1986. "Migratory Movement and Population Structure of Humpback Whales (*Megaptera novaeangliae*) in the Central and Eastern North Pacific." *Marine Ecology Progress Series* 31 (2): 105–19. https://www.jstor.org/stable/24817567.

Barendse, Jaco, Peter B. Best, Inês Carvalho, and Cristina Pomilla. 2013. "Mother Knows Best: Occurrence and Associations of Resighted Humpback Whales Suggest Maternally Derived Fidelity to a Southern Hemisphere Coastal Feeding Ground." *PLOS ONE* 8 (12): e81238. https://doi.org/10.1371/journal.pone.0081238.

Beirão-Campos, Luiza, Mauricio Cantor, Leonardo Flach, and Paulo C. Simões-Lopes. 2016. "Guiana Dolphins Form Social Modules in a Large Population with High Ranging Overlap and Small Demographic Changes." *Behavioral Ecology and Sociobiology* 70 (11): 1821–30. https://doi.org/10.1007/s00265-016-2188-x.

Benoit-Bird, Kelly J., and Whitlow W. L. Au. 2009. "Cooperative Prey Herding by the Pelagic Dolphin, *Stenella longirostris*." *Journal of the Acoustical Society of America* 125 (1): 125–37. https://doi.org/10.1121/1.2967480.

Bergman, Thore J., and Jacinta C. Beehner. 2015. "Measuring Social Complexity." *Animal Behaviour* 103 (May): 203–9. https://doi.org/10.1016/j.anbehav.2015.02.018.

Best, Peter B., Simon H. Elwen, Per J. Palsbøll, Meredith Thornton, Evan Austin, and Katja Vinding. 2015. "Possible Non-Offspring Nursing in the Southern Right Whale, *Eubalaena australis*." *Journal of Mammalogy* 96 (2): 405–16. https://doi.org/10.1093/jmammal/gyv042.

Bräger, Stefan. 1999. "Association Patterns in Three Populations of Hector's Dolphin, *Cephalorhynchus hectori*." *Canadian Journal of Zoology* 77 (1): 13–18. https://doi.org/10.1139/z98-192.

Brent, Lauren J. N., Daniel W. Franks, Emma A. Foster, Kenneth C. Balcomb, Michael A. Cant, and Darren P. Croft. 2015. "Ecological Knowledge, Leadership, and the Evolution of Menopause in Killer Whales." *Current Biology* 25 (6): 746–50. https://doi.org/10.1016/j.cub.2015.01.037.

Briga, Michael, Ido Pen, and Jonathan Wright. 2012. "Care for Kin: Within-Group Relatedness and Allomaternal Care Are Positively Correlated and Conserved Throughout the Mammalian Phylogeny." *Biology Letters* 8 (4): 533–36. https://doi.org/10.1098/rsbl.2012.0159.

Busson, Marine, Matthieu Authier, Christophe Barbraud, et al. 2019. "Role of Sociality in the Response of Killer Whales to an Additive Mortality Event." *Proceedings of the National Academy of Sciences* 116 (24): 11812–17. https://doi.org/10.1073/pnas.1817174116.

Cade, David E., James A. Fahlbusch, William K. Oestreich, et al. 2021. "Social Exploitation of Extensive, Ephemeral, Environmentally Controlled Prey Patches by Supergroups of Rorqual Whales." *Animal Behaviour* 182 (December): 251–66. https://doi.org/10.1016/j.anbehav.2021.09.013.

Cameron, Elissa Z., Trine H. Setsaas, and Wayne L. Linklater. 2009. "Social Bonds Between Unrelated Females Increase Reproductive Success in Feral Horses." *Proceedings of the National Academy of Sciences* 106 (33): 13850–53.

Cantor, Mauricio, and Damien R. Farine. 2018. "Simple Foraging Rules in Competitive Environments Can Generate Socially Structured Populations." *Ecology and Evolution* 8 (10): 4978–91. https://doi.org/10.1002/ece3.4061.

Cantor, Mauricio, and Hal Whitehead. 2015. "How Does Social Behavior Differ Among Sperm Whale Clans?" *Marine Mammal Science* 31 (4): 1275–90. https://doi.org/10.1111/mms.12218.

Carroll, Emma L., Rachel M. Fewster, Simon J. Childerhouse, Nathalie J. Patenaude, Laura Boren, and C. Scott Baker. 2016. "First Direct Evidence for Natal Wintering Ground Fidelity and Estimate of Juvenile Survival in the New Zealand Southern Right Whale *Eubalaena australis*." *PLOS ONE* 11 (1): e0146590. https://doi.org/10.1371/journal.pone.0146590.

Chang, Wei-Lun, Leszek Karczmarski, Shiang-Lin Huang, Glenn Gailey, and Lien-Siang Chou. 2016. "Reproductive Parameters of the Taiwanese Humpback Dolphin (*Sousa chinensis taiwanensis*)." *Regional Studies in Marine Science* 8 (November): 459–65. https://doi.org/10.1016/j.rsma.2016.08.001.

Clutton-Brock, Tim H., and D. Lukas. 2012. "The Evolution of Social Philopatry and Dispersal in Female Mammals." *Molecular Ecology* 21 (3): 472–92. https://doi.org/10.1111/j.1365-294X.2011.05232.x.

Colbeck, Gabriel J., Pierre Duchesne, Lianne D. Postma, Véronique Lesage, Mike O. Hammill, and Julie Turgeon. 2013. "Groups of Related Belugas (*Delphinapterus leucas*) Travel Together During Their Seasonal Migrations in and Around Hudson Bay." *Proceedings of the Royal Society B* 280 (1752): 20122552. https://doi.org/10.1098/rspb.2012.2552.

Connor, Richard C., and Michael Krützen. 2015. "Male Dolphin Alliances in Shark Bay: Changing Perspectives in a 30-Year Study." *Animal Behaviour* 103 (May): 223–35. https://doi.org/10.1016/j.anbehav.2015.02.019.

Connor, Richard, Janet Mann, and Jana Watson-Capps. 2006. "A Sex-Specific Affiliative Contact Behavior in Indian Ocean Bottlenose Dolphins, *Tursiops* sp." *Ethology* 112 (7): 631–38. https://doi.org/10.1111/j.1439-0310.2006.01203.x.

Connor, Richard C., Andrew F. Richards, Rachel A. Smolker, and Janet Mann. 1996. "Patterns of Female Attractiveness in Indian Ocean Bottlenose Dolphins." *Behaviour* 133 (1–2): 37–69. https://doi.org/10.1163/156853996X00026.

Corkeron, Peter J., and Richard C. Connor. 1999. "Why Do Baleen Whales Migrate?" *Marine Mammal Science* 15 (4): 1228–45. https://doi.org/10.1111/j.1748-7692.1999.tb00887.x.

Coscarella, Mariano A., Susana N. Pedraza, and Enrique A. Crespo. 2010. "Behavior and Seasonal Variation in the Relative Abundance of Commerson's Dolphin (*Cephalorhynchus commersonii*) in Northern Patagonia, Argentina." *Journal of Ethology* 28 (3): 463–70. https://doi.org/10.1007/s10164-010-0206-4.

Da Silva, Vera M. F., Priscila M. Silva, Flávia Schlichta, et al. 2021. "Aggression Towards Neonates and Possible Infanticide in the Boto, or Amazon River Dolphin (*Inia geoffrensis*)." *Behaviour* 158 (11): 971–84. https://doi.org/10.1163/1568539X-bja10103.

Diaz-Aguirre, Fernando, Guido J. Parra, Cecilia Passadore, and Luciana Möller. 2020. "Kinship and Reproductive Condition Correlate with Affiliation Patterns in Female Southern Australian Bottlenose Dolphins." *Scientific Reports* 10 (1): 1–12. https://doi.org/10.1038/s41598-020-58800-2.

Dines, James P., Sarah L. Mesnick, Katherine Ralls, Laura May-Collado, Ingi Agnarsson, and Matthew D. Dean. 2015. "A Trade-Off Between Precopulatory and Postcopulatory Trait Investment in Male Cetaceans." *Evolution* 69 (6): 1560–72. https://doi.org/10.1111/evo.12676.

Dudzinski, Kathleen M., and Christine A. Ribic. 2017. "Pectoral Fin Contact as a Mechanism for Social Bonding among Dolphins." *Animal Behavior and Cognition* 4 (1): 30–48. https://doi.org/10.12966/abc.03.02.2017.

Dudzinski, Kathleen M., Manon Themelin, Heather M. Manitzas Hill, and Teresa T. Bolton.

2022. "Allomaternal Care and Allonursing Behaviors by a Primiparous Bottlenose Dolphin." *Aquatic Mammals* 48 (6): 536–40. https://doi.org/10.1578/AM.48.6.2022.536.

Dunphy-Daly, Meagan M., Michael R. Heithaus, and Diane E. Claridge. 2008. "Temporal Variation in Dwarf Sperm Whale (*Kogia sima*) Habitat Use and Group Size off Great Abaco Island, Bahamas." *Marine Mammal Science* 24 (1): 171–82. https://doi.org/10.1111/j.1748-7692.2007.00183.x.

Dwyer, Sarah L., Gabriela Tezanos-Pinto, Ingrid N. Visser, Matthew D. M. Pawley, Anna M. Meissner, Jo Berghan, and Karen A. Stockin. 2014. "Overlooking a Potential Hotspot at Great Barrier Island for the Nationally Endangered Bottlenose Dolphin of New Zealand." *Endangered Species Research* 25 (2): 97–114. https://doi.org/10.3354/esr00613.

Ellis, Samuel, Daniel W. Franks, Stuart Nattrass, et al. 2017. "Mortality Risk and Social Network Position in Resident Killer Whales: Sex Differences and the Importance of Resource Abundance." *Proceedings of the Royal Society B* 284 (1865): 20171313. https://doi.org/10.1098/rspb.2017.1313.

Ellis, Samuel, Daniel W. Franks, Stuart Nattrass, et al. 2018. "Analyses of Ovarian Activity Reveal Repeated Evolution of Post-Reproductive Lifespans in Toothed Whales." *Scientific Reports* 8 (1): 12833. https://doi.org/10.1038/s41598-018-31047-8.

Elwen, Simon H., and Peter B. Best. 2004. "Female Southern Right Whales *Eubalaena australis*: Are There Reproductive Benefits Associated with Their Coastal Distribution off South Africa?" *Marine Ecology Progress Series* 269: 289–95. https://doi.org/10.3354/meps269289.

Engelhaupt, Daniel, A. Rus Hoelzel, Colin Nicholson, et al. 2009. "Female Philopatry in Coastal Basins and Male Dispersion Across the North Atlantic in a Highly Mobile Marine Species, the Sperm Whale (*Physeter macrocephalus*)." *Molecular Ecology* 18 (20): 4193–205. https://doi.org/10.1111/j.1365-294X.2009.04355.x.

Ermak, Jessica, Kristin Brightwell, and Quincy Gibson. 2017. "Multi-Level Dolphin Alliances in Northeastern Florida Offer Comparative Insight into Pressures Shaping Alliance Formation." *Journal of Mammalogy* 98 (4): 1096–104. https://doi.org/10.1093/jmammal/gyx053.

Evacitas, Florence Chan, Wen-Yuan Kao, Graham A. J. Worthy, and Lien-Siang Chou. 2017. "Annual Variability in Dentin δ15N and δ13C Reveal Sex Differences in Weaning Age and Feeding Habits in Risso's Dolphins (*Grampus griseus*)." *Marine Mammal Science* 33 (3): 748–70. https://doi.org/10.1111/mms.12396.

Feyrer, Laura Joan, Shu ting Zhao, Hal Whitehead, and Cory J. D. Matthews. 2020. "Prolonged Maternal Investment in Northern Bottlenose Whales Alters Our Understanding of Beaked Whale Reproductive Life History." *PLOS ONE* 15 (6): e0235114. https://doi.org/10.1371/journal.pone.0235114.

Filatova, Olga A., Filipa I. P. Samarra, Volker B. Deecke, John K. B. Ford, Patrick J. O. Miller, and Harald Yurk. 2015. "Cultural Evolution of Killer Whale Calls: Background, Mechanisms and Consequences." *Behaviour* 152 (15): 2001–38. https://doi.org/10.1163/1568539X-00003317.

Foley, Heather J. 2018. "Spatial Ecology and Movement Patterns of Deep-Diving Odontocetes in the Western North Atlantic." PhD diss., North Carolina State University. https://www.proquest.com/openview/1c10a053edc25b44213db28b54f3b571/1?pq-origsite=gscholar&cbl=18750&diss=y.

Ford, John K. B., and Randall R. Reeves. 2008. "Fight or Flight: Antipredator Strategies of Baleen Whales." *Mammal Review* 38 (1): 50–86. https://doi.org/10.1111/j.1365-2907.2008.00118.x.

Foroughirad, Vivienne, Céline H. Frère, Alexis L. Levengood, Anna M. Kopps, Ewa Krzyszczyk, and Janet Mann. 2023. "Small Effects of Family Size on Sociality despite Strong Kin Preferences in Female Bottlenose Dolphins." *Animal Behaviour* 195 (January): 53–66. https://doi.org/10.1016/j.anbehav.2022.10.011.

Foroughirad, Vivienne, Alexis L. Levengood, Janet Mann, and Céline H. Frère. 2019. "Quality and Quantity of Genetic Relatedness Data Affect the Analysis of Social Structure." *Molecular Ecology Resources* 19 (5): 1181–94. https://doi.org/10.1111/1755-0998.13028.

Fox, Adam G., and Robert F. Young. 2012. "Foraging Interactions Between Wading Birds and Strand-Feeding Bottlenose Dolphins (*Tursiops truncatus*) in a Coastal Salt Marsh." *Canadian Journal of Zoology* 90 (6): 744–52. https://doi.org/10.1139/Z2012-043.

Frère, Céline H., M. Krützen, J. Mann, et al. 2010a. "Home Range Overlap, Matrilineal and Biparental Kinship Drive Female Associations in Bottlenose Dolphins." *Animal Behaviour* 80 (3): 481–86. https://doi.org/10.1016/j.anbehav.2010.06.007.

Frère, Céline H., Michael Krützen, Janet Mann, Richard C. Connor, Lars Bejder, and William B. Sherwin. 2010b. "Social and Genetic Interactions Drive Fitness Variation in a Free-Living Dolphin Population." *Proceedings of the National Academy of Sciences* 107 (46): 19949–54. https://doi.org/10.1073/pnas.1007997107.

Galezo, A. A., E. Krzyszczyk, and J. Mann. 2018. "Sexual Segregation in Indo-Pacific Bottlenose Dolphins Is Driven by Female Avoidance of Males." *Behavioral Ecology* 29 (2): 377–86. https://doi.org/10.1093/beheco/arx177.

Garland, Ellen C., and Emma L. Carroll. 2022. "Culture and Social Learning in Baleen Whales." In *Ethology and Behavioral Ecology of Mysticetes*, edited by Christopher W. Clark and Ellen C. Garland, 177–91. Springer.

Gazda, Stefanie K. 2016. "Driver-Barrier Feeding Behavior in Bottlenose Dolphins (*Tursiops truncatus*): New Insights from a Longitudinal Study." *Marine Mammal Science* 32 (3): 1152–60. https://doi.org/10.1111/mms.12314.

Gerber, Livia, Richard C. Connor, Simon J. Allen, et al. 2022. "Social Integration Influences Fitness in Allied Male Dolphins." *Current Biology* 32 (7): 1664–69. https://doi.org/10.1016/j.cub.2022.03.027.

Gero, Shane, Jonathan Gordon, and Hal Whitehead. 2015. "Individualized Social Preferences and Long-Term Social Fidelity Between Social Units of Sperm Whales." *Animal Behaviour* 102 (April): 15–23. https://doi.org/10.1016/j.anbehav.2015.01.008.

Girvan, M., and M. E. J. Newman. 2002. "Community Structure in Social and Biological Networks." *Proceedings of the National Academy of Sciences* 99 (12): 7821–26. https://doi.org/10.1073/pnas.122653799.

Gowans, Shannon, Bernd Würsig, and Leszek Karczmarski. 2007. "The Social Structure and Strategies of Delphinids: Predictions Based on an Ecological Framework." *Advances in Marine Biology* 53: 195–294. https://doi.org/10.1016/S0065-2881(07)53003-8.

Groenewoud, Frank, Joachim Gerhard Frommen, Dario Josi, Hirokazu Tanaka, Arne Jungwirth, and Michael Taborsky. 2016. "Predation Risk Drives Social Complexity in Cooperative Breeders." *Proceedings of the National Academy of Sciences* 113 (15): 4104–9. https://doi.org/10.1073/pnas.1524178113.

Groot, Nikita E., Rochelle Constantine, Ellen C. Garland, and Emma L. Carroll. 2023. "Phylogenetically Controlled Life History Trait Meta-Analysis in Cetaceans Reveals Unexpected Negative Brain Size and Longevity Correlation." *Evolution* 77 (2): 534–49. https://doi.org/10.1093/evolut/qpac050.

Hartman, Karin L., Fleur Visser, and A. J. E. Hendriks. 2008. "Social Structure of Risso's Dolphins (*Grampus griseus*) at the Azores: A Stratified Community Based on Highly Associated Social Units." *Canadian Journal of Zoology* 86 (4): 294–306. https://doi.org/10.1139/Z07-138.

Heithaus, Michael R., and Lawrence M. Dill. 2002. "Food Availability and Tiger Shark Predation Risk Influence Bottlenose Dolphin Habitat Use." *Ecology* 83 (2): 480–91. https://doi.org/10.1890/0012-9658(2002)083[0480:FAATSP]2.0.CO;2.

Heldstab, Sandra A., Karin Isler, Judith M. Burkart, and Carel P. van Schaik. 2019. "Allomaternal Care, Brains and Fertility in Mammals: Who Cares Matters." *Behavioral Ecology and Sociobiology* 73 (6): 71. https://doi.org/10.1007/s00265-019-2684-x.

Herzing, Denise L. 1997. "The Life History of Free-Ranging Atlantic Spotted Dolphins (*Stenella frontalis*): Age Classes, Color Phases, and Female Reproduction." *Marine Mammal Science* 13 (4): 576–95. https://doi.org/10.1111/j.1748-7692.1997.tb00085.x.

Hill, Russell A., R. Alexander Bentley, and Robin I. M Dunbar. 2008. "Network Scaling Reveals Consistent Fractal Pattern in Hierarchical Mammalian Societies." *Biology Letters* 4 (6): 748–51. https://doi.org/10.1098/rsbl.2008.0393.

Irvine, Lyn G., Michele Thums, Christine E. Hanson, Clive R. McMahon, and Mark A. Hindell. 2017. "Quantifying the Energy Stores of Capital Breeding Humpback Whales and Income Breeding Sperm Whales Using Historical Whaling Records." *Royal Society Open Science* 4 (3): 160290. https://doi.org/10.1098/rsos.160290.

Isbell, Lynne, and Truman Young. 2002. "Ecological Models of Female Social Relationships in Primates: Similarities, Disparities, and Some Directions for Future Clarity." *Behaviour* 139 (February): 177–202. https://doi.org/10.1163/156853902760102645.

Janik, Vincent M., Laela S. Sayigh, and Randall S. Wells. 2006. "Signature Whistle Shape Conveys Identity Information to Bottlenose Dolphins." *Proceedings of the National Academy of Sciences* 103 (21): 8293–97. https://doi.org/10.1073/pnas.0509918103.

Kappeler, Peter M., Tim Clutton-Brock, Susanne Shultz, and Dieter Lukas. 2019. "Social Complexity: Patterns, Processes, and Evolution." *Behavioral Ecology and Sociobiology* 73 (1): 5. https://doi.org/10.1007/s00265-018-2613-4.

Karniski, Caitlin, Ewa Krzyszczyk, and Janet Mann. 2018. "Senescence Impacts Reproduction and Maternal Investment in Bottlenose Dolphins." *Proceedings of the Royal Society B* 285 (1883): 20181123. https://doi.org/10.1098/rspb.2018.1123.

King, Stephanie L., Richard C. Connor, and Stephen H. Montgomery. 2022. "Social and Vocal Complexity in Bottlenose Dolphins." *Trends in Neurosciences* 45 (12): 881–83. https://doi.org/10.1016/j.tins.2022.09.006.

Konrad, Christine M., Timothy R. Frasier, Hal Whitehead, and Shane Gero. 2019. "Kin Selection and Allocare in Sperm Whales." *Behavioral Ecology* 30 (1): 194–201. https://doi.org/10.1093/beheco/ary143.

Konrad, Christine M., Shane Gero, Timothy R. Frasier, and Hal Whitehead. 2018. "Kinship Influences Sperm Whale Social Organization Within, but Generally Not Among, Social Units." *Royal Society Open Science* 5 (8): 180914. https://doi.org/10.1098/rsos.180914.

Kopps, Anna M., Corinne Y. Ackermann, William B. Sherwin, Simon J. Allen, Lars Bejder, and Michael Krützen. 2014. "Cultural Transmission of Tool Use Combined with Habitat Specializations Leads to Fine-Scale Genetic Structure in Bottlenose Dolphins." *Proceedings of the Royal Society B* 281 (1782): 20133245. https://doi.org/10.1098/rspb.2013.3245.

Lawson Handley, Lori J., and Nicolas Perrin. 2007. "Advances in Our Understanding of Mammalian Sex-Biased Dispersal." *Molecular Ecology* 16 (8): 1559–78. https://doi.org/10.1111/j.1365-294X.2006.03152.x.

Lee, Phyllis C., and Cynthia J. Moss. 1986. "Early Maternal Investment in Male and Female African Elephant Calves." *Behavioral Ecology and Sociobiology* 18 (5): 353–61. https://doi.org/10.1007/BF00299666.

Leung, Elaine S., Valeria Vergara, and Lance G. Barrett-Lennard. 2010. "Allonursing in Captive Belugas (*Delphinapterus leucas*)." *Zoo Biology* 29 (5): 633–37. https://doi.org/10.1002/zoo.20295.

Levengood, Alexis L., Kasha Strickland, Vivienne Foroughirad, et al. 2022. "Heterogeneity in Resource Competition Covaries with Individual Variation in Long-Term Social Relationships." Edited by Amanda Ridley. *Behavioral Ecology* 33 (4): 745–57. https://doi.org/10.1093/beheco/arac037.

Lukas, Dieter, and Tim H. Clutton-Brock. 2018. "Social Complexity and Kinship in Animal Societies." *Ecology Letters* 21 (8): 1129–34. https://doi.org/10.1111/ele.13079.

Lusseau, David, Karsten Schneider, Oliver J. Boisseau, Patti Haase, Elisabeth Slooten, and Steve M. Dawson. 2003. "The Bottlenose Dolphin Community of Doubtful Sound Features a Large Proportion of Long-Lasting Associations." *Behavioral Ecology and Sociobiology* 54 (4): 396–405. https://doi.org/10.1007/s00265-003-0651-y.

Lyrholm, Thomas, Olof Leimar, Bo Johanneson, and Ulf Gyllensten. 1999. "Sex-Biased Dispersal in Sperm Whales: Contrasting Mitochondrial and Nuclear Genetic Structure of Global Populations." *Proceedings of the Royal Society B* 266 (1417): 347–54. https://doi.org/10.1098/rspb.1999.0644.

Manlik, Oliver, Jane A. McDonald, Janet Mann, et al. 2016. "The Relative Importance of Reproduction and Survival for the Conservation of Two Dolphin Populations." *Ecology and Evolution* 6 (11): 3496–512. https://doi.org/10.1002/ece3.2130.

Mann, Janet. 2019. "Maternal Care and Calf Development in Odontocetes." In *Ethology and Behavioral Ecology of Toothed Whales and Dolphins*, edited by Bernd Würsig, 95–114. Springer.

Mann, Janet, and H. Barnett. 1999. "Lethal Tiger Shark (*Galeocerdo cuvier*) Attack on Bottlenose Dolphin (*Tursiops* sp.) Calf: Defense and Reactions by the Mother." *Marine Mammal Science* 15 (2): 568–75. https://doi.org/10.1111/j.1748-7692.1999.tb00823.x.

Mann, Janet, R. C. Connor, L. M. Barrè, and M. R. Heithaus. 2000. "Female Reproductive Success in Bottlenose Dolphins (*Tursiops* sp.): Life History, Habitat, Provisioning, and Group-Size Effects." *Behavioral Ecology* 11 (2): 210–19. https://doi.org/10.1093/beheco/11.2.210.

Mann, Janet, and Caitlin Karniski. 2017. "Diving Beneath the Surface: Long-Term Studies of Dolphins and Whales." *Journal of Mammalogy* 98 (3): 621–30. https://doi.org/10.1093/jmammal/gyx036.

Mann, Janet, Brooke L. Sargeant, Jana J. Watson-Capps, et al. 2008. "Why Do Dolphins Carry Sponges?" *PLOS ONE* 3 (12): e3868. https://doi.org/10.1371/journal.pone.0003868.

Mann, Janet, and Barbara Smuts. 1998. "Natal Attraction: Allomaternal Care and Mother-Infant Separations in Wild Bottlenose Dolphins." *Animal Behaviour* 55 (5): 1097–113. https://doi.org/10.1006/anbe.1997.0637.

Mann, Janet, Margaret A. Stanton, Eric M. Patterson, Elisa J. Bienenstock, and Lisa O. Singh. 2012. "Social Networks Reveal Cultural Behaviour in Tool-Using Dolphins." *Nature Communications* 3 (1): 980. https://doi.org/10.1038/ncomms1983.

Mann, Janet, and Jana J. Watson-Capps. 2005. "Surviving at Sea: Ecological and Behavioural Predictors of Calf Mortality in Indian Ocean Bottlenose Dolphins, *Tursiops* sp." *Animal Behaviour* 69 (4): 899–909. https://doi.org/10.1016/j.anbehav.2004.04.024.

Mason, Suzanne, Chandra Salgado Kent, and Kerstin Bilgmann. 2021. "Common Dolphins Form Unexpected Strong Social Bonds: Insights into Social Plasticity of Delphinids." *Marine Mammal Science* 37 (4): 1174–95. https://doi.org/10.1111/mms.12815.

Mastick, Natalie C., David Wiley, David E. Cade, Colin Ware, Susan E. Parks, and Ari S. Friedlaender. 2022. "The Effect of Group Size on Individual Behavior of Bubble-Net Feeding Humpback Whales in the Southern Gulf of Maine." *Marine Mammal Science* 38 (3): 959–74. https://doi.org/10.1111/mms.12905.

McAuliffe, Katherine, and Hal Whitehead. 2005. "Eusociality, Menopause and Information in Matrilineal Whales." *Trends in Ecology and Evolution* 20 (12): 650. https://doi.org/10.1016/j.tree.2005.09.003.

McEntee, Molly H. F., M. MacQueeney, D. C. Alvarado, and Janet Mann. 2023. "Infanticide and Sexual Conflict in Cetaceans." In *Sex in Cetaceans: Morphology, Behavior, and the Evolution of Sexual Strategies*, edited by Bernd Würsig and Dara N. Orbach. Springer.

McSweeney, Daniel J., Robin W. Baird, Sabre D. Mahaffy, Daniel L. Webster, and Gregory S. Schorr. 2009. "Site Fidelity and Association Patterns of a Rare Species: Pygmy Killer Whales (*Feresa attenuata*) in the Main Hawaiian Islands." *Marine Mammal Science* 25 (3): 557–72. https://doi.org/10.1111/j.1748-7692.2008.00267.x.

Mesnick, Sarah L., Lisa T. Ballance, Paul R. Wade, Karen Pryor, and Randall R. Reeves. 2019. "Oceanic Dolphin Societies: Diversity, Complexity, and Conservation." In *Ethology and Behavioral Ecology of Odontocetes*, edited by Bernd Würsig, 183–209. Springer. https://doi.org/10.1007/978-3-030-16663-2_9.

Miketa, Madison L. 2018. "Social and Behavioral Responses to Environmental Stressors in Bottlenose Dolphins in Shark Bay, Australia." *Biology*. Georgetown University.

Mock, Douglas W., and Masahiro Fujioka. 1990. "Monogamy and Long-Term Pair Bonding in Vertebrates." *Trends in Ecology and Evolution* 5 (2): 39–43. https://doi.org/10.1016/0169-5347(90)90045-F.

Möller, Luciana M. 2012. "Sociogenetic Structure, Kin Associations and Bonding in Delphinids." *Molecular Ecology* 21 (3): 745–64. https://doi.org/10.1111/j.1365-294X.2011.05405.x.

Möller, Luciana M., and Robert G. Harcourt. 2008. "Shared Reproductive State Enhances Female Associations in Dolphins." *International Journal of Ecology* 2008 (February): 498390. https://doi.org/10.1155/2008/498390.

Moscovice, Liza R., Cédric Sueur, and Filippo Aureli. 2020. "How Socio-Ecological Factors Influence the Differentiation of Social Relationships: An Integrated Conceptual Framework." *Biology Letters* 16 (9): 20200384. https://doi.org/10.1098/rsbl.2020.0384.

Newsome, Seth D., Michael A. Etnier, Daniel H. Monson, and Marilyn L. Fogel. 2009. "Retrospective Characterization of Ontogenetic Shifts in Killer Whale Diets via δ13C and δ15N Analysis of Teeth." *Marine Ecology Progress Series* 374: 229–42. https://doi.org/10.3354/meps07747.

Nielsen, Mia L. K., Samuel Ellis, Michael N. Weiss, et al. 2023. "Temporal Dynamics of Mother-Offspring Relationships in Bigg's Killer Whales: Opportunities for Kin-Directed Help by Post-Reproductive Females." *Proceedings of the Royal Society B* 290 (2000): 20230139. https://doi.org/10.1098/rspb.2023.0139.

Nishita, Miki, Miki Shirakihara, and Masao Amano. 2017. "Patterns of Association Among Female Indo-Pacific Bottlenose Dolphins (*Tursiops aduncus*) in a Population Forming Large Groups." *Behaviour* 154 (9–10): 1013–28. https://doi.org/10.1163/1568539X-00003453.

Norris, Kenneth S., and T. P. Dohl. 1980. "Behavior of the Hawaiian Spinner Dolphin, Stenella Longirostris." *Fishery Bulletin* 77 (4): 821–49.

O'Brien, Kristin. 2013. "Population Status and Social Structure of Northern Bottlenose Whales (*Hyperoodon ampullatus*) on the Scotian Shelf." Master's thesis, Dalhousie University Halifax, Nova Scotia, Canada. https://library-archives.canada.ca/eng/services/services-libraries/theses/Pages/item.aspx?idNumber=1032932987.

Palmer, Emily I., Emma L. Betty, Sinéad Murphy, Matthew R. Perrott, Adam N. H. Smith, and Karen A. Stockin. 2022. "Reproductive Biology of Female Common Dolphins (*Delphinus delphis*) in New Zealand Waters." *Marine Biology* 169 (12): 158. https://doi.org/10.1007/s00227-022-04139-3.

Palombit, Ryne A. 2015. "Infanticide as Sexual Conflict: Coevolution of Male Strategies and Female Counterstrategies." *Cold Spring Harbor Perspectives in Biology* 7 (6): a017640. https://doi.org/10.1101/cshperspect.a017640.

Parsons, Kim M., Kenneth C. Balcomb III, John K. B. Ford, and John W. Durban. 2009. "The Social Dynamics of Southern Resident Killer Whales and Conservation Implications for This Endangered Population." *Animal Behaviour* 77 (4): 963–71. https://doi.org/10.1016/j.anbehav.2009.01.018.

Pilot, Malgorzata, Marilyn E. Dahlheim, and A. Rus Hoelzel. 2010. "Social Cohesion Among Kin, Gene Flow Without Dispersal and the Evolution of Population Genetic Structure in the Killer Whale (*Orcinus orca*)." *Journal of Evolutionary Biology* 23 (1): 20–31. https://doi.org/10.1111/j.1420-9101.2009.01887.x.

Pitman, Robert L., Lisa T. Ballance, Sarah I. Mesnick, and Susan J. Chivers. 2001. "Killer Whale Predation on Sperm Whales: Observations and Implications." *Marine Mammal Science* 17 (3): 494–507. https://doi.org/10.1111/j.1748-7692.2001.tb01000.x.

Pitman, Robert L., Volker B. Deecke, Christine M. Gabriele, et al. 2017. "Humpback Whales Interfering When Mammal-Eating Killer Whales Attack Other Species: Mobbing Behavior and Interspecific Altruism?" *Marine Mammal Science* 33 (1): 7–58. https://doi.org/10.1111/mms.12343.

Ramp, Christian, Wilhelm Hagen, Per Palsbøll, Martine Bérubé, and Richard Sears. 2010. "Age-Related Multi-Year Associations in Female Humpback Whales (*Megaptera novaeangliae*)." *Behavioral Ecology and Sociobiology* 64 (10): 1563–76. https://doi.org/10.1007/s00265-010-0970-8.

Rankin, Robert William, Vivienne Jilla Foroughirad, Ewa Beata Krzyszczyk, Céline H. Frère, and Janet Mann. 2022. "Changes in Social Position Predict Survival in Bottlenose Dolphins." Preprint. *bioRxiv*. https://doi.org/10.1101/2022.08.25.505273.

Rasmussen, Kristin, Daniel M. Palacios, John Calambokidis, et al. 2007. "Southern Hemisphere Humpback Whales Wintering off Central America: Insights from Water Temperature into the Longest Mammalian Migration." *Biology Letters* 3 (3): 302–5. https://doi.org/10.1098/rsbl.2007.0067.

Reeves, Randall R., Joel Berger, and Phillip J. Clapham. 2006. "Killer Whales as Predators of Large Baleen Whales and Sperm Whales." In *Whales, Whaling, and Ocean Ecosystems*, edited by James A. Estes, Douglas R. DeMaster, Daniel F. Doak, Terri M. Williams, and Robert L. Brownell Jr., 174–87. University of California Press.

Rendall, Drew, Peter S. Rodman, and Roger E. Emond. 1996. "Vocal Recognition of Individuals and Kin in Free-Ranging Rhesus Monkeys." *Animal Behaviour* 51 (5): 1007–15. https://doi.org/10.1006/anbe.1996.0103.

Rendell, Luke, Mauricio Cantor, Shane Gero, Hal Whitehead, and Janet Mann. 2019. "Causes and Consequences of Female Centrality in Cetacean Societies." *Philosophical Transactions of the Royal Society B* 374 (1780): 20180066. https://doi.org/10.1098/rstb.2018.0066.

Riccialdelli, Luciana, Seth D. Newsome, Natalia A. Dellabianca, Ricardo Bastida, Marilyn L. Fogel, and R. Natalie P. Goodall. 2013. "Ontogenetic Diet Shift in Commerson's Dolphin (*Cephalorhynchus commersonii commersonii*) off Tierra Del Fuego." *Polar Biology* 36 (5): 617–27. https://doi.org/10.1007/s00300-013-1289-5.

Robinson, Kevin. 2014. "Agonistic Intraspecific Behavior in Free-Ranging Bottlenose Dolphins: Calf-Directed Aggression and Infanticidal Tendencies by Adult Males." *Marine Mammal Science* 30 (January): 381–88. https://doi.org/10.1111/mms.12023.

Rosenbaum, Stacy, and Lee T. Gettler. 2018. "With a Little Help from Her Friends (and Family) Part I: The Ecology and Evolution of Non-Maternal Care in Mammals." *Physiology and Behavior* 193 (September): 1–11. https://doi.org/10.1016/j.physbeh.2017.12.025.

Ruiz-Cooley, Rocio I., Tim Gerrodette, Susan J. Chivers, and Kerri Danil. 2021. "Cooperative Feeding in Common Dolphins as Suggested by Ontogenetic Patterns in δ15N Bulk and Amino Acids." *Journal of Animal Ecology* 90 (6): 1583–95. https://doi.org/10.1111/1365-2656.13478.

Samuni, Liran, Anna Preis, Tobias Deschner, Catherine Crockford, and Roman M. Wittig. 2018. "Reward of Labor Coordination and Hunting Success in Wild Chimpanzees." *Communications Biology* 1 (1): 138. https://doi.org/10.1038/s42003-018-0142-3.

Santacesaria, Francesca Cornelia, Carmelo Fanizza, Stefano Bellomo, et al. 2022. "Preliminary Analysis of Risso's Dolphin Social Structure in the Gulf of Taranto (Northern Ionian Sea, Central Eastern Mediterranean Sea)." In *2022 IEEE International Workshop on Metrology for the Sea; Learning to Measure Sea Health Parameters (MetroSea)*, 242–46. https://doi.org/10.1109/MetroSea55331.2022.9950996.

Sarano, François, Justine Girardet, Véronique Sarano, et al. 2021. "Kin Relationships in Cultural Species of the Marine Realm: Case Study of a Matrilineal Social Group of Sperm Whales off Mauritius Island, Indian Ocean." *Royal Society Open Science* 8 (2): 201794. https://doi.org/10.1098/rsos.201794.

Sargeant, Brooke L., and Janet Mann. 2009. "Developmental Evidence for Foraging Traditions in Wild Bottlenose Dolphins." *Animal Behaviour* 78 (3): 715–21. https://doi.org/10.1016/j.anbehav.2009.05.037.

Sayigh, Laela S., Randall S. Wells, and Vincent M. Janik. 2017. "What's in a Voice? Dolphins Do Not Use Voice Cues for Individual Recognition." *Animal Cognition* 20: 1067–79. https://doi.org/10.1007/s10071-017-1123-5.

Schulz, Tyler M., and W. Don Bowen. 2005. "The Evolution of Lactation Strategies in Pinnipeds: A Phylogenetic Analysis." *Ecological Monographs* 75 (2): 159–77. https://doi.org/10.1890/04-0319.

Schweitzer, Cécile, Frédéric Lévy, and Cécile Arnould. 2011. "Increasing Group Size Decreases Social Bonding in Young Japanese Quail, *Coturnix japonica*." *Animal Behaviour* 81 (3): 535–42. https://doi.org/10.1016/j.anbehav.2010.11.017.

Scott, Erin M., Janet Mann, Jana J. Watson-Capps, Brooke L. Sargeant, and Richard C. Connor. 2005. "Aggression in Bottlenose Dolphins: Evidence for Sexual Coercion, Male-Male

Competition, and Female Tolerance Through Analysis of Tooth-Rake Marks and Behaviour." *Behaviour* 142 (1): 21–44. https://www.jstor.org/stable/4536227.

Silk, Joan B., Jacinta C. Beehner, Thore J. Bergman, et al. 2010. "Strong and Consistent Social Bonds Enhance the Longevity of Female Baboons." *Current Biology* 20 (15): 1359–61. https://doi.org/10.1016/j.cub.2010.05.067.

Similä, Tiu, and Fernando Ugarte. 1993. "Surface and Underwater Observations of Cooperatively Feeding Killer Whales in Northern Norway." *Canadian Journal of Zoology* 71 (8): 1494–99. https://doi.org/10.1139/z93-210.

Smith, Brian D. 2018. "Irrawaddy Dolphin: *Orcaella brevirostris*." In *Encyclopedia of Marine Mammals*, 3rd ed., edited by Bernd Würsig, Johannes G. M. Thewissen, and Kit M. Kovacs, 525–29. Academic Press. https://doi.org/10.1016/B978-0-12-804327-1.00161-8.

Smith, Tanya M., Christine Austin, Katie Hinde, Erin R. Vogel, and Manish Arora. 2017. "Cyclical Nursing Patterns in Wild Orangutans." *Science Advances* 3 (5): e1601517. https://doi.org/10.1126/sciadv.1601517.

Smolker, Rachel A., Andrew F. Richards, Richard C. Connor, and John W. Pepper. 1992. "Sex Differences in Patterns of Association Among Indian Ocean Bottlenose Dolphins." *Behaviour* 123 (1–2): 38–69. https://doi.org/10.1163/156853992X00101.

Stanton, Margaret A., and Janet Mann. 2012. "Early Social Networks Predict Survival in Wild Bottlenose Dolphins." *PLOS ONE* 7 (10): e47508. https://doi.org/10.1371/journal.pone.0047508.

Strickland, Kasha, Alexis Levengood, Vivienne Foroughirad, Janet Mann, Ewa Krzyszczyk, and Céline H. Frère. 2017. "A Framework for the Identification of Long-Term Social Avoidance in Longitudinal Datasets." *Royal Society Open Science* 4 (8). https://doi.org/10.1098/rsos.170641.

Subramanian, Annamalai, Shinsuke Tanabe, and Ryo Tatsukawa. 1988. "Estimating Some Biological Parameters of Baird's Beaked Whales Using PCBs and DDE as Tracers." *Marine Pollution Bulletin* 19 (6): 284–87. https://doi.org/10.1016/0025-326X(88)90600-5.

Torres Ortiz, Sara, Johanna Stedt, Henrik Skov Midtiby, Henrik Dyrberg Egemose, and Magnus Wahlberg. 2021. "Group Hunting in Harbour Porpoises (*Phocoena phocoena*)." *Canadian Journal of Zoology* 99 (6): 511–20. https://doi.org/10.1139/cjz-2020-0289.

Towers, Jared R., Muriel J. Hallé, Helena K. Symonds, et al. 2018. "Infanticide in a Mammal-Eating Killer Whale Population." *Scientific Reports* 8 (1): 4366. https://doi.org/10.1038/s41598-018-22714-x.

Tsai, Yi-Jiun Jean, and Janet Mann. 2013. "Dispersal, Philopatry, and the Role of Fission-Fusion Dynamics in Bottlenose Dolphins." *Marine Mammal Science* 29 (2): 261–79. https://doi.org/10.1111/j.1748-7692.2011.00559.x.

Vachon, Felicia, Hal Whitehead, and Timothy R. Frasier. 2018. "What Factors Shape Genetic Diversity in Cetaceans?" *Ecology and Evolution* 8 (3): 1554–72. https://doi.org/10.1002/ece3.3727.

Valenzuela, Luciano O., Mariano Sironi, Victoria J. Rowntree, and Jon Seger. 2009. "Isotopic and Genetic Evidence for Culturally Inherited Site Fidelity to Feeding Grounds in Southern Right Whales (*Eubalaena australis*)." *Molecular Ecology* 18 (5): 782–91. https://doi.org/10.1111/j.1365-294x.2008.04069.x.

Van Noordwijk, Maria A., Christopher W. Kuzawa, and Carel P. van Schaik. 2013. "The Evolution of the Patterning of Human Lactation: A Comparative Perspective." *Evolutionary Anthropology: Issues, News, and Reviews* 22 (5): 202–12. https://doi.org/10.1002/evan.21368.

Wade, Paul R., Randall R. Reeves, and Sarah L. Mesnick. 2012. "Social and Behavioural Factors in Cetacean Responses to Overexploitation: Are Odontocetes Less 'Resilient' Than Mysticetes?" *Journal of Marine Biology* 15: e567276. https://doi.org/10.1155/2012/567276.

Wallen, Megan. M., Eric M. Patterson, Ewa Krzyszczyk, and Janet Mann. 2016. "The Ecological Costs to Females in a System with Allied Sexual Coercion." *Animal Behaviour* 115 (May): 227–36. https://doi.org/10.1016/j.anbehav.2016.02.018.

Weidt, Andrea, Sarah E. Hofmann, and Barbara König. 2008. "Not Only Mate Choice Matters: Fitness Consequences of Social Partner Choice in Female House Mice." *Animal Behaviour* 75 (3): 801–8. https://doi.org/10.1016/j.anbehav.2007.06.017.

Weinrich, Mason T. 1991. "Stable Social Associations Among Humpback Whales (*Megaptera novaeangliae*) in the Southern Gulf of Maine." *Canadian Journal of Zoology* 69 (12): 3012–19. https://doi.org/10.1139/z91-425.

Weinrich, Mason T., Howard Rosenbaum, C. Scott Baker, Alexis L. Blackmer, and Hal Whitehead. 2006. "The Influence of Maternal Lineages on Social Affiliations Among Humpback Whales (*Megaptera novaeangliae*) on Their Feeding Grounds in the Southern Gulf of Maine." *Journal of Heredity* 97 (3): 226–34. https://doi.org/10.1093/jhered/esj018.

Weiss, Michael N., Samuel Ellis, and Darren P. Croft. 2021. "Diversity and Consequences of Social Network Structure in Toothed Whales." *Frontiers in Marine Science* 8: 688842. https://doi.org/10.3389/fmars.2021.688842.

Weiss, Michael N., Daniel W. Franks, Darren P. Croft, and Hal Whitehead. 2019. "Measuring the Complexity of Social Associations Using Mixture Models." *Behavioral Ecology and Sociobiology* 73 (1): 8. https://doi.org/10.1007/s00265-018-2603-6.

Weller, David W. 2018. "Predation on Marine Mammals." In *Encyclopedia of Marine Mammals*, 3rd ed., edited by Bernd Würsig, Johannes G. M. Thewissen, and Kit M. Kovacs, 772–80. Academic Press. https://doi.org/10.1016/B978-0-12-804327-1.00206-5.

Whitehead, Hal. 1996. "Babysitting, Dive Synchrony, and Indications of Alloparental Care in Sperm Whales." *Behavioral Ecology and Sociobiology* 38 (4): 237–44. https://doi.org/10.1007/s002650050238.

Whitehead, Hal. 1998. "Cultural Selection and Genetic Diversity in Matrilineal Whales." *Science* 282 (5394): 1708–11. https://doi.org/10.1126/science.282.5394.1708.

Whitehead, Hal. 2008. *Analyzing Animal Societies: Quantitative Methods for Vertebrate Social Analysis*. University of Chicago Press.

Whitehead, Hal, and Shane Gero. 2015. "Conflicting Rates of Increase in the Sperm Whale Population of the Eastern Caribbean: Positive Observed Rates Do Not Reflect a Healthy Population." *Endangered Species Research* 27 (3): 207–18. https://doi.org/10.3354/esr00657.

Whitehead, Hal, F. Vachon, and T. R. Frasier. 2017. "Cultural Hitchhiking in the Matrilineal Whales." *Behavior Genetics* 47 (3): 324–34. https://doi.org/10.1007/s10519-017-9840-8.

Wiszniewski, Joanna, David Lusseau, and Luciana M. Möller. 2010. "Female Bisexual Kinship Ties Maintain Social Cohesion in a Dolphin Network." *Animal Behaviour* 80 (5): 895–904. https://doi.org/10.1016/j.anbehav.2010.08.013.

Wittig, Roman M., Catherine Crockford, Julia Lehmann, Patricia L. Whitten, Robert M. Seyfarth, and Dorothy L. Cheney. 2008. "Focused Grooming Networks and Stress Alleviation in Wild Female Baboons." *Hormones and Behavior* 54 (1): 170–77. https://doi.org/10.1016/j.yhbeh.2008.02.009.

Wittig, Roman M., Catherine Crockford, Anja Weltring, Kevin E. Langergraber, Tobias Deschner, and Klaus Zuberbühler. 2016. "Social Support Reduces Stress Hormone Levels in

Wild Chimpanzees Across Stressful Events and Everyday Affiliations." *Nature Communications* 7 (1): 13361. https://doi.org/10.1038/ncomms13361.

Wrangham, Richard W. 1980. "An Ecological Model of Female-Bonded Primate Groups." *Behaviour* 75 (3–4): 262–300. https://www.jstor.org/stable/4534085.

Wray, Janie, Eric M. Keen, and Éadin N. O'Mahony. 2021. "Social Survival: Humpback Whales (*Megaptera novaeangliae*) Use Social Structure to Partition Ecological Niches Within Proposed Critical Habitat." *PLOS ONE* 16 (6): e0245409. https://doi.org/10.1371/journal.pone.0245409.

Wright, Brianna M., Eva H. Stredulinsky, Graeme M. Ellis, and John K. B. Ford. 2016. "Kin-Directed Food Sharing Promotes Lifetime Natal Philopatry of Both Sexes in a Population of Fish-Eating Killer Whales, *Orcinus orca*." *Animal Behaviour* 115 (May): 81–95. https://doi.org/10.1016/j.anbehav.2016.02.025.

Würsig, Bernd, Nicholas Duprey, and Jody Weir. 2007. "Dusky Dolphins (*Lagenorhynchus obscurus*) in New Zealand Waters." *Present Knowledge and Research Goals. DOC Research and Development Series* 270: 1–28.

Young, Christopher, Bonaventura Majolo, Michael Heistermann, Oliver Schülke, and Julia Ostner. 2014. "Responses to Social and Environmental Stress Are Attenuated by Strong Male Bonds in Wild Macaques." *Proceedings of the National Academy of Sciences* 111 (51): 18195–200. https://doi.org/10.1073/pnas.1411450111.

Zanardo, Nikki, K. Bilgmann, G. J. Parra, and L. M. Möller. 2016. "Socio-Genetic Structure of Short-Beaked Common Dolphins in Southern Australia." *Journal of Zoology* 299 (2): 89–97. https://doi.org/10.1111/jzo.12330.

6

Uncovering Baleen Whale Sociality

REBECCA DUNLOP AND SUSAN PARKS

Abstract

Baleen whales are mammals with superlative traits. They are large; the blue whale is one of the largest species ever to live on the planet with lengths up to 30 m and 200 metric tons in mass. They are long-lived; the bowhead whale's life span exceeds 150 years. They have long migrations; humpback whale annual migrations exceed 16,000 km each year. Their social interactions can involve acoustic exchanges of calls that can travel hundreds of kilometers. These extreme traits make behavioral studies of baleen whale social interactions uniquely challenging. Early studies of baleen whale social behavior on the calving and feeding grounds described short-lived and fluid social interactions beyond the early mother-calf pair bond. Methodological advances have facilitated the development of long-term databases of identified individuals and novel tools to study their social interactions. These have led to exciting new discoveries around the complexity in the behavior of baleen whales and potential long-term social interactions that span decades. However, several fundamental questions remain. This chapter will review the emerging patterns of social complexity in baleen whales and examine the functional benefits and implications of these social structures.

Evolutionary Drivers of Sociality in Baleen Whales

The baleen whales of today encompass 15 different species that have evolved to inhabit all the world's oceans. Their distribution ranges from polar seas to temperate and tropical waters, and from shallow and coastal waters to the deep ocean basins. The most ancient lineage of the living baleen whales is the Balaenidae, which include four extant species, the Arctic bowhead whale (*Balaena balaena*) and the three right whale species including the North Pacific (*Eubalaena japonica*), the North Atlantic (*E. glacialis*) and the southern right whale (*E. australis*; see McGowen et al., this volume). These four species have very similar body plans with broad, stocky bodies, dark gray coloring, and large skulls with arched upper jaws carrying long plates of baleen for filtering their food. The most diverse family of baleen whales is the Balaenopteridae, which includes the largest of the whales, the blue whale (*Balaenoptera musculus*), as well as fin whales (*B. physalus*), the medium-sized humpback whale (*Megaptera novaeangliae*), and the relatively small minke whale (*B. acutorostrata*; see McGowen et al., this volume). Many species within this family are long, flexible, torpedo-shaped whales that can move through the water at high speeds. They also possess conspicuous, grooved throat plates that expand when they fill their mouth with water and prey. This adaptation allows them to engulf large amounts of prey per scoop that they then filter with relatively short plates of baleen. Despite these morphological differences, both families evolved from a common baleen whale ancestor, which split from the odontocete lineage (toothed whales) approximately 34 million years ago.

The evolution of baleen whales can be briefly summarized as "lose teeth and gain size." The late Eocene and early Oligocene periods had high oceanic productivity (Schumacher and Lazarus 2004). High-density patches of phyto- and zooplankton produced high-density patches of small prey, and this abundant prey likely led to the evolution of filter feeding animals (Slater et al. 2017). The increase in oceanic productivity coincided with one of the major steps during the evolution of baleen whales: the progressive loss of teeth (Tsai and Fordyce 2015). As they evolved, they also grew bigger, likely due to the high prey densities and energetic efficiency of bulk feeding (Goldbogen and Madsen 2018). With such a high energetic intake, some species, such as the blue whale, can grow to 30 m in length, and weigh up to 200 metric tons. In addition, their ability to store massive amounts of energy allowed them to survive long periods of fasting, enabling migration over long distances between seasonal highly productive foraging grounds and warmer and more sheltered calving grounds (Boyd 2004; Christiansen et al. 2013; Stern and Friedlander 2018). As a result, the life history of the baleen whale evolved into what we see today: breeding and feeding months separated by long migrations. This large-scale seasonal migration has implications for the social interactions in baleen whale species.

Baleen whale species have large spatiotemporal distributions (Corkeron and Connor 1999). Their breeding and feeding grounds can be extensive, such as the Great Barrier Reef breeding grounds on the eastern coast of Australia (Smith et al. 2012), and Southern Ocean feeding grounds for South Pacific humpback whales (Constantine et al. 2014). Populations can disperse over vast distances during migration (Steel et al. 2018; Stevick et al. 2016). Therefore, there is the potential for individuals to have social interactions at multiple spatial and temporal scales. At a relatively small spatial scale, there can be physical contact between animals, such as between a mother and her calf, or during mating. At a medium scale, vocalizations used for social interactions can propagate hundreds of meters, and at a large scale, low-frequency vocalizations, such as those used in their songs, can propagate kilometers. Temporally, mother and calves are likely to be the longest lasting continuous bond, where in humpback whales, for example, mothers and calves will stay together for approximately one year (Clapham 1996).

In *Cetacean Societies* (Mann et al. 2000), the social system of humpback whales was highlighted as one of the best-studied baleen whale species at that time (Clapham 2000). The breeding and calving grounds of many populations of humpback whales are close to shore, allowing for accessibility of this species for study. For example, the breeding population of North Pacific humpback whales in the Hawaiian Islands have been subject to ongoing research since the 1970s (Wolman and Jurasz 1977; Herman and Antinoja 1977). These early studies noted that groups were mostly small, commonly sighted groups of one to three whales (with or without a newborn calf; Herman and Antinoja 1977; Herman et al. 1980; Glockner and Venus 1983). Larger groups were also present on the breeding grounds, generally consisting of a single female being followed by multiple males in competitive pods (Tyack and Whitehead 1983; Baker and Herman 1984; Clapham et al. 1992). The highest degree of association was between mothers and their nursing calves. Humpback whale mothers were not known to associate with their calves after weaning (Clapham 2000; Connor et al. 2005), though mothers have been observed in association with their yearling calves on the return migration to the breeding and calving grounds the following year (Baker et al. 1987). Other groupings, including competitive groups, consisted of short-term associations that lasted a matter of hours (Mobley and Herman 1985; Clapham 1993).

Humpback whale social interactions on the foraging grounds were also thought to primarily consist of short-term associations. For example, on the foraging grounds in the western North Atlantic, most humpback whales are sighted alone or in pairs over a period of days (Whitehead 1983; Weinrich and Kuhlberg 1991). These short-term associations have been attributed to the formation of coordinated feeding groups to increase foraging capacity (Clapham 1993; Clapham 1996). However, long-term studies of known individuals in this population have shown evidence for repeated short-term associations with the same individuals, suggesting that some whales appear to have preferred associates on the feeding grounds. For example, pairings of non-lactating females have been observed to last up to six feeding seasons (Ramp et al. 2010; Weinrich and Kuhlberg 1991; Weinrich 1991). These repeated associations observed on the feeding grounds suggest that humpback whales are capable of individual recognition, even if the associations may be temporally limited during foraging.

Baleen whales produce low-frequency vocalizations that can propagate over large distances. This enables them to maintain acoustic contact with each other far beyond the immediate group they happen to be found in, and far beyond visual range. Therefore, to understand sociality in baleen whales, it is important to describe both spatial associations and acoustic exchanges that mark a range of social interactions. The large spatial, and long temporal, scales over which baleen whales live mean "standard" association indices, and categorization of interactions, based on physical proximity, may not be suitable for capturing the complexity of baleen whale sociality (Payne and Webb 1971; Clapham 2000). However, having such a large communication space makes it particularly challenging for researchers to observe and document these associations. As a result, evidence for long-term social bonds in baleen whales is often limited to observations of special cases that involve repeated associations of individuals near one another, such as during periods of group foraging, and during within-group social interactions.

For the remainder of this chapter, we will focus on two species, the humpback whale, and the North Atlantic right whale, representing the current knowledge of social complexity within a balaenopterid and a balaenid species. These species are among the best-studied baleen whales, with extensive long-term research of identifiable individuals being carried out on both the breeding and foraging grounds. Given the global distribution of humpback whales and the three right whale species, there are multiple opportunities to assess similarities and differences both within and between species in different ocean basins. The two species have broad similarities in their life histories, with annual migratory cycles. However, differences in their foraging strategies and mating systems provide significant underlying life history differences that may have contributed to the marked differences in the apparent social complexity in the two species.

Repeated associations among individuals can only be documented and understood if individuals can be identified and we can infer something about their relatedness. A revolution in identification of free-ranging wild baleen whales took place in the mid-1970s with the discovery that individuals had unique and stable markings that could be used to document individual identity in the field through photographs. Humpback whales and right whales both have unique stable markings—fluke and dorsal fins for humpback whales, and callosity patterns in right whales—that allow for photographic identification of individuals (Katona and Beard 1990; Clapham 2000; Payne et al. 1983; Kraus et al. 1986). This ability to identify individuals through visual sightings has allowed exploration of associations between, and among, individuals in these species that date back to the 1970s in some populations. Additionally, substantial genetic databases have been collected from both humpback whale populations and right whale species (Frasier et al. 2007; Robbins 2007), which allow the assessment of genetic relatedness of individuals that are observed interacting. Long-term collection of data has allowed for the development of family trees in North Atlantic right whales (Suárez-Menéndez et al. 2023), where maternity, and in many cases, paternity have been determined (Frasier et al. 2007). For humpback whales, maternity associations and, more rarely, paternity assignments, have been made in several populations (Clapham and Palsbøll 1997; Cerchio et al. 2005; Suárez-Menéndez et al. 2023; Eichenberger et al. 2023). These long-term observations of known individuals give us new insights into their social associations. As a result of these long-term studies, research into baleen whale sociality has moved from short-term observations at a group level, to long-term, individual-level observations spanning over four decades, and with it, a deeper understanding of their social structure and social relationships.

Social Relationships on the Breeding Grounds

Humpbacks migrate from cold water feeding grounds to warmer waters to mate and give birth (Chittleborough 1965; Tyack and Whitehead 1983). Right whales, though part of a different family, have a broadly similar life history consisting of an annual migratory cycle ranging between high-latitude foraging grounds and mid- to lower-latitude shallow water breeding grounds (IWC 2001; Clapham et al. 2004). While on the breeding grounds, these species exhibit quite different mating strategies. In both mating systems, males and females likely mate with multiple individuals. Their behavior on the breeding grounds, however, is quite different. Right whales form relatively large social aggregations, referred to as surface-active groups, comprising multiple individuals rolling and touching each other at the surface (Donnelly 1967; Payne 1995; Kraus and Hatch 2001; Best et al. 2003; Parks et al. 2007). In many of these groups, sexual activity is also observed, leading to the early theory that these groups function for courtship or mating (Donnelly 1967). Right whales have proportionally massive testes mass for their body size, and their testes are the largest in the world. They are up to almost a metric ton in size (Omura et al. 1969). This has been taken as morphological evidence of the role of sperm competition in their mating system (Brownell and Ralls 1986). Behavioral observations of these large aggregations of whales suggest multiple males compete to get close to the female to achieve a successful mating, and multiple males that successfully get close to the female can inseminate the same female over a short period (Kraus and Hatch 2001; Mate et al. 2005). Males that successfully inseminate a female face the additional competition for fertilization of the egg through sperm competition.

Humpbacks follow a female defense polygyny (Helweg et al. 1992) or male dominance polygyny (Clapham and Palsbøll 1997; Cerchio et al. 2005; Darling et al. 2006) mating system, which often involves aggressive competitive interactions (Mesnick and Ralls 2009). This competitive behavior results in the formation of large groups ("competitive pods") that appear to function in intrasexual competition (Tyack and Whitehead 1983; Baker and Herman 1984). In general, these groups typically begin with a male joining and escorting a female, including mothers with suckling calves (Craig et al. 2002; Morete et al. 2007). This male, considered the primary escort, may then be joined by more males (termed secondary escorts) that attempt to displace the primary escort through physical competition (Tyack and Whitehead 1983; Baker and Herman 1984; Clapham et al. 1992). Primary escorts tend to be larger than secondary escorts, are typically among the largest animals in the group, and likely to be sexually mature (Spitz et al. 2002). Testosterone levels in breeding humpback whales support these conclusions, with primary escorts tending to have higher testosterone levels than secondary escorts (Mingramm et al. 2020). Although immature animals may also be present in competitive groups, they tend to remain on the periphery of the group (Spitz et al. 2002), suggesting that males undergo temporary dominance sorting via intrasexual competition.

The common denominator for both species is the lack of evidence for long-term associations beyond the mother-calf pair on the calving and breeding grounds. Humpback whales commonly form small groups that associate for a few hours (Mobley and Herman 1985) and do not re-form in successive breeding seasons (Valsecchi et al. 2002). Right whales form social groups consisting of adults and juveniles on the calving and breeding grounds that can range from male-female pairs to larger aggregations of more than three animals (Parks et al. 2007). Observations of copulations in these groups led researchers to conclude that these interactions primarily functioned for reproduction (Donnelly 1967). However, later research found that right whales were commonly sighted in these social surface-active groups when conception was not possible, for example, when the individual was sexually immature or already pregnant (Parks et al. 2007). Potential functions for these non-conceptive groups have been suggested to involve practice for later successful mating attempts or for females to assess the performance of different males to make a choice and mate with the selected male later (Kraus and Hatch 2001). The latter explanation would necessitate the ability of whales to recognize individuals and would require the presence of more complex longer-term relationships among individuals. So far, there has been no evidence that right whales form long-term associations within these surface-active groups. However, unlike humpback whales, no studies have been published exploring the association indices of right whales within these groups. Given the existing genetic and sighting databases available for North Atlantic right whales, this analysis could shed light on the potential for repeated associations through time in this species.

As highlighted earlier in the chapter, baleen whales rely on acoustic communication, and these signals can transmit tens of kilometers. It therefore stands to reason that baleen whales mediate social interactions acoustically. The following section outlines what is currently known about how their acoustic communication is used to mediate social interactions on the breeding grounds.

Male humpback whales sing (Payne and McVay 1971; Winn and Winn 1978; Hersh et al., this volume), and their song is one of the most conspicuous underwater acoustic displays in the animal kingdom. Surprisingly, and despite the large number of studies undertaken, the function of humpback whale song is still debated in the literature. While it is most likely to be a sexual display, there is debate as to whether it is directed at other males (Tyack 1981; Frankel et al. 1995; Darling and Bérubé 2001; Cholewiak et al. 2018), females (Tyack 1981; Smith et al. 2008), or both (Tyack 1981; Frankel et al. 1995; Herman 2017). Males can be alone (i.e., not part of a group) when they sing (Tyack 1981; Darling and Bérubé 2001), presumably advertising to other males or females in the area. Singing males may also "escort" a female (Darling and Bérubé 2001; Smith et al. 2008), which suggests, at least in this context, it is directed at the female. Singers are often approached by other males, and when they come together or "join," the singer usually stops singing (Tyack 1981; Darling and Bérubé 2001; Smith et al. 2008; Dunlop and Noad 2016). Lone singers are sometimes joined by another male and will stop singing as soon as they are joined (Tyack 1981; Frankel et al. 1995; Darling and Bérubé 2001). Theories on the function of these interactions range from dominance sorting by interrupting the male display (Darling et al. 2006), the facilitation of male-male cooperation since these male pairs sometimes remain together and join other groups containing a female (Darling and Bérubé 2001; Darling et al. 2006), or as males prospecting for females (Smith et al. 2008).

Males do not always sing. On the breeding grounds, male humpbacks adopt a conditional mating strategy consisting of two tactics: to sing, or to seek out and join with females without singing (Clapham 1996; Craig et al. 2002). Research on their breeding interactions during migration suggest males choose their tactic according to the density of other nearby males (Dunlop and Noad 2021; Dunlop and Frère 2023). Studies found that when the number of nearby males was low, a higher proportion of males sang, and these singers sometimes joined a group containing a female. As the number of nearby males, and therefore male-male competition, increased, males tended to stop singing and switch to non-singing tactics (Dunlop and Frère 2023). This was found to operate at both a proximate and ultimate level. At a proximate level, males responded to their immediate social environment and selected their breeding tactic according to the number of other males in their surrounding area (Dunlop and Noad 2016; Dunlop and Frère 2023). At a population level, male density increased steadily following the moratorium on whaling (Noad et al. 2019). Consequently, male humpback whales in the eastern Australian population were found to shift their breeding strategy from primarily using singing tactics in the early post-whaling years, to using non-singing tactics in later years (Dunlop and Frère 2023).

To assess their social environment, i.e., the presence of other males and females within their vicinity, they must rely on acoustic information, given other animals are well out of visual range but within acoustic range. This acoustic information is contained within song and social sounds. Singing is a loud broadcast acoustic display and likely contains information related to the fitness of the singer. This information is available to females and other males within acoustic range. Males that sing while escorting a female in the presence of other males are at risk of attracting competition for that female and therefore being displaced as her primary escort (Dunlop and Noad 2016, 2021). In fact, singing males usually stop singing after joining a female if other males are in the area, potentially to reduce this risk (Dunlop and Noad 2016). The alternative tactic is to actively, but more quietly, seek out females to join. By not singing, they are not broadcasting their fitness to other animals in the area and therefore reducing the risk of attracting competition (Dunlop and Noad 2021). Male humpback whales also emit social sounds that could be related to dominance sorting behaviors within competitive groups (Dunlop et al. 2007; Cusano et al. 2021; Cusano et al. 2020). Groups, especially those containing multiple males, can be highly vocal (Dunlop et al. 2007; Cusano et al. 2021), and these vocalizations are likely to be audible to other males (Dunlop 2018). This means males gain acoustic information on the presence of other males (and likely females) in the area.

Therefore, the mating strategy of male humpbacks is behaviorally plastic. As these signals are available to animals within their acoustic network, it is possible both within- and between-group dominance sorting occurs acoustically. In addition, given whales are constantly moving in and out of each other's acoustic network, this creates an unpredictable social environment. To manage this, male humpback whales must continually assess and make tactical trade-off decisions based on animals within this network.

Right whale species have a quite different vocal repertoire compared to humpback whales. Rather than having complex and changing structured male songs, right whales across all three species have been shown to have a graded repertoire of tonal, pulsive, and impulsive signal types that appear to be produced by both sexes (Clark 1982; Clark 1983; Webster et al. 2016; Parks et al. 2011). Unlike humpback whales, the most complex acoustic sequences described for right whales are produced by females within surface-active groups (Parks and Tyack 2005). Females produce a wide variety of loud tonal signals, similar in structure to humpback whale song units, but lacking the repetitive structure of male humpback whale song. Playback experiments of these signals show that these sounds from females are particularly attractive to adult males on the foraging grounds (Parks 2003). Intriguingly, playbacks of these calls from southern right whale surface-active groups recorded on the winter calving grounds attracted both female and male North Atlantic right whales of all ages on the summer foraging grounds. One interpretation of this observation could be that the novelty of the signals, from a "new" right whale, was attractive to right whales in the habitat. Alternatively, the repertoires of these two species, while similar, may differ sufficiently to elicit a curious response from right whales. A future study using calls from another balaenid, the bowhead whale, may be useful in distinguishing between these two hypotheses.

Right whale males also produce signals associated with mating. Although they don't sing, recent recordings found that the gunshot, a loud impulsive sound typically detected during the late summer and autumn months, may be used in a repeating pattern by adult males, potentially akin to song (Parks and Tyack 2005; Crance et al. 2019). This research is in its infancy, largely due to the offshore nature of these displays (Matthews et al. 2014; Crance et al. 2017; Shabangu et al. 2021) and the very low population sizes of North Atlantic and North Pacific right whales, which were decimated centuries before people started exploring the sounds of whales (IWC 2001). Given the potential for female incitation of male-male competition for mating in surface-active groups, and the seasonal acoustic displays by males in the autumn on the higher latitude foraging grounds, right whales appear to have a significantly different use of acoustics for reproduction than humpback whales. More research is needed to better understand the diversity of signals produced by right whales.

Mother-Calf Associations

The mother-calf association remains the longest and most stable social association documented across all baleen whale species. Calves learn their migratory route from their mothers and regularly return to their natal calving grounds and foraging grounds (Carroll et al. 2015). Mother-calf pairs make up the most stable social association in the right whale (Taber and Thomas 1982; Best 1990; Hamilton and Cooper 2010; Hamilton et al. 1995) and humpback whale calving and breeding grounds (Cartwright and Sullivan 2009; Szabo and Duffus 2008; Pack et al. 2017). Right whale mother-calf pairs are commonly sighted spatially separated from other right whales, with relatively few sightings of mother-calf pairs associated with other right whales close by (Taber and Thomas 1982; Rayment et al. 2012). By contrast, humpback calves are often sighted within a breeding group (Cartwright and Sullivan 2009; Smith et al. 2008). This difference facilitates two areas of research. For humpbacks, the focus has been on the audience effect in terms of females with their newborn calf detecting and avoiding harassing males. For right whales, mother-calf habitat selection and acoustic behavior suggests attempts to hide acoustically, though whether this is from harassing males or potential predators is yet to be determined. Both areas of research go toward understanding how social interactions are mediated acoustically, as both help to elucidate what information is contained within their acoustic signals, and how whales might use this information socially.

With the overlap in breeding and calving grounds in humpback whales, males can sometimes join and harass a mother with her calf (Cartwright and Sullivan 2009). When this occurs, the calf may be at higher risk of separation from its mother and, because the group tends to move faster, is also subject to a higher energetic output (Cartwright and Sullivan 2009; Craig et al. 2014). Because of this, humpback whale females with their calf have evolved a way of minimizing contact with males. Mothers with a calf tend to avoid these males behaviorally, by moving into shallow water (Kavanagh et al. 2017; Craig et al. 2014), as well as acoustically, by producing vocal sounds at lower intensities, which is a form of vocal crypsis (Dunlop 2016; Videsen et al. 2021; Indeck et al. 2020).

Right whale mother-calf pairs also show evidence for attention to audience effects in their communication by using acoustically cryptic behaviors when calves are young and potentially vulnerable to predation. Right whale females appear to select shallow habitat areas on the calving grounds that have poor acoustic propagation characteristics, allowing them to minimize the detection likelihood of their acoustic communication signals from potential eavesdropping predators or conspecifics (Zeh et al. 2022). Studies of North Atlantic (Cusano et al. 2019; Parks et al. 2019a, 2019b) and southern (Dombroski et al. 2020) right whale mother-calf pairs found that novel, extremely quiet, pulsed signals are produced preferentially by lactating females with calves. These signals are rarely produced by any other right whales on the calving grounds. In addition, lactating female right whales have rarely been documented producing higher amplitude call types on the calving grounds (Nielsen et al. 2019; Parks et al. 2019a; Dombroski et al. 2020). Though these findings are not directly related to sociality, they show that female humpback and right whales are responding to their social environment. Like males, their communication space is an important factor in determining their behavior, adding strength to the argument that social interactions extend beyond their immediate group.

Social Relationships on the Foraging Grounds

It has been a long-standing question of how baleen whales are able to find their prey, given the limited visibility in the ocean waters, and whether this involves cooperative interactions between group members. Humpback social interactions on the foraging grounds are broadly like associations on the breeding grounds, that is, short-term associations that last in the order of minutes to days. Most humpback whales are sighted alone or in pairs over a period of days (Whitehead 1983; Weinrich and Kuhlberg 1991). Right whale social associations on the foraging grounds are like the social associations observed on the calving grounds. They are generally sighted alone or in social surface-active groups, with the mother-calf association in the first year being the highest degree of association observed (Hamilton and Cooper 2010). Like breeding interactions, these findings suggest that baleen whale social structure on the feeding grounds is not complex, but consists of short-term, group-based interactions that can result from aggregation in areas with dense prey availability.

Larger associations of ephemeral and changing group compositions, often associated with active foraging, are becoming increasingly common across all known foraging grounds in humpback whales. This is likely due to their increased numbers following the moratorium on whaling (Jackson et al. 2016; Noad et al. 2019). Group sizes now range from four to eight in the western

North Atlantic and North Pacific feeding areas (Sharpe 2001; Mastick et al. 2022), to "super groups" of 20 to 200 whales in the Southern Hemisphere (Findlay et al. 2017; Pirotta et al. 2021). Often the density of humpback whales in an area is related to the density of available prey, meaning whales can be found in large numbers in geographically limited areas. The size of associated feeding groups has been shown to vary with prey type and the size of the prey patch, with larger groups of humpbacks observed near larger patches of fish prey (Whitehead 1983). Within dense aggregations of prey, extremely high densities of whales have been observed in the order of 100 whales per 2.5 km^2. These observations do not necessarily indicate that these animals have stable social associations in these large groups, only that they are aggregated in an area with abundant food resources. Similarly, in right whales, surface-active groups described for the calving grounds, are now also seen on the foraging grounds, with larger groups typically seen in the late summer and early autumn (Parks et al. 2007). Like humpback whales, right whales are known to aggregate in high densities in areas with high concentrations of prey availability. For example, up to 50% of the known population of the endangered North Atlantic right whale has been sighted in their springtime feeding habitat in Massachusetts (Mayo et al. 2018).

Turning back to acoustic communication, humpback whales are highly vocal on the feeding grounds suggesting that there are social interactions occurring. A wide range of social calls have been detected across multiple foraging grounds in the North Pacific (Cerchio and Dahlheim 2001; Wild and Gabriele 2014; Fournet et al. 2015), North Atlantic (Stimpert et al. 2011; Parks et al. 2014; Videsen et al. 2021), and Southern Ocean (Van Opzeeland et al. 2013). In other baleen whale species, there is evidence of increased signal production in areas of high prey availability where active foraging is occurring (Croll et al. 2002; Oleson et al. 2007). These acoustic behaviors may function to advertise resource availability to conspecifics, through a specific feeding call, attract conspecifics to areas of high prey concentration, or as a way of coordinating feeding activity to increase the energetic intake per individual.

Group-based coordination of behaviors is also an indicator of social complexity within a species (Freeberg and Krams 2015). Several studies have highlighted high levels of coordinated foraging behavior in different humpback whale behaviors. This includes timing and role coordination in group surfacing feeding (Sharpe 2001; Wiley et al. 2011; Mastick et al. 2022), as well as timing and coordination of bottom foraging among temporarily stable social groups that last on the order of hours to days (Ware et al. 2014; see also Wild and Rendell, this volume). Interestingly, on the feeding grounds, detailed observations suggest that coordinated humpback whale group feeding behaviors may not all be cooperative. In some situations, whales produce "trumpet blows" when new individuals join stable groups. These "trumpet blows" are also produced in competitive groups on the breeding grounds during periods of aggressive physical contact between competing males (Cusano et al. 2021; Silber 1986). Data from tagged individuals indicate that not all individuals contribute bubbles to the corralling of prey in bubble-net feeding, but rather simply surface inside the bubble nets created by other whales (Wiley et al. 2011). This suggests that different whales may play different roles in these foraging groups, and exploration of whether the degree of contribution affects the stability of a short-term association would be an interesting avenue to explore (Mastick et al. 2022).

Humpback whales are known to learn novel acoustic signals through their changing songs (see Hersh et al., this volume). There is additional evidence of social learning occurring on the foraging grounds. A novel foraging behavior, where whales used lobtailing at the surface to create a surface bubble cloud, started in the 1980s in the western North Atlantic humpback whale population (Weinrich et al. 1992). This behavior was tracked using social network analyses as it spread rapidly through the population (Allen et al. 2013) demonstrating that humpback whales possess an ability for rapid social learning for other behaviors (Wild and Rendell, this volume). This ability for social learning indicates that social interactions with conspecifics are important, and individuals notice, and can mimic, the behaviors of other conspecifics.

For right whales, the only evidence of stable long-term associations for North Atlantic right whales during migration and on the foraging grounds is the association between mother-calf pairs (Hamilton and Cooper 2010). The average time of weaning of North Atlantic right whales is thought to be in the middle of the summer feeding season, about 8–9 months after birth, when the mother-calf pair are on the foraging grounds (Hamilton et al. 2022). There is strong evidence in all three right whale species that there is matrilineally determined site fidelity as the calf travels with its mother during its first migration (Malik et al. 1999; Best 2000; Burnell 2001; Valenzuela et al. 2009; Carroll et al. 2015; Garland and Carroll 2022). This site fidelity includes both return to the natal calving areas and for species where data are available, often fidelity to specific feeding areas.

Historically, foraging grounds used by all three species are known to be more widespread than observed today, with the suggestion that the memories of those foraging grounds transmitted through social learning were lost as individual whales were removed from the species through whaling (Clapham et al. 2004; Jackson et al. 2016). Unlike humpback whales, right whales are not known to feed in a coordinated manner to entrap their prey but feed primarily on very small zooplankton with limited mobility (Kenney et al. 1986). There have been observations of coordinated feeding activities in right whales in the form of feeding echelons (Arguelles et al. 2023). In these feeding associations, the right whales feed with their mouths open in a formation like a V of flying geese. One whale leads the group while the others feed in the slipstream of the whale in front of them. This formation functions to concentrate the zooplankton (Fish et al. 2013), efficiently "mowing" down large areas of high zooplankton abundance. However, there have been no studies exploring the social associations of whales in these feeding echelons, the stability of these associations, or whether these ephemeral groups arise spontaneously from whales co-located in the same feeding patch or if whales have preferred associates for this foraging behavior. This is a potentially fruitful research area in right whale sociality.

Future Avenues for Study

In this chapter we have reviewed advances in our understanding of social interactions, with a focus on the two baleen whale species for which we have sufficient data to do so. Exploration of the social associations and interactions among other baleen whale species should be a high priority. The relatively small number of species, with significant differences in their prey, migratory patterns, and reproductive strategies (Eichenberger et al. 2023) suggest that variations in the selective pressures for social behavior must vary across these families. Being able to track individuals throughout their migratory range will also aid our understanding of their social structure (e.g., Abrahms et al. 2019). This is likely the next step that we can take to better understand how whales mediate their close associations with other whales.

With the advent of more advanced genetic (see Foote et al., this volume; Parsons et al., this volume) and imaging techniques, it is becoming possible to identify more individuals and build more catalogs to track social interactions in other baleen species (Attard et al. 2018; Blount et al. 2022; Cheeseman et al. 2022). For both right whales and humpback whales, individual identifying characteristics make it possible for researchers to document when individuals are co-located in space and time during visual surveys. The addition of genetic information for several right whale and humpback whale populations means there is now a possibility to reexamine previous interactions to assess the relatedness of individuals involved (Pomilla and Rosenbaum 2006; Rew et al. 2011). In right whales, assessment of the genetic relatedness of individuals in surface-active groups in different habitats may help shed light on the functions of these social interactions that likely vary seasonally.

Another likely avenue for study is the potential for an individually distinctive vocal recognition system in baleen whale species. Based on available observations, it has been suggested that social associations, and potentially dominance hierarchies, could be mediated through individual recognition of others acoustically (Darling and Bérubé 2001; Tyack 2022). In humpback whales, individually distinctive traits have been documented in performance of shared songs (Lamoni et al. 2023). McCordic et al. (2016) found that there is measurable individual distinctiveness in the production of a single call type, the contact calls in right whales, indicating that there is the potential for "speaker recognition" through voice cues in baleen whales. Playback studies are an obvious tool with which to test for differential responses to signals from different individuals. This research could be used to compare behavioral responses to signals from close associates or kin versus individuals rarely observed to associate within the population, to signals from other populations. Indeed, a playback study of North Atlantic right whales elicited significantly different responses to signals from surface-active groups played from North Atlantic versus southern right whales (Parks 2003) showing at least the ability to discriminate species-level differences in acoustic signaling from two closely related species.

Summary

Studying large, long-lived nomadic baleen whales presents a range of challenges to researchers. Despite a long history of human interactions with baleen whales, there are still fundamental gaps in our knowledge about their basic biology. For understanding their social complexity, the scale and scope of their social interactions exceed our current technological abilities to observe them. Increased technological advances in imaging and genetics for individual identification, combined with global databases that can integrate observations from multiple groups, are starting to provide glimpses into the social lives of these enigmatic animals. As these species and populations continue to grow, with the reduced pressure from active large-scale hunting, it is likely that new types of interactions will be observed.

References

Abrahms, Briana, Elliott L. Hazen, Ellen O. Aikens, et al. 2019. "Memory and Resource Tracking Drive Blue Whale Migrations." *Proceedings of the National Academy of Sciences* 116 (12): 5582–87. https://doi.org/10.1073/pnas.1819031116.

Allen, Jenny A., Mason T. Weinrich, William Hoppitt, and Luke Rendell. 2013. "Network-Based

Diffusion Analysis Reveals Cultural Transmission of Lobtail Feeding in Humpback Whales." *Science* 340 (6131): 485–88. https://doi.org/10.1126/science.1231976.

Arguelles, M., Carla Fiorito, Mariano Alberto Coscarella, Ana Fazio, and Marcelo Bertellotti. 2023. "First Observations of Cooperative Circle Feeding in Southern Right Whales (*Eubalaena australis*)." *Aquatic Mammals* 49 (1): 1–6. https://doi.org/10.1578/am.49.1.2023.1.

Attard, Catherine, Luciano B. Beheregaray, Jonathan Sandoval-Castillo, et al. 2018. "From Conservation Genetics to Conservation Genomics: A Genome-Wide Assessment of Blue Whales (*Balaenoptera musculus*) in Australian Feeding Aggregations." *Royal Society Open Science* 5 (1): 170925. https://doi.org/10.1098/rsos.170925.

Baker, C. Scott, and Louis M. Herman. 1984. "Aggressive Behavior Between Humpback Whales (*Megaptera novaeangliae*) Wintering in Hawaiian Waters." *Canadian Journal of Zoology* 62 (10): 1922–37. https://doi.org/10.1139/z84-282.

Baker, C. Scott, Anjanette Perry, and Louis. M. Herman. 1987. "Reproductive Histories of Female Humpback Whales *Megaptera novaeangliae* in the North Pacific." *Marine Ecology Progress Series* 41 (2): 103–14. https://doi.org/10.3354/meps041103.

Best, Peter B. 1990. "Trend in the Inshore Right Whale Population of South Africa, 1969–87." *Marine Mammal Science* 6 (2): 93–108. https://doi.org/10.1111/j.1748-7692.1990.tb00232.x.

Best, Peter B. 2000. "Coastal Distribution, Movements and Site Fidelity of Right Whales *Eubalaena australis* off South Africa, 1969–98." *South African Journal of Marine Science* 22: 43–55. https://doi.org/10.2989/025776100784125618.

Best, Peter B., Cathy M. Schaeff, Desray Reeb, and Per J. Palsbøll. 2003. "Composition and Possible Function of Social Groupings of Southern Right Whales in South African Waters." *Behaviour* 140 (11/12): 1469–94. https://www.jstor.org/stable/4536102.

Blount, Drew, Shane Gero, Jon Van Oast, et al. 2022. "Flukebook: An Open-Source AI Platform for Cetacean Photo Identification." *Mammalian Biology* 102 (3): 1005–23. https://doi.org/10.1007/s42991-021-00221-3.

Boyd, Ian. L. 2004. "Migration of Marine Mammals." In *Biological Resources and Migration*, edited by Dietrich Werner, 203–10. Springer. https://doi.org/10.1007/978-3-662-06083-4_20.

Brownell, Robert L., and Katherine Ralls. 1986. "Potential for Sperm Competition in Baleen Whales." *Reports of the International Whaling Commission* 8: 97–112.

Burnell, Stephen R. 2001. "Aspects of the Reproductive Biology, Movements and Site Fidelity of Right Whales off Australia." *Journal of Cetacean Research and Management* 2: 89–102. https://doi.org/10.47536/jcrm.vi.272.

Carroll, Emma L., C. Scott Baker, Mandy Watson, et al. 2015. "Cultural Traditions Across a Migratory Network Shape the Genetic Structure of Southern Right Whales Around Australia and New Zealand." *Scientific Reports* 5 (1). https://doi.org/10.1038/srep16182.

Cartwright, Rachel, and Matthew Sullivan. 2009. "Associations with Multiple Male Groups Increase the Energy Expenditure of Humpback Whale (*Megaptera novaeangliae*) Female and Calf Pairs on the Breeding Grounds." *Behaviour* 146 (11): 1573–600. https://doi.org/10.1163/156853909x458377.

Cerchio, Salvatore, and Marilyn Dahlheim. 2001. "Variation in Feeding Vocalizations of Humpback Whales *Megaptera novaeangliae* from Southeast Alaska." *Bioacoustics* 11 (4): 277–95. https://doi.org/10.1080/09524622.2001.9753468.

Cerchio, Salvatore, Jeff K. Jacobsen, Danielle Cholewiak, Erin A. Falcone, and D. Andrew Merriwether. 2005. "Paternity in Humpback Whales, *Megaptera novaeangliae*: Assessing Polygyny and Skew in Male Reproductive Success." *Animal Behaviour* 70 (2): 267–77. https://doi.org/10.1016/j.anbehav.2004.10.028.

Cheeseman, Ted, Ken Southerland, Jinmo Park, et al. 2022. "Advanced Image Recognition: A Fully Automated, High-Accuracy Photo-Identification Matching System for Humpback Whales." *Mammalian Biology* 102 (3): 915–29. https://doi.org/10.1007/s42991-021-00180-9.

Chittleborough, R. Graham. 1965. "Dynamics of Two Populations of the Humpback Whale, *Megaptera novaeangliae* (Borowski)." *Australian Journal of Marine and Freshwater Research* 16: 33–128. https://doi.org/10.1071/MF9650033.

Cholewiak, Danielle, Salvatore Cerchio, Jeff K. Jacobsen, Jorge Urbán-R., and Christopher W. Clark. 2018. "Songbird Dynamics Under the Sea: Acoustic Interactions Between Humpback Whales Suggest Song Mediates Male Interactions." *Royal Society Open Science* 5 (2): 171298. https://doi.org/10.1098/rsos.171298.

Christiansen, Fredrik, Gísli A. Víkingsson, Marianne H. Rasmussen, and David Lusseau. 2013. "Minke Whales Maximise Energy Storage on Their Feeding Grounds." *Journal of Experimental Biology* 216 (3): 427–36. https://doi.org/10.1242/jeb.074518.

Clapham, Phillip J. 1993. "Social Organization of Humpback Whales on a North Atlantic Feeding Ground." *Symposium of the Zoological Society of London* 66: 131–45. https://doi.org/10.1093/oso/9780198540694.003.0008.

Clapham, Phillip J. 1996. "The Social and Reproductive Biology of Humpback Whales: An Ecological Perspective." *Mammal Review* 26 (1): 27–49. https://doi.org/10.1111/j.1365-2907.1996.tb00145.x.

Clapham, Phillip J. 2000. "The Humpback Whale: Seasonal Feeding and Breeding in a Baleen Whale." In *Cetacean Societies: Field Studies of Dolphins and Whales*, edited by Janet Mann, Richard C. Connor, Peter L. Tyack, and Hal Whitehead, 173–96. University of Chicago Press.

Clapham, Phillip J., Caroline Good, Sara E. Quinn, Randall R. Reeves, James E. Scarff, and Robert L. Brownell Jr. 2004. "Distribution of North Pacific Right Whales (*Eubalaena japonica*) as Shown by 19th and 20th Century Whaling Catch and Sighting Records." *Journal of Cetacean Research and Management* 6 (1): 1–6. https://doi.org/10.47536/jcrm.v6i1.783.

Clapham, Phillip J., and Per J. Palsbøll. 1997. "Molecular Analysis of Paternity Shows Promiscuous Mating in Female Humpback Whales (*Megaptera novaeangliae*, Borowski)." *Proceedings of the Royal Society B* 264 (1378): 95–98. https://doi.org/10.1098/rspb.1997.0014.

Clapham, Phillip J., Per J. Palsbøll, David K. Mattila, and Oswaldo Vásquez. 1992. "Composition and Dynamics of Humpback Whale Competitive Groups in the West Indies." *Behaviour* 122 (3–4): 182–94. https://doi.org/10.1163/156853992x00507.

Clark, Christopher W. 1982. "The Acoustic Repertoire of the Southern Right Whale, a Quantitative Analysis." *Animal Behaviour* 30 (4): 1060–71. https://doi.org/10.1016/s0003-3472(82)80196-6.

Clark, Christopher W. 1983. "Acoustic Communication and Behavior of the Southern Right Whale." In *Communication and Behavior of Whales*, edited by Roger Payne, Westview Press.

Connor, Richard C., Mason T. Weinrich, and Katherine A. Sardi. 2005. "Social Interactions of Humpback Whale (*Megaptera novaeangliae*) Mother/Calf Pairs on a North Atlantic Feeding Ground." *Behaviour* 142 (6): 731–50. https://doi.org/10.1163/1568539054729132.

Constantine, Rochelle, Debbie Steel, Judith Allen, et al. 2014. "Remote Antarctic Feeding Ground Important for East Australian Humpback Whales." *Marine Biology* 161 (5): 1087–93. https://doi.org/10.1007/s00227-014-2401-2.

Corkeron, Peter J., and Richard C. Connor. 1999. "Why Do Baleen Whales Migrate?" *Marine Mammal Science* 15 (4): 1228–45. https://doi.org/10.1111/j.1748-7692.1999.tb00887.x.

Craig, Alison S., Louis M. Herman, and Adam A. Pack. 2002. "Male Mate Choice and Male-Male Competition Coexist in the Humpback Whale (*Megaptera novaeangliae*)." *Canadian Journal of Zoology* 80 (4): 745–55. https://doi.org/10.1139/z02-050.

Craig, Alison S., Louis M. Herman, Adam A. Pack, and James O. Waterman. 2014. "Habitat Segregation by Female Humpback Whales in Hawaiian Waters: Avoidance of Males?" *Behaviour* 151 (5): 613–31. https://doi.org/10.1163/1568539x-00003151.

Crance, Jessica L., Catherine L. Berchok, and Jennifer L. Keating. 2017. "Gunshot Call Production by the North Pacific Right Whale *Eubalaena japonica* in the Southeastern Bering Sea." *Endangered Species Research* 34 (September): 251–67. https://doi.org/10.3354/esr00848.

Crance, Jessica L., Catherine L. Berchok, Dana Wright, Arial Brewer, and Daniel Woodrich. 2019. "Song Production by the North Pacific Right Whale, *Eubalaena japonica*." *Journal of the Acoustical Society of America* 145 (6): 3467–79. https://doi.org/10.1121/1.5111338.

Croll, Donald A., Christopher W. Clark, Alejandro Acevedo, et al. 2002. "Only Male Fin Whales Sing Loud Songs." *Nature* 417 (6891): 809. https://doi.org/10.1038/417809a.

Cusano, Dana A., Lisa Conger, Sofie M. Van Parijs, and Susan E. Parks. 2019. "Implementing Conservation Measures for the North Atlantic Right Whale: Considering the Behavioral Ontogeny of Mother-Calf Pairs." *Animal Conservation* 22 (3): 228–37. https://doi.org/10.1111/acv.12457.

Cusano, Dana A., Katherine L. Indeck, Michael J. Noad, and Rebecca A. Dunlop. 2020. "Humpback Whale (*Megaptera novaeangliae*) Social Call Production Reflects Both Motivational State and Arousal." *Bioacoustics* 31 (1): 17–40. https://doi.org/10.1080/09524622.2020.1858450.

Cusano, Dana A., David Paton, Michael J. Noad, and Rebecca A. Dunlop. 2021. "Socially Complex Breeding Interactions in Humpback Whales Are Mediated Using a Complex Acoustic Repertoire." *Frontiers in Marine Science* 8: 665186. https://doi.org/10.3389/fmars.2021.665186.

Darling, James D., and Martine Bérubé. 2001. "Interactions of Singing Humpback Whales with Other Males." *Marine Mammal Science* 17 (3): 570–84. https://doi.org/10.1111/j.1748-7692.2001.tb01005.x.

Darling, James D., Meagan E. Jones, and Charles P. Nicklin. 2006. "Humpback Whale Songs: Do They Organize Males During the Breeding Season?" *Behaviour* 143 (9): 1051–101. https://doi.org/10.1163/156853906778607381.

Dombroski, Julia R. G., Susan E. Parks, Paulo A. C. Flores, et al. 2020. "Animal-Borne Tags Provide Insights into the Acoustic Communication of Southern Right Whales (*Eubalaena australis*) on the Calving Grounds." *Journal of the Acoustical Society of America* 147 (6): EL498–503. https://doi.org/10.1121/10.0001391.

Donnelly, B. G. 1967. "Observations on the Mating Behaviour of the Southern Right Whale *Eubalaena australis*." *South African Journal of Science* 64 (5): 176–81. https://hdl.handle.net/10520/AJA00382353_3933.

Dunlop, Rebecca A. 2016. "Changes in Vocal Parameters with Social Context in Humpback Whales: Considering the Effect of Bystanders." *Behavioral Ecology and Sociobiology* 70 (6): 857–70. https://doi.org/10.1007/s00265-016-2108-0.

Dunlop, Rebecca A. 2018. "The Communication Space of Humpback Whale Social Sounds in Wind-Dominated Noise." *Journal of the Acoustical Society of America* 144 (2): 540–51. https://doi.org/10.1121/1.5047744.

Dunlop, Rebecca A., and Céline Frère. 2023. "Post-Whaling Shift in Mating Tactics in Male Humpback Whales." *Communications Biology* 6 (1). https://doi.org/10.1038/s42003-023-04509-7.

Dunlop, Rebecca A., and Michael J. Noad. 2016. "The 'Risky' Business of Singing: Tactical Use of Song During Joining by Male Humpback Whales." *Behavioral Ecology and Sociobiology* 70 (12): 2149–60. https://doi.org/10.1007/s00265-016-2218-8.

Dunlop, Rebecca A., and Michael J. Noad. 2021. "The Eavesdropping Risk of Conspicuous Sexual Signaling in Humpback Whales." *Behavioral Ecology and Sociobiology* 75 (8). https://doi.org/10.1007/s00265-021-03048-7.

Dunlop, Rebecca A., Michael J. Noad, Douglas H. Cato, and Dale Stokes. 2007. "The Social Vocalization Repertoire of East Australian Migrating Humpback Whales (*Megaptera novaeangliae*)." *Journal of the Acoustical Society of America* 122 (5): 2893–905. https://doi.org/10.1121/1.2783115.

Eichenberger, Franca, Ellen C. Garland, and Emma L. Carroll. 2023. "Reproductive Tactics in Baleen Whales." In *Sex in Cetaceans: Morphology, Behavior, and the Evolution of Sexual Strategies*, edited by Bernd Würsig and Dara N. Orbach, 469–98. Springer.

Findlay, Ken P., S. Mduduzi Seakamela, Michael A. Meÿer, et al. 2017. "Humpback Whale 'Super-Groups': A Novel Low-Latitude Feeding Behaviour of Southern Hemisphere Humpback Whales (*Megaptera novaeangliae*) in the Benguela Upwelling System." *PLOS ONE* 12 (3): e0172002. https://doi.org/10.1371/journal.pone.0172002.

Fish, Frank E., Kimberly T. Goetz, David J. Rugh, and Linda Vate Brattström. 2013. "Hydrodynamic Patterns Associated with Echelon Formation Swimming by Feeding Bowhead Whales (*Balaena mysticetus*)." *Marine Mammal Science* 29 (4). https://doi.org/10.1111/mms.12004.

Fournet, Michelle E. H., Andy Szabo, and David K. Mellinger. 2015. "Repertoire and Classification of Non-Song Calls in Southeast Alaskan Humpback Whales (*Megaptera novaeangliae*)." *Journal of the Acoustical Society of America* 137 (1): 1–10. https://doi.org/10.1121/1.4904504.

Frankel, Adam S., Christopher G. Clark, Louis M. Herman, and Christine M. Gabriele. 1995. "Spatial Distribution, Habitat Utilization, and Social Interactions of Humpback Whales, *Megaptera novaeangliae*, off Hawai'i, Determined Using Acoustic and Visual Techniques." *Canadian Journal of Zoology* 73 (6): 1134–46. https://doi.org/10.1139/z95-135.

Frasier, Timothy R., Philip K. Hamilton, Moira W. Brown, et al. 2007. "Patterns of Male Reproductive Success in a Highly Promiscuous Whale Species: The Endangered North Atlantic Right Whale." *Molecular Ecology* 16 (24): 5277–93. https://doi.org/10.1111/j.1365-294x.2007.03570.x.

Freeberg, Todd M., and Indriķis Krams. 2015. "Does Social Complexity Link Vocal Complexity and Cooperation?" *Journal of Ornithology* 156 (S1): 125–32. https://doi.org/10.1007/s10336-015-1233-2.

Garland, Ellen C., and Emma L. Carroll. 2022. "Culture and Social Learning in Baleen Whales." In *Ethology and Behavioral Ecology of Mysticetes*, edited by Christopher W. Clark and Ellen C. Garland, 177–91. Springer. https://doi.org/10.1007/978-3-030-98449-6_8.

Glockner, Deborah A., and Spearous C. Venus. 1983. "Identification, Growth Rate and Behavior of Humpback Whale (*Megaptera novaeangliae*) Cows and Calves in the Waters off Maui, Hawaii, 1977–79." In *Communication and Behavior of Whales*, edited by Roger Payne, 223–58. Westview Press.

Goldbogen, Jeremy A., and Peter T. Madsen. 2018. "The Evolution of Foraging Capacity and

Gigantism in Cetaceans." *Journal of Experimental Biology* 221 (11): jeb166033. https://doi.org/10.1242/jeb.166033.

Hamilton, Philip K., and Lindsay A. Cooper. 2010. "Changes in North Atlantic Right Whale (*Eubalaena glacialis*) Cow-Calf Association Times and Use of the Calving Ground: 1993–2005." *Marine Mammal Science* 26 (4): 896–916. https://doi.org/10.1111/j.1748-7692.2010.00378.x.

Hamilton, Philip K., Brenna A. Frasier, Lisa Conger, R. Clay George, Katharine Jackson, and Timothy R. Frasier. 2022. "Genetic Identifications Challenge Our Assumptions of Physical Development and Mother-Calf Associations and Separation Times: A Case Study of the North Atlantic Right Whale (*Eubalaena glacialis*)." *Mammalian Biology* 102 (4): 1389–408. https://doi.org/10.1007/s42991-021-00177-4.

Hamilton, Philip K., Marilyn K. Marx, and Scott D. Kraus. 1995. "Weaning in North Atlantic Right Whales." *Marine Mammal Science* 11 (3): 386–90. https://doi.org/10.1111/j.1748-7692.1995.tb00293.x.

Helweg, David A., Adam S. Frankel, Joseph R. Mobley, and Louis M. Herman. 1992. "Humpback Whale Song—Our Current Understanding." In *Marine Mammal Sensory Systems*, edited by Jeanette A. Thomas, Ronald A. Kastelein, and Alexander Ya. Supin, 459–83. https://doi.org/10.1007/978-1-4615-3406-8_32.

Herman, Louis M. 2017. "The Multiple Functions of Male Song Within the Humpback Whale (*Megaptera novaeangliae*) Mating System: Review, Evaluation, and Synthesis." *Biological Reviews* 92 (3): 1795–818. https://doi.org/10.1111/brv.12309.

Herman, Louis M., and Ronald C. Antinoja. 1977. "Humpback Whales in the Hawaiian Breeding Waters: Population and Pod Characteristics." *Scientific Reports of the Whales Research Institute* (Tokyo) 29: 59–85.

Herman, Louis M., Paul H. Forestell, and Ronald C. Antinoja. 1980. "The 1976/77 Migration of Humpback Whales into Hawaiian Waters: Composite Description." In *Final Report to the U.S. Marine Mammal Commission, Report # MMC-77/19. National Technical Information Service PBSO-162332.* United States National Technical Information Services, 55.

Indeck, Katherine L., Elisa Girola, Maëlle Torterotot, Michael J. Noad, and Rebecca A. Dunlop. 2020. "Adult Female-Calf Acoustic Communication Signals in Migrating East Australian Humpback Whales." *Bioacoustics* 30 (3): 341–65. https://doi.org/10.1080/09524622.2020.1742204.

IWC (International Whaling Commission). 2001. "Report of the Workshop on the Comprehensive Assessment of Right Whales: A Worldwide Comparison." *Journal of Cetacean Research and Management* 2: 1–60. https://doi.org/10.47536/jcrm.vi.270.

Jackson, Jennifer A., Emma L. Carroll, Tim D. Smith, Alexandre N. Zerbini, Nathalie J. Patenaude, and C. Scott Baker. 2016. "An Integrated Approach to Historical Population Assessment of the Great Whales: Case of the New Zealand Southern Right Whale." *Royal Society Open Science* 3 (3): 150669. https://doi.org/10.1098/rsos.150669.

Katona, Steven K., and Judith A. Beard. 1990. "Population Size, Migrations and Feeding Aggregations of the Humpback Whale in the Western North Atlantic Ocean." *Reports of the International Whaling Commission* 12: 295–305.

Kavanagh, Ailbhe S., Michael J. Noad, Simon P. Blomberg, et al. 2017. "Factors Driving the Variability in Diving and Movement Behavior of Migrating Humpback Whales (*Megaptera novaeangliae*): Implications for Anthropogenic Disturbance Studies." *Marine Mammal Science* 33 (2): 413–39. https://doi.org/10.1111/mms.12375.

Kenney, Robert D., Martin A. M. Hyman, Ralph E. Owen, Gerald P. Scott, and Howard E. Winn.

1986. “Estimation of Prey Densities Required by Western North Atlantic Right Whales.” *Marine Mammal Science* 2 (1): 1–13. https://doi.org/10.1111/j.1748-7692.1986.tb00024.x.

Kraus, Scott D., and Jeremy J. Hatch. 2001. “Mating Strategies in the North Atlantic Right Whale (*Eubalaena glacialis*).” *Journal of Cetacean Research and Management* 2: 237–44.

Kraus, Scott D., Karen E. Moore, Carol A. Price, et al. 1986. “The Use of Photographs to Identify Individual North Atlantic Right Whales (*Eubalaena glacialis*).” *Reports of the International Whaling Commission* 10: 145–51.

Lamoni, Luca, Ellen C. Garland, Jenny A. Allen, Jennifer Coxon, Michael J. Noad, and Luke Rendell. 2023. “Variability in Humpback Whale Songs Reveals How Individuals Can Be Distinctive When Sharing a Complex Vocal Display.” *Journal of the Acoustical Society of America* 153 (4): 2238–50. https://doi.org/10.1121/10.0017602.

Malik, Sobia, Moira W. Brown, Scott D. Kraus, Amy R. Knowlton, Philip K. Hamilton, and Bradley N. White. 1999. “Assessment of Mitochondrial DNA Structuring and Nursery Use in the North Atlantic Right Whale (*Eubalaena glacialis*).” *Canadian Journal of Zoology* 77 (8): 1217–22. https://doi.org/10.1139/cjz-77-8-1217.

Mastick, Natalie, David N. Wiley, David E. Cade, Colin Ware, Susan E. Parks, and Ari S. Friedlaender. 2022. “The Effect of Group Size on Individual Behavior of Bubble-Net Feeding Humpback Whales in the Southern Gulf of Maine.” *Marine Mammal Science* 38 (3): 959–74. https://doi.org/10.1111/mms.12905.

Mate, Bruce, Peter Duley, Barbara Lagerquist, Frederick Wenzel, Alison Stimpert, and Phil Clapham. 2005. “Observations of a Female North Atlantic Right Whale (*Eubalaena glacialis*) in Simultaneous Copulation with Two Males: Supporting Evidence for Sperm Competition.” *Aquatic Mammals* 31 (2): 157. https://doi.org/10.1578/AM.31.2.2005.157.

Matthews, Leanna P., Jessica A. McCordic, and Susan E. Parks. 2014. “Remote Acoustic Monitoring of North Atlantic Right Whales (*Eubalaena glacialis*) Reveals Seasonal and Diel Variations in Acoustic Behavior.” *PLOS ONE* 9 (3): e91367. https://doi.org/10.1371/journal.pone.0091367.

Mayo, Charles A., Laura C. Ganley, Christine A. Hudak et al. 2018. “Distribution, Demography, and Behavior of North Atlantic Right Whales (*Eubalaena glacialis*) in Cape Cod Bay, Massachusetts, 1998–2013.” *Marine Mammal Science* 34 (4): 979–96. https://doi.org/10.1111/mms.12511.

McCordic, Jessica A., Holly Root-Gutteridge, Dana A. Cusano, Samuel L. Denes, and Susan E. Parks. 2016. “Calls of North Atlantic Right Whales *Eubalaena glacialis* Contain Information on Individual Identity and Age Class.” *Endangered Species Research* 30 (May): 157–69. https://doi.org/10.3354/esr00735.

Mesnick, Sarah L., and Katherine Ralls. 2009. “Mating Systems.” In *Encyclopedia of Marine Mammals*, edited by W. F. Perrin, B. Würsig, and Johannes G. M. Thewissen. Academic Press, 712–19. https://doi.org/10.1016/B978-0-12-373553-9.00165-6.

Mingramm, Fletcher M. J., Tamara Keeley, Deanne J. Whitworth, and Rebecca A. Dunlop. 2020. “The Influence of Physiological Status on the Reproductive Behavior of Humpback Whales (*Megaptera novaeangliae*).” *Hormones and Behavior* 117 (January): 104606. https://doi.org/10.1016/j.yhbeh.2019.104606.

Mobley, Joseph R., and Louis M. Herman. 1985. “Transience of Social Affiliations Among Humpback Whales (*Megaptera novaeangliae*) on the Hawaiian Wintering Grounds.” *Canadian Journal of Zoology* 63 (4): 762–72. https://doi.org/10.1139/z85-111.

Morete, Maria E., Tatiana L. Bisi, and Sérgio Rosso. 2007. “Temporal Pattern of Humpback

Whale (*Megaptera novaeangliae*) Group Structure Around Abrolhos Archipelago Breeding Region, Bahia, Brazil." *Journal of the Marine Biological Association of the United Kingdom* 87 (1): 87–92. https://doi.org/10.1017/s0025315407054616.

Nielsen, Mia L. K., Kate R. Sprogis, Lars Bejder, Peter T. Madsen, and Fredrik Christiansen. 2019. "Behavioural Development in Southern Right Whale Calves." *Marine Ecology Progress Series* 629 (October): 219–34. https://doi.org/10.3354/meps13125.

Noad, Michael J., Eric Kniest, and Rebecca A. Dunlop. 2019. "Boom to Bust? Implications for the Continued Rapid Growth of the Eastern Australian Humpback Whale Population Despite Recovery." *Population Ecology* 61 (2): 198–209. https://doi.org/10.1002/1438-390x.1014.

Oleson, Erin M., John Calambokidis, William C. Burgess, Mark A. McDonald, Carrie A. LeDuc, and John A. Hildebrand. 2007. "Behavioral Context of Call Production by Eastern North Pacific Blue Whales." *Marine Ecology Progress Series* 330 (January): 269–84. https://doi.org/10.3354/meps330269.

Omura, Hideo, S. Ohsumi, T. Nemoto, K. Nasu, and T. Kasuya. 1969. "Black Right Whales in the North Pacific." *Scientific Reports of the Whales Research Institute* (Tokyo) 21: 1–78.

Pack, Adam A., Louis M. Herman, Alison S. Craig, et al. 2017. "Habitat Preferences by Individual Humpback Whale Mothers in the Hawaiian Breeding Grounds Vary with the Age and Size of Their Calves." *Animal Behaviour* 133 (November): 131–44. https://doi.org/10.1016/j.anbehav.2017.09.012.

Parks, Susan E. 2003. "Response of North Atlantic Right Whales (*Eubalaena glacialis*) to Playback of Calls Recorded from Surface Active Groups in Both the North and South Atlantic." *Marine Mammal Science* 19 (3): 563–80. https://doi.org/10.1111/j.1748-7692.2003.tb01321.x.

Parks, Susan E., Moira W. Brown, Lisa Conger, et al. 2007. "Occurrence, Composition, and Potential Functions of North Atlantic Right Whale (*Eubalaena glacialis*) Surface Active Groups." *Marine Mammal Science* 23 (4): 868–87. https://doi.org/10.1111/j.1748-7692.2007.00154.x.

Parks, Susan E., Dana A. Cusano, Alison K. Stimpert, Mason T. Weinrich, Ari S. Friedlaender, and David N. Wiley. 2014. "Evidence for Acoustic Communication Among Bottom Foraging Humpback Whales." *Scientific Reports* 4 (1). https://doi.org/10.1038/srep07508.

Parks, Susan E., Dana A. Cusano, Sofie M. Van Parijs, and Douglas P. Nowacek. 2019a. "Acoustic Crypsis in Communication by North Atlantic Right Whale Mother-Calf Pairs on the Calving Grounds." *Biology Letters* 15 (10): 20190485. https://doi.org/10.1098/rsbl.2019.0485.

Parks, Susan E., Dana A. Cusano, Sofie M. Van Parijs, and Douglas P. Nowacek. 2019b. "North Atlantic Right Whale (*Eubalaena glacialis*) Acoustic Behavior on the Calving Grounds." *Journal of the Acoustical Society of America* 146 (1): EL15–21. https://doi.org/10.1121/1.5115332.

Parks, Susan E., Mark Johnson, Douglas P. Nowacek, and Peter L. Tyack. 2011. "Individual Right Whales Call Louder in Increased Environmental Noise." *Biology Letters* 7 (1): 33–35. https://doi.org/10.1098/rsbl.2010.0451.

Parks, Susan E., and Peter L. Tyack. 2005. "Sound Production by North Atlantic Right Whales (*Eubalaena glacialis*) in Surface Active Groups." *Journal of the Acoustical Society of America* 117 (5): 3297–306. https://doi.org/10.1121/1.1882946.

Payne, Roger. 1995. *Among Whales*. Vol. 95. Scribner.

Payne, Roger, Oliver Brazier, Eleanor M. Dorsey, Judith S. Perkins, Victoria J. Rowntree, and A. Titus. 1983. "External Features in Southern Right Whales (*Eubalaena australis*) and Their Use in Identifying Individuals." In *Communication and Behavior of Whales*, edited by Roger Payne, xii, 371–445. Westview Press.

Payne, Roger S., and Scott McVay. 1971. "Songs of Humpback Whales: Humpbacks Emit Sounds in Long, Predictable Patterns Ranging over Frequencies Audible to Humans." *Science* 173 (3997): 585–97. https://doi.org/10.1126/science.173.3997.585.

Payne, Roger, and Douglas C. Webb. 1971. "Orientation by Means of Long Range Acoustic Signaling in Baleen Whales." *Annals of the New York Academy of Sciences* 188 (1): 110–41. https://doi.org/10.1111/j.1749-6632.1971.tb13093.x.

Pirotta, Vanessa, Kylie Owen, David Donnelly, M. Brasier, and Robert Harcourt. 2021. "First Evidence of Bubble-Net Feeding and the Formation of 'Super-Groups' by the East Australian Population of Humpback Whales During Their Southward Migration." *Aquatic Conservation: Marine and Freshwater Ecosystems* 31 (9): 2412–19. https://doi.org/10.1002/aqc.3621.

Pomilla, Cristina, and Howard C. Rosenbaum. 2006. "Estimates of Relatedness in Groups of Humpback Whales (*Megaptera novaeangliae*) on Two Wintering Grounds of the Southern Hemisphere." *Molecular Ecology* 15 (9): 2541–55. https://doi.org/10.1111/j.1365-294x.2006.02943.x.

Ramp, Christian, Martine Bérubé, Per Palsbøll, Wilhelm Hagen, and Richard Sears. 2010. "Sex-Specific Survival in the Humpback Whale *Megaptera novaeangliae* in the Gulf of St. Lawrence, Canada." *Marine Ecology Progress Series* 400 (February): 267–76. https://doi.org/10.3354/meps08426.

Rayment, William, Anthony R. Davidson, Stephen M. Dawson, Elisabeth Slooten, and Trudi Webster. 2012. "Distribution of Southern Right Whales on the Auckland Islands Calving Grounds." *New Zealand Journal of Marine and Freshwater Research* 46 (3): 431–36. https://doi.org/10.1080/00288330.2012.697072.

Rew, Mary Beth, Jooke Robbins, David K. Mattila, Per J. Palsbøll, and Martine Bérubé. 2011. "How Many Genetic Markers to Tag an Individual? An Empirical Assessment of False Matching Rates Among Close Relatives." *Ecological Applications* 21 (3): 877–87. https://doi.org/10.1890/10-0348.1.

Robbins, Julie. 2007. "Structure and Dynamics of the Gulf of Maine Humpback Whale Population." PhD diss., University of St. Andrews.

Schumacher, Stefanie, and David Lazarus. 2004. "Regional Differences in Pelagic Productivity in the Late Eocene to Early Oligocene: A Comparison of Southern High Latitudes and Lower Latitudes." *Palaeogeography, Palaeoclimatology, Palaeoecology* 214 (3): 243–63. https://doi.org/10.1016/j.palaeo.2004.06.018.

Shabangu, Fannie W., Rex K. Andrew, and Ken P. Findlay. 2021. "Acoustic Occurrence, Diel-Vocalizing Pattern, and Detection Ranges of Southern Right Whale Gunshot Sounds off South Africa's West Coast." *Marine Mammal Science* 37 (2): 733–50. https://doi.org/10.1111/mms.12760.

Sharpe, Fred A. 2001. "Social Foraging of the Southeast Alaskan Humpback Whale, *Megaptera novaeangliae*." PhD diss., Simon Fraser University.

Silber, Gregory K. 1986. "The Relationship of Social Vocalizations to Surface Behavior and Aggression in the Hawaiian Humpback Whale (*Megaptera novaeangliae*)." *Canadian Journal of Zoology* 64 (10): 2075–80. https://doi.org/10.1139/z86-316.

Slater, Graham J., Jeremy A. Goldbogen, and Nicholas D. Pyenson. 2017. "Independent Evolution of Baleen Whale Gigantism Linked to Plio-Pleistocene Ocean Dynamics." *Proceedings of the Royal Society B* 284 (1855): 20170546. https://doi.org/10.1098/rspb.2017.0546.

Smith, Joshua N., Anne W. Goldizen, Rebecca A. Dunlop, and Michael J. Noad. 2008. "Songs of Male Humpback Whales, *Megaptera novaeangliae*, Are Involved in Intersexual Interactions." *Animal Behaviour* 76 (2): 467–77. https://doi.org/10.1016/j.anbehav.2008.02.013.

Smith, Joshua N., Hedley S. Grantham, Nick Gales, Michael C. Double, Michael J. Noad, and David Paton. 2012. "Identification of Humpback Whale Breeding and Calving Habitat in the Great Barrier Reef." *Marine Ecology Progress Series* 447 (February): 259–72. https://doi.org/10.3354/meps09462.

Spitz, Scott S., Louis M. Herman, Adam A. Pack, and Mark H. Deakos. 2002. "The Relation of Body Size of Male Humpback Whales to Their Social Roles on the Hawaiian Winter Grounds." *Canadian Journal of Zoology* 80 (11): 1938–47. https://doi.org/10.1139/z02-177.

Steel, Debbie, M. Eric Anderson, Claire Garrigue, et al. 2018. "Migratory Interchange of Humpback Whales (*Megaptera novaeangliae*) Among Breeding Grounds of Oceania and Connections to Antarctic Feeding Areas Based on Genotype Matching." *Polar Biology* 41 (4): 653–62. https://doi.org/10.1007/s00300-017-2226-9.

Stern, Seth J., and Ari S. Friedlander. 2018. "Migrating and Movement." In *Encyclopedia of Marine Mammals*, 3rd ed., edited by Bernd Würsig, Johannes G. M. Thewissen, and Kit M. Kovacs, 602–6. Academic Press.

Stevick, Peter T., Simon Berrow, Martine Bérubé, et al. 2016. "There and Back Again: Multiple and Return Exchange of Humpback Whales Between Breeding Habitats Separated by an Ocean Basin." *Journal of the Marine Biological Association of the United Kingdom* 96 (4): 885–90. https://doi.org/10.1017/s0025315416000321.

Stimpert, Alison K., Whitlow W. L. Au, Susan E. Parks, Thomas P. Hurst, and David N. Wiley. 2011. "Common Humpback Whale (*Megaptera novaeangliae*) Sound Types for Passive Acoustic Monitoring." *Journal of the Acoustical Society of America* 129 (1): 476–82. https://doi.org/10.1121/1.3504708.

Suárez-Menéndez, Marcos, Martine Bérubé, Fabrício Furni, et al. 2023. "Wild Pedigrees Inform Mutation Rates and Historic Abundance in Baleen Whales." *Science* 381 (6661): 990–95. https://doi.org/10.1126/science.adf2160.

Szabo, Andy, and David A. Duffus. 2008. "Mother-Offspring Association in the Humpback Whale, *Megaptera novaeangliae*: Following Behaviour in an Aquatic Mammal." *Animal Behaviour* 75 (3): 1085–92. https://doi.org/10.1016/j.anbehav.2007.08.019.

Taber, Sara M., and Peter O. Thomas. 1982. "Calf Development and Mother-Calf Spatial Relationships in Southern Right Whales." *Animal Behaviour* 30 (4): 1072–83. https://doi.org/10.1016/s0003-3472(82)80197-8.

Tsai, Ching-Hui, and R. Ewan Fordyce. 2015. "The Earliest Gulp-Feeding Mysticete (*Cetacea: mysticeti*) from the Oligocene of New Zealand." *Journal of Mammalian Evolution* 22 (4): 535–60. https://doi.org/10.1007/s10914-015-9290-0.

Tyack, Peter. 1981. "Interactions Between Singing Hawaiian Whales and Conspecifics Nearby." *Behavioral Ecology and Sociobiology* 8 (2): 105–16. https://doi.org/10.1007/bf00300822.

Tyack, Peter L. 2022. "Social Organization of Baleen Whales." In *Ethology and Behavioral Ecology of Mysticetes*, edited by Christopher W. Clark and Ellen C. Garland, 147–75. Springer. https://doi.org/10.1007/978-3-030-98449-6_7.

Tyack, Peter, and Hal Whitehead. 1983. "Male Competition in Large Groups of Wintering Humpback Whales." *Behaviour* 83: 132–54. https://doi.org/10.1163/156853982x00067.

Valenzuela, Luciano O., Mariano Sironi, Victoria J. Rowntree, and Jon Seger. 2009. "Isotopic and Genetic Evidence for Culturally Inherited Site Fidelity to Feeding Grounds in Southern

Right Whales (*Eubalaena australis*)." *Molecular Ecology* 18 (5): 782–91. https://doi.org/10.1111/j.1365-294x.2008.04069.x.

Valsecchi, Elena, Peter T. Hale, Peter J. Corkeron, and William Amos. 2002. "Social Structure in Migrating Humpback Whales (*Megaptera novaeangliae*)." *Molecular Ecology* 11 (3): 507–18. https://doi.org/10.1046/j.0962-1083.2001.01459.x.

Van Opzeeland, Ilse, Sofie Van Parijs, Lars Kindermann, Elke Burkhardt, and Olaf Boebel. 2013. "Calling in the Cold: Pervasive Acoustic Presence of Humpback Whales (*Megaptera novaeangliae*) in Antarctic Coastal Waters." *PLOS ONE* 8 (9): e73007. https://doi.org/10.1371/journal.pone.0073007.

Videsen, Simone K. A., Malene Simon, Mark Johnson, Peter T. Madsen, and Fredrik Christiansen. 2021. "Cryptic Vocal Behavior of Foraging Humpback Whales on Feeding Grounds in West Greenland." *Journal of the Acoustical Society of America* 150 (4): 2879–87. https://doi.org/10.1121/10.0006735.

Ware, Colin, David N. Wiley, Ari S. Friedlaender, et al. 2014. "Bottom Side-Roll Feeding by Humpback Whales (*Megaptera novaeangliae*) in the Southern Gulf of Maine, U.S.A." *Marine Mammal Science* 30 (2): 494–511. https://doi.org/10.1111/mms.12053.

Webster, Trudi, Stephen M. Dawson, William Rayment, Susan E. Parks, and Sofie M. Van Parijs. 2016. "Quantitative Analysis of the Acoustic Repertoire of Southern Right Whales in New Zealand." *Journal of the Acoustical Society of America* 140 (1): 322–33. https://doi.org/10.1121/1.4955066.

Weinrich, Mason T. 1991. "Stable Social Associations Among Humpback Whales (*Megaptera novaeangliae*) in the Southern Gulf of Maine." *Canadian Journal of Zoology* 69 (12): 3012–19. https://doi.org/10.1139/z91-425.

Weinrich, Mason T., and Anne E. Kuhlberg. 1991. "Short-Term Association Patterns of Humpback Whale (*Megaptera novaeangliae*) Groups on Their Feeding Grounds in the Southern Gulf of Maine." *Canadian Journal of Zoology* 69 (12): 3005–11. https://doi.org/10.1139/z91-424.

Weinrich, Mason T., Mark R. Schilling, and Cynthia R. Belt. 1992. "Evidence for Acquisition of a Novel Feeding Behaviour: Lobtail Feeding in Humpback Whales, *Megaptera novaeangliae*." *Animal Behaviour* 44 (6): 1059–72. https://doi.org/10.1016/s0003-3472(05)80318-5.

Whitehead, Hal. 1983. "Structure and Stability of Humpback Whale Groups off Newfoundland." *Canadian Journal of Zoology* 61 (6): 1391–97. https://doi.org/10.1139/z83-186.

Wild, Lauren A., and Christine Gabriele. 2014. "Putative Contact Calls Made by Humpback Whales (*Megaptera novaeangliae*) in Southeastern Alaska." *Canadian Acoustics* 42 (4): 23–31.

Wiley, David N., Colin Ware, Alessandro Bocconcelli, et al. 2011. "Underwater Components of Humpback Whale Bubble-Net Feeding Behaviour." *Behaviour* 148 (5–6): 575–602. https://doi.org/10.1163/000579511x570893.

Winn, Howard E., and Lois K. Winn. 1978. "The Song of the Humpback Whale *Megaptera novaeangliae* in the West Indies." *Marine Biology* 47 (2): 97–114. https://doi.org/10.1007/bf00395631.

Wolman, Allen A., and Charles M. Jurasz. 1977. "Humpback Whales in Hawaii—Vessel Census, 1976." *Marine Fisheries Review* 39 (7): 1–5.

Zeh, Julia, Julia R. G. Dombroski, and Susan E. Parks. 2022. "Preferred Shallow-Water Nursery Sites Provide Acoustic Crypsis to Southern Right Whale Mother-Calf Pairs." *Royal Society Open Science* 9 (5). https://doi.org/10.1098/rsos.220241.

7

Cetacean Social Structure and Identity Signaling

JULIE N. OSWALD, VINCENT M. JANIK, MADALINA MATEI, VOLKER B. DEECKE, AND PETER L. TYACK

Abstract

An important function of many communication signals is broadcasting the identity of the signaler. Identity signals can encode information at the level of the individual, group, population or species and can provide fitness benefits to both sender and receiver. In this chapter we discuss the current state of knowledge regarding identity signaling in cetaceans and how identity signaling has coevolved with social structure. As most cetacean communication is acoustic, we focus on acoustic identity signaling. Group stability is highly diverse in cetaceans, varying from fluid groups in delphinid fission-fusion societies to closed matrilineal groups with almost no immigration and emigration through life. Identity signaling maps onto these social differences with individual identity being reliably encoded in learned signature whistles in many dolphin species, group identity reflected in call repertoires that may permit an assessment of relatedness in species such as killer whales and sperm whales, and species identity being most apparent in baleen whale mate attraction where animals travel long distances to find partners and errors in species recognition would lead to wasted mating opportunities. Future studies are needed to broaden our understanding of how identity is encoded and decoded, and how this influences social decision-making in each species.

Introduction

Variation in communication signals between individuals can have many causes and, if it correlates with a form of identity, can be used by a receiver to extract identity information. However, it can be useful to distinguish identity *signals*, where variation was selected to broadcast identity, from identity *cues*, which may vary for other reasons (Maynard Smith and Harper 2003). For example, there has been considerable debate about whether variations in dialects in songbirds result from cultural or genetic drift, have been selected as signals for population identification, or were selected for other functions (Slater 1989). Different factors can contribute to variation in vocalizations, with optimizing sound transmission in different environments, differences in vocal anatomy, and vocal learning being prominent ones (Vernes et al. 2021). For example, Dowling et al. (2012) report significant increases in the minimum frequency of songbirds recorded along a gradient from quiet rural habitats to urban habitats corresponding to increasing low-frequency noise levels. Differences in vocal anatomy between individuals may lead to vocal differences that allow individual recognition based on voices (Aubin and Jouventin 2002; Carlson et al. 2020). Finally, vocal production learning (Janik and Slater 2000; Tyack 2008; Vernes et al. 2021) can lead to variation when animals either copy signals or copy the way in which other animals use signals, with copying errors leading to cultural drift over time between individuals or groups (Boughman 1998; Slater and Lachlan 2003). Another example of vocal learning involves animals actively modifying existing signals or inventing new ones, leading to signals unique to individuals, social groups, or populations (Farabaugh et al. 1994; Boughman 1998).

In many contexts, communicating identity can provide fitness benefits for both sender and receiver, leading to selection for more effective encoding and decoding of individual-, group-, or species-specific signal variation (Tibbetts and Dale 2007). The most direct fitness benefit of recognizing others involves interactions of related animals, since in kin groups, benefits to any member of the group increase the inclusive fitness of all individuals (Hamilton 1964). Mother-infant interactions are the most obvious case, but closed family groups benefit in a similar way. However, even among unrelated individuals, the ability of members of social species to base behavioral decisions on past experiences with specific individuals confers considerable advantages (Connor and King, this volume). Recognition can be helpful when trying to stay in contact with a mate or during mate guarding, but also when trying to assess whether a conspecific or a member of another species is a known or unknown cooperator or threat. For example, male songbirds with neighboring territories may share a repertoire of song types (Beecher et al. 2000). This allows them to recognize the presence of a stranger who sings different songs and may allow them to direct a threat by repeating the song type of a neighbor (Beecher et al. 1997). Within a social group, relationships often depend on past cooperation and here individual recognition provides the ability to identify individuals that are more or less likely to cooperate (e.g., Dolivo and

Taborsky 2015), which is deemed to be a key prerequisite for the evolution of reciprocal altruism (Trivers 1971).

Due to its inclusive fitness benefits, kin recognition is common among animals with parental care. Mechanisms of kin recognition vary in nature and complexity from genetic variation at single alleles (e.g., Keller and Ross 1998) or multiple gene loci (e.g., the mammalian major histocompatibility complex; Brown and Eklund 1994) to complex learned cues to identify kin groups (e.g., Sharp et al. 2005; Deecke et al. 2010). Kin recognition also frequently functions in mate choice: avoiding mating with close relatives can prevent fitness costs from inbreeding depression caused by deleterious recessive mutations (e.g., Charlesworth and Willis 2009).

Where groups are stable, signaling and decoding group identity can provide fitness benefits by maintaining group cohesion and coordinating collective behaviors. Long-range signals allow group members to maintain contact over large distances and to relocate each other after periods of separation. Effective individual and group recognition is particularly important where multiple groups form temporary aggregations, or in dense breeding colonies (e.g., Aubin and Jouventin 2002; Charrier et al. 2002, 2003).

At the species level, identity signaling can provide fitness advantages by allowing animals to direct courtship and mating at conspecifics thus avoiding investing time and resources in cross-species matings that are either not viable or result in hybrid offspring with reduced fitness. Identity signaling is therefore an important mechanism in the evolution of assortative mating, reproductive isolation, and ultimately speciation (Mallet 2005).

In this chapter we summarize what is known about identity signaling in cetaceans and how this has coevolved with social structure. In the aquatic environment sound travels much farther than light and much faster than chemicals. This leads most cetacean communication to be acoustic. We therefore focus on acoustic identity signaling, although we also touch on other modalities. Cetaceans are a particularly interesting taxon in the study of identity signaling because they are capable of vocal learning, adjust their vocal output to optimize signal transmission, and have unique constraints on vocal production due to changing water pressure affecting their air-driven production apparatus (Janik and Slater 1997; Tyack and Sayigh 1997; Tyack 2008). Here, we discuss the production and perception of signals that communicate information about species, population, relatedness, or kin as well as social group and individual identity. Previous reviews of these subjects were provided by Tyack (1986b) and Connor et al. (1998), and identity signaling has also featured in broader reviews on cetacean communication such as those of Tyack (2000) and Janik (2014). These reviews described patterns of vocal variation and vocal learning in different cetacean species, and how identity signaling maps onto their social systems. Since then, scientists have expanded our knowledge on this topic considerably, finding the same patterns but providing more detail on their variety, development, and functions.

Species Identity

The most discussed identity signals in animal communication are those used to select mates of the same species. If species are defined as groups of organisms within which males and females can produce fertile offspring, then reproductive effort directed toward other species is wasted, which may create a selection pressure for males and females to evolve systems to identify members of the same species. Well-established examples of species identity signaling in animals include acoustic signals in túngara frogs (*Physalaemus pustulosus*; Ryan and Rand 1995), damselfish (*Dascyllus albisella*; Kihslinger and Klimley 2002), and cowbirds (*Molothrus ater*; Hauber et al. 2001), and visual signals in fireflies (family Lampyridae; Goh et al. 2022).

Many cetaceans have distinctive morphologies and coloration patterns that allow human observers to identify them to species with confidence. However, to date no studies have determined whether cetaceans use these cues for species recognition. In fact, most cetaceans live in low-visibility environments suggesting that visual signaling of species identity would not function reliably beyond close range. Instead, it has been suggested that many coloration patterns function to camouflage species in the aquatic environment (Mitchell 1970). The most efficient modality for communication of species identity in the aquatic environment is sound (see Dunlop and Parks, this volume).

Some sounds produced by cetaceans are distinctive and easily identified to species by humans. For example, the low-frequency downsweeps produced by fin whales (*Balaenoptera physalus*; Watkins et al. 1987), tonal calls produced by blue whales (*B. musculus*; Mellinger and Clark 2003), and clicks produced by sperm whales (*Physeter macrocephalus*; Goold and Jones 1995) and each known species of beaked whale (family Ziphiidae, Baumann-Pickering et al. 2013) are characteristic of those species. Many baleen whales produce songs with the potential to attract mating partners over long distances. Long-range propagation can degrade the information in signals, and the cost of errors in distant species identification likely contributed to the evolution of clear encoding of species identity in these animals. In contrast, the whistles and clicks produced by many delphinid species exhibit considerable overlap in the time-frequency characteristics typically measured by bioacousticians and

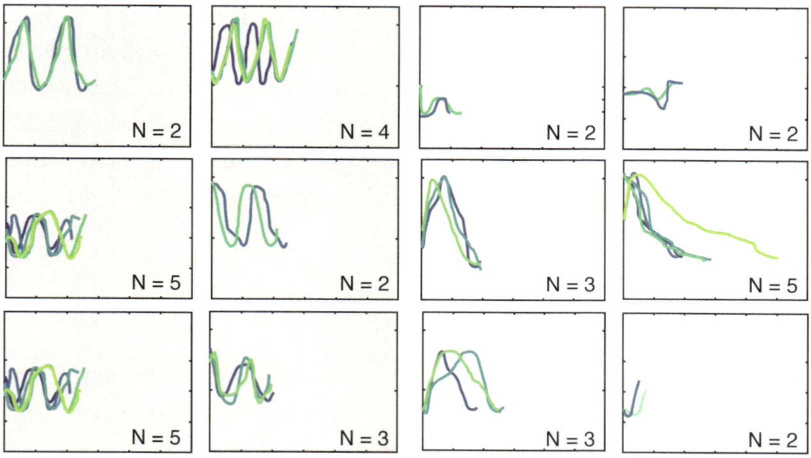

FIGURE 7.1. Examples of frequency contours from oscillatory whistle types (gray backgrounds) and non-oscillatory whistle types (white backgrounds) produced exclusively by long-beaked common dolphins (*Delphinus delphis bairdii*) that were detected during multiple encounters in the Southern California Bight. Colors represent individual whistle contours. Numbers in each plot are the number of whistles in each whistle type. Axes for each plot are the same, with the *x*-axis showing 0–3 s and the *y*-axis showing 0–23 kHz. Oscillatory whistle types (i) consist of at least two cycles of similar magnitude with two or more maxima; (ii) the maximum and minimum of each cycle is separated by at least 1 kHz; and (iii) at least 50% of the contour exhibits this pattern.
Source: Modified from Oswald et al. 2021.

can therefore be much more difficult to ascribe to a particular species (Oswald et al. 2007). If the primary selective advantage of these signals is to broadcast species identity, we would not expect so much overlap between species. Perhaps general time-frequency characteristics of sounds are not being used by delphinids to signal species identity. Instead, this information may be encoded in specific whistle contours, akin to the individually identifying signature whistles in bottlenose dolphins (*Tursiops truncatus* and *T. aduncus*). Oswald et al. (2021) showed that short-beaked common dolphins (*Delphinus delphis delphis*) produce whistles with oscillatory shapes (Gruden et al. 2016) much more commonly than do sympatric long-beaked common dolphins (*D. d. bairdii*; fig. 7.1). The differential usage of specific whistle contours such as oscillatory whistles could be a method for conveying species or subspecies identity. Whether and how cetaceans use acoustic signals to identify conspecifics are still open questions and key to answering them will be experiments to determine whether animals perceive and respond to conspecific calls differentially compared to those from a different species.

Signals evolved to broadcast identity are subject to eavesdropping, and the other context in which species recognition is important is therefore predator or prey recognition. Predators can eavesdrop on signals produced by their prey, allowing them to identify and locate it (e.g., Gannon et al. 2005). Similarly, some potential prey species have evolved complex recognition mechanisms, e.g., discriminating calls from threatening killer whales (*Orcinus orca*) from those of nonthreatening ones or other species (e.g., Deecke et al. 2002; Bowers et al. 2018; Curé et al. 2019). The specific acoustic parameters used in predator and prey recognition are not currently understood and require further investigation.

Population Identity

A variety of cetacean species show clear variation in their vocal signals that delineates populations and social groups. This variation can occur over a range of spatial scales: from populations separated by continents down to differences in vocal dialects of sympatric subpopulations within a species. Hypotheses for the causes of this variation differ depending on scale. Andrew (1962) and Slater (1989) elaborated the hypothesis that geographic variation may simply be an epiphenomenon of differentiation caused by genetic or cultural drift, especially for populations that are isolated from one another (and therefore act as identity cues rather than signals). Studying regional dialects in a songbird, Nottebohm (1969) hypothesized that differences between populations in reproductive advertisement displays could function in the selection of mates adapted to local conditions and thus provide selective advantages. Several functions proposed for vocal dialects focus on social association rather than location, proposing benefits for animals that match one another's vocalizations (Podos and Warren 2007). Here, we will start by discussing geographic variation of populations that may be isolated, followed by vocal dialects of populations that may be sympatric.

VARIATION BETWEEN GEOGRAPHICALLY ISOLATED POPULATIONS

Širović and Oleson (2022) reviewed the global distribution of 11 regional blue whale song types and argued that several can be used to identify populations but that song types are not always clear indicators of genetic differentiation. They pointed out that only one song type is recognized in the North Atlantic but more than one genetically defined population of blue whale occurs there, while more than one song type can be found off Chile where there is little evidence for genetic variation in this area. Hatch and Clark (2004) found significant differences in acoustic features of songs of fin whales from 11 regions in the North Atlantic and North Pacific, but these acoustic differences were not reflected in genetic differences. They suggested that acoustic differences may reflect more recent patterns of movement and social behavior rather than time spans required for genetic differentiation of populations to accrue. All these results raise questions about how reliably acoustic variation can be assumed to indicate variation sufficient to define genetically distinct populations.

The long propagation ranges of low-frequency sounds in the ocean and the mobility of migratory cetaceans in oceanic habitats with few barriers to movement make it difficult to know which populations are truly isolated. For example, the low-frequency songs of fin and blue whales can be heard over hundreds of kilometers (Širović et al. 2007). The higher-frequency components of songs of humpback whales (*Megaptera novaeangliae*) are often only detectable up to tens of kilometers away (Helble et al. 2013). However, humpback whale songs can be shared across populations separated by 6,000 km in the South Pacific (Garland et al. 2011), probably due to whales hearing one another as they are close together on migration or on shared feeding grounds (Garland et al. 2013; Garrigue et al. 2015). The speed with which new songs are taken up by populations across the South Pacific and the repeated uptake of songs from a different population when a few whales migrate from one population to another (Noad et al. 2000) argue against acoustic differentiation being selected to maintain local genetic differentiation. However, it does provide strong evidence that the songs are learned as described in the vocal culture chapter of this book (Hersh et al., this volume).

The acoustic adaptation hypothesis of Wiley and Richards (1982) suggests that differences in sound propagation in different habitats may select for different vocal patterns, leading to habitat-specific dialects. For an aquatic example, underwater sound will not propagate well in water that is shallower than the wavelength of the sound and one might therefore expect that in shallower habitats, sounds would be selected to be higher in frequency. However, most cetaceans live in a broad range of water depths and use a wide frequency range, suggesting a need for more flexible ways to adapt to local acoustic conditions. In fact, clear differences in vocal patterns of cetaceans have been described in different habitats and acoustic conditions. For example, Parks et al. (2007) showed that the contact calls of the North Atlantic right whale (*Eubalaena glacialis*) were higher in frequency than in the southern right whale (*E. australis*). Rather than being a function of water depth, however, these differences appear to compensate for low-frequency vessel noise, which is more prevalent in the North than South Atlantic, and call frequencies have increased over time in both habitats as vessel noise has increased (Parks et al. 2007; Hotchkin and Parks 2013). Recordings of animals in habitats with different noise conditions might appear to represent regional call types adapted to the noise, but rather, they could be a product of short-term vocal plasticity in all exposed individuals. This review highlights the dangers of assuming that geographic variation in calls has been selected to identify populations rather than resulting from drift between isolated populations or resulting from other factors such as the acoustic soundscape.

VARIATION AMONG SYMPATRIC SUBPOPULATIONS OR ECOTYPES

Distinctive acoustic repertoires have been reported for some populations or subpopulations of cetaceans that are not isolated from one another. For example, several sympatric killer whale ecotypes in the eastern North Pacific Ocean are genetically, acoustically, and behaviorally distinct and have specialized foraging strategies (Barrett-Lennard et al. 1996; Ford and Ellis 2014; Moura et al. 2014). Members of the fish-eating resident ecotype feed exclusively on fish, primarily on salmon (*Oncorhynchus* spp.). Transients (also known as Bigg's killer whales) feed on marine mammals. Ecotypes show striking differences in vocal behavior including the use of echolocation, pulsed calls, and whistles (Barrett-Lennard et al. 1996; Deecke et al. 2005; Riesch and Deecke 2011). Each ecotype has structural variation in its pulsed calls (see, e.g., fig. 7.2) that allow researchers to identify them from acoustic recordings and are also decoded by other marine mammal species. For example, playback experiments have shown that harbor seals (*Phoca vitulina*) react much more strongly to calls from mammal-eating killer whales than they do to sympatric fish-eating killer whales (Deecke et al. 2002). Killer whales that feed on mammal prey with good hearing use more cryptic echolocation than those feeding on fish that hear poorly in the frequency range of echolocation (Barrett-Lennard et al. 1996). Decoding cues and signals of ecotype identity may allow killer whales to identify suitable social and hunting partners. Studies have demonstrated striking morphological differences between sympatric mammal-eating and fish-eating ecotypes (e.g., skull and tooth morphology [Fung 2016], body size and proportion [Kotik et al. 2023]). If offspring resulting from inter-ecotype matings have suboptimal intermediate phenotypes resulting in reduced fitness, this would create additional selective pressure for improved identity signaling for ecotypes. However, these hypotheses about the functions of killer whale dialects for conspecific communication have not been well tested. We know more about selective pressures from prey that lead to differences in the calling behavior of ecotypes than we do of selective pressures from conspecific communication.

Sperm whales produce clicks both for echolocation and for communication as rhythmic patterns called codas (Watkins and Schevill 1977). Categorization of coda types based on the number and rhythmic pattern of clicks suggests that many different groups of sperm whales produce the same repertoires of coda types, forming a vocal clan (Rendell and Whitehead 2003a). Some codas are used by many clans, while others, called identity codas, are better predictors of clan membership (Hersh et al. 2021). In the Pacific Ocean, vocal clans may overlap geographically but do not mix at close range (Rendell and Whitehead 2003b). Vocal clans appear to have different movement patterns and foraging strategies (Whitehead and Rendell 2003; see Parsons et al., this volume). Coda repertoires correlate with mitochondrial haplotype variation (Whitehead et al. 1998) and are a better predictor of mtDNA variation than geographic distance (Rendell et al. 2012). One interpretation of this could be that variation in coda type is caused by genetic variation but Whitehead et al. (1998) argue that sperm whales learn their codas from a matrilineal group that has the same mtDNA as their mother. Despite claims for vocal culture in sperm whales, we know of no direct evidence for vocal production learning in this species. However, usage learning has been suggested to play a role in the development of sperm whale click behavior (Janik and Slater 2000).

Rendell and Whitehead (2003b) argue that matrilineal family units of sperm whales only associate with family units from the same vocal clan, creating larger groups that produce the same clan-specific coda repertoire. They hypothesize that codas function in inter-unit interactions, mediating whether units join or avoid one another. This would predict a differential response with whales approaching groups producing codas from the same vocal clan and avoiding groups producing codas from different clans. However, playback experiments of codas to separate groups (Rendell 2003) have failed to show differential responses to codas from the same or different vocal clans. Jacobs et al. (2024) measured the intensity of codas, concluding that they are "not used for long-range acoustic communication between units and clans, but likely serve to mediate . . . intra-unit communication." Codas tend to be produced in exchanges by whales within the same group (Weilgart and Whitehead 1993) at ranges of <500 m (Schulz et al. 2008), which supports the conclusions of Jacobs et al. (2024).

A null hypothesis for geographic or social variation in sound signals posits that this is a by-product of drift or of characteristics such as dispersal and learning that are driven by selection pressures unrelated to dialects. Several functional hypotheses for vocal dialects focus on adaptation to local habitat conditions or social groups. The social functions of vocal dialects in killer and sperm whales can be directly tested using behavioral observation and well-designed playback experiments. Further, spatially explicit agent-based models can also help test these factors (including male-male aggression, female choice, dispersal patterns, and biases in song learning) affecting vocal convergence within a region (e.g., Lachlan et al. 2004) and divergence between populations (e.g., Ellers and Slabbekoorn 2003).

Group Identity

Most of the evidence for signaling of group identity in cetaceans comes in the form of much greater variation in signals between closed social groups compared to within these groups. Experimental evidence that this variation is decoded by conspecifics to obtain information about group identity is currently limited.

Some of the best-studied examples of group-specific variation in vocal signals come from killer whales. In the northeast Pacific Ocean, members of the fish-eating resident ecotype live in stable matrilineal groups with virtually no dispersal of individuals between groups (Bigg et al. 1990). Matrilines have group-specific repertoires of 5–15 highly stereotyped pulsed call types, and recordings from captive individuals suggest that all individuals in a matriline produce all call types in its repertoire (Ford 1991). Some call types are shared with other, presumably related, matrilines, but many resident populations contain matrilines that have no call types in common yet frequently interact socially (fig. 7.2). Ford (1991) grouped all matrilines that had at least one call type in common into a common acoustic clan. Call types that are shared by several matrilines usually exhibit group-specific differences in call structure (Deecke et al. 1999; Miller and Bain 2000) and usage of shared call types also differs between matrilines with some groups producing shared call types much more often, giving rise to the notion of family signature calls (Weiß et al. 2006). Filatova et al. (2011) investigated responses of Kamchatkan resident killer whales to playbacks of calls from the same social group compared to a different social group. The results suggest differences in the response (calling in response to all playbacks of calls from the same six social groups versus no vocal responses to six playbacks of calls from a different group) but should be considered preliminary since each playback lasted two minutes, and therefore reactions could have been due to an increased chance of recognizing individuals in the playback rather than an overall group identity or a mechanism to match known calls.

While the composition of repertoires of resident killer whale matrilines remains stable over many decades (Filatova et al. 2015), the structure of stereotyped call types undergoes subtle changes over time, and these changes

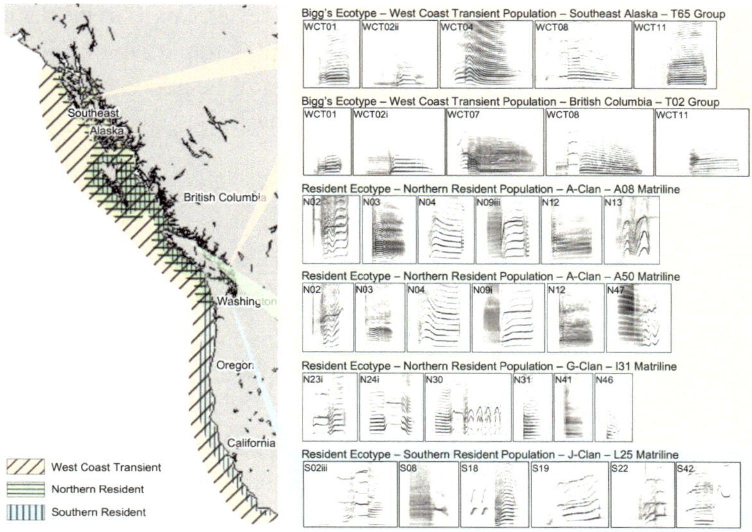

FIGURE 7.2. Variation in vocal repertoires of killer whales in the northeast Pacific. The left panel shows the home ranges of three killer whale populations belonging to two ecotypes (fish-eating residents and mammal-hunting transient or Bigg's killer whales). The right panel shows common call types for different social groups. T65 and T02 groups are part of the West Coast Transient population. All members of this population share some call types (WCT01, WCT02, and WCT11) but share no call types with resident killer whales. A08 and A50 matrilines belong to the same acoustic clan (A-Clan) in the Northern Resident population and share many call types. However, some shared call types show distinct structural differences (e.g., N09iii subtype in A08 compared to N09i subtype in A50). I31 matriline belongs to the same population but is a member of G-Clan and shares no call types with A08 or A50, although it regularly associates with these groups. L25 matriline belongs to a different resident population and is not known to associate with any of the Northern Resident matrilines and shares no call types with them. Pointers indicate the location where each recording was made. Spectrograms show a frequency range of 0–15 kHz and vary in length between 1.0 s (WCT01 produced by T02 group) and 2.2 s (WCT08 produced by T02 group). Recordings of T02 group and L25 matriline were supplied by Paul Spong.

can be transmitted between matrilines demonstrating a role of vocal learning and cultural transmission in determining call structure (Deecke et al. 2000). Structural variation of shared call types is correlated with association behaviors of matrilines, as well as the relatedness of the groups' matriarchs (Deecke et al. 2010). In captivity, killer whales have been shown to copy stereotyped call types from social companions (Crance et al. 2014). Group-specific repertoires and structural variation in shared call types may therefore be an epiphenomenon of vocal learning processes. However, in Northeast Pacific resident killer whales, mating usually occurs between individuals with different vocal repertoires (Barrett-Lennard 2000) suggesting that vocal variation may function in mate choice, acting as a learned kin recognition signal as has been suggested for some bird species (Sharp et al. 2005). Kin recognition has not been studied specifically in cetaceans, but Sayigh et al. (1999) found that bottlenose dolphins turn more toward the whistles of closely related individuals such as mothers or their own offspring. Bottlenose dolphins develop signature whistles that allow individual recognition (see next section), but it is possible that general kin recognition occurs as well.

Individual Identity

The textbook example for individual recognition signals in cetaceans is signature whistles of bottlenose dolphins (Caldwell et al. 1990; Janik and Sayigh 2013). Signature whistles are individually distinctive whistles that appear to be acquired by vocal production learning where infants either copy a whistle and then modify it (Sayigh et al. 1995; Fripp et al. 2005) or invent a whistle that does not overlap with those they hear in their environment (Sayigh et al. 2022). The exact mechanism of acquisition is still unclear, but signature whistles emerge in the first few months of life and can remain stable for decades (Sayigh et al. 2022). Signature whistles can also change, either abruptly or gradually (Sayigh et al. 2022); in some males, these changes occur when they enter an alliance with one or more other males. Signature whistles of alliance members in the Sarasota, Florida population of *Tursiops truncatus* are often more similar to each other than they are to those of other alliances but still retain distinctive features that allow individual recognition (Watwood et al. 2004). By contrast, King et al. (2018) found no evidence that allied male *T. aduncus* in Shark Bay, Western Australia, change their signature whistles to become more similar to one another. Social factors such as the tendency for pairs of males to ally in Sarasota, versus larger nested alliances in Shark Bay, may influence the variation and stability of whistles. Playback experiments with natural signature whistles have shown that mothers and offspring react more strongly to each other than to other known individuals (Sayigh et al. 1999) and that bottlenose dolphins remember signature whistles of close associates for decades (Bruck 2013). Playback experiments with synthetic signals (Janik et al. 2006) have demonstrated that the individual identity information is encoded in the frequency modulation pattern of whistles and not in general voice features (Sayigh et al. 2017). Individuals use signature whistles primarily when separated from group members (Janik and Slater 1998), and signature whistles account for 30–70% of all whistles produced by wild bottlenose dolphins (Buckstaff 2004; Cook et al. 2004; Watwood et al. 2005). Signature whistles stand out from other recognition signals in that they consist of unique frequency modulation patterns that are effectively new whistle types rather than modifications of existing ones (fig. 7.3). Even when dolphins ap-

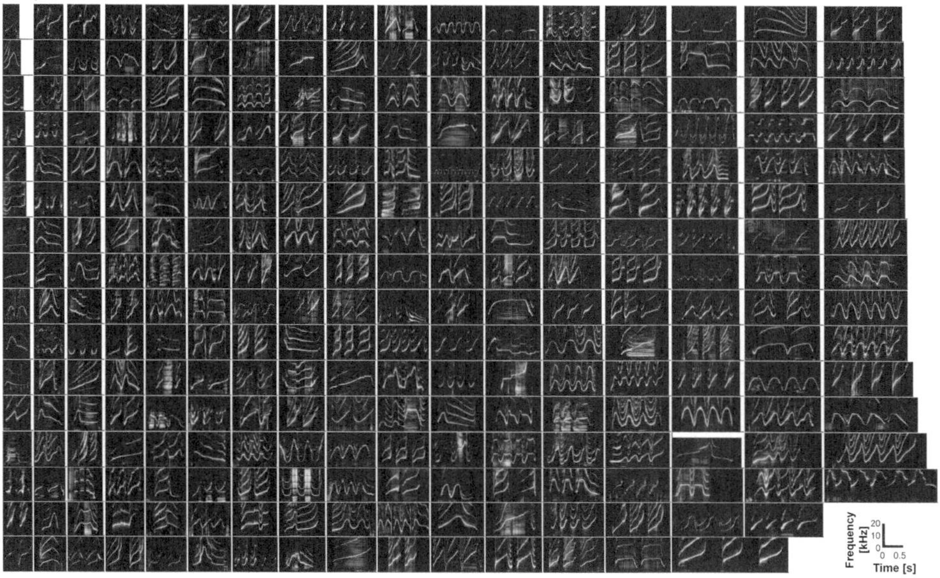

FIGURE 7.3. Example spectrograms (48 kHz sample rate, 1024 pt Hanning window, 75% overlap) of signature whistles recorded from 269 individual bottlenose dolphins in Sarasota Bay, Florida. Whistles are sorted into columns based on increasing duration, and time and frequency axes are standardized. *Source*: Sayigh et al. 2022.

pear to use another's whistle as a template, changes are usually substantial. Such changes are made possible by their vocal production learning skills, and Janik (2014) argues that effective identity signaling may have been the selective advantage that led to the evolution of vocal learning in delphinids.

Vocal learning allows bottlenose dolphins to use identity signals in an unusual way. While copying voice features of others (i.e., subtle differences that affect all call types of an individual and thereby indicate identity) is difficult and has not been documented in nonhuman animals, bottlenose dolphins regularly match whistle types of others (Janik 2000). Signature whistle copying was first reported from two captive animals that provided the first evidence that dolphins learn whistles from each other (Tyack 1986b). Tyack (1986b) hypothesized that copying is used to address or find the animal that is copied. King et al. (2013) found that copying of signature whistles occurs primarily between closely associated individuals such as mothers and their calves as well as members of male alliances. A playback study confirmed that individuals react to hearing copies of their own signature whistle by whistling in response, further supporting an addressing or labeling function (King and Janik 2013). The only other nonhuman taxon in which such copying and addressing has been reported are parrots, with the best data coming from spectacled parrotlets (*Forpus conspicillatus*, Wanker et al. 2005).

Cognitive studies have examined the extent to which dolphins might understand the use of whistles as labels for objects or other dolphins in their environment. Training studies show that dolphins can learn to copy novel human-synthesized sounds and use them to label objects in social interactions (Richards et al. 1984). A cross-modal study on individual identification demonstrated that individual identity is a modality-independent concept in dolphins, since dolphin subjects showed different reactions to simultaneous presentation of whistles and urine samples depending on whether they came from the same individual or not (Bruck et al. 2022).

Most of these studies of identity signaling in the bottlenose dolphin focus on how well the receiver can distinguish between learned calls and how the receiver responds to these calls. We have no evidence of this kind of response to specific signals for other cetacean species, and therefore suggestions of individual identity being encoded in vocalizations of other species are yet to be confirmed. It is not yet known whether other species produce signals evolved to carry signature information, whether individual identity is encoded in more subtle acoustic cues, or neither. Nousek et al. (2006) reported that individual killer whales have distinctive voice differences and McCordic et al. (2016) reported that North Atlantic right whale upcalls differ between individuals. These differences are subtle, suggesting that they may stem from by-product distinctiveness caused by morphological differences or perhaps requiring enhanced discrimination abilities of conspecific listeners. There have also been suggestions that sperm whale click patterns may be individually distinctive. Each sperm whale click is made up of several pulses (Møhl 2001), with interpulse intervals (IPIs) thought to be generated by reflections within the spermaceti organ (Norris and Harvey 1972). IPIs scale with the size of the individual's spermaceti organ and therefore differ between individuals of different sizes (Møhl et al. 1981; Antunes et al. 2011). Gero et al. (2016) argue that if whales recognize these IPI differences among group mates, they may be used to mediate social interactions within the group. However, Bøttcher et al. (2018) suggest that the IPIs of sperm whales vary too much within individuals to be useful for individual identification and Zimmer et al. (2005) show that much of the variation in IPIs depends on the orientation of the clicking whale relative to the receiver. In addition to suggestions about distinctiveness of IPIs, Antunes et al. (2011) report individually distinctive timing of clicks in one type of sperm whale coda. These differences are thought to result from neural control of the interclick timing, which may enable each member of a group to develop distinctive timing.

How Does Social Structure Relate to Identity Signaling?

Tyack (1986a) highlights that the form of social associations is likely to be reflected by the identity signals of cetaceans. Vocal learning appears to be used in different ways by different species, resulting in different patterns of variation (Janik 2014). Bottlenose dolphins, for example, live in dynamic social systems where they interact with many other individuals, and while they typically have preferred social partners, group composition can change on a minute-by-minute basis. This social system requires clear signals for individual recognition, and vocal learning appears to be used to produce novel sounds that are individually distinctive and stand out from the whistles of their conspecifics. When stable groups of unrelated animals form, vocal learning can enable convergence on stable group-distinctive calls within weeks to months (Tyack 2008). In contrast to fluid bottlenose dolphin groups, resident killer whales live in stable family groups where the presence of close relatives is almost always a given. Killer whales learn the calls of their family group when young, and these calls may change slowly over years and decades (Deecke et al. 2000). As groups grow, they may split with each subgroup continuing independent processes of change, resulting in group-distinctive repertoires of signals that relate to association and thereby relatedness (Deecke et al. 2010).

Among baleen whales, humpback and possibly bowhead whales (*Balaena mysticetus*) use vocal learning to create substantially different song displays in the same oceanic basin and change them over time (Garland et al. 2011; Stafford et al. 2018). Songs of humpback whales recorded within a population are similar at the same time but can change on a month-by-month basis (Payne et al. 1983). Songs of bowhead whales may also change over time, with as many as 76 different song types recorded in one year by Stafford et al. (2018). There appears to be a selective pressure for novelty in both species, but more pressure for conformity to a changing consensus song in humpbacks. By contrast, the songs of fin and blue whales are less diverse within a population and change more slowly over time (McDonald et al. 2009; Miksis-Olds et al. 2018). One factor that may affect this is the remarkable range over which the low-frequency sounds of fin and blue whale propagate in deep ocean (Payne and Webb 1971). Slow changes in acoustic signals result in population differences (e.g., McDonald et al. 2006) but as in killer whales it is unclear to what extent this information is used in group or population recognition or how well they map onto populations as defined genetically (Tyack 2022).

In contrast, the overlap in acoustic features of different delphinid species suggests that their vocalizations are not strongly selected for species identification (Lammers and Oswald 2015). It is possible that long-term relationships between individuals are more important in making mating choices, and individual identity signals would help to keep track of these relationships. When mating occurs during encounters between groups, unfamiliar animals may have a multitude of other sensory inputs that give information on whether another individual is a suitable mating partner. New research into the potential use of specific whistle types for species recognition is still in its infancy and may reveal another route for dolphins to identify species.

Looking Ahead

Considerable progress has been made in the study of cetacean identity signaling since the reviews by Tyack (1986b) and Connor et al. (1998). However, key questions remain unanswered, including the taxonomic breadth of vocal learning in identity signaling, the specific types of identity information communicated by signals, how these signals are perceived, and how different populations compare in social structure and identity signaling. Here we identify specific topics and questions where we think evidence is weak or lacking and suggest how new studies may improve our understanding.

While vocal learning has been clearly demonstrated in several cetacean species (Janik 2014), only bottlenose dolphins have been studied in detail for the role of vocal learning in individual recognition and addressing others (Janik and Sayigh 2013). However, the exact role that learning plays in the development of individually distinctive whistles is still unknown, and many cetaceans appear to use vocal learning primarily in other contexts (e.g., repertoire development in killer whales [Filatova et al. 2015] and reproductive advertisement in baleen whales [Garland et al. 2011]). New machine learning methods may enable clearer descriptions and categorizations of vocalizations produced by different groups and species and in different contexts (Oswald et al. 2022). These analyses require identifying which animal has made a sound and in what behavioral context. This information will allow us to investigate situations in which identity signals are being used, filling another knowledge gap, particularly for animals in the wild. For studies of vocal learning, it is also important to record the sounds an animal hears as potential models for learning. Interindividual distances and background noise have been hypothesized to be major drivers for the evolution of clear identity signals, yet few studies have been able to address their role directly.

How receivers react to signals and what meaning they can extract from them is key to understanding the evolution and functionality of identity signals. Few cetacean studies have been able to address these questions. Research into whether animals can distinguish between different sound types to extract identity information is in its infancy, currently comprising a handful of training and playback studies (e.g., Sayigh et al. 1999; Bruck et al. 2022; see King et al., this volume). The context of mate selection and reproductive advertisement provides little-explored opportunities to follow individuals and observe their displays and behavior. For example, aspects such as the complexity of courtship displays and the behavior of receivers in the presence of displaying individuals could be monitored and compared to mating success and subsequent genetic assessments of paternity. Such assessments may give insight into the signal characteristics that cetaceans use to identify species and choose mates. Observational studies could be supplemented by playback experiments using cues from known individuals of different species and conspecifics with different acoustic displays and status as well as unknown individuals to validate candidate identity characteristics.

It is somewhat surprising that the roles of signal-to-noise ratio and distance between signaler and receiver have not received more attention. Mammals may increase the amplitude of acoustic signals or use other compensation techniques when ambient noise levels are elevated (e.g., Holt et al. 2009; Hotchkin and Parks 2013). Questions remain, however, regarding the contexts in which cetaceans are likely to adapt call characteristics such as source level, and the potential function of such changes (e.g., maintaining a critical detection range to minimize eavesdropping by predators/conspecifics or maximizing source levels during vocal advertisements). To examine these questions, more studies need to collect acoustic recordings using calibrated hydrophones to estimate signal source levels, measure ambient noise levels, and investigate signal propagation. Additionally, animal-borne tags can measure sound levels of vocalizations at the tagged animal and the noise levels it receives, revealing how source levels may change as a function of ambient noise (e.g., Parks et al. 2011). Last, playback experiments can determine the probability of cetaceans responding to acoustic signals as a function of received level or signal-to-noise ratio (e.g., Miller et al. 2014).

Finally, understanding how identity signals vary as a function of differences in social behavior, social structure, and habitat will require studying the structure and use of these signals across populations, communities, and species that vary in habitat use and social structure. Comparison of the acoustic structures of identity signals is hampered by inadequate standardization of acoustic measures and nomenclature. Comparative studies will require closer collaboration between scientists to agree on standardized methods to collect comparable data from different populations and species. Some aspects often overlooked by field studies are how to systematically categorize strength of bonds between individuals, forms of dependency between mother and young, definitions of group and community structure, and frequency of interactions with unknown individuals. Standardized cross-taxa comparative studies would allow us to test hypotheses regarding the evolution of identity signals and social behavior across cetaceans.

Our review demonstrates that cetacean identity signaling is complex, with many interesting and unique features. The cognitive versatility and learning abilities of these animals are likely products of social challenges they faced in their evolutionary return to a fully aquatic lifestyle, since maintaining contact with social partners and distinguishing friend from foe is difficult in low visibility. Comparing identity signaling between these animals and other taxa will help us understand how biological systems develop effective communication and cohesion strategies in demanding environments.

References

Andrew, Richard J. 1962. "Evolution of Intelligence and Vocal Mimicking: Studies of Large-Brained Mammals Promise to Elucidate Some Problems of Human Evolution." *Science* 137: 585–89. https://doi.org/10.1126/science.137.3530.585.

Antunes, Ricardo, Tyler Schulz, Shane Gero, Hal Whitehead, Jonathan Gordon, and Luke Rendell. 2011. "Individually Distinctive Acoustic Features in Sperm Whale Codas." *Animal Behaviour* 81: 723–30. https://doi.org/10.1016/j.anbehav.2010.12.019.

Aubin, Thierry, and Pierre Jouventin. 2002. "How to Vocally Identify Kin in a Crowd: The Penguin Model." *Advances in the Study of Behavior* 31: 243–77. https://doi.org/10.1016/s0065-3454(02)80010-9.

Barrett-Lennard, Lance G. 2000. "Population Structure and Mating Patterns of Killer Whales (*Orcinus orca*) as Revealed by DNA Analysis." PhD diss., University of British Columbia.

Barrett-Lennard, Lance G., John K. B. Ford, and Kathy A. Heise. 1996. "The Mixed Blessing of Echolocation: Differences in Sonar Use by Fish-Eating and Mammal-Eating Killer Whales." *Animal Behaviour* 51: 553–65. https://doi.org/10.1006/anbe.1996.0059.

Baumann-Pickering, Simone, Mark A. McDonald, Anne E. Simonis, et al. 2013. "Species-Specific Beaked Whale Echolocation Signals." *Journal of the Acoustical Society of America* 134: 2293–301. https://doi.org/10.1121/1.4817832.

Beecher, Michael D., S. Elizabeth Campbell, and J. Cully Nordby. 2000. "Territory Tenure in Song Sparrows is Related to Song Sharing with Neighbours, but Not to Repertoire Size." *Animal Behaviour* 59: 29–37. https://doi.org/10.1006/anbe.1999.1304.

Beecher, Michael D., J. Cully Nordby, S. Elizabeth Campbell, John M. Burt, Christopher E. Hill, and Adrian L. O'Loghlen. 1997. "What Is the Function of Song Learning in Songbirds?" In *Perspectives in Ethology*, vol. 12, *Communication*, edited by Donald H. Owings, Michael D. Beecher, and Nicholas S. Thompson, 77–97. Springer. https://doi.org/10.1007/978-1-4899-1745-4_4.

Bigg, Michael A., Peter F. Olesiuk, Graeme M. Ellis, John K. B. Ford, and Kenneth C. Balcomb. 1990. "Social Organization and Genealogy of Resident Killer Whales (*Orcinus orca*) in the Coastal Waters of British Columbia and Washington State." *Report of the International Whaling Commission* 12: 383–405.

Bøttcher, Anne, Shane Gero, Kristian Beedholm, Hal Whitehead, and Peter T. Madsen. 2018. "Variability of the Inter-Pulse Interval in Sperm Whale Clicks with Implications for Size Estimation and Individual Identification." *Journal of the Acoustical Society of America* 144: 365–74. https://doi.org/10.1121/1.5047657.

Boughman, Janette W. 1998. "Vocal Learning by Greater Spear-Nosed Bats." *Proceedings of the Royal Society B* 265: 227–33. https://doi.org/10.1098/rspb.1998.0286.

Bowers, Matthew T., Ari S. Friedlaender, Vincent M. Janik, et al. 2018. "Selective Reactions to Different Killer Whale Call Categories in Two Delphinid Species." *Journal of Experimental Biology* 221 (11): jeb162479. https://doi.org/10.1242/jeb.162479.

Brown, Jerram L., and Amy Eklund. 1994. "Kin Recognition and the Major Histocompatibility Complex: An Integrative Review." *American Naturalist* 143: 435–61. https://doi.org/10.1086/285612.

Bruck, Jason N. 2013. "Decades-Long Social Memory in Bottlenose Dolphins." *Proceedings of the Royal Society B* 280: 20131726. https://doi.org/10.1098/rspb.2013.1726.

Bruck, Jason N., Sam F. Walmsley, and Vincent M. Janik. 2022. "Cross-Modal Perception of Identity by Sound and Taste in Bottlenose Dolphins." *Science Advances* 8 (20): eabm7684. https://doi.org/10.1126/sciadv.abm76.

Buckstaff, Kara C. 2004. "Effects of Watercraft Noise on the Acoustic Behavior of Bottlenose Dolphins, *Tursiops truncatus*, in Sarasota Bay, Florida." *Marine Mammal Science* 20: 709–25. https://doi.org/10.1111/j.1748-7692.2004.tb01189.x.

Caldwell, Melba C., David K. Caldwell, and Peter L. Tyack. 1990. "Review of the Signature-Whistle-Hypothesis for the Atlantic Bottlenose Dolphin." In *The Bottlenose Dolphin*, edited by S. Leatherwood, Randall R. Reeves, 199–234. Academic Press. https://www.doi.org/10.1016/b978-0-12-440280-5.50014-7.

Carlson, Nora V., E. McKenna Kelly, and Iain Couzin. 2020. "Individual Vocal Recognition Across Taxa: A Review of the Literature and a Look into the Future." *Philosophical Transactions of the Royal Society B* 375: 20190479. https://doi.org/10.1098/rstb.2019.0479.

Charlesworth, Deborah, and John H. Willis. 2009. "The Genetics of Inbreeding Depression." *Nature Reviews Genetics* 10: 783–96. https://doi.org/10.1038/nrg2664.

Charrier, Isabelle, Nicolas Mathevon, and Pierre Jouventin. 2002. "How Does a Fur Seal Mother Recognize the Voice of her Pup? An Experimental Study of *Arctocephalus tropicalis*." *Journal of Experimental Biology* 205 (5): 603–12. https://doi.org/10.1242/jeb.205.5.603.

Charrier, Isabelle, Nicolas Mathevon, and Pierre Jouventin. 2003. "Vocal Signature Recognition of Mothers by Fur Seal Pups." *Animal Behaviour* 65: 543–50. https://doi.org/10.1006/anbe.2003.2073.

Connor, Richard C., Janet Mann, Peter L. Tyack, and Hal Whitehead. 1998. "Social Evolution in Toothed Whales." *Trends in Ecology and Evolution* 13: 228–32. https://doi.org/10.1016/S0169-5347(98)01326-3.

Cook, Mandy L. H., Laela S. Sayigh, James E. Blum, and Randall S. Wells. 2004. "Signature-Whistle Production in Undisturbed Free-Ranging Bottlenose Dolphins (*Tursiops truncatus*)." *Proceedings of the Royal Society B* 271: 1043–49. https://doi.org/10.1098/rspb.2003.2610.

Crance, Jessica L., Anne E. Bowles, and Alan Garver. 2014. "Evidence for Vocal Learning in Juvenile Male Killer Whales, *Orcinus orca*, from an Adventitious Cross-Socializing Experiment." *Journal of Experimental Biology* 217 (8): 1229–37. https://doi.org/10.1242/jeb.094300.

Curé, Charlotte, Saana Isojunno, Heike I. Vester, et al. 2019. "Evidence for Discrimination Between Feeding Sounds of Familiar Fish and Unfamiliar Mammal-Eating Killer Whale Ecotypes by Long-Finned Pilot Whales." *Animal Cognition* 22: 863–82. https://doi.org/10.1007/s10071-019-01282-1.

Deecke, Volker B., Lance G. Barrett-Lennard, Paul Spong, and John K. B. Ford. 2010. "The Structure of Stereotyped Calls Reflects Kinship and Social Affiliation in Resident Killer Whales (*Orcinus orca*)." *Naturwissenschaften* 97: 513–18. https://doi.org/10.1007/s00114-010-0657-z.

Deecke, Volker B., John K. B. Ford, and Peter J. B. Slater. 2005. "The Vocal Behaviour of Mammal-Eating Killer Whales: Communicating with Costly Calls." *Animal Behaviour* 69 (2): 395–405. https://doi.org/10.1016/j.anbehav.2004.04.014.

Deecke, Volker B., John K. B. Ford, and Paul Spong. 1999. "Quantifying Complex Patterns of Bioacoustic Variation: Use of a Neural Network to Compare Killer Whale (*Orcinus orca*) Dialects." *Journal of the Acoustical Society of America* 105: 2499–507. https://doi.org/10.1121/1.426853.

Deecke, Volker B., John K. B. Ford, and Paul Spong. 2000. "Dialect Change in Resident Killer Whales: Implications for Vocal Learning and Cultural Transmission." *Animal Behaviour* 60: 629–38. https://doi.org/10.1006/anbe.2000.1454.

Deecke, Volker B., Peter J. B. Slater, and John K. B. Ford. 2002. "Selective Habituation Shapes Acoustic Predator Recognition in Harbour Seals." *Nature* 420: 171–73. https://doi.org/10.1038/nature01030.

Dolivo, Vassilissa, and Michael Taborsky. 2015. "Norway Rats Reciprocate Help According to the Quality of Help They Received." *Biology Letters* 11: 20140959. https://doi.org/10.1098/rsbl.2014.0959.

Dowling, Jenélle L., David A. Luther, and Peter P. Marra. 2012. "Comparative Effects of Urban Development and Anthropogenic Noise on Bird Songs." *Behavioral Ecology* 23: 201–9. https://doi.org/10.1093/beheco/arr176.

Ellers, Jacintha, and Hans Slabbekoorn. 2003. "Song Divergence and Male Dispersal Among Bird Populations: A Spatially Explicit Model Testing the Role of Vocal Learning." *Animal Behaviour* 65: 671–81. https://doi.org/10.1006/anbe.2003.2081.

Farabaugh, Susan M., Alison Linzenbold, and Robert J. Dooling. 1994. "Vocal Plasticity in Budgerigars (*Melopsittacus undulatus*): Evidence for Social Factors in the Learning of Contact Calls." *Journal of Comparative Psychology* 108: 81–92. https://doi.org/10.1037/0735-7036.108.1.81.

Filatova, Olga A., Filipa I. P. Samarra, Volker B. Deecke, John K. B. Ford, Patrick J. O. Miller, and Harald Yurk. 2015. "Cultural Evolution of Killer Whale Calls: Background, Mechanisms and Consequences." *Behaviour* 152 (15): 2001–38. https://doi.org/10.1163/1568539X-00003317.

Ford, John K. B. 1991. "Vocal Traditions Among Resident Killer Whales (*Orcinus orca*) in Coastal Waters of British Columbia." *Canadian Journal of Zoology* 69 (6): 1454–83. https://doi.org/10.1139/z91-206.

Ford, John K. B., and G. M. Ellis. 2014. "You Are What You Eat: Ecological Specializations and Their Influence on the Social Organization and Behaviour of Killer Whales." In *Primates and Cetaceans: Field Research and Conservation of Complex Mammalian Societies*, edited by J. Yamagiwa and L. Karczmarski, 75–98. Springer. https://doi.org/10.1007/978-4-431-54523-1_4.

Fripp, Deborah, Caryn Owen, Ester Quintana-Rizzo, et al. 2005. "Bottlenose Dolphin (*Tursiops truncatus*) Calves Appear to Model Their Signature Whistles on the Signature Whistles of Community Members." *Animal Cognition* 8: 17–26. https://doi.org/10.1007/s10071-004-0225-z.

Fung, Charissa W. 2016. "Cranial Shape Correlates with Diet Specialization in Northeast Pacific Killer Whale (*Orcinus orca*) Ecotypes." Master's thesis, University of British Columbia. https://doi.org/10.14288/1.0314137.

Gannon, Damon P., Nélio B. Barros, Douglas P. Nowacek, Andrew J. Read, Danielle M. Waples, and Randall S. Wells. 2005. "Prey Detection by Bottlenose Dolphins, *Tursiops truncatus*: An Experimental Test of the Passive Listening Hypothesis." *Animal Behaviour* 69: 709–20. https://doi.org/10.1016/j.anbehav.2004.06.020.

Garland, Ellen C., Jason Gedamke, Melinda L. Rekdahl, Michael J. Noad, Claire Garrigue, and Nick Gales. 2013. "Humpback Whale Song on the Southern Ocean Feeding Grounds: Implications for Cultural Transmission." *PLOS ONE* 8: e79422. https://doi.org/10.1371/journal.pone.0079422.

Garland, Ellen C., Anne W. Goldizen, Melinda L. Rekdahl, et al. 2011. "Dynamic Horizontal Cultural Transmission of Humpback Whale Song at the Ocean Basin Scale." *Current Biology* 21 (8): 687–91. https://doi.org/10.1016/j.cub.2011.03.019.

Garrigue, Claire, Phillip J. Clapham, Ygor Geyer, Amy S. Kennedy, and Alexandre N. Zerbini. 2015. "Satellite Tracking Reveals Novel Migratory Patterns and the Importance of Seamounts for Endangered South Pacific Humpback Whales." *Royal Society Open Science* 2: e150489. https://doi.org/10.1098/rsos.150489.

Gero, Shane, Hal Whitehead, and Luke Rendell. 2016. "Individual, Unit and Vocal Clan Level Identity Cues in Sperm Whale Codas." *Royal Society Open Science* 3: 150372. https://doi.org/10.1098/rsos.150372.

Goh, King-Siang, Chia-Ming Lee, and Tzi-Yuan Wang. 2022. "Species-Specific Flash Patterns Track the Nocturnal Behavior of Sympatric Taiwanese Fireflies." *Biology* 11: 58. https://doi.org/10.3390/biology11010058.

Goold, John C., and Sarah E. Jones. 1995. "Time and Frequency Domain Characteristics of Sperm Whale Clicks." *Journal of the Acoustical Society of America* 98: 1279–91. https://doi.org/10.1121/1.413465.

Gruden, Pina, Paul R. White, Julie N. Oswald, et al. 2016. "Differences in Oscillatory Whistles Produced by Spinner (*Stenella longirostris*) and Pantropical Spotted (*Stenella attenuata*) Dolphins." *Marine Mammal Science* 32: 520–34. https://doi.org/10.1111/mms.12280.

Hamilton, William D. 1964. "The Genetical Evolution of Social Behaviour. I." *Journal of Theoretical Biology* 7 (1): 1–16. https://doi.org/10.1016/0022-5193(64)90038-4.

Hatch, Leila T., and Christopher W. Clark. 2004. "Acoustic Differentiation Between Fin Whales in Both the North Atlantic and North Pacific Oceans, and Integration with Genetic Estimates of Divergence." Paper SC/56/SD6 presented to the IWC Scientific Committee, July, Sorrento, Italy.

Hauber, Mark E., Stefani A. Russo, and Paul W. Sherman. 2001. "A Password for Species Recognition in a Brood-Parasitic Bird." *Proceedings of the Royal Society B* 268: 1041–48. https://doi.org/10.1098/rspb.2001.1617.

Helble, Tyler A., Gerald L. D'Spain, John A. Hildebrand, Gregory S. Campbell, Richard L. Campbell, and Kevin D. Heaney. 2013. "Site Specific Probability of Passive Acoustic Detection of

Humpback Whale Calls from Single Fixed Hydrophones." *Journal of the Acoustical Society of America* 134: 2556–70. https://doi.org/10.1121/1.4816581.

Hersh, Taylor A., Shane Gero, Luke Rendell, and Hal Whitehead. 2021. "Using Identity Calls to Detect Structure in Acoustic Datasets." *Methods in Ecology and Evolution* 12: 1668–78. https://doi.org/10.1111/2041-210X.13644.

Holt, Marla M., Dawn P. Noren, Val Veirs, Candice K. Emmons, and Scott Veirs. 2009. "Speaking Up: Killer Whales (*Orcinus orca*) Increase their Call Amplitude in Response to Vessel Noise." *Journal of the Acoustical Society of America* 125: EL27–32. https://doi.org/10.1121/1.3040028.

Hotchkin, Cara, and Susan Parks. 2013. "The Lombard Effect and other Noise-Induced Vocal Modifications: Insight from Mammalian Communication Systems." *Biological Reviews* 88: 809–24. https://doi.org/10.1111/brv.12026.

Jacobs, Ellen R., Shane Gero, Chloe E. Malinka, et al. 2024. "The Active Space of Sperm Whale Codas: Inter-Click Information for Intra-Unit Communication." *Journal of Experimental Biology* 227 (4): jeb246442. https://doi.org/10.1242/jeb.246442.

Janik, Vincent M. 2000. "Whistle Matching in Wild Bottlenose Dolphins (*Tursiops truncatus*)." *Science* 289: 1355–57. https://doi.org/10.1126/science.289.5483.1355.

Janik, Vincent M. 2014. "Cetacean Vocal Learning and Communication." *Current Opinion in Neurobiology* 28: 60–65. https://doi.org/10.1016/j.conb.2014.06.010.

Janik, Vincent M., and Laela S. Sayigh. 2013. "Communication in Bottlenose Dolphins: 50 Years of Signature Whistle Research." *Journal of Comparative Physiology A* 199: 479–89. https://doi.org/10.1007/s00359-013-0817-7.

Janik, Vincent M., Laela S. Sayigh, and Randall S. Wells. 2006. "Signature Whistle Shape Conveys Identity Information to Bottlenose Dolphins." *Proceedings of the National Academy of Sciences* 103 (21): 8293–97. https://doi.org/10.1073/pnas.0509918103.

Janik, Vincent M., and Peter J. B. Slater. 1997. "Vocal Learning in Mammals." *Advances in the Study of Behaviour* 26: 59–99. https://doi.org/10.1016/S0065-3454(08)60377-0.

Janik, Vincent M., and Peter J. B. Slater. 1998. "Context-Specific Use Suggests That Bottlenose Dolphin Signature Whistles Are Cohesion Calls." *Animal Behaviour* 56: 829–38. https://doi.org/10.1006/anbe.1998.0881.

Janik, Vincent M., and Peter J. B. Slater. 2000. "The Different Roles of Social Learning in Vocal Communication." *Animal Behaviour* 60: 1–11. https://doi.org/10.1006/anbe.1998.0881.

Keller, Laurent, and Kenneth G. Ross. 1998. "Selfish Genes: A Green Beard in the Red Fire Ant." *Nature* 394: 573–75. https://doi.org/10.1038/29064.

Kihslinger, Rebecca L., and A. Peter Klimley. 2002. "Species Identity and the Temporal Characteristics of Fish Acoustic Signals." *Journal of Comparative Psychology* 116: 210–14. https://doi.org/10.1037/0735-7036.116.2.210.

King, Stephanie L., Whitney R. Friedman, Simon J. Allen, et al. 2018. "Bottlenose Dolphins Retain Individual Vocal Labels in Multi-Level Alliances." *Current Biology* 28: 1993–99. https://doi.org/10.1016/j.cub.2018.05.013.

King, Stephanie L., and Vincent Michael Janik. 2013. "Bottlenose Dolphins Can Use Learned Vocal Labels to Address Each Other." *Proceedings of the National Academy of Sciences* 110: 13216–21. https://doi.org/10.1073/pnas.1304459110.

King, Stephanie L., Laela S. Sayigh, Randall S. Wells, Wendi Fellner, and Vincent M. Janik. 2013.

"Vocal Copying of Individually Distinctive Signature Whistles in Bottlenose Dolphins." *Proceedings of the Royal Society B* 280: e20130053. https://doi.org/10.1098/rspb.2013.0053.

Kotik, Chloe, John W. Durban, Holly Fearnbach, and Lance G. Barrett-Lennard. 2023. "Morphometrics of Mammal-Eating Killer Whales from Drone Photogrammetry, with Comparison to Sympatric Fish-Eating Killer Whales in the Eastern North Pacific." *Marine Mammal Science* 39: 42–58. https://doi.org/10.1111/mms.12965.

Lachlan, Robert F., Vincent M. Janik, and Peter J. B. Slater. 2004. "The Evolution of Conformity Enforcing Behaviour in Cultural Communication Systems." *Animal Behaviour* 68: 561–70. https://doi.org/10.1016/j.anbehav.2003.11.015.

Lammers, Marc O., and Julie N. Oswald. 2015. "Analyzing the Acoustic Communication of Dolphins." In *Dolphin Communication and Cognition: Past, Present, and Future*, edited by Denise L. Herzing and Christine M. Johnson, 107–37. MIT Press. https://doi.org/10.7551/mitpress/9939.003.0008.

Mallet, James. 2005. "Hybridization as an Invasion of the Genome." *Trends in Ecology and Evolution* 20: 229–37. https://doi.org/10.1016/j.tree.2005.02.010.

Maynard Smith, John, and David Harper. 2003. *Animal Signals*. Oxford University Press.

McCordic, Jessica A., Holly Root-Gutteridge, Dana A. Cusano, Samuel L. Denes, and Susan E. Parks. 2016. "Calls of North Atlantic Right Whales *Eubalaena glacialis* Contain Information on Individual Identity and Age Class." *Endangered Species Research* 30: 157–69. https://doi.org/10.3354/esr00735.

McDonald, Mark A., John A. Hildebrand, and Sarah L. Mesnick. 2009. "Worldwide Decline in Tonal Frequencies of Blue Whale Songs." *Endangered Species Research* 9 (1): 13–21. https://doi.org/10.3354/esr00217.

McDonald, Mark A., Sarah L. Mesnick, and Hildebrand, John A. 2006. "Biogeographic Characterization of Blue Whale Song Worldwide: Using Song to Identify Populations." *Journal of Cetacean Research and Management* 8: 55–65. https://doi.org/10.47536/jcrm.v8i1.702.

Mellinger, David K., and Christopher W. Clark. 2003. "Blue Whale (*Balaenoptera musculus*) Sounds from the North Atlantic." *Journal of the Acoustical Society of America* 114: 1108–19. https://doi.org/10.1121/1.1593066.

Miksis-Olds, Jennifer L., Sharon L. Nieukirk, and Danielle V. Harris. 2018. "Two Unit Analysis of Sri Lankan Pygmy Blue Whale Song over a Decade." *Journal of the Acoustical Society of America* 144: 3618–26. https://doi.org/10.1121/1.5084269.

Miller, Patrick J. O., Ricardo N. Antunes, Paul J. Wensveen, et al. 2014. "Dose-Response Relationships for the Onset of Avoidance of Sonar by Free-Ranging Killer Whales." *Journal of the Acoustical Society of America* 135: 975–93. https://doi.org/10.1121/1.4861346.

Miller, Patrick J. O., and David E. Bain. 2000. "Within-Pod Variation in Killer Whale Calls." *Animal Behaviour* 60: 617–28. https://doi.org/10.1006/anbe.2000.1503.

Mitchell, Edward. 1970. "Pigmentation Pattern Evolution in Delphinid Cetaceans: An Essay in Adaptive Coloration." *Canadian Journal of Zoology* 48: 717–40. https://doi.org/10.1139/z70-129.

Møhl, Bertel. 2001. "Sound Transmission in the Nose of the Sperm Whale *Physeter catodon*. A Post Mortem Study." *Journal of Comparative Physiology A* 187: 335–40. https://doi.org/10.1007/s003590100205.

Møhl, Bertel, E. Larsen, and Mats Amundin. 1981. "Sperm Whale Size Determination: Outlines of an Acoustic Approach." *Mammals in the Sea* 3: 327–31.

Moura, Andre E., John G. Kenny, Roy Chaudhuri, et al. 2014. "Population Genomics of the Killer Whale Indicates Ecotype Evolution in Sympatry Involving Both Selection and Drift." *Molecular Ecology* 23: 5179–5192. https://doi.org/10.1111/mec.12929.

Noad, Michael J., Douglas H. Cato, Michael M. Bryden, Micheline-Nicole Jenner, and K. Curt S. Jenner. 2000. "Cultural Revolution in Whale Songs." *Nature* 408 (6812): 537. https://doi.org/10.1038/35046199.

Norris, Kenneth S., and George W. Harvey. 1972. "A Theory for the Function of the Spermaceti Organ of the Sperm Whale (*Physeter catodon* L.)." In *Animal Orientation and Navigation*, edited by Sidney R. Galler, Klaus Schmidt-Koenig, George J. Jacobs, and Richard E. Belleville, NASA Document ID 19720017437. National Aeronautics and Space Administration.

Nottebohm, Fernando. 1969. "The Song of the Chingolo, *Zonotrichia capensis*, in Argentina: Description and Evaluation of a System of Dialects." *Condor* 71: 299–315. https://doi.org/10.2307/1366306.

Nousek, Anna E., Peter J. B. Slater, Chao Wang, and Patrick J. O. Miller. 2006. "The Influence of Social Affiliation on Individual Vocal Signatures of Northern Resident Killer Whales (*Orcinus orca*)." *Biology Letters* 2: 481–84. https://doi.org/10.1098/rsbl.2006.0517.

Oswald, Julie N., Christine Erbe, William L. Gannon, Shyam Madhusudhana, and Jeanette A. Thomas. 2022. "Detection and Classification Methods for Animal Sounds." In *Exploring Animal Behavior Through Sound*, vol. 1, edited by Christine Erbe, and Jeanette A. Thomas, 269–317. Springer. https://doi.org/10.1007/978-3-030-97540-1_8.

Oswald, Julie N., Shannon Rankin, Jay Barlow, and Marc O. Lammers. 2007. "A Tool for Real-Time Acoustic Species Identification of Delphinid Whistles." *Journal of the Acoustical Society of America* 122: 587–95. https://doi.org/10.1121/1.2743157.

Oswald, Julie N., Sam F. Walmsley, Caroline Casey, Selene Fregosi, Brandon Southall, and Vincent M. Janik. 2021. "Species Information in Whistle Frequency Modulation Patterns of Common Dolphins." *Philosophical Transactions of the Royal Society B* 376: e20210046. https://doi.org/10.1098/rstb.2021.0046.

Parks, Susan E., Christopher W. Clark, and Peter L. Tyack. 2007. "Short- and Long-Term Changes in Right Whale Calling Behavior: The Potential Effects of Noise on Acoustic Communication." *Journal of the Acoustical Society of America* 122: 3725–31. https://doi.org/10.1121/1.2799904.

Parks, Susan E., Mark Johnson, Douglas Nowacek, and Peter L. Tyack. 2011. "Individual Right Whales Call Louder in Increased Environmental Noise." *Biology Letters* 7: 33–35. https://doi.org/10.1098/rsbl.2010.0451.

Payne, Katharine, Peter L. Tyack, and Roger S. Payne. 1983. "Progressive Changes in the Songs of Humpback Whales *Megaptera novaeangliae*: A Detailed Analysis of Two Seasons in Hawaii." In *Communication and Behavior of Whales*, edited by Roger Payne, 9–57. Westview Press.

Payne, Roger, and Douglas Webb. 1971. "Orientation by Means of Long Range Acoustic Signaling in Baleen Whales." *Annals of the New York Academy of Sciences* 188 (1): 110–41. https://doi.org/10.1111/j.1749-6632.1971.tb13093.x.

Podos, Jeffrey, and Paige S. Warren. 2007. "The Evolution of Geographic Variation in Birdsong." *Advances in the Study of Behavior* 37: 403–58. https://doi.org/10.1016/S0065-3454(07)37009-5.

Rendell, Luke E. 2003. "Cultural Variation and Usage of Coda Vocalisations by Sperm Whales, *Physeter macrocephalus*." PhD diss., Dalhousie University.

Rendell, Luke, Sarah L. Mesnick, Merel L. Dalebout, Jessica Burtenshaw, and Hal Whitehead. 2012. "Can Genetic Differences Explain Vocal Dialect Variation in Sperm Whales, *Physeter macrocephalus*?" *Behavior Genetics* 42: 332–43. https://doi.org/10.1007/s10519-011-9513-y.

Rendell, Luke E., and Hal Whitehead. 2003a. "Comparing Repertoires of Sperm Whale Codas: A Multiple Methods Approach." *Bioacoustics* 14: 61–81. https://doi.org/10.1080/09524622.2003.9753513.

Rendell, Luke E., and Hal Whitehead. 2003b. "Vocal Clans in Sperm Whales (*Physeter macrocephalus*)." *Proceedings of the Royal Society B* 270 (1512): 225–31. https://doi.org/10.1098/rspb.2002.2239.

Richards, Douglas G., James P. Wolz, and Louis M. Herman. 1984. "Vocal Mimicry of Computer-Generated Sounds and Vocal Labeling of Objects by a Bottlenosed Dolphin, *Tursiops truncatus*." *Journal of Comparative Psychology* 98: 10–28. https://doi.org/10.1037/0735-7036.98.1.10.

Riesch, Rüdiger, and Volker B. Deecke. 2011. "Whistle Communication in Mammal-Eating Killer Whales (*Orcinus orca*): Further Evidence for Acoustic Divergence Between Ecotypes." *Behavioral Ecology and Sociobiology* 65: 1377–87. https://doi.org/10.1007/s00265-011-1148-8.

Ryan, Michael J., and A. Stanley Rand. 1995. "Female Responses to Ancestral Advertisement Calls in Túngara Frogs." *Science* 269: 390–92. https://doi.org/10.1126/science.269.5222.390.

Sayigh, Laela S., Vincent M. Janik, Frants Havmand Jensen, Michael D. Scott, Peter L. Tyack, and Randall S. Wells. 2022. "The Sarasota Dolphin Whistle Database: A Unique Long-Term Resource for Understanding Dolphin Communication." *Frontiers in Marine Science* 9: 923046. https://doi.org/10.3389/fmars.2022.923046.

Sayigh, Laela S., Peter L. Tyack, Randall S. Wells, Michael D. Scott and A. Blair Irvine. 1995. "Sex Differences in Signature Whistle Production of Free-Ranging Bottlenose Dolphins, *Tursiops truncatus*." *Behavioral Ecology and Sociobiology* 36: 171–77. https://doi.org/10.1007/BF00177793.

Sayigh, Laela S., Peter L. Tyack, Randall S. Wells, Andrew R. Solow, Michael D. Scott and A. Blair Irvine. 1999. "Individual Recognition in Wild Bottlenose Dolphins: A Field Test Using Playback Experiments." *Animal Behaviour* 57: 41–50. https://doi.org/10.1006/anbe.1998.0961.

Sayigh, Laela S., Randall S. Wells, and Vincent M. Janik. 2017. "What's in a Voice? Dolphins Do Not Use Voice Cues for Individual Recognition." *Animal Cognition* 20: 1067–79. https://doi.org/10.1007/s10071-017-1123-5.

Schulz, Tyler M., Hal Whitehead, Shane Gero, and Luke Rendell. 2008. "Overlapping and Matching of Codas in Vocal Interactions Between Sperm Whales: Insights into Communication Function." *Animal Behaviour* 76: 1977–88. https://doi.org/10.1016/j.anbehav.2008.07.032.

Sharp, Stuart P., Andrew McGowan, Matthew J. Wood, and Ben J. Hatchwell. 2005. "Learned Kin Recognition Cues in a Social Bird." *Nature* 434: 1127–30. https://doi.org/10.1038/nature03522.

Širović, Ana, John A. Hildebrand, and Sean M. Wiggins. 2007. "Blue and Fin Whale Call Source Levels and Propagation Range in the Southern Ocean." *Journal of the Acoustical Society of America* 122 (2): 1208–15. https://doi.org/10.1121/1.2749452.

Širović, Ana, and Erin M. Oleson. 2022. "The Bioacoustics of Blue Whales: Global Diversity and Behavioral Variability in a Foraging Specialist." In *Ethology and Behavioral Ecology of*

Mysticetes, edited by Christopher W. Clark and Ellen C. Garland, 195–221. Springer. https://doi.org/10.1007/978-3-030-98449-6_9.

Slater, Peter J. B. 1989. "Bird Song Learning: Causes and Consequences." *Ethology Ecology and Evolution* 1: 19–46. https://doi.org/10.1080/08927014.1989.9525529.

Slater, Peter J. B., and R. F. Lachlan. 2003. "Is Innovation in Bird Song Adaptive?" In *Animal Innovation*, edited by Simon M. Reader and Kevin N. Laland, 117–35. Oxford University Press. https://doi.org/10.1093/acprof:oso/9780198526223.003.0005.

Stafford, Kathleen M., Christian Lydersen, Øystein Wiig, and Kit M. Kovacs. 2018. "Extreme Diversity in the Songs of Spitsbergen's Bowhead Whales." *Biology Letters* 14: 20180056. https://doi.org/10.1098/rsbl.2018.0056.

Tibbetts, Elizabeth A., and James Dale. 2007. "Individual Recognition: It Is Good to Be Different." *Trends in Ecology and Evolution* 22: 529–37. https://doi.org/10.1016/j.tree.2007.09.001.

Trivers, Robert L. 1971. "The Evolution of Reciprocal Altruism." *Quarterly Review of Biology* 46: 35–57. https://doi.org/10.1086/406755.

Tyack, Peter L. 1986a. "Population Biology, Social Behavior and Communication in Whales and Dolphins." *Trends in Ecology and Evolution* 1: 144–50. https://doi.org/10.1016/0169-5347(86)90042-X.

Tyack, Peter L. 1986b. "Whistle Repertoires of Two Bottlenosed Dolphins, *Tursiops truncatus*: Mimicry of Signature Whistles?" *Behavioral Ecology and Sociobiology* 18: 251–57. https://doi.org/10.1007/BF00300001.

Tyack, Peter L. 2000. "Functional Aspects of Cetacean Communication." In *Cetacean Societies: Field Studies of Dolphins and Whales*, edited by Janet Mann, Richard C. Connor, Peter L. Tyack, and Hal Whitehead, 270–307. University of Chicago Press.

Tyack, Peter L. 2008. "Convergence of Calls as Animals Form Social Bonds, Active Compensation for Noisy Communication Channels, and the Evolution of Vocal Learning in Mammals." *Journal of Comparative Psychology* 122: 319. https://doi.org/10.1037/a0013087.

Tyack, Peter L. 2022. "Social Organization of Baleen Whales." In *Ethology and Behavioral Ecology of Mysticetes*, edited by Christopher W. Clark and Ellen C. Garland, 147–75. Springer. https://doi.org/10.1007/978-3-030-98449-6_7.

Tyack, Peter L., and Laela S. Sayigh. 1997. "Vocal Learning in Cetaceans." In *Social Influences on Vocal Development*, edited by Charles T. Snowdon and Martine Hausberger, 208–33. Cambridge University Press. https://doi.org/10.1017/CBO9780511758843.011.

Vernes, Sonja C., Buddhamas Pralle Kriengwatana, Veronika C. Beeck, et al. 2021. "The Multi-Dimensional Nature of Vocal Learning." *Philosophical Transactions of the Royal Society B* 376: e20200236. https://doi.org/10.1098/rstb.2020.0236.

Wanker, Ralf, Yasuko Sugama, and Sabine Prinage. 2005. "Vocal Labelling of Family Members in Spectacled Parrotlets, *Forpus conspicillatus*." *Animal Behaviour* 70: 111–18. https://doi.org/10.1016/j.anbehav.2004.09.022.

Watkins, William A., and William E. Schevill. 1977. "Sperm Whale Codas." *Journal of the Acoustical Society of America* 62 (6): 1485–90. https://doi.org/10.1121/1.381678.

Watkins, William A., Peter Tyack, Karen E. Moore, and James E. Bird. 1987. "The 20-Hz Signals of Finback Whales (*Balaenoptera physalus*)." *Journal of the Acoustical Society of America* 82 (6): 1901–12. https://doi.org/10.1121/1.395685.

Watwood, Stephanie L., Edward C. G. Owen, Peter L. Tyack, and Randall S. Wells. 2005. "Signature Whistle Use by Temporarily Restrained and Free-Swimming Bottlenose Dolphins, *Tursiops truncatus*." *Animal Behaviour* 69: 1373–86. https://doi.org/10.1016/j.anbehav.2004.08.019.

Watwood, Stephanie. L., Peter L. Tyack, and Randall S. Wells. 2004. "Whistle Sharing in Paired Male Bottlenose Dolphins, *Tursiops truncatus*." *Behavioural Ecology and Sociobiology* 55: 531–43. https://doi.org/10.1007/s00265-003-0724-y.

Weilgart, Linda, and Hal Whitehead. 1993. "Coda Communication by Sperm Whales (*Physeter macrocephalus*) off the Galapagos Islands." *Canadian Journal of Zoology* 71: 744–52. https://doi.org/10.1139/z93-098.

Weiß, Brigitte M., Friedrich Ladich, Paul Spong, and Helena Symonds. 2006. "Vocal Behavior of Resident Killer Whale Matrilines with Newborn Calves: The Role of Family Signatures." *Journal of the Acoustical Society of America* 119: 627–35. https://doi.org/10.1121/1.2130934.

Whitehead, Hal, Mary Dillon, Susan Dufault, Linda Weilgart, and Jonathan Wright. 1998. "Non-Geographically Based Population Structure of South Pacific Sperm Whales: Dialects, Fluke-Markings and Genetics." *Journal of Animal Ecology* 67: 253–62. https://doi.org/10.1046/j.1365-2656.1998.00187.x.

Whitehead, Hal, and Luke Rendell. 2004. "Movements, Habitat Use and Feeding Success of Cultural Clans of South Pacific Sperm Whales." *Journal of Animal Ecology* 73 (1): 190–96. https://doi.org/10.1111/j.1365-2656.2004.00798.x.

Wiley, R. Haven, and Douglas G. Richards. 1982. "Adaptations for Acoustic Communication in Birds: Sound Transmission and Signal Detection." In *Acoustic Communication in Birds: Production, Perception and Design Features of Sounds*, edited by Donald E. Kroodsma, and Edward H. Miller, 131–81. Academic Press. https://doi.org/10.1016/b978-0-08-092416-8.50014-0.

Zimmer, Walter M. X., Peter T. Madsen, Valeria Teloni, Mark P. Johnson, and Peter L. Tyack. 2005. "Off-Axis Effects on the Multipulse Structure of Sperm Whale Usual Clicks with Implications for Sound Production." *Journal of the Acoustical Society of America* 118: 3337–45. https://doi.org/10.1121/1.2082707.

PART III

Reproduction, Cooperation, and Conflict in Cetacean Societies

8

Sociality, Life History, and the Evolution of Menopause in Toothed Whales

SAMUEL ELLIS,
MIA LYBKÆR KRONBORG NIELSEN,
AND DARREN P. CROFT

Abstract

Toothed whales are an ideal group in which to study the coevolution of sociality and life history because they show remarkable variability in both. For example, they are the only mammal group where menopause—prolonged female post-reproductive life—has evolved multiple times. This chapter reviews the literature and performs original analysis to explore how: (i) group living; (ii) social relationships; (iii) intrasexual conflict; and (iv) interactions among kin may be linked to life history evolution in toothed whales. The chapter then presents an in-depth review of the literature concerning the evolution of menopause in resident-ecotype killer whales as a case study of how intergenerational help and harm can drive life history evolution. Building on killer whales as a case study, the chapter then places the evolution of menopause in toothed whales in a comparative context by drawing social and life history comparisons across species with and without menopause. The chapter aims to highlight that toothed whales offer a fascinating and important group in which to explore and understand the coevolution of sociality and life history.

Introduction

Historically, life history theory has focused on population demography and paid less attention to the interplay between the social environment and life history and how the two can coevolve. Sociality (here taken to encompass group living, social relationships, mating systems, and kin structure; Kappeler 2019) can have fundamental implications for survival and reproductive success (Hamilton 1964; Krause and Ruxton 2002; Snyder-Mackler et al. 2020). Thus, sociality and life history are tightly linked. For social species, life history traits evolve in, and can be directly influenced by, the social environment. These effects, however, are not necessarily only in one direction. Life history traits also have the potential to shape the social environment due to the impact of reproduction and survival on social group structure. In this chapter, we present a general overview of the potential for the coevolution of life history and sociality in toothed whales with a focus on how patterns of cooperation and conflict in social groups can shape life history evolution.

Toothed whale species exhibit marked diversity in both sociality and life history, making them an ideal group in which to investigate the coevolution of sociality and life histories. Sociality is ubiquitous yet highly variable in toothed whales. Accompanying this social variation is an equally impressive variation in life history. Perhaps most notably, they are the only genera in which menopause—prolonged female post-reproductive life—has evolved multiple times (Ellis et al. 2018a, 2018b). Here, we outline four different areas where sociality and life history evolution can interact in toothed whales. First, we consider how living in a group can influence survival and reproduction and the potential resulting consequences for life history evolution. Second, we consider the social relationships individuals form within and between groups and how these relationships can shape survival and reproductive success. Third, we consider sexual conflict and, in particular, how male-male competition may shape life history evolution. Finally, we examine how intergenerational patterns of cooperation and conflict can shape life history evolution and, in particular, the evolution of menopause.

Group Living and Life History Evolution

Almost all toothed whales are usually observed in close association with one or more conspecifics: they are group living. Looking at life history patterns across mammals generally, recent work has found that group-living mammals have a longer lifespan than solitary species, which may relate to several factors, including a lower risk of predation and starvation and the opportunity to gain inclusive fitness benefits by providing parental or grandparental care (Zhu et al. 2023). However, to date, no studies have looked at how group size is related to lifespan in social mammals. Here we explore the relationship between group size and lifespan in toothed whales. We define "group size" as the number of whales typically in close association when encountered, we use this definition because the key benefits of living in larger social groups, such as avoiding predation and cooperative hunting, are likely to have important effects at these very local scales.

We undertook an analysis of the relationship between group size and lifespan in toothed whales. We found that in toothed whales there is some evidence (an 87% probability) that species living in smaller groups (<10 individuals) live relatively longer as adults (adult lifespan / age at maturity) than species inhabiting large (>61 individuals) or medium-sized (10–60 individuals) groups, given their body size (fig. 8.1; app. 8A). One potential explanation for why toothed whales living in smaller groups live longer is the interplay between group size and predation risk on juveniles. In toothed whales the two key benefits of group living are likely protection from predation, particularly predation by killer whales (Baird 2000), and access to resources (Connor 2000). Both of these benefits may decrease the mortality risk of juveniles more than of adults. Demographic theory demonstrates that an increase in extrinsic (age-independent) mortality risk cannot increase rate of aging, but age-dependent changes in mortality risk can (Wensink et al. 2017; Moorad et al. 2019). The reason for this can be conceptualized in terms of reproductive value (Wensink et al. 2017). When juvenile mortality risk increases, the reproductive value of older mature individuals also increases, because having survived that early-life mortality becomes a key predictor of likelihood of contributing to future generations (Wensink et al. 2017). Higher reproductive value of older individuals will select for increased late-life survival and slower senescence (Wensink et al. 2017). Therefore, if mortality risk to juvenile toothed whales decreases with increasing group size relative to the risk to adults, then we expect that whales inhabiting smaller groups will have longer adult lifespans than species inhabiting larger groups. The theoretical links between group living and life history evolution are only just being developed; there are other factors likely to be influencing the relationship between group size and lifespan, such as habitat and age-correlated costs and benefits of group living, that are not considered in our analysis. Nevertheless, given their ubiquitous yet varied sociality, toothed whales present an important system in which to study this question as the field develops.

Social Relationships and Life History Evolution

Social structure—the pattern of social interactions, social relationships, and dominance interactions in a population (Kappeler 2019)—has the potential to drive life history differences between species, populations, and individuals. One pathway by which social structure could influence life history evolution in toothed whales is through modulating the risk of pathogen transmission (e.g., cetacean morbillivirus; Weiss et al. 2020). The emergent social structure of a population is likely to influence the size and severity of disease outbreaks.

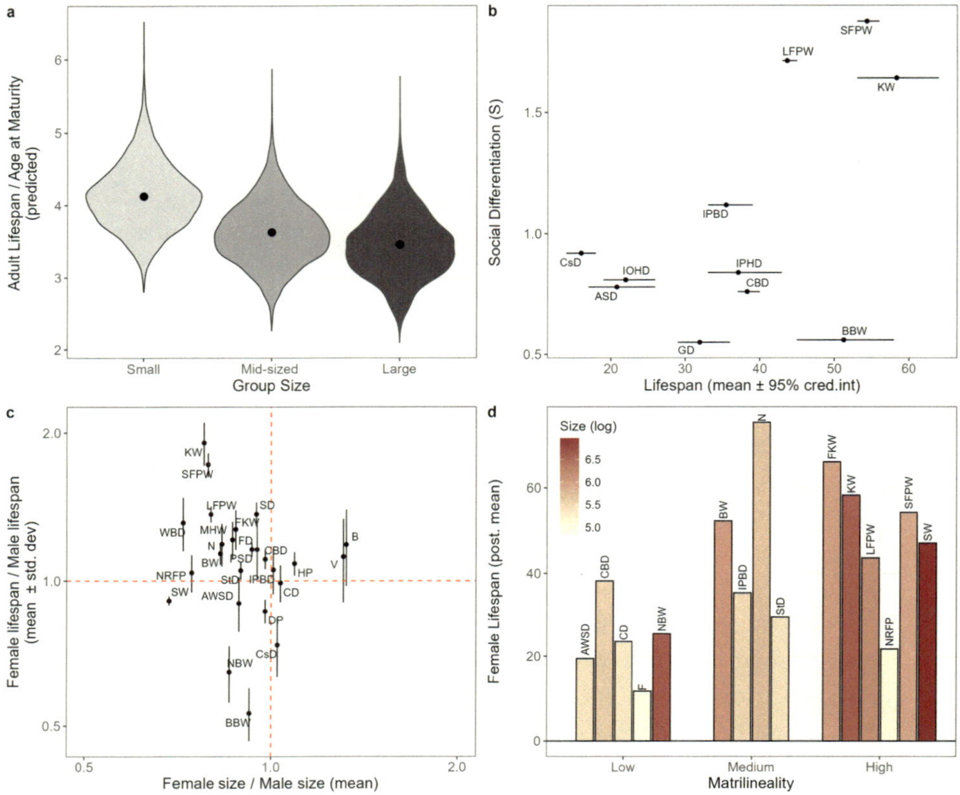

FIGURE 8.1. Sociality and life history coevolution in toothed whales. (a) Adult lifespan / age at maturity ratio for small, midsized, and large toothed whale groups. Violins show the posterior distributions from a Bayesian model. The distributions show the predicted adult divided by immature lifespan for a 230 cm long toothed whale (median toothed whale size in sample). The model predicts the adult/immature distribution as the mean of a lognormal distribution: $\mu_i \sim \alpha + \beta_{\text{groupsize}_i} + \gamma\text{SIZE}_i$, fitted with weakly informative priors, and carrying through errors in lifespan estimation and size estimation. (b) Toothed whale social differentiation given predicted ordinary maximum lifespan (posterior mean ±95% credible interval). Social differentiation (*S*) describes the variation in association indices (Whitehead 2008). Species indicated by initials (app. 8A). (c) Relationship between female-male size differences and female-male lifespan differences for 25 species of toothed whale (app. 8A). Dashed lines show the axis where the female/male ratio = 1; that is, females and males are the same size or have the same lifespan. (d) Matrilineality and female lifespan in toothed whales. Bars show the mean estimate of maximum lifespan for females (app. 8A), and bar fill color indicates log female size, with redder colors representing larger sizes. Abbreviations above the bars show species (app. 8A). Using the same model structure and priors as panel (a) demonstrates that in species with low matrilineality females have shorter lifespans than females in species with either medium (P[low < medium] = 0.95) or high (P[low < high] = 0.95) matrilineality, given their size.

Simulation studies suggest that in highly modular social structures outbreaks tend to be less severe because the disease mostly spreads *within* rather than *between* subgroups, limiting the size of the outbreak (Silk and Fefferman 2021). Where the social structure of a population or species leaves them particularly vulnerable to disease outbreaks, we might predict increased investment in early-life reproduction rather than long lifespan due to the risks of dying of disease. However, there are some caveats to this prediction. One is that social decisions (and therefore emergent social structure) are also under selection—and the emergent social structure in high-disease-risk societies is presumably adaptive despite the increased risk of disease-induced mortality. Highly modular networks in toothed whales could, for example, allow individuals to have easier access to vertically transmitted (from one generation to another) social information (Weiss et al. 2021). Considerably more work is needed to understand if and how social structure can influence life history in toothed whales.

As well as varying between species, there is growing evidence that life history can also vary between individuals within the same population (Dammhahn et al. 2018). The position of individuals within their social structure—their social position—has the potential to affect their life history trajectory. A body of theoretical and empirical work in a variety of animals has shown that an individual's social position affects their exposure to both disease and socially transmitted information. This is clearly demonstrated in Indo-Pacific bottlenose dolphins where the spread of both disease (tattoo skin disease; Powell et al. 2020) and foraging behaviors ("shelling"; Wild et al. 2020) have been linked to social network structure. Further, studies in toothed whales (and other species) have shown that social position can correlate with fitness (Snyder-Mackler et al. 2020). For example, southern resident killer whale males in less central social positions have higher annual mortality risks than males in more central social positions (Ellis et al. 2017). Similarly, in Indo-Pacific bottlenose dolphins, young males in a more central social position in their social network have an increased probability of reaching sexual maturity (Stanton and Mann 2012), and on reaching adulthood more central males secure more paternities than less central males (Gerber et al. 2022). Links between fitness and social position raise the possibility that individual life history in toothed whales could be affected by relative social position. For example, an increased mortality risk driven by a peripheral social position could select for an increased "pace-of-life" with individuals investing more in reproduction and less on long-term survival—but to date this has not been tested in toothed whales.

The direction of causality in the relationship between social structure and life history could also be reversed, that is, life history could drive variation in social structure. Recently, Silk and Hodgson (2021) hypothesized that species with a slow pace-of-life (longer lifespans, slower reproduction) should, in general, benefit more from strongly differentiated social bonds because both partners are more likely to live long enough to receive the beneficial payoff from their relationship (Silk and Hodgson 2021; but see Aubier and Kokko 2022). Correlating toothed whale lifespan and social differentiation we find no clear evidence for this hypothesis: there does not appear to be a simple relationship between social differentiation and lifespan in toothed whales (fig. 8.1b). While some long-lived toothed whales such as killer whales, long-finned pilot whales, and short-finned pilot whales show very high social differentiation (social differentiation [variation in association index; Whitehead 2008]; $S = 1.64$, 1.71, and 1.81, respectively), similarly long-lived Baird's beaked whales have very low levels of social differentiation ($S = 0.56$). Social differentiation in toothed whales with short and mid-length lifespans are similarly very variable with no clear relationship between social differentiation and lifespan: For example, common bottlenose dolphins have nearly double the mean ordinary maximum lifespan of Atlantic spotted dolphins (app. 8A), yet their social differentiation values are very similar ($S = 0.76$ and 0.78, respectively).

Intrasexual Conflict and Life History Evolution

It has generally been argued that higher male reproductive competition increases selection for investment in reproduction and decreases selection for a long lifespan in male mammals (Trivers 1972). Overall, this male reproductive competition hypothesis is supported in mammals with polygynous mating systems where, in general, the degree of reproductive competition is associated with relatively shorter male lifespans (Lemaître et al. 2020). The intensity of male reproductive competition is an important driver of sex size dimorphism in mammals—relatively larger males suggest more intense male-male reproductive competition (Isaac 2005). A comparative prediction of the male reproductive competition hypothesis is, therefore, that sex size dimorphism should be associated with sex differences in lifespan in toothed whales. Comparing the ratio of female to male lifespan and size in toothed whales, we find that in most toothed whale species, males are larger than females, and females live longer than males (fig. 8.1c; app. 8A), which is consistent with the male reproductive competition hypothesis. Although the general pattern is consistent with male reproductive competition, there are several species where males are both larger and live longer than females (fig. 8.1c; app. 8A). This inconsistency is perhaps unsurprising because there are other forms of sexual competition that are likely to influence sex size and lifespan differences, such as trade-offs between pre- and postcopulatory competition (Dines et al. 2015) and the extent and intensity of female-female reproductive competition (Isaac 2005). In order to fully understand the role of mating systems in explaining cetacean life history, more data on mating patterns and levels of reproductive competition in a greater variety of species are required.

Intergenerational Cooperation, Conflict, and the Evolution of Menopause

Matrilineality—the extent to which individuals associate with their maternal relatives as adults—is perhaps the key axis of cetacean sociality (Rendell et al. 2019). High matrilineality provides the opportunity for intergenerational help and harm and therefore, the opportunity for kin selection to drive life history evolution (see Foroughirad et al. and Parsons et al., this volume). Here we first take a comparative approach to explore the correlation between matrilineality and lifespan in toothed whales, and then we focus on the impact of intergenerational help and harm on the evolution of menopause.

There is considerable variation in the degree of matrilineality in toothed whales, even in the handful of species where kinship structure is known in detail (Rendell et al. 2019; Weiss et al. 2021). In a comparative analysis we found that, in toothed whale populations, the degree of matrilineality is linked to life history evolution. We found that female toothed whale species inhabiting societies with high or medium matrilineality (i.e., they are either in close association or in the same social unit as some of their matrilineal relatives as adults) have longer lifespans than species not in association with their relatives—given their size (fig. 8.1d; app. 8A). However, the causal direction in this relationship is unclear. It is plausible that only in species with relatively long lifespans do the generation times of matrilineal kin overlap for long enough to regularly inhabit the same social group. This view is challenged, however, by the high matrilineality in the relatively short-lived narrow-ridged finless porpoise (fig. 8.1d; app. 8A), which have been observed in groups of three matrilineal generations (Chen et al. 2016)—suggesting that it is possible to have a short lifespan and high matrilineality. An alternative explanation is that the presence of close relatives selects for longer lifespans (Péron et al. 2019). This explanation is based on theoretical results suggesting that intergenerational transfer of fitness from older generations to younger generations can select for longer lifespans (Lee 2008), which is supported by empirical results in mammals more generally (Péron et al. 2019). In toothed whales there is evidence of intergenerational transfer of both knowledge and resources (Brent et al. 2015; Wright et al. 2016; Wild et al. 2019), which is clearly only possible when close relatives are nearby. A third hypothesis for the link between matrilineality and life history in toothed whales is that they coevolve in the same selective environment. Toothed whales remain an important group in which to investigate this question and differentiate these hypotheses as the theoretical and empirical basis of this field develops.

One of the best examples of the coevolution between kin structure and life history evolution in mammals is the evolution of extended female post-reproductive lifespans in humans and toothed whales (Ellis et al. 2024). Hereafter, for simplicity, we use the term "menopause" to refer to physiological termination of reproduction before the end of life resulting in females regularly living for an extended period of time after their last reproductive event, even though in toothed whales there is no menstruation, sensu stricto, as the uterine lining is reabsorbed. Menopause is a rare life history strategy and, outside of humans, it has only been reported as a species-level trait under natural conditions in short-finned pilot whales (Kasuya and Marsh 1984), killer whales (Olesiuk et al. 1990; Nielsen et al. 2021), false killer whales (Photopoulou et al. 2017), narwhals (Ellis et al. 2018a), and beluga whales (Ellis et al. 2018a). Although evidence for menopause in the latter three species has sometimes been conflicting (Photopoulou et al. 2017; Ellis et al. 2018a; Péron et al. 2019). Ellis et al. (2018a) used counts of ovarian corpora from deceased whales to measure patterns of ovarian activity across the lifetimes of several species of toothed whale. They found no evidence of slowing of reproduction in false killer whales, but in a detailed analysis, including of pregnancy rates in deceased whales, Photopoulou et al. (2017) suggested that females of that species cease reproduction well before the end of life; on balance, we consider it likely that menopause is present in false killer whales. In beluga whales, evidence from multiple populations suggests that ovarian activity ceases in the midforties, although pregnancy data remain scarce (Suydam 2009; Ellis et al. 2018a; Ferguson et al. 2020). Narwhals may continue to reproduce relatively later in life than other species with menopause (Garde et al. 2015) but have very long lifespans (regularly living into their eighties and nineties; Ellis et al. 2025), resulting in significant post-reproductive periods (Ellis et al. 2018a). Overall, we suggest there is strong evidence of menopause in all three species, but long-term individual-based studies are needed for unambiguous confirmation.

In the rest of this section, we outline our current understanding of how cooperation and conflict in family groups have shaped the evolution of menopause in toothed whales. We first focus on resident killer whales in the northeast Pacific Ocean, which provide the best-studied example of the evolution of menopause in the wild. We then take a comparative approach to explore how sociality drives the evolution of menopause—and when it does not—in toothed whales by (i) comparing killer whale populations; (ii) contrasting killer whales and sperm whales; and (iii) exploring differences between short-finned and long-finned pilot whales.

THE EVOLUTION OF MENOPAUSE IN RESIDENT KILLER WHALES

Two populations of resident-ecotype killer whales reside in the northeast Pacific Ocean: the southern and the northern resident killer whales. They are exclusively fish-eating, and in the summer months, the majority of their diet is Chinook salmon (*Oncorhynchus tshawytscha*). The populations move into the inshore waters to feed on the migrating salmon during the summer, returning to the same areas year on year, allowing births and deaths to be recorded on an annual basis. Long-term observation has established that female resident killer whales cease reproduction at age 41 (age at which 95% of population fecundity has been completed); have a greater than 40% chance of reaching this age given they survive to adulthood; and, for those that reach that age, can expect to live for another 20 years after the cessation of reproduction (Nielsen et al. 2021; fig. 8.2a). This observed life history trajectory has been validated by hormone profiling, which shows the physiological termination of reproduction in midlife (Wasser et al. 2017). The outcome of this life history pattern is that, at any time, >25% of adult female years are being lived by post-reproductive females (Nielsen et al. 2021). This life history profile is very similar to that exhibited by humans without access to modern medical care (Alberts et al. 2013). Research in resident killer whales has been instrumental in establishing that the evolution of menopause in toothed whales is driven by social factors.

INCLUSIVE FITNESS BENEFITS OF HELPING KIN IN LATE LIFE

When social groups are composed of overlapping generations of close kin, post-reproductive females may continue to gain inclusive fitness benefits by helping kin survive and reproduce rather than reproducing themselves directly. Two routes have been proposed through which older females can help

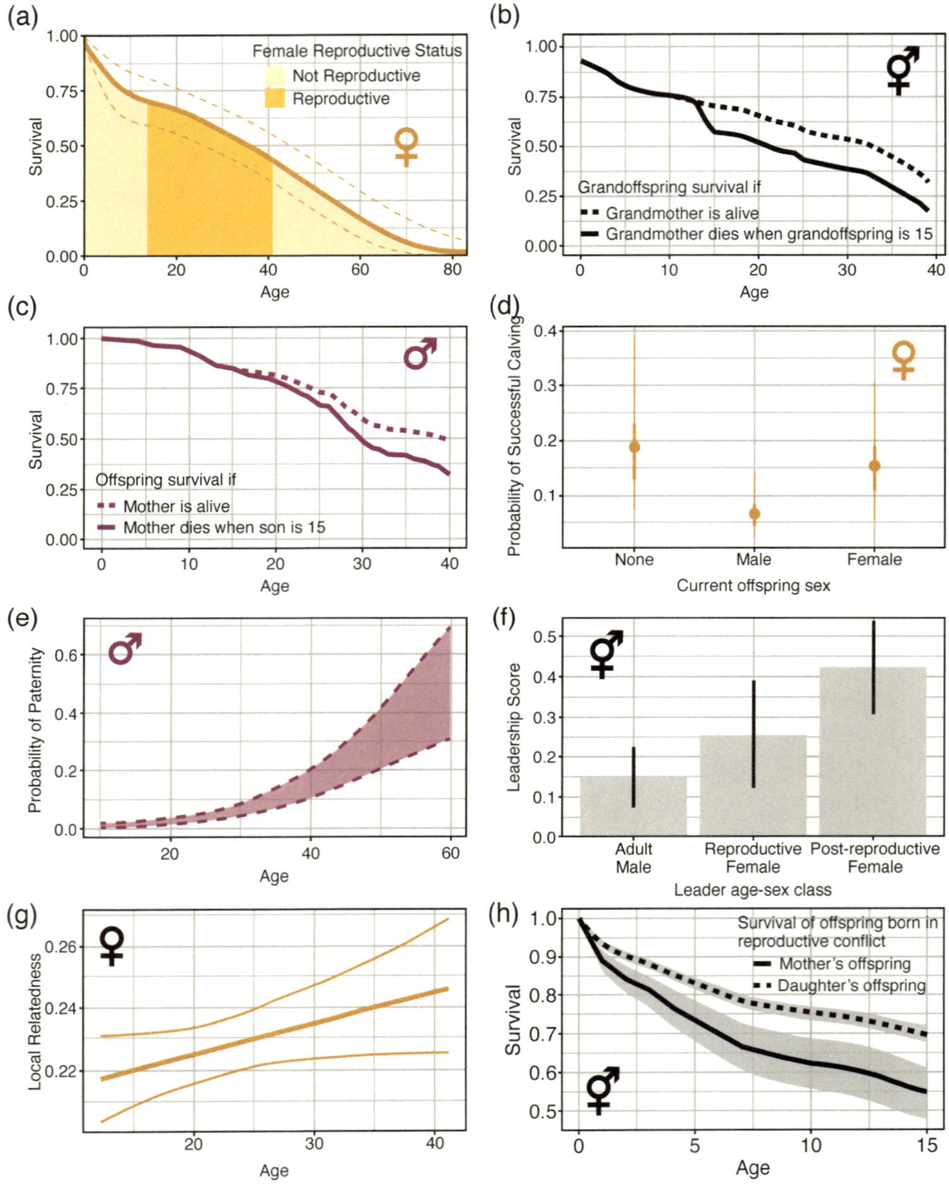

FIGURE 8.2. Aspects of menopause in resident killer whales. In all figures, dark yellow shows effects on females, maroon shows effects on males, and black/gray shows effects on both sexes. (a) Representation of female resident killer whale life history, derived from Nielsen et al. (2021). Survival represents probability of reaching a given age (solid line posterior mean, dashed lines 95% credible interval). Area fill shows whether females of that age are expected to be reproductively active (dark yellow) or not reproductively active (light yellow) which represent both immaturity and post-reproductive lifespan. (b) Grandmother effect in resident killer whales, from Nattrass et al. (2019). Predicted survival of whales (of both sexes combined) of whales whose grandmother is alive and post-reproductive (dashed line) compared to those

their younger kin: from mothers to their offspring (cf. the mother hypothesis; Williams 1957) and from grandmothers to their grandoffspring (cf. the grandmother hypothesis; Hawkes et al. 1998).

The social organization of resident killer whales is unusual, neither males nor females disperse from their family group (Bigg et al. 1990; Ellis et al. 2021), which results in mothers interacting with their sons, their daughters, and their daughters' offspring (grandoffspring) for their entire life. The presence of grandmothers in these matrilineal groups provides a significant fitness benefit to their grandoffspring: grandmothers increase the survival probability of both their male and female grandoffspring (Nattrass et al. 2019). The annual probability of dying (mortality hazard) for a grandoffspring that loses their grandmother is 4.5 times higher than a grandoffspring whose grandmother is still alive (Nattrass et al. 2019; fig. 8.2b). An interesting consequence of this is that the interbirth interval is increased for daughters whose mother was still alive, probably because it is more likely their offspring will survive infancy, and females will tend not to resume reproduction while their calves are very young (Nattrass et al. 2019).

Resident killer whale mothers continue to have a substantial impact on offspring survival well past weaning and into adulthood. The survival benefits,

---

whose grandmother dies when they are aged 15 (solid line). Survival before 15 is identical in both conditions. Predictions are derived from a fitted survival model. (c) Mother effect in resident killer whale males, from Foster et al. (2012). Survival of males whose mother is alive (dashed line) compared to those whose mother dies when they are 15 (solid line). Survival before 15 is identical in both conditions. Predictions are derived from a fitted survival model. (d) Resident killer whale sons impose a reproductive cost on their mothers, from Weiss et al. (2023). Categories represent females who have a living son (Male condition), no offspring (None condition), or a daughter (Female condition). Points and thick and thin error bars represent the posterior mean and 50% and 95% credible intervals from a Bayesian model, respectively. (e) Probability of males achieving paternity given their age, from Ford et al. (2018). Ribbon shows the 95% credible interval of a model fitted to genetic paternity data and shows the relationship including males with no known offspring. (f) Leadership in resident killer whales, derived from data in Brent et al. (2015). A whale's leadership score is the proportion of times they were observed leading their group. Bars show the mean leadership score for whales of a given age-sex class, and error bars show the standard deviation around that mean. (g) Resident female kinship dynamics during reproductively active years, from Ellis et al. (2022). Local relatedness is the average relatedness of a female to her social group, here defined as those with whom she shares a strong social connection. Thick and thin yellow lines show the posterior mean and 95% credible interval of a model fitted to real data. (h) The outcome of reproductive conflict in resident killer whales, from Croft et al. (2017). Reproductive conflict occurs when a mother and her daughter both reproduce at the same time (±2 years). Lines show the survival (± standard error, gray ribbon) of the offspring of the mother (solid line) and her daughter (dashed line) born into reproductive conflict. Lines are derived from a fitted survival model.

Sources: All figures are redrawn, adapted, or derived from other publications; full details for each can be found in the indicated publications.

however, are not equal for sons and daughters; post-reproductive females disproportionally increase the survival of their sons compared to their daughters. Males under 30 are 3.1 times more likely to die if their mother is dead than if they are alive, while there is no effect in females (Foster et al. 2012; fig. 8.2c). Even more dramatically, if a post-reproductive mother dies when their offspring are over 30, the chance of dying in the next year is increased by a factor of 13.9 for sons and 5.4 for daughters (Foster et al. 2012). Further, in the southern resident killer whale mothers exhibit lifelong maternal investment in their sons, and this investment comes at a significant cost, with mothers less likely to reproduce again following the birth of a son but not following the birth of a daughter (Weiss et al. 2023; fig. 8.2d).

Preferential investment in male offspring could be linked to the reproductive differences between the sexes in resident killer whales. Adult killer whale males have a higher potential rate of reproduction than adult females and, importantly, in resident killer whales, paternity in a given year tends to be dominated by the oldest male(s) in the population (Ford et al. 2018; fig. 8.2e). Investing in the survival of sons, therefore, has the potential to provide females with a very large inclusive fitness payoff. A further consideration is that because mating in the resident killer whales almost always occurs between individuals from different social groups (Ford et al. 2018) when a son reproduces, the cost of rearing a calf is not carried by the social group of the son (Cant and Johnstone 2008). In contrast, when a daughter reproduces, the calf is born into the group of the daughter that reproduced, increasing within-group competition. In summary, females can increase their inclusive fitness by preferentially investing in male offspring and continuing that investment even when their sons are mature.

HOW DO POST-REPRODUCTIVE FEMALES HELP?

Three key pathways have been identified by which older females provide intergenerational help. First, female resident killer whales disproportionally share fish they catch with others in their social group, particularly their offspring (Wright et al. 2016). However, the propensity to share with offspring is dependent on both offspring age and sex: females regularly share with their adult sons but rarely with their adult daughters (Wright et al. 2016). Second, older females may be able to provide valuable ecological knowledge related to fluctuations in prey abundance. Salmon abundance varies in space and time and predicting the timing of this variation likely impacts the survival and reproduction of the killer whale population (Ward et al. 2009; Ford et al.

2010). In the resident killer whales it is the older post-reproductive mothers that act as leaders of the group movement around the foraging grounds especially in times of low salmon abundance (Brent et al. 2015; fig. 8.2f). Third, post-reproductive females provide social support to their sons. Resident killer whale males whose mother is alive and post-reproductive receive fewer socially inflicted injuries than males whose mother is either dead or is still reproducing (Grimes et al. 2023). There may be other pathways yet to be investigated by which post-reproductive females can help their kin—such as passing on important social connections or facilitating mating opportunities.

INTERGENERATIONAL REPRODUCTIVE CONFLICT AND THE EVOLUTION OF MENOPAUSE

Intergenerational benefits from older females are not exclusive to species where females have an extended post-reproductive life (Péron et al. 2019). To explain the evolution of menopause in resident killer whales, as well as explaining the benefits of long life, we also need to explain why females stop reproducing before the end of life. In addition to the fitness benefits of helping kin late in life, menopause may have evolved to avoid the costs of continued reproduction. In humans, various costs have been considered, such as the risks related to childbirth in late life (Williams 1957) or the limitations that having your own dependent offspring may place on your ability to help your other kin (Kim et al. 2019). These costs, however, are unlikely to be strong enough to explain why females stop reproduction even under elevated mortality (Lahdenperä et al. 2011).

In multigenerational groups, there is another potential cost of late-life reproduction—the cost of reproductive conflict between generations (Cant and Johnstone 2008). Because of the dispersal and mating patterns in resident killer whales (bisexual philopatry with out-group mating) females are born into a group without their father. As females reproduce, however, they recruit their sons to their local group, which will increase a female's local relatedness to males with age—driving an overall increase in mean relatedness between a female and her local group with age (Johnstone and Cant 2010; Ellis et al. 2022; fig. 8.2g). The consequence for life history evolution in killer whales of this age-linked change in relatedness (kinship dynamics) is that younger females will be under weaker selection not to harm the group (to whom they are less related) compared to older females, who will be under stronger selection to help the group (Johnstone and Cant 2010). Reproduction within a group is an indiscriminate form of harming as it increases the local competition among breeders (Johnstone and Cant 2010), and younger females are therefore predicted to invest more effort in reproductive competition than older females (Croft et al. 2017). In resident killer whales, this intergenerational competition is manifested when reproductive competition occurs. When mothers reproduce at the same time as their daughters—and are therefore in competition—the survival probability of the older mother's offspring is significantly lower than the survival probability of her daughter's offspring (Croft et al. 2017; fig. 8.2h). This is expected if the daughters are investing more in competition for reproductive resources than their mothers (Croft et al. 2017). Menopause provides a route out of this conflict for older female resident killer whales: by stopping reproduction, females avoid the costs of this reproductive competition.

Taken together, this body of work provides unique insight into the evolution of menopause and provides a clear link between sociality and life history evolution. In resident killer whales, menopause evolved through both the inclusive fitness benefits of late-life helping, particularly of sons, and intergenerational reproductive conflict. The inclusive fitness benefits of grandmothers and mothers supporting their kin can explain the extended female lifespan and the costs of intergenerational reproductive conflict can explain why reproduction does not continue into later life.

MENOPAUSE IN OTHER TOOTHED WHALES

Resident killer whales are not the only toothed whale species with menopause. In fact, toothed whales are the only mammal taxa where menopause has evolved multiple times (fig. 8.3), providing a unique opportunity to examine the evolution of menopause in a comparative context. Exploiting this opportunity has demonstrated that females in species with menopause have a longer lifespan, but the same reproductive lifespan, as expected for a whale of their size (Ellis et al. 2024). Further, females in species with menopause spend longer alive at the same time as their grandoffspring than species without menopause, but do not spend any longer in reproductive competition than species without menopause (Ellis et al. 2024). These results concur with the studies in resident killer whales: both highlight important roles for both grandmother effects in promoting extended female lifespans and reproductive conflict selecting against extended female reproductive lifespans (Ellis et al. 2024). In the final section of this chapter we use three comparisons as case studies to highlight the role of sociality in the evolution of menopause in a comparative context.

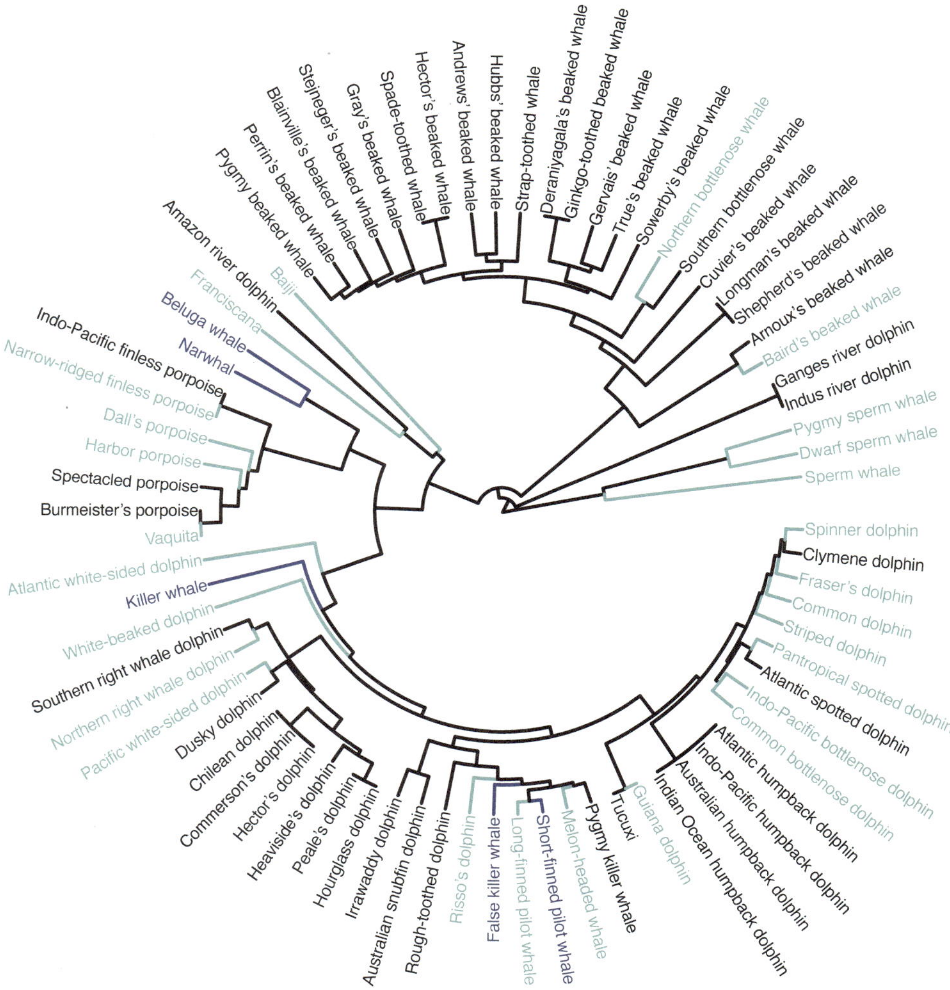

FIGURE 8.3. Toothed whale phylogenetic tree derived from McGowen et al. (2020). Menopause has evolved at least four times independently (dark blue branches and text). Light blue branches are species known not to have menopause. Black branches are species where the menopause state is unknown.

1. Resident-Ecotype Versus Bigg's-Ecotype Killer Whales

Like the sympatric resident-ecotype killer whales, mammal-eating Bigg's-ecotype killer whales in the northeast Pacific Ocean (also referred to as transient-ecotype killer whales) exhibit menopause (Nielsen et al. 2021). This is perhaps surprising because Bigg's killer whales differ from resident killer whales in several important respects, including in diet, hunting behavior, and—importantly—social organization (Baird and Whitehead 2000; Nielsen et al. 2023). In contrast to the bisexual philopatry of resident killer whales, Bigg's killer whales exhibit a form of female-biased dispersal (Baird and Whitehead 2000; Nielsen et al. 2023). The probability of female Bigg's killer whales sharing a strong social bond with their mother decreases steadily across early adulthood, reaching about 10% by age 30—indicating that, by middle age, few females are in close association with their mothers (Nielsen et al. 2023). In contrast, most male Bigg's killer whales share a strong social bond with their mother for as long as both of them are alive, although a few males do exhibit an alternative "roving" strategy (Nielsen et al. 2023). Given the importance of bisexual philopatry to the evolution of menopause in resident killer whales this is at first glance something of a puzzle.

There are three factors likely to be important in explaining the occurrence of menopause in Bigg's killer whales. First, as the probability of having dispersed increases gradually during adulthood most females will have at least their first offspring while in close association with their mother, there is therefore still ample scope for intergenerational concurrent reproduction and therefore reproductive conflict between mothers and their daughters (Nielsen et al. 2023). Second, investment in sons is a key benefit of extended lifespans in resident killer whales, and the absence of male dispersal in Bigg's killer whales shows there is potential for this key benefit of post-reproductive life to also be operating in this population. Third, over the last 40 years the Bigg's killer whales have experienced and are experiencing rapid population growth and space-use change (Ford et al. 2012)—likely driven by the recovery of key mammalian prey stocks (Muto et al. 2020). This recovery may have led to changes in behavior—potentially increasing the relative benefit of dispersal compared to philopatry for female Bigg's killer whales. Another possibility is that menopause is an ancestral trait in killer whales and is not currently under positive or negative selection in Bigg's killer whales, persisting as a relic trait. Future research into the demographics of other populations and ecotypes and the costs of reproductive competition will shed light on the evolutionary history of menopause in killer whales.

The benefits, if any, older female Bigg's killer whales provide their group remain unknown but may be similar to those of residents—they may perhaps lead the group to productive areas, play important roles in hunting, or play a role in initiating, or providing protection from, infanticide (Towers et al. 2018). The comparison between Bigg's and resident killer whales demonstrates that even under very different ecologies and apparently different social organizations the conditions necessary for the evolution of menopause can still be present in a toothed whale species.

2. Killer Whales Versus Sperm Whales

Despite some clear differences in their ecological niche, life history, and social systems, killer whales and sperm whales share a number of important similarities. Both killer whales and sperm whales are apex predators in their environment, are relatively immune to predation as adults (Jefferson et al. 1991) with a long lifespan (Ellis et al. 2025), have a female-centered or matrilineal social structure of closely related females and their offspring (Bigg et al. 1990; Whitehead 2003), and have considerable sexual size dimorphism with very large males (Dines et al. 2015). In both species, there is evidence that only older males are the most significant contributors to annual reproduction (Best et al. 1984; Ford et al. 2018). However, whereas killer whales have menopause, sperm whales do not (Ellis et al. 2018a).

An important difference in the social organization of sperm whales and killer whales that may explain this life history difference is the social behavior of males. Unlike male killer whales, male sperm whales permanently disperse from their natal social unit before sexual maturity (Whitehead 2003). This has two consequences pertinent to the evolution of menopause. First, as kinship dynamics are driven by changing relatedness to the opposite sex (Johnstone and Cant 2010; Ellis et al. 2022), female sperm whales will not experience an increase in relatedness to their group as they age. Therefore, and unlike killer whales, there is no increased selection for helping with age in female sperm whales or selection to harm when young—reducing the selective pressures for the evolution of menopause (Johnstone and Cant 2010; Ellis et al. 2022). Second, even though male sperm whale reproductive success is likely to be linked to age, due to male dispersal females have no mechanism by which they can help the survival or reproduction of their adult sons. The combination of these effects means that the key selective pressures for extended post-reproductive female lifespan are absent in sperm whales. The comparison of sperm whales and killer whales demonstrates the importance of males—when considered as part of the extended phenotype of their mothers—in selecting for the evolution of menopause.

3. Short- Versus Long-Finned Pilot Whales

Female short-finned pilot whales live, on average, 25% of their adult life years as post-reproductive (Ellis et al. 2018a). Long-finned pilot whale females, in contrast, will only live 1% of their adult life post-reproductive (Ellis et al. 2018a). This difference in life history between two closely related species is all the more striking because short- and long-finned pilot whales are so morphologically and behaviorally similar (Olson 2018). Importantly, current evidence suggests that both species share a "resident killer whale"–like social organization based on bisexual philopatry with out-group mating (Amos et al. 1993; Van Cise et al. 2017)—which will (and does in long-finned pilot whales) lead to kinship dynamics characterized by increasing female relatedness with age (Nichols et al. 2016; Ellis et al. 2022). Further, despite not having menopause, long-finned pilot whales show evidence of intergenerational reproductive conflict (Nichols et al. 2020). The reason for the differences in life history between the two species remains a mystery, however some hints may come from a recent large-scale comparative analysis of toothed whale life history. Ellis et al. (2024) found that while females from both pilot whale species reproduce for as long as expected for an odontocete of their size, short-finned pilot whale females live for considerably longer than expected given their size (54 years; app. 8A), whereas long-finned pilot whale females live for approximately as long as expected given their size (43 years; app. 8A). This suggests that selection has acted to extend the lifespan of female short-finned, but not long-finned, pilot whales.

The most notable difference between short- and long-finned pilot whales is in their ecology: short-finned pilot whales largely inhabit tropical and subtropical waters whereas long-finned pilot whales inhabit temperate and polar waters in both hemispheres (Olson 2018). In killer whales, selection is likely to have acted to extend female lifespan due to the balance of costs and benefits of extended post-reproductive life. It may be that differences in ecology between the tropics and the "antitropical" regions change this balance in short- compared to long-finned pilot whales. For example, features of the tropical and subtropical ocean may mean that the knowledge of older females has more value for species inhabiting this region. Tropical and subtropical waters are less productive than temperate and polar waters, and therefore the experience of older female short-finned pilot whales may allow them to significantly benefit their kin using their ecological knowledge. The lower abundance of killer whales in the tropics may also be a factor (Ford 2018); with the high abundance of killer whales in temperate and polar waters, and the entailed increased predation risk, it may simply be too difficult for long-finned pilot whales to evolve an extended lifespan. Supporting this, the mortality doubling rate of long-finned pilot whales is twice that of short-finned pilot whales: suggesting that the mortality rate of long-finned pilot whales is considerably higher than that of short-finned pilot whales (Foote 2008). The result of this is that the efficacy of late-life selection in long-finned pilot whales will be much lower than that of short-finned pilot whales, and therefore any benefits that older females can provide will be under much weaker selection, making it much harder for menopause to evolve (Foote 2008). While both of these explanations are at this stage speculative, the pilot whales illustrate how similarities in social systems are not enough in and of themselves to drive the evolution of menopause. Social systems can predispose a species to menopause, but ecology and behavior are important to the story as well.

Conclusions

Since the publication of *Cetacean Societies*, our understanding of the links between sociality and life history has advanced considerably. In the original book, both female and male life histories were described in detail, particularly with respect to reproductive strategies (Connor et al. 2000; Whitehead and Mann 2000). These chapters have been very influential in our understanding of cetacean sociality and life history, particularly by placing reproductive strategies, life history, and social decisions within the same framework. Three developments have driven changes in our ability to understand the links between sociality and life history in the last two decades. First, and most importantly, long-term studies of toothed whale populations have matured and expanded. The importance of long-term data is evident throughout this chapter but is particularly highlighted by the step change in our understanding of the evolution of menopause in resident killer whales since the publication of the original edition, which has been driven by data from long-term studies in the northeast Pacific Ocean. Second, our ability to characterize and study cetacean social structures has been revolutionized by the widespread application of social network analysis in the taxa (Weiss et al. 2021). Indeed, work in toothed whales has been influential in establishing social network analysis as a general tool for understanding animal societies (Whitehead 2008). And third, theoretical models linking sociality to the evolution of menopause (e.g., Johnstone and Cant 2010; Chan et al. 2016), and life history more generally, have begun to be published, generating testable predictions that can be applied to toothed whales.

Although we are just at the beginning of understanding how sociality and life history evolve, there is growing evidence that the two can and do coevolve. Here, we have shown that group living, social relationships, and sexual conflict all have at least the potential to influence life history. The clearest example of how sociality and life history can coevolve is how intergenerational cooperation and conflict can shape life history evolution—specifically the evolution of menopause in resident killer whales. The complementary hypotheses of inclusive fitness benefits of late-life helping and intergenerational reproductive conflict provide a more complete understanding of how and why menopause has evolved. The evolution of menopause in toothed whales offers a clear window into how sociality can drive life history evolution and will hopefully act as an important starting point for further insights into this important topic, in both toothed whales and other animals.

Acknowledgments

S. E. was funded for this work by a Leverhulme Trust Early Career Research Fellowship. M. L. K. N. acknowledges funding from a NERC PhD studentship, and D. P. C. from a NERC Standard Grant (no. NE S010327/1). The authors would additionally like to thank Dr. Micheal Weiss, and other members of the Centre for Research in Animal Behaviour at the University of Exeter, for valuable insights and discussions during the writing of this chapter.

Appendix 8A

TABLE 8.1. Toothed whale social and life history data

Species	*Female lifespan (post. mean [95% credible interval])*	*Male lifespan (post. mean [95% credible interval])*	*Female/male lifespan (post. mean ± post. std. dev.)*	*Mean female / mean male size (length in m)*	*Typical group size*	*Social differentiation*	*Matrilineality*
Atlantic spotted dolphin (*Stenella frontalis*) [ASD]	20.78 [17–26]	31.14 [27–37]				0.78	
Atlantic white-sided dolphin (*Lagenorhynchus acutus*) [AWSD]	19.44 [18–21]	20.75 [19–23]	0.9 ± 0.12	2.14 / 2.41	Midsized		Low
Baiji (*Lipotes vexillifer*) [B]	21.23 [19–24]	17.02 [14–21]	1.19 ± 0.21	2.3 / 1.74	Small		
Baird's beaked whale (*Berardius bairdii*) [BBW]	51.28 [45–58]	80.81 [73–89]	0.53 ± 0.07	9.51 / 10.31	Small	0.56	
Beluga whale (*Delphinapterus leucas*) [BW]	52.36 [51–54]	50.23 [48–53]	1.14 ± 0.06	4.17 / 5.04	Small		Medium
Chilean dolphin (*Cephalorhynchus eutropia*)	19.87 [17–24]	17.96 [15–22]					
Commerson's dolphin (*Cephalorhynchus commersonii*) [CsD]	16.02 [14–18]	20.31 [18–23]	0.73 ± 0.1	1.52 / 1.49	Small	0.92	
Common bottlenose dolphin (*Tursiops truncatus*) [CBD]	38.26 [37–40]	36.42 [35–38]	1.11 ± 0.05	2.6 / 2.66	Small	0.76	Low
Common dolphin (*Delphinus delphis*) [CD]	23.57 [23–24]	25.84 [24–28]	0.99 ± 0.09	1.94 / 1.88	Large		Low
Cuvier's beaked whale (*Ziphius cavirostris*)	22.73 [20–27]	49.78 [42–59]					
Dall's porpoise (*Phocoenoides dalli*) [DP]	11.12 [11–12]	12.46 [12–13]	0.87 ± 0.05	1.83 / 1.87	Midsized		
Dwarf sperm whale (*Kogia sima*)	19.06 [16–23]	16.64 [14–20]			Small		
False killer whale (*Pseudorca crassidens*) [FKW]	66.02 [62–71]	58.07 [53–64]	1.28 ± 0.12	3.74 / 4.26	Midsized		High
Franciscana (*Pontoporia blainvillei*) [F]	11.69 [11–12]	11.08 [10–12]			Small		Low
Fraser's dolphin (*Lagenodelphis hosei*) [FD]	18.92 [17–21]	19.17 [18–21]	1.16 ± 0.15	2.35 / 2.48	Large		
Guiana dolphin (*Sotalia guianensis*) [GD]	31.96 [29–36]	27.16 [25–30]			Small	0.55	
Harbor porpoise (*Phocoena phocoena*) [HP]	15.87 [15–17]	14.82 [14–16]	1.09 ± 0.06	1.57 / 1.44	Small		
Hector's dolphin (*Cephalorhynchus hectori*)	21.18 [18–25]	22.13 [19–27]					
Indian Ocean humpback dolphin (*Sousa plumbea*) [IOHD]	21.99 [19–26]	26.52 [24–29]			Small	0.81	

(*continued*)

TABLE 8.1. *(continued)*

Species	*Female lifespan (post. mean [95% credible interval])*	*Male lifespan (post. mean [95% credible interval])*	*Female/male lifespan (post. mean ± post. std. dev.)*	*Mean female / mean male size (length in m)*	*Typical group size*	*Social differentiation*	*Matrilineality*
Indo-Pacific bottlenose dolphin (*Tursiops aduncus*) [IPBD]	35.44 [33–39]	34.35 [32–37]	1.05 ± 0.11	2.41 / 2.39	Midsized	1.12	Medium
Indo-Pacific finless porpoise (*Neophocaena phocaenoides*)	28.71 [24–34]	19.95 [17–24]					
Indo-Pacific humpback dolphin (*Sousa chinensis*) [IPHD]	37.07 [33–43]	31.27 [28–35]			Midsized	0.84	
Killer whale (*Orcinus orca*) [KW]	58.31 [53–64]	38.65 [36–42]	1.91 ± 0.19	6.78 / 8.72	Midsized	1.64	High
Long-finned pilot whale (*Globicephala melas*) [LFPW]	43.68 [43–45]	37.97 [37–39]	1.37 ± 0.05	4.41 / 5.52	Midsized	1.71	High
Melon-headed whale (*Peponocephala electra*) [MHW]	39.86 [37–43]	40.09 [37–43]	1.22 ± 0.11	2.15 / 2.48	Large		
Narrow-ridged finless porpoise (*Neophocaena asiaeorientalis*) [NRFP]	21.76 [20–24]	21.16 [20–23]	1.04 ± 0.09	1.37 / 1.84	Small		High
Narwhal (*Monodon monoceros*) [N]	75.34 [68–83]	72.77 [66–80]	1.19 ± 0.12	2.95 / 3.54	Small		Medium
Northern bottlenose whale (*Hyperoodon ampullatus*) [NBW]	25.45 [23–28]	32.41 [29–36]	0.64 ± 0.09	7.5 / 8.75	Small		Low
Northern right whale dolphin (*Lissodelphis borealis*)	27.02 [25–29]	22.45 [21–24]			Large		
Pacific white-sided dolphin (*Lagenorhynchus obliquidens*)	39.03 [36–42]	37.41 [35–41]			Large		
Pantropical spotted dolphin (*Stenella attenuata*) [PSD]	33.63 [33–34]	33.28 [32–34]	1.16 ± 0.04	1.97 / 2.12	Large		
Peale's dolphin (*Lagenorhynchus australis*)	18.69 [15–24]	19.24 [15–25]					
Pygmy sperm whale (*Kogia breviceps*)	21.11 [19–24]	26.08 [22–31]			Small		

Risso's dolphin (*Grampus griseus*)	32.27 [30–35]	26.47 [24–30]			Midsized		
Rough-toothed dolphin (*Steno bredanensis*)	35.74 [31–42]	33.15 [30–37]			Midsized		
Short-finned pilot whale (*Globicephala macrorhynchus*) [SFPW]	54.33 [53–56]	40.16 [38–42]	1.72 ± 0.1	3.58 / 4.53	Midsized	1.87	High
Sperm whale (*Physeter macrocephalus*) [SW]	47.21 [46–48]	58.02 [57–59]	0.91 ± 0.02	10.7 / 15.65	Small		High
Spinner dolphin (*Stenella longirostris*) [SD]	21.83 [21–22]	17.87 [17–19]	1.37 ± 0.08	1.83 / 1.93	Large		
Striped dolphin (*Stenella coeruleoalba*) [StD]	29.65 [29–30]	30.64 [30–32]	1.05 ± 0.04	2.1 / 2.36	Midsized		Medium
Vaquita (*Phocoena sinus*) [V]	21.7 [19–26]	20.16 [17–24]	1.12 ± 0.22	1.4 / 1.07	Small		
White-beaked dolphin (*Lagenorhynchus albirostris*) [WBD]	34.25 [31–38]	31.04 [28–35]	1.32 ± 0.16	2.02 / 2.81	Small		

Notes: Species are included if lifespan data are available for both sexes. Where necessary, square brackets show the shortened species name used for point labels in figure 8.1. Lifespan data from Ellis et al. (2023). Female/male lifespan is calculated from 40,000 posterior draws; methods and details in Ellis et al. (2025). Female and male size based on length, derived from the published expert consensus, details in references (Ellis et al. 2025), and data are included in marinelifehistdata R package (https://github.com/samellisq/marinelifehistdata). Typical group size based on the number of whales typically observed swimming together; data from relevant entries in Würsig et al. (2018). Groups are categorized as "small" if they contain fewer than 10 whales, "midsized" if they contain 10–60 whales, and "large" if they contain >60 whales. These categories reflect biology and the structure of the data (there are no species with typical group sizes between 56 and 100). Full dataset and derivation are included in the marinelifehistdata R package. Social differentiation (*S*) as collated in Weiss et al. (2021). Matrilineality refers to the extent to which adults associate with their close matrilineal kin. High matrilineality indicates that adults of at least one sex are almost always found in close association with their relatives. Medium matrilineality indicates that whales of at least one sex inhabit the same wider social group as some of their matrilineal kin but are not necessarily always in close association or do not share particularly strong social bonds. Low matrilineality indicates that whales are rarely observed in the same wider social group or in close association with their matrilineal kin as adults. These categories are necessarily imprecise but aim to capture the continuum of matrilineal kinship structure hypothesized in other studies (Rendell et al. 2019). Matrilineality data and references in the marinelifehistdata R package. References for matrilineality categorization are as follows: AWSD (Pugliares-Bonner et al. 2021); BW (Colbeck et al. 2013); CBD (Louis et al. 2018; Nykänen et al. 2018); CD (Viricel et al. 2008); FKW (Martien et al. 2019); F (Mendez et al. 2010); IPBD (Möller et al. 2006; Foroughirad et al. 2019; Diaz-Aguirre et al. 2020); KW (Bigg et al. 1990; Parsons et al. 2009); LFPW (Amos et al. 1993; Nichols et al. 2016); NRFP (Chen et al. 2016); N (Palsbøll et al. 1997); NBW (Gowans et al. 2001); SFPW (Alves et al. 2013; Van Cise et al. 2017); SW (Gero et al. 2008); StD (Gaspari et al. 2007). In all columns, empty cells indicate that no data are available.

References

Alberts, Susan C., Jeanne Altmann, Diane K. Brockman, et al. 2013. "Reproductive Aging Patterns in Primates Reveal That Humans Are Distinct." *Proceedings of the National Academy of Sciences* 110 (33): 13440–45. https://doi.org/10.1073/pnas.1311857110.

Alves, Filipe, Sophie Quérouil, Ana Dinis, et al. 2013. "Population Structure of Short-Finned Pilot Whales in the Oceanic Archipelago of Madeira Based on Photo-Identification and Genetic Analyses: Implications for Conservation." *Aquatic Conservation: Marine and Freshwater Ecosystems* 23 (5): 758–76. https://doi.org/10.1002/aqc.2332.

Amos, Bill, Christian Schlötterer, and Diethard Tautz. 1993. "Social Structure of Pilot Whales Revealed by Analytical DNA Profiling." *Science* 260 (5108): 670–72. https://doi.org/10.1126/science.8480176.

Aubier, Thomas G., and Hanna Kokko. 2022. "Volatile Social Environments Can Favour Investments in Quality Over Quantity of Social Relationships." *Proceedings of the Royal Society B* 289 (1973): 20220281. https://doi.org/10.1098/rspb.2022.0281.

Baird, Robin W. 2000. "The Killer Whale: Foraging Specializations and Group Hunting." In *Cetacean Societies: Field Studies of Dolphins and Whales*, edited by Janet Mann, Richard C. Connor, Peter L. Tyack, and Hal Whitehead, 127–53. University of Chicago Press.

Baird, Robin W., and Hal Whitehead. 2000. "Social Organization of Mammal-Eating Killer Whales: Group Stability and Dispersal Patterns." *Canadian Journal of Zoology* 78 (12): 2096–105. https://doi.org/10.1139/z00-155.

Best, Peter B., P. A. S. Canham, and N. Macleod. 1984. "Patterns of Reproduction in Sperm Whales, *Physeter macrocephalus*." *Report of the International Whaling Commission* 6: 51–80.

Bigg, Michael A., Peter F. Olesiuk, Graeme M. Ellis, John K. B. Ford, and Kenneth C. Balcomb. 1990. "Social Organization and Genealogy of Resident Killer Whales (*Orcinus orca*) in the Coastal Waters of British Columbia and Washington State." *Report of the International Whaling Commission* 12: 383–405.

Brent, Lauren J. N., Daniel W. Franks, Emma A. Foster, Kenneth C. Balcomb, Michael A. Cant, and Darren P. Croft. 2015. "Ecological Knowledge, Leadership, and the Evolution of Menopause in Killer Whales." *Current Biology* 25 (6): 746–50. https://doi.org/10.1016/j.cub.2015.01.037.

Cant, Michael A., and Rufus A. Johnstone. 2008. "Reproductive Conflict and the Separation of Reproductive Generations in Humans." *Proceedings of the National Academy of Sciences* 105 (14): 5332–36. https://doi.org/10.1073/pnas.0711911105.

Chan, Matthew H., Kristen Hawkes, and Peter Kim. 2016. "Evolution of Longevity, Age at Last Birth and Sexual Conflict with Grandmothering." *Journal of Theoretical Biology* 393 (March): 145–57. https://doi.org/10.1016/j.jtbi.2015.12.014.

Chen, Minmin, Yusheng Zheng, Yiming Hao, et al. 2016. "Parentage-Based Group Composition and Dispersal Pattern Studies of the Yangtze Finless Porpoise Population in Poyang Lake." *International Journal of Molecular Sciences* 17 (8): 1268. https://doi.org/10.3390/ijms17081268.

Colbeck, Gabriel J., Pierre Duchesne, Lianne D. Postma, Véronique Lesage, Mike O. Hammill, and Jean Turgeon. 2013. "Groups of Related Belugas (*Delphinapterus leucas*) Travel Together During Their Seasonal Migrations in and Around Hudson Bay." *Proceedings of the Royal Society B* 280 (1752): 20122552. https://doi.org/10.1098/rspb.2012.2552.

Connor, Richard C. 2000. "Group Living in Whales and Dolphins." In *Cetacean Societies: Field*

Studies of Dolphins and Whales, edited by Janet Mann, Richard C. Connor, Peter L. Tyack, and Hal Whitehead, 199–219. University of Chicago Press.

Connor, Richard C., Andrew J. Read, and Richard W. Wrangham. 2000. "Male Reproductive Strategies and Social Bonds." In *Cetacean Societies: Field Studies of Dolphins and Whales*, edited by Janet Mann, Richard C. Connor, Peter L. Tyack, and Hal Whitehead, 247–69. University of Chicago Press.

Croft, Darren P., Rufus A. Johnstone, Samuel Ellis, et al. 2017. "Reproductive Conflict and the Evolution of Menopause in Killer Whales." *Current Biology* 27 (2): 298–304. https://doi.org/10.1016/j.cub.2016.12.015.

Dammhahn, Melanie, Niels Jeroen Dingemanse, Petri T. Niemelä, and Denis Réale. 2018. "Pace-of-Life Syndromes: A Framework for the Adaptive Integration of Behaviour, Physiology and Life History." *Behavioral Ecology and Sociobiology* 72 (3). https://doi.org/10.1007/s00265-018-2473-y.

Diaz-Aguirre, Fernando, Guido J. Parra, Cecilia Passadore, and Luciana M. Möller. 2020. "Kinship and Reproductive Condition Correlate with Affiliation Patterns in Female Southern Australian Bottlenose Dolphins." *Scientific Reports* 10 (1). https://doi.org/10.1038/s41598-020-58800-2.

Dines, James P., Sarah L. Mesnick, Katherine Ralls, Laura J. May-Collado, Ingi Agnarsson, and Matthew D. Dean. 2015. "A Trade-Off Between Precopulatory and Postcopulatory Trait Investment in Male Cetaceans." *Evolution* 69 (6): 1560–72. https://doi.org/10.1111/evo.12676.

Ellis, Samuel, Darren P. Croft, Mia L. K. Nielsen, Daniel W. Franks, and Michael N. Weiss. 2025. "Bayesian Inference of Toothed Whale Lifespans." *Biological Journal of the Linnean Society* 145 (1). https://doi.org/10.1093/biolinnean/blaf022.

Ellis, Samuel, Daniel W. Franks, Stuart Nattrass, et al. 2017. "Mortality Risk and Social Network Position in Resident Killer Whales: Sex Differences and the Importance of Resource Abundance." *Proceedings of the Royal Society B* 284 (1865): 20171313. https://doi.org/10.1098/rspb.2017.1313.

Ellis, Samuel, Daniel W. Franks, Stuart Nattrass, et al. 2018a. "Analyses of Ovarian Activity Reveal Repeated Evolution of Post-Reproductive Lifespans in Toothed Whales." *Scientific Reports* 8 (1). https://doi.org/10.1038/s41598-018-31047-8.

Ellis, Samuel, Daniel W. Franks, Stuart Nattrass, et al. 2018b. "Postreproductive Lifespans Are Rare in Mammals." *Ecology and Evolution* 8 (5): 2482–94. https://doi.org/10.1002/ece3.3856.

Ellis, Samuel, Daniel W. Franks, Mia L. K. Nielsen, Michael N. Weiss, Darren P. Croft. 2024. The Evolution of Menopause in Toothed Whales. *Nature* 627 579–85.

Ellis, Samuel, Daniel W. Franks, Michael N. Weiss, et al. 2021. "Mixture Models as a Method for Comparative Sociality: Social Networks and Demographic Change in Resident Killer Whales." *Behavioral Ecology and Sociobiology* 75 (4). https://doi.org/10.1007/s00265-021-03006-3.

Ellis, Samuel, Rufus A. Johnstone, Michael A. Cant, et al. 2022. "Patterns and Consequences of Age-Linked Change in Local Relatedness in Animal Societies." *Nature Ecology and Evolution* 6 (11): 1766–76. https://doi.org/10.1038/s41559-022-01872-2.

Ferguson, Steven H., Cornelia Willing, Trish C. Kelley, David A. Boguski, David J. Yurkowski, and Cortney A. Watt. 2020. "Reproductive Parameters for Female Beluga Whales (*Delphinapterus leucas*) of Baffin Bay and Hudson Bay, Canada." *Arctic* 73 (4): 405–20. https://doi.org/10.14430/arctic71435.

Foote, Andrew D. 2008. "Mortality Rate Acceleration and Post-Reproductive Lifespan in Matrilineal Whale Species." *Biology Letters* 4 (2): 189–91. https://doi.org/10.1098/rsbl.2008.0006.

Ford, John K. B. 2018. "Killer Whale *Orcinus orca*." In *Encyclopedia of Marine Mammals*, 3rd ed., edited by Bernd Würsig, Johannes G. M. Thewissen, and Kit M. Kovacs, 531–37. Academic Press.

Ford, John K. B., Graeme M. Ellis, Peter F. Olesiuk, and Kenneth C. Balcomb. 2010. "Linking Killer Whale Survival and Prey Abundance: Food Limitation in the Oceans' Apex Predator?" *Biology Letters* 6 (1): 139–42. https://doi.org/10.1098/rsbl.2009.0468.

Ford, John Kenneth Baker, Graeme Ellis, and Eva H. Stredulinsky. 2012. *Information in Support of the Identification of Critical Habitat for Transient Killer Whales (Orcinus orca) off the West Coast of Canada*. Canadian Science Advisory Secretariat Research Document 2012/155.

Ford, Michael J., Kim M. Parsons, Eric J. Ward, et al. 2018. "Inbreeding in an Endangered Killer Whale Population." *Animal Conservation* 21 (5): 423–32. https://doi.org/10.1111/acv.12413.

Foroughirad, Vivienne, Alexis L. Levengood, Janet Mann, and Céline H. Frère. 2019. "Quality and Quantity of Genetic Relatedness Data Affect the Analysis of Social Structure." *Molecular Ecology Resources* 19 (5): 1181–94. https://doi.org/10.1111/1755-0998.13028.

Foster, Emma A., Daniel W. Franks, Sonia Mazzi, et al. 2012. "Adaptive Prolonged Postreproductive Life Span in Killer Whales." *Science* 337 (6100): 1313. https://doi.org/10.1126/science.1224198.

Garde, Eva, Steen Honoré Hansen, Susanne Ditlevsen, et al. 2015. "Life History Parameters of Narwhals (*Monodon monoceros*) from Greenland." *Journal of Mammalogy* 96 (4): 866–79. https://doi.org/10.1093/jmammal/gyv110.

Gaspari, Stefania, Arianna Azzellino, Sabina Airoldi, and A. Rus Hoelzel. 2007. "Social Kin Associations and Genetic Structuring of Striped Dolphin Populations (*Stenella coeruleoalba*) in the Mediterranean Sea." *Molecular Ecology* 16 (14): 2922–33. https://doi.org/10.1111/j.1365-294x.2007.03295.x.

Gerber, Livia, Richard C. Connor, Simon J. Allen, et al. 2022. "Social Integration Influences Fitness in Allied Male Dolphins." *Current Biology* 32 (7): 1664–69. https://doi.org/10.1016/j.cub.2022.03.027.

Gero, Shane, Dan Engelhaupt, and Hal Whitehead. 2008. "Heterogeneous Social Associations Within a Sperm Whale, *Physeter macrocephalus*, Unit Reflect Pairwise Relatedness." *Behavioral Ecology and Sociobiology* 63 (1): 143–51. https://doi.org/10.1007/s00265-008-0645-x.

Gowans, Shannon, Hal Whitehead, and Sascha K. Hooker. 2001. "Social Organization in Northern Bottlenose Whales, *Hyperoodon ampullatus*: Not Driven by Deep-Water Foraging?" *Animal Behaviour* 62 (2): 369–77. https://doi.org/10.1006/anbe.2001.1756.

Grimes, Charli, Lauren J. N. Brent, Samuel Ellis, et al. 2023. "Postreproductive Female Killer Whales Reduce Socially Inflicted Injuries in Their Male Offspring." *Current Biology* 33 (15): 3250–56. https://doi.org/10.1016/j.cub.2023.06.039.

Hamilton, William D. 1964. "The Genetical Evolution of Social Behaviour. I." *Journal of Theoretical Biology* 7 (1): 1–16. https://doi.org/10.1016/0022-5193(64)90038-4.

Hawkes, Kristen, James F. O'Connell, Nicholas Jones, Helen Alvarez, and Eric L. Charnov. 1998. "Grandmothering, Menopause, and the Evolution of Human Life Histories." *Proceedings of the National Academy of Sciences* 95 (3): 1336–39. https://doi.org/10.1073/pnas.95.3.1336.

Isaac, Joanne L. 2005. "Potential Causes and Life-History Consequences of Sexual Size Dimorphism in Mammals." *Mammal Review* 35 (1): 101–15. https://doi.org/10.1111/j.1365-2907.2005.00045.x.

Jefferson, Thomas A., Pam Joyce Stacey, and Robin W. Baird. 1991. "A Review of Killer Whale

Interactions with Other Marine Mammals: Predation to Co-Existence." *Mammal Review* 21 (4): 151–80. https://doi.org/10.1111/j.1365-2907.1991.tb00291.x.

Johnstone, Rufus A., and Michael A. Cant. 2010. "The Evolution of Menopause in Cetaceans and Humans: The Role of Demography." *Proceedings of the Royal Society B* 277 (1701): 3765–71. https://doi.org/10.1098/rspb.2010.0988.

Kappeler, Peter M. 2019. "A Framework for Studying Social Complexity." *Behavioral Ecology and Sociobiology* 73 (1). https://doi.org/10.1007/s00265-018-2601-8.

Kasuya, T., and H. Marsh. 1984. "Life History and Reproductive Biology of the Short-Finned Pilot Whale, *Globicephala macrorhynchus*, off the Pacific Coast of Japan." *Report of the International Whaling Commission* 6: 259–310.

Kim, Peter, John S. McQueen, and Kristen Hawkes. 2019. "Why Does Women's Fertility End in Mid-Life? Grandmothering and Age at Last Birth." *Journal of Theoretical Biology* 461 (January): 84–91. https://doi.org/10.1016/j.jtbi.2018.10.035.

Krause, Jens, and Graeme D. Ruxton. 2002. *Living in Groups*. Oxford University Press.

Lahdenperä, Mirkka, Andrew F. Russell, Marc Adélard Tremblay, and Virpi Lummaa. 2011. "Selection on Menopause in Two Premodern Human Populations: No Evidence for the Mother Hypothesis." *Evolution* 65 (2): 476–89. https://doi.org/10.1111/j.1558-5646.2010.01142.x.

Lee, Ronald. 2008. "Sociality, Selection, and Survival: Simulated Evolution of Mortality with Intergenerational Transfers and Food Sharing." *Proceedings of the National Academy of Sciences* 105 (20): 7124–28. https://doi.org/10.1073/pnas.0710234105.

Lemaître, Jean-François, Victor Ronget, Morgane Tidière, et al. 2020. "Sex Differences in Adult Lifespan and Aging Rates of Mortality Across Wild Mammals." *Proceedings of the National Academy of Sciences* 117 (15): 8546–53. https://doi.org/10.1073/pnas.1911999117.

Louis, Marie, Benoit Simon-Bouhet, Amélia Viricel, et al. 2018. "Evaluating the Influence of Ecology, Sex and Kinship on the Social Structure of Resident Coastal Bottlenose Dolphins." *Marine Biology* 165: 1–12. https://doi.org/10.1007/s00227-018-3341-z.

Martien, Karen K., Barbara L. Taylor, Susan J. Chivers, Sabre D. Mahaffy, Antoinette M. Gorgone, and Robin W. Baird. 2019. "Fidelity to Natal Social Groups and Mating Within and Between Social Groups in an Endangered False Killer Whale Population." *Endangered Species Research* 40 (November): 219–30. https://doi.org/10.3354/esr00995.

McGowen, Michael R., Georgia Tsagkogeorga, Sandra Álvarez-Carretero, et al. 2019. "Phylogenomic Resolution of the Cetacean Tree of Life Using Target Sequence Capture." *Systematic Biology* 69 (3): 479–501. https://doi.org/10.1093/sysbio/syz068.

Mendez, Martín, Howard C. Rosenbaum, Randall S. Wells, Andrew Stamper, and Pablo Bordino. 2010. "Genetic Evidence Highlights Potential Impacts of By-Catch to Cetaceans." *PLOS ONE* 5 (12): e15550. https://doi.org/10.1371/journal.pone.0015550.

Möller, Luciana M., Luciano B. Beheregaray, S. J. Allen, and Robert Harcourt. 2006. "Association Patterns and Kinship in Female Indo-Pacific Bottlenose Dolphins (*Tursiops aduncus*) of Southeastern Australia." *Behavioral Ecology and Sociobiology* 61 (1): 109–17. https://doi.org/10.1007/s00265-006-0241-x.

Moorad, Jacob A., Daniel Promislow, and Jonathan Silvertown. 2019. "Evolutionary Ecology of Senescence and a Reassessment of Williams' 'Extrinsic Mortality' Hypothesis." *Trends in Ecology and Evolution* 34 (6): 519–30. https://doi.org/10.1016/j.tree.2019.02.006.

Muto, M. M., V. T. Helker, B. J. Delean, et al. 2020. *Alaska Marine Mammal Stock Assessments, 2019*. NOAA.

Nattrass, Stuart, Darren P. Croft, Samuel Ellis, et al. 2019. "Postreproductive Killer Whale Grandmothers Improve the Survival of Their Grandoffspring." *Proceedings of the National Academy of Sciences* 116 (52): 26669–73. https://doi.org/10.1073/pnas.1903844116.

Nichols, Hazel J., Kevin Arbuckle, Karen Fullard, and William Amos. 2020. "Why Don't Long-Finned Pilot Whales Have a Widespread Postreproductive Lifespan? Insights From Genetic Data." *Behavioral Ecology* 31 (2): 508–18. https://doi.org/10.1093/beheco/arz211.

Nichols, Hazel J., Lilith J. Zecherle, and Kevin Arbuckle. 2016. "Patterns of Philopatry and Longevity Contribute to the Evolution of Post-Reproductive Lifespan in Mammals." *Biology Letters* 12 (2): 20150992. https://doi.org/10.1098/rsbl.2015.0992.

Nielsen, Mia L. K., Samuel Ellis, Jared R. Towers, et al. 2021. "A Long Postreproductive Life Span Is a Shared Trait Among Genetically Distinct Killer Whale Populations." *Ecology and Evolution* 11 (13): 9123–36. https://doi.org/10.1002/ece3.7756.

Nielsen, Mia L. K., Samuel Ellis, Michael N. Weiss, et al. 2023. "Temporal Dynamics of Mother-Offspring Relationships in Bigg's Killer Whales: Opportunities for Kin-Directed Help by Post-Reproductive Females." *Proceedings of the Royal Society B* 290 (2000): 20230139. https://doi.org/10.1098/rspb.2023.0139.

Nykänen, Milaja, Eileen Dillane, Anneli Englund, et al. 2018. "Quantifying Dispersal Between Marine Protected Areas by a Highly Mobile Species, the Bottlenose Dolphin, *Tursiops truncatus*." *Ecology and Evolution* 8 (18): 9241–58. https://doi.org/10.1002/ece3.4343.

Olesiuk, P. F., M. A. Bigg, and G. M. Ellis. 1990. "Life History and Population Dynamics of Resident Killer Whales (*Orcinus orca*) in the Coastal Waters of British Colombia and Washington State." *Report of the International Whaling Commission* 12: 209–43.

Olson, P. A. 2018. "Pilot Whales *Globicephala melas* and *G. macrorhynchus*." In *Encyclopedia of Marine Mammals*, 3rd ed., edited by Bernd Würsig, Johannes G. M. Thewissen, and Kit M. Kovacs. Academic Press.

Palsbøll, Per, Mads Peter Heide-Jørgensen, and Rune Dietz. 1997. "Population Structure and Seasonal Movements of Narwhals, Monodon Monoceros, Determined From mtDNA Analysis." *Heredity* 78 (3): 284–92. https://doi.org/10.1038/hdy.1997.43.

Parsons, Kim M., Kenneth C. Balcomb III, John K. B. Ford, and John W. Durban. 2009. "The Social Dynamics of Southern Resident Killer Whales and Conservation Implications for This Endangered Population." *Animal Behaviour* 77 (4): 963–71. https://doi.org/10.1016/j.anbehav.2009.01.018.

Péron, Guillaume, Christophe Bonenfant, Jean-François Lemaître, Victor Ronget, Morgane Tidière, and Jean-Michel Gaillard. 2019. "Does Grandparental Care Select for a Longer Lifespan in Non-Human Mammals?" *Biological Journal of the Linnean Society* 128 (92): 360–72. https://doi.org/10.1093/biolinnean/blz078.

Photopoulou, Theoni, Ines Maria Ferreira, Peter B. Best, Toshio Kasuya, and Helene Marsh. 2017. "Evidence for a Postreproductive Phase in Female False Killer Whales *Pseudorca crassidens*." *Frontiers in Zoology* 14 (1). https://doi.org/10.1186/s12983-017-0208-y.

Powell, Sarah N., Megan M. Wallen, Madison L. Miketa, et al. 2020. "Sociality and Tattoo Skin Disease Among Bottlenose Dolphins in Shark Bay, Australia." *Behavioral Ecology* 31 (2): 459–66. https://doi.org/10.1093/beheco/arz207.

Pugliares-Bonner, Katie R., Kate LaSpina, Kathryn S. Rose, Steven E. Travis, and Kristina M. Cammen. 2021. "Strandings Provide Insight into Social Group Structure of Atlantic White-Sided Dolphins." *Marine Mammal Science* 37 (3): 901–18. https://doi.org/10.1111/mms.12783.

Rendell, Luke, Mauricio Cantor, Shane Gero, Hal Whitehead, and Janet Mann. 2019. "Causes and Consequences of Female Centrality in Cetacean Societies." *Philosophical Transactions of the Royal Society B* 374 (1780): 20180066. https://doi.org/10.1098/rstb.2018.0066.

Silk, Matthew J., and Nina H. Fefferman. 2021. "The Role of Social Structure and Dynamics in the Maintenance of Endemic Disease." *Behavioral Ecology and Sociobiology* 75 (8). https://doi.org/10.1007/s00265-021-03055-8.

Silk, Matthew J., and David J. Hodgson. 2021. "Differentiated Social Relationships and the Pace-of-Life-History." *Trends in Ecology and Evolution* 36 (6): 498–506. https://doi.org/10.1016/j.tree.2021.02.007.

Snyder-Mackler, Noah, Joseph R. Burger, Lauren Gaydosh, et al. 2020. "Social Determinants of Health and Survival in Humans and Other Animals." *Science* 368 (6493). https://doi.org/10.1126/science.aax9553.

Stanton, Margaret A., and Janet Mann. 2012. "Early Social Networks Predict Survival in Wild Bottlenose Dolphins." *PLOS ONE* 7 (10): e47508. https://doi.org/10.1371/journal.pone.0047508.

Suydam, R. S. 2009. "Age, Growth, Reproduction, and Movements of Beluga Whales (*Delphinapterus leucas*) from the Eastern Chukchi Sea." PhD diss., University of Washington.

Towers, Jared R., Muriel J. Hallé, Helena Symonds, et al. 2018. "Infanticide in a Mammal-Eating Killer Whale Population." *Scientific Reports* 8 (1). https://doi.org/10.1038/s41598-018-22714-x.

Trivers, Robert L. 1972. "Parental Investment and Sexual Selection." In *Sexual Selection and the Descent of Man*, edited by B. Campbell, 136–79. Aldine.

Van Cise, Amy M., Karen K. Martien, Sabre D. Mahaffy, et al. 2017. "Familial Social Structure and Socially Driven Genetic Differentiation in Hawaiian Short-Finned Pilot Whales." *Molecular Ecology* 26 (23): 6730–41. https://doi.org/10.1111/mec.14397.

Viricel, Amélia, Allan E. Strand, Patricia E. Rosel, Vincent Ridoux, and Pascale Garcia. 2008. "Insights on Common Dolphin (*Delphinus delphis*) Social Organization from Genetic Analysis of a Mass-Stranded Pod." *Behavioral Ecology and Sociobiology* 63 (2): 173–85. https://doi.org/10.1007/s00265-008-0648-7.

Ward, Eric J., Elizabeth E. Holmes, and Ken C. Balcomb. 2009. "Quantifying the Effects of Prey Abundance on Killer Whale Reproduction." *Journal of Applied Ecology* 46 (3): 632–40. https://doi.org/10.1111/j.1365-2664.2009.01647.x.

Wasser, Samuel K., Jessica I. Lundin, Katherine L. Ayres, et al. 2017. "Population Growth Is Limited by Nutritional Impacts on Pregnancy Success in Endangered Southern Resident Killer Whales (*Orcinus orca*)." *PLOS ONE* 12 (6): e0179824. https://doi.org/10.1371/journal.pone.0179824.

Weiss, Michael N., Samuel Ellis, and Darren P. Croft. 2021. "Diversity and Consequences of Social Network Structure in Toothed Whales." *Frontiers in Marine Science* 8: 688842. https://doi.org/10.3389/fmars.2021.688842.

Weiss, Michael N., Samuel Ellis, Daniel W. Franks, et al. 2023. "Costly Lifetime Maternal Investment in Killer Whales." *Current Biology* 33 (4): 744–48.e3. https://doi.org/10.1016/j.cub.2022.12.057.

Weiss, Michael N., Daniel W. Franks, Kenneth C. Balcomb, et al. 2020. "Modelling Cetacean Morbillivirus Outbreaks in an Endangered Killer Whale Population." *Biological Conservation* 242 (February): 108398. https://doi.org/10.1016/j.biocon.2019.108398.

Wensink, Maarten Jan, Hal Caswell, and Annette Baudisch. 2017. "The Rarity of Survival to Old Age Does Not Drive the Evolution of Senescence." *Evolutionary Biology* 44 (1): 5–10. https://doi.org/10.1007/s11692-016-9385-4.

Whitehead, Hal. 2003. *Sperm Whales: Social Evolution in the Ocean*. University of Chicago Press.

Whitehead, Hal. 2008. *Analyzing Animal Societies: Quantitative Methods for Vertebrate Social Analysis*. University of Chicago Press.

Whitehead, Hal, and Janet Mann. 2000. "Female Reproductive Strategies of Cetaceans: Life Histories and Calf Care." In *Cetacean Societies: Field Studies of Dolphins and Whales*, edited by Janet Mann, Richard C. Connor, Peter L. Tyack, and Hal Whitehead, 219–46. University of Chicago Press.

Wild, Sonja, Simon J. Allen, Michael Krützen, Stephanie L. King, Livia Gerber, and William J. E. Hoppitt. 2019. "Multi-Network-Based Diffusion Analysis Reveals Vertical Cultural Transmission of Sponge Tool Use Within Dolphin Matrilines." *Biology Letters* 15 (7): 20190227. https://doi.org/10.1098/rsbl.2019.0227.

Wild, Sonja, William J. E. Hoppitt, Simon J. Allen, and Michael Krützen. 2020. "Integrating Genetic, Environmental, and Social Networks to Reveal Transmission Pathways of a Dolphin Foraging Innovation." *Current Biology* 30 (15): 3024–30. https://doi.org/10.1016/j.cub.2020.05.069.

Williams, George C. 1957. "Pleiotropy, Natural Selection, and the Evolution of Senescence." *Evolution* 11 (4): 398–411. https://doi.org/10.1111/j.1558-5646.1957.tb02911.x.

Wright, Brianna M., Eva H. Stredulinsky, Graeme M. Ellis, and John K. B. Ford. 2016. "Kin-Directed Food Sharing Promotes Lifetime Natal Philopatry of Both Sexes in a Population of Fish-Eating Killer Whales, *Orcinus orca*." *Animal Behaviour* 115 (May): 81–95. https://doi.org/10.1016/j.anbehav.2016.02.025.

Würsig, Bernd, Johannes G. M. Thewissen, Kit M. Kovacs. 2018. *Encyclopedia of Marine Mammals*. Academic Press.

Zhu, Pingfen, Weiqiang Liu, Xiaoxiao Zhang, et al. 2023. "Correlated Evolution of Social Organization and Lifespan in Mammals." *Nature Communications* 14 (1). https://doi.org/10.1038/s41467-023-35869-7.

9

Us Against Them: Alliance Formation in Toothed Whales

RICHARD CONNOR AND STEPHANIE L. KING

Abstract

For forty years, an extraordinary population of Indo-Pacific dolphins in Shark Bay, Western Australia, has proven to be the Rosetta Stone for understanding dolphin intelligence in the wild, most prominently in the study of multilevel male alliances. The Shark Bay dolphins have the largest known multilevel alliance network outside humans and have converged with humans in the formation of strategic intergroup alliances. Importantly, these findings show that dolphins, like nonhuman primates, are valuable model systems for understanding human social and cognitive evolution. Here we summarize what we have learned about alliances in toothed whales, using Shark Bay as a case study, before discussing evidence for coalitions and alliances in other cetacean species, where progress has been slower but with notable discoveries in several species. We show how observations of alliances in odontocetes have led to new theory on alliance formation in mammals and provided new insights into human social evolution. We expand the discussion of coalitions and alliances into the realm of interspecific interactions, then reflect on why the study of alliances and coalitions in odontocetes and nonhuman primates followed different trajectories. From this comparative perspective, we make recommendations for the future study of alliances and coalitions in toothed whales.

Introduction

When *Cetacean Societies* was first published in 2000, the bottlenose dolphin research in Shark Bay had been underway for 18 years. By then much had been discovered about this extraordinary population of dolphins. We knew that strongly bonded pairs and trios of males, "first-order" alliances, would herd estrus females, and multiple pairs and trios would join forces in "second-order" alliances to take females aggressively from other alliances or defend against such attacks (Connor et al. 1992b). This was a major discovery, as it was the first case of multiple-level male alliances within a social group outside of humans, with significant implications for the "social brain" hypothesis for the evolution of large brains and intelligence (Connor et al. 1992a, 1992b). We had also discovered a "super alliance" of 14 males that comprised more labile pairs and trios, which at the time was thought to be unusual (Connor et al. 1999). Affiliative contact behavior, petting, and rubbing within and between the sexes had been documented, as had physical synchrony, particularly among the males in the form of synchronous displays (Connor et al. 2000b). Yet we knew little about the function of synchrony nor how affiliative contact was used to mediate relationships. We had yet to examine genetic relationships among the males and we knew little of their communication system. Over two decades later, and four decades since the Shark Bay dolphin project's inception, our understanding of the Shark Bay alliances has increased dramatically. For example, we now know that there are three alliance levels and that Shark Bay males have the largest known multilevel alliance network outside humans (Connor et al. 2022).

Multilevel Alliances in Shark Bay

At the first level, pairs or trios work together to herd single estrus females in events termed "consortships" that can last hours to weeks (Connor and Krützen 2015; Connor et al. 1992b; Connor et al. 1996; fig. 9.1). The stability of these first-order alliances can be highly variable, with some males having consistent first-order allies and others consorting females with numerous males from within their second-order alliance in a given mating season. Because of this continuous variation in first-order stability we did not adopt the coalition/alliance dichotomy from primatology (De Waal and Harcourt 1992), in which coalitions are instances of two or more conspecifics joining forces against another, and individuals who repeatedly form coalitions constitute an alliance (see Connor et al. 2022).

Within second-order alliances, males with more stable first-order alliances spend more time consorting females (Connor et al. 2001; Connor and Krützen 2015). Finding such covariation between consortship success, along with rare observations of one first-order alliance taking a female from another from the same second-order alliance, might reflect dominance relationships. While first-order alliances predominantly consist of trios in Shark Bay, the percentage of first-order trios (versus pairs), consortship rate, and

FIGURE 9.1. Alliance formation in Shark Bay, Western Australia. (*Top left*) Researchers conducting a focal follow on allied male dolphins in Shark Bay. (*Top right*) Three allied males ("Urchin," "Cookie," and "Smokey"), members of a second-order alliance known as the "Red Cliff Rascals," herding female "Lick." "Cookie" featured in *Cetacean Societies* in 2000 as a newborn calf (fig. 4.15). (*Bottom left*) Three allied males expressing their social bonds through synchronous movement. (*Bottom right*) Three second-order alliances ("Hooligans," "Space Cadets," and "Lissos Loonies").
Source: Photos by Simon J. Allen and Shark Bay Dolphin Research.

the occurrence of fresh rake marks (indicative of recent competition), all increase from south to north along the east side of Peron Peninsula in Shark Bay (Connor et al. 2017; Hamilton et al. 2019; Sørensen et al. 2024). Such habitat-related differences in alliance behavior within a single social network are an ideal and unique setting for testing hypotheses on ecological impacts on male alliance formation.

There is a continuous range of second-order alliance sizes from 4 to 14 males (the lower end is defined by the minimum size of two first-order alliances). The original "super alliance" of 14 males was not an anomaly; three alliances achieved and maintained a size of 14 members for three years before losing members (Connor et al. 1999; Connor and Krützen 2015). It is an impressive sight when all members of a large second-order alliance are together, with several consorted females and their calves (fig. 9.1).

The formation of second-order alliances is a protracted process, with social bonds that form between unrelated, but similarly aged males during the juvenile period predicting future alliance membership (Gerber et al. 2020, 2021). Kinship can explain some associations during adolescence (8–14 years), but not adulthood (Gerber et al. 2020). We know of at least three cases in which successive maternal brothers are in different second-order alliances (Gerber et al. unpublished data). Thus, while kin and nonkin may be important for social bond and skill development during adolescence, males form adult alliances with similar aged peers (Gerber et al. 2020, 2021). Similar aged kin might be favored allies if they were available, but they are not; with a four-or-more year interval between surviving offspring, and because female dolphins give birth to a single infant, they cannot produce "ready-made" alliances of relatives, like litter-bearing lions who often synchronize births with female relatives (Connor 2007; Packer and Pusey 1987).

If not kinship, on what basis do males select and compete for allies? Possibilities include resource holding potential, compatible personalities, dominance rank, and foraging homophily. For example, there is considerable variation in learned foraging tactics among individuals in Shark Bay, including those that range in the same areas (Wild et al. 2019; Mann et al. 2012; Sargeant et al. 2005). This variation includes an alliance of males that use marine sponges as a foraging tool, a specialized behavior more common in females (Bizzozzero et al. 2019), and differences in preferred foraging habitat among alliances with broadly overlapping ranges (O'Brien et al. 2020). Males with shared foraging habitats, prey types, or hunting techniques may be able to coordinate their lives together more efficiently than would be the case if they differed in these attributes.

Once established, membership in second-order alliances is largely closed, as adults rarely join or leave to join other second-order alliances, thus the second-order alliance is the core male social unit in Shark Bay (Connor and Krützen 2015; Connor et al. 2022). Changes in second-order alliance membership are closer in nature to the "secondary transfers" that occur in primates when, for example, a male disperses a second time from the group he joined after leaving his natal group (Connor et al. 2022). Although unusual, such transfers appear highly strategic, as when the addition of a new member in an older declining alliance allowed members to resume consorting females in their strongly preferred trios (Connor and Krützen 2015). In general, however, older alliances become gradually smaller as individuals disappear and are presumed to have died (Randić et al. 2012; King et al. 2021; Connor et al. 2022). In 2024, we were fortunate to observe the male Gripe. Now on his own, he is thought to be the only remaining member of the original 14-member second-order alliance that was discovered in 1994, a.k.a. the "super alliance" (Connor et al. 1999, 2001).

Third-order or "intergroup" alliances occur when second-order alliances associate preferentially and cooperate against other groups (Connor et al. 2011). Given that bond strengths (as measured by association indices) among some second- and third-order allies are similar, and second- and third-order alliances share the same function, why does a third alliance level exist and how is it distinguished from the second? Five criteria clearly distinguish second- and third-order alliances (Connor et al. 2022). First, cluster analysis clearly distinguishes the two levels and males are rarely with third-order allies without the presence of second-order allies. Second, males consort females almost exclusively with their second-order allies. Third, once formed, second-order alliances are largely closed. Fourth, male dolphins respond more strongly to signature whistle playbacks of second- than third-order allies irrespective of bond strength (King et al. 2021; see King et al., this volume). Finally, signature whistle exchanges can be used by males as a replacement for physical bonding for allies they spend less time with and are significantly more likely to occur between second-order allies than third-order allies, which would not be the case if second- and third-order allies were interchangeable (Chereskin et al. 2022). We suspect the third level of alliance in Shark Bay is insurance for occasions when second-order alliances are threatened by larger groups or second-order allies are widely dispersed and unavailable to assist in conflicts (Connor et al. 2011). Supporting this idea, we now know that third-order alliance relationships increase male access to females, thereby increasing their reproductive success (Connor et al. 2022).

The long-term persistence of second-order alliances led to the realization that they are akin to stable primate groups that are "closed" or "semi-closed" (Connor et al. 2022). Yet we had previously described the overall society as "open"; lacking the social boundary found in most terrestrial mammals (like a chimpanzee community, baboon troop, or wolf pack) and characterized instead by a continuous mosaic of overlapping female and male alliance ranges (Randić et al. 2012). These two perspectives are not incompatible; the male alliances are "semi-closed" social groups, but their social relationships are not (fig. 9.2). Not only do they interact with the females and juveniles in their range, they also are found in social groups with other allied and unallied males, forming the largest known continuous alliance network outside of humans (Connor et al. 2022). Indeed, social interactions with non-allied males may form the strategic basis of those unusual alliance shifts (Connor et al. 2022). Exceptionally among nonhuman primates, bonobos (*Pan paniscus*) have a variety of social interactions with members of other groups that include simple coalitions but not intergroup alliances (Furuichi 2020; Tokuyama et al. 2019; Samuni and Surbeck 2023). The discovery of third-order

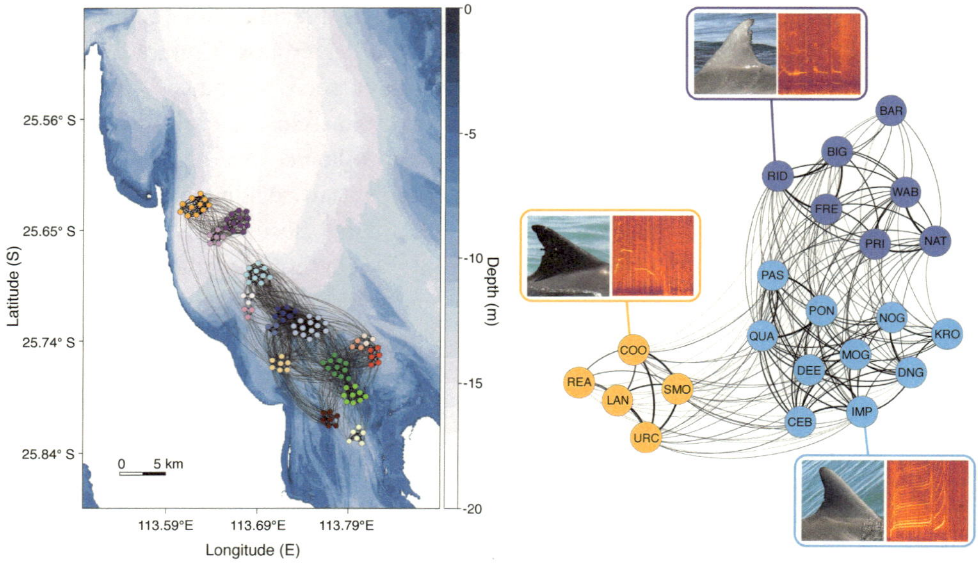

FIGURE 9.2. Multilevel dolphin alliances in Shark Bay. The left panel shows a social network plot of 12 second-order alliances and five trios (121 males). Lines (edge weights) represent association strength (i.e., social bond strength), and circles (nodes) represent individual males. Node colors denote alliance membership (second-order alliance or trio), with third-order allies sharing similar colors. Alliance locations are geo-referenced based on their approximate mating season core home range. The right panel shows a network plot of three second-order alliances, who also formed third-order alliances, with males color coded by alliance membership. Individual males are identified by both their unique dorsal fins and their individually distinctive signature whistles.
Source: (*Left*) Adapted from Connor et al. (2022). (*Right*) Friedman et al. (2023).

or between-group alliances in dolphins is therefore important, because it shows that only humans and dolphins have converged in forming strategic intergroup alliances between nonkin (Connor et al. 2022). It is truly astonishing that an animal whose body plan and habitat could not be more different from humans have shown us that intergroup male alliances can arise directly from a chimpanzee-like promiscuous mating system without key features implicated in cooperation between human groups, such as male-female pair bonds and male parental care (Connor et al. 2022).

Mediating Alliance Relationships in Shark Bay

Observations of social interactions and vocalizations have revealed large variation in relationships among males both within and between the three alliance levels; such differentiated relationships add another dimension of social complexity (see Weiss et al., this volume). This additional complexity is consequential as it has been shown that well-integrated males with more homogeneous social bonds with their second-order allies secure more paternities (Gerber et al. 2022). Differentiated relationships within and between alliances are mediated with affiliative contact behaviors, such as petting and rubbing, as well as through vocalizations (King et al. 2018, 2019, 2021; Moore et al. 2020; Chereskin et al. 2022; Friedman et al. 2023). Unsurprisingly, petting frequency increases with association within second-order alliances, but it is striking that even weakly bonded third-order allies participate in intergroup petting following group fusions (Chereskin et al. 2022; Friedman et al. 2023).

Males have individual identity or "signature" whistles that are distinct from their allies (King et al. 2018), and which play a key role in the maintenance of their multilevel alliance relationships (fig. 9.2), rather than utilizing a shared alliance whistle as was proposed previously for allied males in Shark Bay (Smolker and Pepper 1999) and occurs for males in Sarasota Bay, Florida (Watwood et al. 2004; see Oswald et al., this volume). As noted earlier, males use signature whistles to maintain alliance bonds in lieu of opportunities for physical contact (Chereskin et al. 2022).

Bond formation and maintenance among allies is also supported through synchronous behavior, which appears to play an important role in promoting cooperation and signaling alliance unity (Connor 2007; Connor et al. 2006; McCue et al. 2020; Moore et al. 2020). Males will surface side by side synchronously, usually less than a meter apart, breaking the surface within milliseconds of each other (Connor et al. 2006; McCue et al. 2020; fig. 9.1). Synchrony is more common among closely associated males, including first-order allies and the more closely associating pair in first-order trios (Connor et al. 2006). Side-by-side synchronous surfacing between males from different first-order alliances occurs more when second-order allies are excited around a consorted female, suggesting synchrony may reduce tension in situations of potential conflict (Connor et al. 2006). Surprisingly, frame-by-frame analysis showed that males who associate less in second-order alliances engage in more precise synchrony, as if attending more to bonds with males with whom they spend less time (McCue et al. 2020). This result, combined with the greater use of vocal exchanges among allied males with weaker bonds, demonstrates the value males place on all second-order relationships, regardless of time spent together. Males will also perform elaborate displays consisting of synchronous turns and aerial leaps when consorting individual females, indicating that female choice may still be important in this coercive mating system (Connor et al. 2000a). Given the lack of a leader in synchrony, and displays that are too complex for simple entrainment, the only mechanism we are aware of that can explain all dolphin synchrony is the "mirror game" where both males simultaneously imitate each other's moves (Noy et al. 2011; McCue et al. 2020).

Allied males will also synchronize their vocalizations. In Shark Bay, males, but not females, produce narrowband, low-frequency pulsed calls termed "pops" (Connor and Smolker 1996; Vollmer et al. 2015). Pops are produced by males almost exclusively during consortships and function as an agonistic "come-hither" signal, inducing the female to remain close to the popping male (Connor and Smolker 1996; Vollmer et al. 2015). Males popping simultaneously will match each other's pop tempo and also synchronize their pop production (Moore et al. 2020). This acoustic synchrony may strengthen social bonds, promote cooperation among the males, and signal alliance unity to females and rival males. As in humans, synchronous behavior in dolphins may have evolved as a coalition signaling system (King et al. 2022).

Prosocial Dolphins

Male dolphins in Shark Bay help each other. Their cooperation is not based solely on an increase in the probability of immediate benefits for participants (known as a by-product mutualism; see Connor 1995; King et al., this volume), but second- and third-order alliance cooperation involves males helping each other keep or acquire females, which is best explained by delayed benefits models such as reciprocity or possibly pseudo-reciprocity (Connor 2010). The prosocial nature of dolphin alliance relationships is also reflected in their affiliative behavior, including efforts to maintain bonds with more weakly bonded second- and third-order allies with petting, synchrony, and whistle exchanges (Chereskin et al. 2022; Friedman et al. 2023).

Alliances in Other *Tursiops* Populations

Owing to their widespread distribution in temperate and tropical inshore habitats, bottlenose dolphins (the common bottlenose dolphin, *T. truncatus* and the Indo-Pacific bottlenose dolphin, *T. aduncus*) are studied more than any other cetacean. In *Cetacean Societies*, alliance formation was contrasted in three populations; *T. aduncus* in Shark Bay, with first-order pairs and trios and second-order alliances; *T. truncatus* in Sarasota, Florida, USA, with first-order pairs only, and *T. truncatus* in the Moray Firth, Scotland, with no alliances (Connor et al. 2000b). The picture remains the same in Sarasota and the Moray Firth, but we have since discovered the third alliance level in Shark Bay (Connor et al. 2011; Connor et al. 2022). Association data from several additional populations, supplemented by behavioral observations in a few cases, have emerged since. Brightwell and Gibson (2023) reviewed data from 18 widely distributed *Tursiops* studies. Alliances appear to be absent in several populations, especially those at high latitudes. Second, association data indicate two alliance levels in only one population outside of Shark Bay; in St. John's River in Florida, where stable pairs are considered the core unit and second-order alliances are less stable and pervasive (Brightwell et al. 2025). Shark Bay dolphins are thus exceptional in forming mostly first-order trios, large and stable second-order alliances of up to 14 individuals, and third-order alliances.

Bottlenose dolphin populations also vary in fundamental characteristics such as body size and sexual size dimorphism. At the highest latitudes, bottlenose dolphins are twice as long as the Shark Bay dolphins, and sexual size dimorphism is much more pronounced. Both factors should impact alliance formation (Connor et al. 2000a, 2000b). First, variance in body size increases with mean body size in mammals and birds (Hallgrímsson and Maiorana 2000). To the extent that body size is correlated with resource holding potential (RHP), a greater variance in RHP in large-bodied populations may oppose alliance formation because larger males may more easily dominate smaller ones and be targeted by female choice for larger males. Second, greater sexual size dimorphism could impact the coercion of females by allowing males to dominate females more easily or make it more challenging to do so if maneuverability is key.

Given interpopulation variation in key variables such as body size, sexual size dimorphism (SSD), interbirth intervals (impacting operational sex ratios), as well as resources and predators, *Cetacean Societies* could do no better than recommend broad interpopulation studies where these variables could be compared. Hence, the discovery of intrapopulation variation in alliance behavior in Shark Bay assumes great importance (Connor et al. 2017). While factors such as body size and SSD may be important on a global scale, we can conclude that alliance variation (e.g., first-order pairs versus trios) does not require variation in body size and SSD and instead we can, in Shark Bay, focus on ecological variables.

Alliance Formation in Other Odontocetes

Surprisingly, the alliance system that may be most similar to Shark Bay is not found in other bottlenose dolphin populations, but among Risso's dolphins (*Grampus griseus*) in deep-water habitat off the Azores. Risso's dolphins are medium-sized delphinids, twice as long as the Shark Bay *Tursiops*, and dive to hundreds of meters to hunt for squid. Adult male Risso's associate in "clusters" of 3–13 mostly unrelated individuals that persist across years and appear very similar to the stability and size range of second-order alliances in Shark Bay (3–14 including "lone trios" that are remnants of formerly larger second-order alliances; Hartman et al. 2023b). The Risso's male clusters have been observed chasing each other and in mating associations with females, but their function remains unclear (Hartman et al. 2008; Hartman et al. 2020). Like the Shark Bay males, Risso's males often surface side by side synchronously in pairs and occasionally trios, but rarely in larger groups (Hartman et al. 2020). A drone study of synchrony and position within a group of 13 males hints at alliances within groups, revealing two pairs, a trio, and six less closely associated individuals. The individuals in these pairs and trio, who also had higher rates of surfacing synchronously with others in general, were also found more often in the front of the group (Hartman et al. 2020).

In Shark Bay, it is common for males to swim in "formation" with consorted females, a pair of males on either side of and slightly behind the female or a trio side by side behind her (Connor et al. 1992b) and for first-order alliances to travel apart (>10 m) from the rest of the group. Neither of these phenomena, so striking and ubiquitous in Shark Bay, are reported from groups containing a Risso's male cluster and females. Instead, Hartman et al. (2023a) report that a male-female pair may swim side by side, or the female may travel in the center of the larger group. The "bull chases" described in Hartman et al. resemble the behavior during female captures in Shark Bay, but as they often involve males from more than one cluster, may also reflect competition for females. The lack of clear associations between single females and male pairs and trios, in and away from the rest of the male cluster, suggest that access to females in Risso's male clusters might operate under a different set of rules than among allied males in Shark Bay. These differences may owe, in part, to the greater cohesion of Risso's male clusters. The comparison with Shark Bay will grow richer as we learn more about the function of male-male associations in the Risso's and whether there are alliance relationships between male clusters.

Alliance behavior has also been documented in the spotted dolphin (*Stenella frontalis*). In the Bahamas, two separate long-term studies on spotted dolphins in Bimini and Little Bahama Bank (LBB) revealed clear same-sex preferences, with long-lasting associations among adult males (Danaher-Garcia et al. 2020; Elliser and Herzing 2014; Herzing et al. 2017). The function of the strong social bonds among males off Bimini remains unclear (Danaher-Garcia et al. 2022; Danaher-Garcia et al. 2020). In contrast, up to three levels of alliances have been proposed in the LBB study, with long-term stable pairs and trios also forming temporary clusters with other males (Elliser and Herzing 2014). Three to four male spotted dolphins in LBB have been observed chasing, surrounding, and following a female before mating with her (Herzing 1996), and the associations between different first-order alliances (possible second-order alliances) have been observed "during consortship activities" (Herzing 1996). A third alliance level was invoked to explain associations among males during interspecific encounters with bottlenose dolphins (Elliser and Herzing 2014; Herzing and Elliser 2013; Herzing and Johnson 1997; Elliser and Herzing 2016). Missing from the LBB study is quantitative data on functional alliance behavior against conspecifics. The nature of the pairs and trios still rests on a few descriptive accounts from the 1990s (Herzing 1996; Herzing and Johnson 1997) and there is no evidence of aggression against other spotted alliances. The coalitionary interactions against *Tursiops* may be the sole reason for associations among males outside of pairs and trios in this population. This would explain the lack of higher-level associations in the Bimini spotted dolphins, which do not interact with *Tursiops* to the same extent. Remarkably, the majority of the LBB society migrated to Bimini and formed not only associations with Bimini dolphins, but the strongest associations between Bimini and formerly LBB individuals, which included affiliative contact behavior, were between adult males (Danaher-Garcia et al. 2022).

Coalitions or alliances may also be found in the Australian humpback dolphins (*Sousa sahulensis*). Adult male pairs with strong levels of association have been observed engaging in coordinated approaches to adult females (Allen et al. 2017), and males have been documented presenting sponges to receptive females and performing physical displays or postures in the presence of females, although the function of these sexual displays and adult male associations remains unknown (Allen et al. 2017).

Three odontocetes, two ziphiids, and a monodontid that were predicted to live in stable matrilineal groups have been discovered to have societies closer to bottlenose dolphins; with a fission-fusion grouping pattern and possible male alliances. A genetic and behavioral study on belugas did not find the expected stable matrilineal associations, but rather a fission-fusion grouping pattern with bisexual philopatry, at least within the larger community, and extensive interactions with non-relatives (O'Corry-Crowe et al. 2020). Closely associating males who moved together for extended periods indicated male alliances (O'Corry-Crowe et al. 2020). The first observed mating interaction in wild belugas was reminiscent of mating-related aggression by male bottlenose dolphins during consortships but was quite violent: the males moved synchronously as they aggressively herded and mated with the female, who was bleeding around the head and genitals (Lydersen et al. 2023).

Gowans et al. (2001) examined associations among deep-diving bottlenose whales (*Hyperoodon ampullatus*) in a canyon off Nova Scotia to examine the idea, derived from sperm whale studies, that deep diving favors communal care of vulnerable young and thus a matrilineal society with strong bonds among females. Instead, they found association patterns closer to that of bottlenose dolphins: a fission-fusion grouping pattern with weak associations among females and stronger male-male bonds that persisted over years in some cases (Gowans et al. 2001). There are also hints of male alliances in two other ziphiids. Baird (2019) documented an association across years in one maturing male pair of Cuvier's beaked whale (*Ziphius cavirostris*) and Cioffi et al. (2021) recorded highly synchronous dives between associating tagged male pairs that endured for up to 17 days. The first study of the social organization of Baird's beaked whale (*Berardius bairdii*) was limited to 28 encounters of groups averaging about eight whales across 6 years (Fedutin et al. 2015). They found high levels of association across years in pairs and trios of heavily scarred individuals of unknown sex (some have since been genetically confirmed as males; Olga Filatova, pers. comm.).

What is the potential for alliances in odontocetes that live in stable matrilineal social groups of related females and, in some cases, related males? With their huge ranges and deep three-dimensional habitat, territory and resource defense seem an unlikely explanation for such groups. Predator defense is an obvious candidate as odontocetes must deal with sharks and, the most formidable predator in the ocean, the highly intelligent, group-hunting killer whale (*Orcinus orca*). Connor et al. (2000a) suggested a third possibility: stable social groups are alliances against infanticidal conspecifics (see also Möller 2012). They noted that odontocetes have the characteristics that led Van Schaik and Kappeler (1997) to suggest that year-round association between male and female primates was primarily driven by infanticide risk. Those characteristics are broadly seasonal breeding that would allow a female who loses a calf to conceive again and a lactation period that exceeds gestation duration, which excludes a postpartum estrus that would eliminate the advantage of infanticide (Hrdy 1979; Van Schaik and Kappeler 1997).

The most well-known odontocete with year-round association between males and females is the killer whale (*O. orca*). "Resident" or fish-eating killer whales off the West Coast of North America live sympatrically with the mammal-eating "Bigg's" killer whales, in stable matrilineal groups with bisexual philopatry (Baird 2000). In *Cetacean Societies*, Connor et al. (2000a) suggested that observations of residents indicated closer bonds between post-reproductive mothers with adult sons than daughters and predicted that such associations were based on benefits provided by the mother to her sons; in foraging or as allies in contests with other adult males. It is clear now the general hypothesis is correct; post-reproductive females invest heavily in sons (Weiss et al. 2023; Ellis et al., this volume). The first behavioral evidence supported the foraging hypothesis; post-reproductive females lead their sons (more than daughters) during foraging and share more food with adult sons than daughters (Brent et al. 2015; Wright et al. 2016). Just as it seemed the alliance hypothesis was moot, it resurfaced with a remarkable twist in Bigg's killer whales, which also have strong mother-son bonds (Nielsen et al. 2023). A mother, at considerable cost, aided her son during infanticide (Towers et al. 2018). More recently (Grimes et al. 2023) showed that socially inflicted injuries (tooth rake marks) in residents were reduced in adult sons but not daughters with post-reproductive mothers. Such injuries are likely to occur in conflicts with other males or females during attempted infanticide. Given these observations in killer whales, we should be alert to mixed sex, kin-based alliances in other matrilineal species.

Extending the Concept: Interspecies Alliances and Coalitions in Odontocetes

The use of the terms "coalition" and "alliance" in the study of terrestrial mammals has been generally restricted to interactions between conspecifics (De Waal and Harcourt 1992). However, mixed-species groups are common in odontocetes (Syme et al. 2021), and the same coalitionary and alliance behaviors that are directed against conspecifics may be directed against individuals of other species. Male bottlenose dolphins (*Tursiops* spp.) have been observed performing herding-like behavior that includes chasing, flanking, and mounting male and female individuals of other species (*Sousa* [Stensland 2004; Cerchio et al. 2015] and *Sotalia* [May-Collado 2010]), and occasionally the other species performs herding behaviors toward *Tursiops* (Stensland 2004; Cerchio et al. 2015). While infanticidal attacks on calves have been reported in common bottlenose dolphins (Perrtree et al. 2016; Díaz López et al. 2018; Ramos et al. 2022) and Indo-Pacific humpback dolphins (Zheng et al. 2016), groups of male bottlenose dolphins may also "practice" infanticide on harbor porpoises (*Phocoena phocoena*) or common dolphins (*Delphinus delphis*), who are similar in size to bottlenose dolphin calves (Cotter et al. 2012; Methion and Díaz López 2021). Indeed, lethal coalitionary behavior, especially by bottlenose dolphins, may be directed at individuals of a wide range of species (Barnett et al. 2009; Crespo-Picazo et al. 2021).

The most detailed observations of mixed-species groups with coalitionary behavior have been made in the Bahamas, where Atlantic spotted dolphins are found with the larger common bottlenose dolphins (Herzing and Johnson 1997; Elliser and Herzing 2016; Eierman et al. 2019). Aggression was initiated by males of both species, but more often by *Tursiops*, and included sociosexual behavior by male *Tursiops* directed at male and female *Stenella*, and by male *Stenella* who mounted female but not male *Tursiops* (Herzing and Johnson 1997; Cusick and Herzing 2014; Herzing and Elliser 2013). Elliser and Herzing (2014) argued that the interspecific interaction should be considered alliance behavior since it "mirrors" intraspecific alliance behavior. The size difference likely accounts for why aggression could be initiated by single male *Tursiops*, but a coalition of spotted males was required to chase away a single bottlenose male (Herzing and Johnson 1997). Interestingly, mixed-species coalitions have been observed where male bottlenose dolphins teamed up with spotted dolphins against other bottlenose or spotted dolphins (Herzing and Johnson 1997). Regular social interactions between these species include affiliative contact behavior with the pectoral fin; in fact, individuals did not show a preference for petting conspecifics over heterospecifics in mixed-species groups (Eierman et al. 2019). These interspecific affiliative interactions may be the basis for interspecific social bonds (Herzing and Johnson 1997; Eierman et al. 2019) that support the formation of interspecific coalitions in the same manner that affiliative interactions support coalitions and alliances in dolphin and primate groups. Strong long-term associations between individual *Stenella* and *Tursiops* are uncommon in the Bahamas (Herzing et al. 2017), but interspecies coalitions and other forms of social support do not always require high association levels (Connor et al. 2006; Fox et al. 2023). On the other hand, long-term interspecific associations have been found in mixed-species groups of false killer whales and bottlenose dolphins in New Zealand, where the same heterospecifics have been observed in groups together five years and 650 km apart. We suggest that, going forward, researchers should be explicit about the occurrence of coalitionary behavior during interspecific interactions. For example, a statement such as "in mixed groups of bottlenose and humpback dolphins, male bottlenose dolphins often chased individual male *Sousa*" does not distinguish between action by single male bottlenose dolphins at different times versus joint action by more than one bottlenose dolphin.

Coalitions and Alliances in Cetaceans: Looking Forward

Many of the theories, concepts, and definitions used in studies of social mammals were derived from primatology, a field that had a 20-year head start on studies of cetacean societies. However, we have learned that cetacean societies do not fit cleanly within the social parameter space of primate societies; for example, the Shark Bay Indo-Pacific bottlenose dolphins that have multilevel male alliances in an open society or the "resident" killer whales in which adult sons and daughters remain with their mother. Neither of these phenomena has an equivalent in nonhuman primates or other terrestrial mammals.

The study of alliances and coalitions in primates and cetaceans, including efforts to develop predictive models, developed along very different trajectories, so there is very little "cross talk" between the fields. Why is this? In Shark Bay we know the function of all three alliance levels, but in most populations and species alliances are inferred based on high association levels only. The primary cooperative act that defines first-order order alliances in Shark Bay, consorting a female, typically lasts days to weeks, although some pre-mating season events may be quite brief (Connor et al. 1996). The situation in primate studies is very different: the stable bisexual groups of primates (e.g., a baboon troop, a chimpanzee community) do not form alliances with other groups so primate studies and models have focused on who intervenes in squabbles between two individuals within the same social group. These are typically brief events but common enough to study the key variables that determine who intervenes, and thus who forms a coalition, with whom. Studies of female coalitionary interactions in Old World monkeys revealed stable alliances between closely related females in matrilines such that dominance rank is "inherited" and is effectively between matrilines rather than individuals (Chapais 1992). Matrilineal alliances are also found in the most complex non-primate system of alliances among terrestrial mammals, the spotted hyena (East et al. 2009; Ilany et al. 2021).

Coalitions in primates are more strategic where interventions are not based on close relatedness, as in male common chimpanzees (Nishida 1983; Feldblum et al. 2021). Dominance rank features prominently in almost all models of primate coalition formation, which, in the simplest case of three individuals, focus on when one should find top-down (top two ranking against bottom ranking), bottom-up, or bridging coalitions (top and bottom against middle ranking individual; reviewed in Bissonnette et al. 2015). In contrast, the key models of alliance formation in cetaceans have focused not on interventions within groups but the formation of groups based on alliance benefits (Connor and Whitehead 2005; Whitehead and Connor 2005). Roving odontocetes may encounter others in competition over resources, favoring cooperative alliances for resource defense. The *Encounter Rates Model* has been used to explain alliance size and sex differences in odontocete alliance formation (as males with estrus females will be more likely to encounter same-sex competitors than females with quickly consumed prey items; Connor and

Whitehead 2005; Whitehead and Connor 2005) and could be extended to the formation of permanent female or bisexual groups based on infant defense. A number of factors may impact encounter rates other than population density, including home range, detection distance, and operational sex ratio (Connor et al. 2000a, 2000b; Sørensen et al. 2024).

The study of odontocete alliances will progress as association data are supplemented with direct observations of coalitionary behaviors or indirect evidence such as tooth rake marks (Grimes et al. 2023). Behavioral evidence is important, as there can be other explanations for sexual segregation, especially in sexually dimorphic species (Wearmouth and Sims 2008). For example, different foraging requirements may lead to spatial segregation of the sexes and then male groupings could be favored due to predation risk or by-product benefits derived from prey location (e.g., sperm whale bachelor groups). In theory, even without spatial segregation, different foraging requirements leading males and females to pursue different prey that differ in depth and patchiness could lead to differences in male and female grouping patterns in fission-fusion societies. An important lesson from the Towers et al. (2018) observation of infanticide in killer whales is that some behaviors and events that are critical for shaping behavior, and have large fitness impacts, may be quite rare or rarely observed given the limited time humans spend observing any given population. Another key point is that with a dynamic fission-fusion grouping pattern where individuals have considerable association freedom, without behavioral evidence, strong male associations assume a greater evidentiary role. If two monkeys in a cohesive troop attack a third, one can question whether they acted together or both simply responded to some vulnerability in the victim at the same time. The coalition interpretation is stronger if the two attacking monkeys were regular associates and grooming partners within the troop and there was no other known benefit to their association.

Is the study of coalitionary interventions within or independently of alliances possible in odontocetes? We think so. In Shark Bay we often see brief agonistic interactions, "tiffs," between allies, where two males line up head-to-head while vocalizing; but lineups may feature more than one individual on either or both sides. Such spats can occur in the context of normal alliance behavior, such as when two males in a trio line up against the third while the trio is consorting a female. The role of rank in coalitions will need to be considered but our understanding of rank relationships in dolphins has unfortunately not advanced beyond Samuels and Gifford's (1997) groundbreaking study. The correlation between first-order alliance stability and consortship rate (Connor et al. 2001; Connor and Krützen 2015) is suggestive of dominance in Shark Bay second-order alliances, with Wiszniewski et al. (2012) suggesting that alliance partner choice could be influenced by rank. However, given that allies are ubiquitous in a Shark Bay male dolphin's life and the cast of allies in his company varies, often hourly, constructing an "intrinsic" dominance ranking based on fighting ability in dyadic interactions is not a reasonable expectation, but learning about power relationships in the context of alliances is (see Watts 2010). In other populations with simpler alliances, such as the male pairs in Sarasota, it should be possible to detect dominance ranks between alliances. Drones, perhaps in combination with localized vocal recordings, and whistle playbacks to look at approaches and avoidances, offer hope of advancement in the study of coalitions and alliances within and between species, especially in locations with calm clear water or where regular underwater observations are made and videotaped. In addition to the new technology, arguably the most important resource that will lead to new insight and greater resolution is time. As long-term studies on identifiable individuals continue to mature, documenting the rise and fall of new and old alliances, our understanding of the advantages and costs of these associations will surely grow.

References

Allen, Simon J., Stephanie L. King, Michael Krützen, and Alex M. Brown. 2017. "Multi-Modal Sexual Displays in Australian Humpback Dolphins." *Scientific Reports* 7 (1): 1–8. https://doi.org/10.1038/s41598-017-13898-9.

Baird, Robin W. 2000. "The Killer Whale: Foraging Specializations and Group Hunting." In *Cetacean Societies: Field Studies of Dolphins and Whales*, edited by Janet Mann, Richard C. Connor, Peter L. Tyack, and Hal Whitehead, 127–53. University of Chicago Press.

Baird, Robin W. 2019. "Behavior and Ecology of Not-So-Social Odontocetes: Cuvier's and Blainville's Beaked Whales." In *Ethology and Behavioral Ecology of Odontocetes*, edited by Bernd Würsig, 305–29. Springer. https://doi.org/10.1007/978-3-030-16663-2_14.

Barnett, J., N. Davison, R. Deaville, et al. 2009. "Postmortem Evidence of Interactions of Bottlenose Dolphins (*Tursiops truncatus*) with Other Dolphin Species in South-West England." *Veterinary Record* 165 (15): 441–44. https://doi.org/10.1136/vr.165.15.441.

Bissonnette, Annie, Susan Perry, Louise Barrett, et al. 2015. "Coalitions in Theory and Reality: A Review of Pertinent Variables and Processes." *Behaviour* 152 (1): 1–56. https://doi.org/10.1163/1568539X-00003241.

Bizzozzero, Manuela R., Simon J. Allen, Livia Gerber, et al. 2019. "Tool Use and Social Homophily Among Male Bottlenose Dolphins." *Proceedings of the Royal Society B* 286 (1904): 20190898. https://doi.org/https://doi.org/10.1098/rspb.2019.0898.

Brent, Lauren J. N., Daniel W. Franks, Emma A. Foster, Kenneth C. Balcomb, Michael A. Cant, and Darren P. Croft. 2015. "Ecological Knowledge, Leadership, and the Evolution of Menopause in Killer Whales." *Current Biology* 25 (6): 746–50. https://doi.org/10.1016/j.cub.2015.01.037.

Brightwell, Kristin, and Quincy Gibson. 2023. "Inter- and Intrapopulation Variation in Bottlenose Dolphin Mating Strategies." In *Sex in Cetaceans: Morphology, Behavior, and the Evolution of Sexual Strategies*, edited by Bernd Würsig and Dara N. Orbach, 251–78. Springer. https://doi.org/10.1007/978-3-031-35651-3_12.

Brightwell, Kristin, Ewa B. Krzyszczyk, and Quincy A. Gibson. 2025. "Dynamics of Multilevel Alliances in St. Johns River, Florida, Tamanend's Bottlenose Dolphins (*Tursiops erebennus*) with Respect to an Epizootic Unusual Mortality Event." *Marine Mammal Science* 41 (1): e13165. https://doi.org/10.1111/mms.13165.

Cerchio, Salvatore, Norbert Andrianarivelo, and Boris Andrianantenaina. 2015. "Ecology and Conservation Status of Indian Ocean Humpback Dolphins (*Sousa plumbea*) in Madagascar." In *Advances in Marine Biology*, edited by T. A. Jefferson and B. E. Curry, 163–99. Academic Press. https://doi.org/10.1016/bs.amb.2015.09.004.

Chapais, Bernard. 1992. "The Role of Alliances in Social Inheritance of Rank Among Female Primates." In *Coalitions and Alliances in Humans and Other Animals*, edited by F. B. M. de Waal and A. Harcourt, 29–60. Oxford University Press.

Chereskin, Emma, Richard C. Connor, Whitney F. Friedman, et al. 2022. "Allied Male Dolphins Use Vocal Exchanges to 'Bond at a Distance.'" *Current Biology* 32 (7): 1657–63. https://doi.org/10.1016/j.cub.2022.02.019.

Cioffi, William R., Nicola J. Quick, Heather J. Foley, et al. 2021. "Adult Male Cuvier's Beaked Whales (*Ziphius cavirostris*) Engage in Prolonged Bouts of Synchronous Diving." *Marine Mammal Science* 37 (3): 1085–100. https://doi.org/10.1111/mms.12799.

Connor, Richard C. 1995. "The Benefits of Mutualism: A Conceptual Framework." *Biological Reviews* 70: 427–57. https://doi.org/10.1111/j.1469-185X.1995.tb01196.x.

Connor, Richard C. 2007. "Dolphin Social Intelligence: Complex Alliance Relationships in Bottlenose Dolphins and a Consideration of Selective Environments for Extreme Brain Size Evolution in Mammals." *Philosophical Transactions of the Royal Society B* 362 (1480): 587–602. https://doi.org/10.1098/rstb.2006.1997.

Connor, Richard C. 2010. "Cooperation Beyond the Dyad: On Simple Models and a Complex Society." *Philosophical Transactions of the Royal Society B* 365 (1553): 2687–97. https://doi.org/10.1098/rstb.2010.0150.

Connor, Richard C., William R. Cioffi, Srđan Randić, Simon J. Allen, Jana Watson-Capps, and Michael Krützen. 2017. "Male Alliance Behaviour and Mating Access Varies with Habitat in a Dolphin Social Network." *Scientific Reports* 7. https://doi.org/10.1038/srep46354.

Connor, Richard C., Michael R. Heithaus, and Lynne M. Barrè. 1999. "Superalliance of Bottlenose Dolphins." *Nature* 397: 571–72. https://doi.org/10.1038/17501.

Connor, Richard C., Michael R. Heithaus, and Lynne M. Barrè. 2001. "Complex Social Structure, Alliance Stability and Mating Access in a Bottlenose Dolphin 'Super-Alliance.'" *Proceedings of the Royal Society B* 268 (1464): 263–67. https://doi.org/10.1098/rspb.2000.1357.

Connor, Richard C., and Michael Krützen. 2015. "Male Dolphin Alliances in Shark Bay: Changing Perspectives in a 30-Year Study." *Animal Behaviour* 103 (May): 223–35. https://doi.org/10.1016/j.anbehav.2015.02.019.

Connor, Richard C., Michael Krützen, Simon J. Allen, William B. Sherwin, and Stephanie L. King. 2022. "Strategic Intergroup Alliances Increase Access to a Contested Resource in Male Bottlenose Dolphins." *Proceedings of the National Academy of Sciences* 119 (36): e212172311. https://doi.org/10.1073/pnas.2121723119.

Connor, Richard C., Andrew J. Read, and Richard W. Wrangham. 2000a. "Male Reproductive Strategies and Social Bonds." In *Cetacean Societies: Field Studies of Dolphins and Whales*, edited by Janet Mann, Richard C. Connor, Peter L. Tyack, and Hal Whitehead, 247–69. University of Chicago Press.

Connor, Richard C., Andrew F. Richards, Rachel A. Smolker, and J. Mann. 1996. "Patterns of Female Attractiveness in Indian Ocean Bottlenose Dolphins." *Behaviour* 133 (1): 37–69. https://doi.org/10.1163/156853996x00026.

Connor, Richard C., and Rachel A. Smolker. 1996. "'Pop' Goes the Dolphin: A Vocalization Male Bottlenose Dolphins Produce During Consortships." *Behaviour* 133: 643–62. https://doi.org/10.1163/156853996X00404.

Connor, Richard C., Rachel Smolker, and Lars Bejder. 2006. "Synchrony, Social Behaviour and Alliance Affiliation in Indian Ocean Bottlenose Dolphins, *Tursiops aduncus*." *Animal Behaviour* 72 (6): 1371–78. https://doi.org/10.1016/j.anbehav.2006.03.014.

Connor, Richard C., Rachel A. Smolker, and Andrew F. Richards. 1992a. "Dolphin Alliances and Coalitions." In *Coalitions and Alliances in Humans and Other Animals*, edited by Alexander H. Harcourt and Frans de Waal, 415–43. Oxford University Press.

Connor, Richard C., Rachel A. Smolker, and Andrew F. Richards. 1992b. "Two Levels of Alliance Formation Among Male Bottlenose Dolphins (*Tursiops* sp.)." *Proceedings of the National Academy of Sciences* 89 (3): 987–90. https://doi.org/10.1073/pnas.89.3.987.

Connor, Richard C., Jana J. Watson-Capps, William B. Sherwin, and Michael Krützen. 2011. "A New Level of Complexity in the Male Alliance Networks of Indian Ocean Bottlenose Dolphins (*Tursiops* sp.)." *Biology Letters* 7 (4): 623–26. https://doi.org/10.1098/rsbl.2010.0852.

Connor, Richard C., Randall S. Wells, Janet Mann, and Andrew Read. 2000b. "The Bottlenose Dolphin: Social Relationships in a Fission-Fusion Society." In *Cetacean Societies: Field Studies of Dolphins and Whales*, edited by Janet Mann, Richard C. Connor, Peter Tyack, and Hal Whitehead, 91–126. University of Chicago Press.

Connor, Richard, and Hal Whitehead. 2005. "Alliances II. Rates of Encounter During Resource Utilization: A General Model of Intrasexual Alliance Formation in Fission-Fusion Societies." *Animal Behaviour* 69 (1): 127–32. https://doi.org/10.1016/j.anbehav.2004.02.022.

Cotter, Mark P., Daniela Maldini, and Thomas A. Jefferson. 2012. "'Porpicide' in California: Killing of Harbor Porpoises (*Phocoena phocoena*) by Coastal Bottlenose Dolphins (*Tursiops truncatus*)." *Marine Mammal Science* 28 (1). https://doi.org/10.1111/j.1748-7692.2011.00474.x.

Crespo-Picazo, J. L., C. Rubio-Guerri, M. A. Jiménez, et al. 2021. "Bottlenose Dolphins (*Tursiops truncatus*) Aggressive Behavior Towards Other Cetacean Species in the Western Mediterranean." *Scientific Reports* 11 (1): 1–9. https://doi.org/10.1038/s41598-021-00867-6.

Cusick, Jessica A., and Denise L. Herzing. 2014. "The Dynamic of Aggression: How Individual and Group Factors Affect the Long-Term Interspecific Aggression Between Two Sympatric Species of Dolphin." *Ethology* 120 (3): 287–303. https://doi.org/10.1111/eth.12204.

Danaher-Garcia, Nicole A., Richard Connor, Gavin Fay, Kelly Melillo-Sweeting, and Kathleen M. Dudzinski. 2022. "The Partial Merger of Two Dolphin Societies." *Royal Society Open Science* 9 (8). https://doi.org/10.1098/rsos.211963.

Danaher-Garcia, Nicole A., Kelly Melillo-Sweeting, and Kathleen M. Dudzinski. 2020. "Social Structure of Atlantic Spotted Dolphins (*Stenella frontalis*) off Bimini, The Bahamas (2003–2016): Alternate Reasons for Preferential Association in Delphinids." *Acta Ethologica* 23 (1): 9–21. https://doi.org/10.1007/s10211-019-00329-3.

De Waal, Frans B. M., and Alexander H. Harcourt. 1992. "Coalitions and Alliances: A History of Ethological Research." In *Coalitions and Alliances in Humans and Other Animals*, edited by Alexander H. Harcourt and Frans de Waal, 3–19. Oxford University Press.

Díaz López, Bruno, Alfredo López, Séverine Methion, and Pablo Covelo. 2018. "Infanticide Attacks and Associated Epimeletic Behaviour in Free-Ranging Common Bottlenose Dolphins (*Tursiops truncatus*)." *Journal of the Marine Biological Association of the United Kingdom* 98 (5): 1159–67. https://doi.org/10.1017/S0025315417001266.

East, Marion L., Oliver P. Höner, Bettina Wachter, Kerstin Wilhelm, Terry Burke, and Heribert Hofer. 2009. "Maternal Effects on Offspring Social Status in Spotted Hyenas." *Behavioral Ecology* 20 (3): 478–83. https://doi.org/10.1093/beheco/arp020.

Eierman, Laura E., Kristen Laccetti, Kelly Melillo-Sweeting, and J. Daisy Kaplan. 2019. "Interspecies Pectoral Fin Contact Between Bottlenose Dolphins and Atlantic Spotted Dolphins off Bimini, The Bahamas." *Animal Behaviour* 157: 167–76. https://doi.org/10.1016/j.anbehav.2019.09.002.

Elliser, Cindy R., and Denise L. Herzing. 2014. "Long-Term Social Structure of a Resident Community of Atlantic Spotted Dolphins, *Stenella frontalis*, in the Bahamas 1991–2002." *Marine Mammal Science* 30 (1): 308–28. https://doi.org/10.1111/mms.12039.

Elliser, Cindy R., and Denise L. Herzing. 2016. "Changes in Interspecies Association Patterns of Atlantic Bottlenose Dolphins, *Tursiops truncatus*, and Atlantic Spotted Dolphins, *Stenella frontalis*, After Demographic Changes Related to Environmental Disturbance." *Marine Mammal Science* 32 (2): 602–18. https://doi.org/10.1111/mms.12289.

Fedutin, Ivan D., Olga A. Filatova, Evgeny G. Mamaev, Alexandr M. Burdin, and Erich Hoyt. 2015. "Occurrence and Social Structure of Baird's Beaked Whales, *Berardius bairdii*, in the Commander Islands, Russia." *Marine Mammal Science* 31 (3): 853–65. https://doi.org/10.1111/mms.12204.

Feldblum, Joseph T., Christopher Krupenye, Joel Bray, Anne E. Pusey, and Ian C. Gilby. 2021. "Social Bonds Provide Multiple Pathways to Reproductive Success in Wild Male Chimpanzees." *IScience* 24 (8): 102864. https://doi.org/10.1016/j.isci.2021.102864.

Fox, Stephanie A., Martin N. Muller, Nicole Thompson González, et al. 2023. "Weak, but Not Strong, Ties Support Coalition Formation Among Wild Female Chimpanzees." *Philosophical Transactions of the Royal Society B* 378 (1868): 20210427. https://doi.org/10.1098/rstb.2021.0427.

Friedman, Whitney R., Michael Krützen, Stephanie L. King, Simon J. Allen, Livia Gerber, Samuel Wittwer, and Richard C. Connor. 2023. "Inter-Group Alliance Dynamics in Indo-Pacific Bottlenose Dolphins (*Tursiops aduncus*)." *Animal Cognition* 26: 1601–12 https://doi.org/10.1007/s10071-023-01804-y.

Furuichi, Takeshi. 2020. "Variation in Intergroup Relationships Among Species and Among and Within Local Populations of African Apes." *International Journal of Primatology* 41 (2): 203–23. https://doi.org/10.1007/s10764-020-00134-x.

Gerber, Livia, Richard C. Connor, Simon J. Allen, et al. 2022. "Social Integration Influences Fitness in Allied Male Dolphins." *Current Biology* 32 (7): 1664–69. https://doi.org/10.1016/j.cub.2022.03.027.

Gerber, Livia, Richard C. Connor, Stephanie L. King, et al. 2020. "Affiliation History and Age Similarity Predict Alliance Formation in Adult Male Bottlenose Dolphins." *Behavioral Ecology* 31 (2): 361–70. https://doi.org/10.1093/beheco/arz195.

Gerber, Livia, Samuel Wittwer, Simon J. Allen, et al. 2021. "Cooperative Partner Choice in

Multi-Level Male Dolphin Alliances." *Scientific Reports* 11 (1): 1–10. https://doi.org/10.1038/s41598-021-85583-x.

Gowans, Shannon, Hal Whitehead, and Sascha K. Hooker. 2001. "Social Organization in Northern Bottlenose Whales, *Hyperoodon ampullatus*: Not Driven by Deep-Water Foraging?" *Animal Behaviour* 62 (2): 369–77. https://doi.org/10.1006/anbe.2001.1756.

Grimes, Charli, Lauren J. N. Brent, Samuel Ellis, et al. 2023. "Postreproductive Female Killer Whales Reduce Socially Inflicted Injuries in Their Male Offspring." *Current Biology* 33 (15): 3250–56. https://doi.org/10.1016/j.cub.2023.06.039.

Hallgrímsson, Benedikt, and Virginia Maiorana. 2000. "Variability and Size in Mammals and Birds." *Biological Journal of the Linnean Society* 70 (4): 571–95. https://doi.org/10.1006/bijl.1999.0417.

Hamilton, Rebecca A., Teresa Borcuch, Simon J. Allen, et al. 2019. "Male Aggression Varies with Consortship Rate and Habitat in a Dolphin Social Network." *Behavioral Ecology and Sociobiology* 73: 141. https://doi.org/10.1007/s00265-019-2753-1.

Hartman, Karin L., Ing Chen, Pieter A. van der Harst, et al. 2023a. "Kinship Study Reveals Non-Kin-Based Stable Associations in a Medium-Sized Delphinid." *Behavioural Ecology and Sociobiology* 77 (137). https://doi.org/10.1007/s00265-023-03411-w.

Hartman, Karin L., Pieter A. van der Harst, and Raul Vilela. 2020. "Continuous Focal Group Follows Operated by a Drone Enable Analysis of the Relation Between Sociality and Position in a Group of Male Risso's Dolphins (*Grampus griseus*)." *Frontiers in Marine Science* 7: 283. https://doi.org/10.3389/fmars.2020.00283.

Hartman, Karin L., Pieter A. van der Harst, and Raul Vilela. 2023b. "Sex and Sexual Strategies in Deep-Diving Risso's Dolphins." In *Sex in Cetaceans: Morphology, Behavior, and the Evolution of Sexual Strategies*, edited by Bernd Würsig and Dara N. Orbach, 279–306. Springer. https://doi.org/10.1007/978-3-031-35651-3_13.

Hartman, Karin L., Fleur Visser, and A. J. Hendriks. 2008. "Social Structure of Risso's Dolphins (*Grampus griseus*) at the Azores: A Stratified Community Based on Highly Associated Social Units." *Canadian Journal of Zoology* 86 (4): 294–306. https://doi.org/10.1139/Z07-138.

Herzing, Denise L. 1996. "Vocalizations and Associated Underwater Behavior of Free-Ranging Atlantic Spotted Dolphins, *Stenella frontalis* and Bottlenose Dolphins, *Tursiops truncatus*." *Aquatic Mammals* 22: 61–79.

Herzing, Denise L., Bethany N. Augliere, Cindy R. Elliser, Michelle L. Green, and Adam A. Pack. 2017. "Exodus! Large-Scale Displacement and Social Adjustments of Resident Atlantic Spotted Dolphins (*Stenella frontalis*) in the Bahamas." *PLOS ONE* 12. https://doi.org/10.1371/journal.pone.0180304.

Herzing, Denise L., and Cindy R. Elliser. 2013. "Directionality of Sexual Activities During Mixed-Species Encounters Between Atlantic Spotted Dolphins (*Stenella frontalis*) and Bottlenose Dolphins (*Tursiops truncatus*)." *International Journal of Comparative Psychology* 26 (2). https://doi.org/10.46867/ijcp.2013.26.02.04.

Herzing, Denise L., and C. M. Johnson. 1997. "Interspecific Interaction Between Atlantic Spotted Dolphins and Bottlenose Dolphins in the Bahamas, 1985–95." *Aquatic Mammals* 23 (2): 85–99.

Hrdy, Sarah Blaffer. 1979. "Infanticide Among Animals: A Review, Classification, and Examination of the Implications for the Reproductive Strategies of Females." *Ethology and Sociobiology* 1 (1): 13–40. https://doi.org/10.1016/0162-3095(79)90004-9.

Ilany, Amiyaal, Kay E. Holekamp, and Erol Akçay. 2021. "Rank-Dependent Social Inheritance

Determines Social Network Structure in Spotted Hyenas." *Science* 373 (6552): 348–52. https://doi.org/10.1126/science.abc196.

King, Stephanie L., Simon J. Allen, Michael Krützen, and Richard C. Connor. 2019. "Vocal Behaviour of Allied Male Dolphins During Cooperative Mate Guarding." *Animal Cognition* 22 (6): 991–1000. https://doi.org/10.1007/s10071-019-01290-1.

King, Stephanie L., Richard C. Connor, Michael Krützen, and Simon J. Allen. 2021. "Cooperation-Based Concept Formation in Male Bottlenose Dolphins." *Nature Communications* 12 (1): 2373. https://doi.org/10.1038/s41467-021-22668-1.

King, Stephanie L., Richard C. Connor, and Stephen H. Montgomery. 2022. "Social and Vocal Complexity in Bottlenose Dolphins." *Trends in Neurosciences* 45 (12): 881–83. https://doi.org/10.1016/j.tins.2022.09.006.

King, Stephanie L., Whitney F. Friedman, Simon J. Allen, et al. 2018. "Bottlenose Dolphins Retain Individual Vocal Labels in Multi-Level Alliances." *Current Biology* 28 (12): 1993–99. https://doi.org/10.1016/j.cub.2018.05.013.

Lydersen, Christian, Kerstin Langenberger, and Kit M. Kovacs. 2023. "An Observation of White Whale (*Delphinapterus leucas*) Mating Behaviour in the Wild." *Polar Research* 42: 1–9. https://doi.org/10.33265/polar.v42.8875.

Mann, Janet, Margaret A. Stanton, Eric M. Patterson, Elisa J. Bienenstock, and Lisa O. Singh. 2012. "Social Networks Reveal Cultural Behaviour in Tool-Using Dolphins." *Nature Communications* 3. https://doi.org/10.1038/ncomms1983.

May-Collado, Laura J. 2010. "Changes in Whistle Structure of Two Dolphin Species During Interspecific Associations." *Ethology* 116 (11): 1065–74. https://doi.org/10.1111/j.1439-0310.2010.01828.x.

McCue, Laura M., William R. Cioffi, Michael R. Heithaus, Lynne Barrè, and Richard C. Connor. 2020. "Synchrony, Leadership, and Association in Male Indo-Pacific Bottlenose Dolphins (*Tursiops aduncus*)." *Ethology* 126: 741–50. https://doi.org/10.1111/eth.13025.

Methion, Séverine, and Bruno Díaz López. 2021. "Spatial Segregation and Interspecific Killing of Common Dolphins (*Delphinus delphis*) by Bottlenose Dolphins (*Tursiops truncatus*)." *Acta Ethologica* 24 (2): 95–106. https://doi.org/10.1007/s10211-021-00363-0.

Möller, Luciana M. 2012. "Sociogenetic Structure, Kin Associations and Bonding in Delphinids." *Molecular Ecology* 21 (3): 745–64. https://doi.org/10.1111/j.1365-294X.2011.05405.x.

Moore, Bronte M., Richard C. Connor, Simon J. Allen, Michael Krützen, and Stephanie L. King. 2020. "Acoustic Coordination by Allied Male Dolphins in a Cooperative Context." *Proceedings of the Royal Society B* 1924: 20192944. https://doi.org/10.1098/rspb.2019.2944.

Nielsen, Mia L. K., Samuel Ellis, Michael N. Weiss, et al. 2023. "Temporal Dynamics of Mother-Offspring Relationships in Bigg's Killer Whales: Opportunities for Kin-Directed Help by Post-Reproductive Females." *Proceedings of the Royal Society B* 290 (2000): 20230139. https://doi.org/10.1098/rspb.2023.0139.

Nishida, Toshisada. 1983. "Alpha Status and Agonistic Alliance in Wild Chimpanzees (*Pan troglodytes schweinfurthii*)." *Primates* 24: 318–36. https://doi.org/10.1007/BF02381978.

Noy, Lior, Erez Dekel, and Uri Alon. 2011. "The Mirror Game as a Paradigm for Studying the Dynamics of Two People Improvising Motion Together." *Proceedings of the National Academy of Sciences* 108 (52): 20947–52. https://doi.org/10.1073/pnas.1108155108.

O'Brien, Orla, Simon J. Allen, Michael Krützen, and Richard C. Connor. 2020. "Alliance-Specific Habitat Selection by Male Indo-Pacific Bottlenose Dolphins in Shark Bay, Western Australia." *Animal Behaviour* 164: 39–49. https://doi.org/10.1016/j.anbehav.2020.03.014.

O'Corry-Crowe, Greg, Robert Suydam, Lori Quakenbush, et al. 2020. "Group Structure and Kinship in Beluga Whale Societies." *Scientific Reports* 10 (1): 1–21. https://doi.org/10.1038/s41598-020-67314-w.

Packer, Craig, and Anne Pusey. 1987. "Intrasexual Cooperation and the Sex Ratio in African Lions." *American Naturalist* 130: 636–42. https://doi.org/10.1086/284735.

Perrtree, Robin M., Laela S. Sayigh, Allison Williford, Alessandro Bocconcelli, Mary Carla Curran, and Tara M. Cox. 2016. "First Observed Wild Birth and Acoustic Record of a Possible Infanticide Attempt on a Common Bottlenose Dolphin (*Tursiops truncatus*)." *Marine Mammal Science* 32 (1): 376–85. https://doi.org/10.1111/mms.12248.

Ramos, Eric A., Isidore D. Szczepaniak, J. Daisy Kaplan, and Diana Reiss. 2022. "Potential Infanticide Attempt of Common Bottlenose Dolphins (*Tursiops truncatus*) on a Young Calf in a Tropical Caribbean Atoll." *Aquatic Mammals* 48 (2): 132–41. https://doi.org/10.1578/AM.48.2.2022.132.

Randić, Srđan, Richard C. Connor, William B. Sherwin, and Michael Krützen. 2012. "A Novel Mammalian Social Structure in Indo-Pacific Bottlenose Dolphins (*Tursiops* sp.): Complex Male Alliances in an Open Social Network." *Proceedings of the Royal Society B* 279 (1740): 3083–90. https://doi.org/10.1098/rspb.2012.0264.

Samuels, Amy, and Tara Gifford. 1997. "A Quantitative Assessment of Dominance Relations Among Bottlenose Dolphins." *Marine Mammal Science* 13: 70–99. https://doi.org/10.1111/j.1748-7692.1997.tb00613.x.

Samuni, Liran, and Martin Surbeck. 2023. "Cooperation Across Social Borders in Bonobos." *Science* 382 (6672): 805–9. https://doi.org/10.1126/science.adg0844.

Sargeant, Brooke. L., Janet Mann, Per Berggren, and Michael Krützen. 2005. "Specialization and Development of Beach Hunting, a Rare Foraging Behavior, by Wild Bottlenose Dolphins (*Tursiops* sp.)." *Canadian Journal of Zoology* 83: 1400–1410. https://doi.org/10.1139/z05-136.

Smolker, Rachel, and John W. Pepper. 1999. "Whistle Convergence Among Allied Male Bottlenose Dolphins (*Delphinidae Tursiops* sp.)." *Ethology* 105 (7): 595–617. https://doi.org/10.1046/j.1439-0310.1999.00441.x.

Sørensen, Pernille M., Richard C. Connor, Simon J. Allen, et al. 2024. "Communication Range Predicts Dolphin Alliance Size in a Cooperative Mating System." *Current Biology* 34 (20): 4774–80. https://doi.org/10.1016/j.cub.2024.08.032.

Stensland, E. 2004. "Behavioural Ecology of Indo-Pacific Bottlenose and Humpback Dolphins." PhD diss., Stockholm University.

Syme, Jonathan, Jeremy J. Kiszka, and Guido J. Parra. 2021. "Dynamics of Cetacean Mixed-Species Groups: A Review and Conceptual Framework for Assessing Their Functional Significance." *Frontiers in Marine Science* 8: 678173. https://doi.org/10.3389/fmars.2021.678173.

Tokuyama, Nahoko, Tetsuya Sakamaki, and Takeshi Furuichi. 2019. "Inter-Group Aggressive Interaction Patterns Indicate Male Mate Defense and Female Cooperation Across Bonobo Groups at Wamba, Democratic Republic of the Congo." *American Journal of Physical Anthropology* 170 (4): 535–50. https://doi.org/10.1002/ajpa.23929.

Towers, Jared R., Muriel J. Hallé, Helena K. Symonds, et al. 2018. "Infanticide in a Mammal-Eating Killer Whale Population." *Scientific Reports* 8 (1): 1–8. https://doi.org/10.1038/s41598-018-22714-x.

Van Schaik, Carel P., and Peter M. Kappeler. 1997. "Infanticide Risk and the Evolution of Male-Female Association in Primates." *Proceedings of the Royal Society B* 264 (1388): 1687–94. https://doi.org/10.1098/rspb.1997.0234.

Vollmer, Nicole. L., Lee-Ann C. Hayek, Michael R. Heithaus, and Richard C. Connor. 2015. "Further Evidence of a Context-Specific Agonistic Signal in Bottlenose Dolphins: The Influence of Consortships and Group Size on the Pop Vocalization." *Behaviour* 152: 1979–2000. https://doi.org/10.1163/1568539X-00003311.

Watts, David. P. 2010. "Dominance, Power, and Politics in Nonhuman and Human Primates." In *Mind the Gap: Tracing the Origins of Human Universals*, edited by Peter M Kappeler and Joan Silk, 109–38. Springer.

Watwood, Stephanie. L., Peter L. Tyack, and Randall S. Wells. 2004. "Whistle Sharing in Paired Male Bottlenose Dolphins, *Tursiops truncatus*." *Behavioural Ecology and Sociobiology* 55: 531–43. https://doi.org/10.1007/s00265-003-0724-y.

Wearmouth, Victoria. J., and David W. Sims. 2008. "Sexual Segregation in Marine Fish, Reptiles, Birds and Mammals Behaviour Patterns, Mechanisms and Conservation Implications." *Advances in Marine Biology* 54: 107–70.

Weiss, Michael N., Samuel Ellis, Daniel W. Franks, et al. 2023. "Costly Lifetime Maternal Investment in Killer Whales." *Current Biology* 33 (4): 744–48.e3. https://doi.org/10.1016/j.cub.2022.12.057.

Whitehead, Hal, and Richard Connor. 2005. "Alliances I. How Large Should Alliances Be?" *Animal Behaviour* 69 (1): 117–26. https://doi.org/10.1016/j.anbehav.2004.02.021.

Wild, Sonja, Simon J. Allen, Michael Krützen, Stephanie L. King, Livia Gerber, and William J. E. Hoppitt. 2019. "Multi-Network-Based Diffusion Analysis Reveals Vertical Cultural Transmission of Sponge Tool Use Within Dolphin Matrilines." *Biology Letters* 15 (7): 20190227. https://doi.org/10.1098/rsbl.2019.0227.

Wiszniewski, Joanna, Shannon Corrigan, Luciano B. Beheregaray, and Luciana M. Möller. 2012. "Male Reproductive Success Increases with Alliance Size in Indo-Pacific Bottlenose Dolphins (*Tursiops aduncus*)." *Journal of Animal Ecology* 81 (2): 423–31. https://doi.org/10.1111/j.1365-2656.2011.01910.x.

Wright, Brianna M., Eva H. Stredulinsky, Graeme M. Ellis, and John K. B. Ford. 2016. "Kin-Directed Food Sharing Promotes Lifetime Natal Philopatry of Both Sexes in a Population of Fish-Eating Killer Whales, *Orcinus orca*." *Animal Behaviour* 115 (May): 81–95. https://doi.org/10.1016/j.anbehav.2016.02.025.

Zheng, Ruiqiang, Leszek Karczmarski, Wenzhi Lin, Stephen C. Y. Chan, Wei-Lun Chang, and Yuping Wu. 2016. "Infanticide in the Indo-Pacific Humpback Dolphin (*Sousa chinensis*)." *Journal of Ethology* 34 (3): 299–307. https://doi.org/10.1007/s10164-016-0475-7.

10

Communicative and Cognitive Skills Underlying Cooperation in Cetaceans

STEPHANIE L. KING, PERNILLE M. SØRENSEN, AND KELLY JAAKKOLA

Abstract

Cooperation among cetaceans is ubiquitous, with individuals of some species forming long-term cooperative relationships in the form of alliances as a means of enhancing reproductive success, or engaging in cooperative feeding strategies that appear to require coordinated action. While the term "cooperation" has a long history of being used to describe a variety of mutually beneficial foraging, reproductive, and caregiving behaviors among two or more individuals, it is only in recent years that we have begun to critically reflect on the mechanisms underlying such apparent acts of cooperation. Indeed, we have long romanticized the notion of cetacean cooperation as occurring among cognitively advanced and charismatic animals, yet studies have demonstrated that animals can behave in ways that function cooperatively without actively and intentionally cooperating. By contrast, recent experiments with bottlenose dolphins in both human care and the wild have provided compelling new evidence of actively coordinated cooperation, where cetaceans can use vocal signals to facilitate the successful execution of coordinated cooperative actions. This chapter will review the state of the field and describe the suite of cognitive and communicative skills that cetaceans have been shown to possess in a cooperative context.

Introduction

Cooperation pervades all levels of biological organization, from eukaryotic cells to complex societies (West et al. 2007). Consequently, the evolutionary roots of cooperation have long intrigued and challenged researchers across disciplines. At the ultimate level, cooperation can be explained by mechanisms such as kin selection, where individuals engage in self-sacrificial behavior that benefits their genetic fitness by helping relatives (Hamilton 1964, 1963); by-product mutualisms, where individuals behave in ways that increase the receipt of by-product benefits from others, but by doing so confer by-product benefits themselves (Connor 2010); or by reciprocity, where individuals help one another with the expectation that this help will be reciprocated in the future (Trivers 1971). Yet, the cognitive processes underlying the proximate mechanisms of cooperation, explaining *how* animals cooperate, have received far less attention. These mechanisms are important, however, because while many animals appear to accomplish impressive feats of teamwork and collective action, different neural and psychological mechanisms can lead to similar behaviors, and collective behavior can arise from simple processes where individuals perform the same behavior but act independently of each other (Muro et al. 2011; Herbert-Read et al. 2016). Over the years, cooperation experiments with animals in human care have helped shed important light on the cognitive skills underlying coordination toward a shared goal, with current evidence suggesting that only a few species actually understand their partner's role in a cooperative situation and are therefore capable of monitoring their partner's behavior and responding accordingly.

While cooperation is well-documented in wild cetaceans, we still know very little about how these cooperative behaviors are achieved. Here, we review the state of the field and critically evaluate the cognitive mechanisms that could explain the cooperative acts observed within cetacean species. We discuss the experiments that have been conducted with cetaceans in human care, which in recent years have revealed a suite of communicative and cognitive skills that cetaceans use during cooperative behavior, including understanding their partner's role in a cooperative context and using vocal signals to facilitate coordinated cooperative actions. We then move from individual cooperative acts to long-term cooperative relationships and reflect on the cognitive skills that individuals use to form and maintain these cooperative partnerships.

Proximate Mechanisms That Can Theoretically Underlie Cooperative Acts

The proximate mechanisms that can lead to cooperative behavior and coordination can be quite varied (Duguid and Melis 2020; Noë 2006), ranging in extremes from cases where individuals obtain benefits from being in the proximity of others without really coordinating actions to where individuals actively share a joint goal and work toward it together (Duguid and Melis 2020). At the simplest level, by-product collaboration (or by-product mutualism; fig. 10.1a) occurs when increased success is a by-product of individuals simultaneously performing the same behavior at the same time in response to the same environmental stimuli (e.g., prey), but with no evidence for individual roles or spatial coordination (Duguid and Melis 2020). Socially influenced collaboration is similar, except here an individual's decision to pursue a specific goal or perform a specific behavior is influenced by conspecific presence or behavior (fig. 10.1b). For example, an individual may be more likely to engage in hunting behavior when it perceives others engaging in hunting, but with no evidence of intentional strategies to facilitate coordination with partners (Duguid and Melis 2020). In chimpanzees (*Pan troglodytes*; Mine et al. 2022) and wolves (*Canis lupus*; Bailey et al. 2013), for example, group hunting and even role specialization can be explained by simple movement rules (Muro et al. 2011; Herbert-Read et al. 2016) that do not require awareness about the causal role of others (Duguid and Melis 2020). A more complex form of cooperative behavior is actively coordinated collaboration (fig. 10.1c), in which individuals are attempting to coordinate in time and space by monitoring their partner's behavior and responding accordingly, i.e., they have some knowledge of the need for a partner (Duguid and Melis 2020). Evidence of this type of cooperation could be waiting for a partner before engaging in a cooperative task, recruiting a partner to a cooperative task, and supporting a partner by communicating to coordinate actions (Duguid and Melis 2020). Finally, the most sophisticated cognitive and communicative mechanisms underpinning cooperative behavior are those reflecting shared intentionality (fig. 10.1d). Characteristic of many forms of human cooperation, shared intentionality (sometimes called "we-intentionality") requires the ability not only to coordinate behavior but also to represent and engage in a shared goal, as well as a shared motivation to pursue that goal together. Although difficult to identify empirically, evidence for this type of cooperation might include things like using specialized forms of communication to enter into and maintain collaborations, flexibly supporting each other to achieve the shared goal, and having a mutual willingness to share the resulting spoils (Duguid and Melis 2020; Tomasello 2022).

Observing Cetacean Cooperation in the Wild

Cetaceans are well-known for cooperating extensively in the wild, including feeding strategies that appear to require coordinated action and the formation of long-term cooperative alliances. Here, we consider evidence for and against the proximate mechanisms that may explain some of the best-studied cooperative behaviors exhibited by cetaceans.

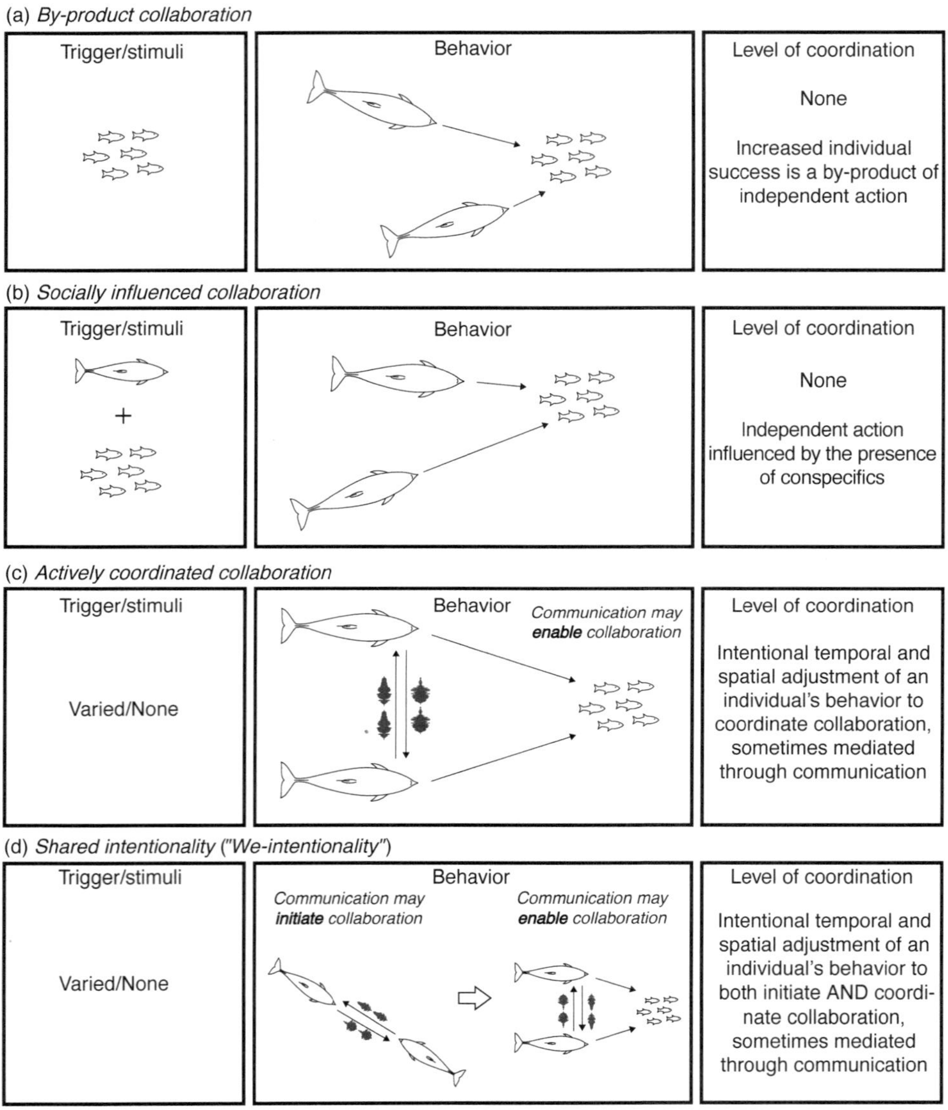

FIGURE 10.1. Schematic of the proximate mechanisms that may explain cooperative acts. (a) By-product collaboration, where individuals perform the same behavior at the same time (i.e., pursuing prey) in response to the same external trigger (i.e., prey) but with no intentional coordination. (b) Socially influenced collaboration, where either the presence of a conspecific or a conspecific engaging in the target behavior influences an individual's decision to pursue a specific goal but with no intentional coordination. (c) Actively coordinated collaboration, where an individual temporally and spatially coordinates actions with others in order to reach a common goal, which can be mediated through communication. (d) Shared intentionality, where an individual temporally and spatially coordinates actions with others with whom they also share knowledge of their intentions and goals. Shared intentionality is difficult to identify empirically, but convincing evidence from wild animals may include a demonstration that individuals use intentional communication to decide whether to engage in a cooperative act together and to coordinate

COOPERATIVE FEEDING

Cetaceans engage in a variety of group-level foraging behaviors that certainly appear cooperative in nature. For example, humpback whales (*Megaptera novaeangliae*) engage in bubble-net feeding during which groups of whales will produce bubbles (fig. 10.2), creating columns of ascending air that will trap their prey, allowing the whales to feed in the center (D'Vincent et al. 1985; Wiley et al. 2011; Wild and Rendell, this volume). Group hunting has been documented in harbor porpoises (*Phocoena phocoena*), where individuals seem to favor different behavioral positions (quantified by swimming position and orientation with respect to the fish school) when hunting in groups (Torres Ortiz et al. 2021), and also in killer whales (*Orcinus orca*; Similä and Ugarte 1993). Bottlenose dolphins (*Tursiops truncatus*) in the Bahamas will swim rapidly in a dispersed line abreast, with dolphins at the end of the line swimming faster and diving synchronously to form a large, loose circle, presumably to herd fish into (Rossbach 1999). This mirrors cooperative prey herding by spinner dolphins (*Stenella longirostris*), where individuals swim in a line and then form a circle, allowing them to increase the density of their prey and per capita feeding success (Benoit-Bird and Au 2009). Yet, for all these behaviors, a causal understanding of the role that others are playing is not necessarily required for successful foraging behavior to occur. An analysis of the vocalizations produced by cooperatively foraging spinner dolphins revealed that typical social calls, i.e., whistles, were not used when dolphins were foraging, but that echolocation rates increased during behavioral transitions (Benoit-Bird and Au 2009). This suggests that perhaps animals are using echolocation clicks to infer the positions of conspecifics or the location of prey, instead of engaging in active communication to coordinate their behavior. If this is true, these behaviors could be explained by socially influenced (non-intentional) collaboration.

Interestingly, cetaceans are one of the few groups of animals capable of division of labor with role specialization (DLRS), which occurs when individuals specialize in a subtask during repetitions of a group task (Gazda et al. 2005; Hamilton et al. 2022). This is relatively rare outside of humans (*Homo sapiens*) and eusocial animals, and, so far, has only been documented in a limited number of species, including lions (*Panthera leo*; Stander 1992), mice

---

the cooperative act (Duguid and Melis 2020; Tomasello 2022). We propose that communication behavior provides good evidence for the occurrence of (c) and (d); however, it is important to differentiate communication for the purpose of manipulating another individual from communication for the purpose of coordinating collaboratively with that individual.

FIGURE 10.2. Cooperative behavior among cetacean species. (*Top left*) Humpback whales (*Megaptera novaeangliae*) engaging in bubble-net feeding in southeast Alaska. (*Top right*) Driver-barrier bottlenose dolphins (*Tursiops truncatus*) in Cedar Key, Florida. (*Bottom left*) Three B1 pack ice killer whales (*Orcinus orca*) spy-hop next to a crabeater seal (*Lobodon carcinophaga*) hauled out on an ice floe, in the coastal waters of the western Antarctic Peninsula. (*Bottom right*) Bottlenose dolphins with sound and movement recording tags (DTAG-3) engaging in a cooperative button-pressing task.
Sources: Photos courtesy of (*top left*) van Aswegen, Alaska Whale Foundation, and Marine Mammal Research Program, University of Hawai'i; (*top right*) Cedar Key Dolphin Project, taken under permit no. 14450/21938-04; (*bottom left*) Leigh S. Hickmott; (*bottom right*) Dolphin Research Center.

(*Mus spicilegus*; Hurtado et al. 2013), bottlenose dolphins (Gazda et al. 2005; Gazda 2016; Hamilton et al. 2022), and potentially chimpanzees (Boesch 2002) and humpback whales (D'Vincent et al. 1985; Wiley et al. 2011). In Cedar Key, Florida, bottlenose dolphins are well-known for engaging in role-specialized "driver-barrier feeding," where a "driver" dolphin herds mullet (*Mugil* spp.) toward "barrier" dolphins, resulting in the mullet leaping out of the water and the dolphins then catching them in the air (fig. 10.2). It has been shown that driver dolphins are more successful with barrier dolphins present, and that barrier dolphins also do better when there are more of them (Gazda 2016). It remains to be shown, however, whether individuals are intentionally coordinating with each other. Recent work has revealed that only the driver echolocates during driver-barrier feeding events, and that barrier dolphins may use the driver dolphin's echolocation to infer either the relative position and movement of the driver or the location and movement of the prey (Hamilton et al. 2022). Like group hunting in other species, this cooperative behavior does not necessarily require an understanding of the causal roles of individuals. Indeed, drivers will drive mullet even if no other dolphins are present (Gazda et al. 2005), and therefore barrier dolphins (or "cooperative partners") are not required for the driver to be successful. Nonetheless, there are clear by-product benefits to barrier dolphins being present. Thus, at the simplest level, driver-barrier behavior could be another example of socially influenced collaboration, where dolphins may be responding to what each other are doing without specifically attempting to coordinate their actions. While there is evidence that whistle use increases immediately prior to driver-barrier events, which offers up a potential avenue for determining whether individuals might be actively coordinating during this behavior, the function of these whistles remains unknown and other explanations, such as arousal, cannot currently be ruled out (Hamilton et al. 2022).

Another form of cooperative foraging behavior among cetaceans requires synchronized coordination rather than role specialization. Bottlenose dolphins in South Carolina have been shown to simultaneously swim side by side toward the shore and then surge out of the water in unison onto the mud banks to feed on small fish stranded by the surge wave before moving back into the water (Hoese 1971; Petricig 1995; Duffy-Echevarria et al. 2008). Unlike driver-barrier feeding, there is no evidence to date that individuals occupy particular stranding positions (Duffy-Echevarria et al. 2008). Intentional stranding has been observed in killer whales in a number of locations (Lopez and Lopez 1985; Guinet and Bouvier 1995), and similar behavior is also evident in pack ice killer whales (*Orcinus orca* type B) in Antarctic Peninsula waters (Pitman and Durban 2012), where they engage in a "cooperative wave-washing behavior" (fig. 10.2). Upon detecting a seal (i.e., Weddell seals, *Leptonychotes weddellii*) hauled out on the ice, members of the group initiate a wave-wash attack that involves synchronously charging underwater side by side toward the ice floe, lifting their tails and diving under the ice, creating a wave that breaks over the floe and typically washes the seal into the water (Pitman and Durban 2012). Typically, the majority, if not all, of the whales immediately turn back toward the floe after surfacing on the opposite side and spy-hop to check on the seal location (Pitman and Durban 2012). While, in general terms, performing the same behavior at the same time could be explained by socially influenced collaboration, the tight behavioral synchronization requires the animals to coordinate their behaviors in time and space, suggesting actively coordinated collaboration.

INTERSPECIES COOPERATION

Another example of cooperative foraging in cetaceans that should not be overlooked is the dolphin-fisher mutualism (see Cantor et al., this volume). In the most studied case, in Laguna, southern Brazil, bottlenose dolphins are known to cooperate with artisanal fishers to catch mullet (Pryor et al. 1990; Simões-Lopes 1991; Simões-Lopes et al. 1998; Cantor et al. 2023). This appears to be coordinated by the dolphins approaching fishers closely and providing a behavioral cue (e.g., a sudden deep dive, or a head or tail slap) at which time the fishers cast their nets (Cantor et al. 2023). Both fishers and dolphins benefit from this interaction and are more likely to catch mullet when synchronizing their foraging behavior (Cantor et al. 2023). The study demonstrates that both dolphins and humans are paying attention to each other's behavior, that the dolphins' behavior provides a cue to fishers of when nets should be cast, and that the ability of fishers to interpret dolphin behavior is a key element of this foraging mutualism (Cantor et al. 2023). However, unlike at least one other case of human-wildlife cooperation (honeyguides and humans; Spottiswoode et al. 2016), there is currently no evidence that dolphins and fishers are intentionally communicating with each other. Rather, it may simply be that both the fishers and dolphins are adept at timing their actions based on the others' behavior, without any overt attempts at communication to influence the others' behavior. Thus, while active communication remains an interesting topic for future work, for now this cooperative foraging interaction could theoretically be explained by socially influenced collaboration; where the presence or behavior of one species influences the behavior of the other, without necessarily requiring intentional strategies to facilitate cooperation.

REPRODUCTIVE COOPERATION

Another famous example of cetacean cooperation can be found in Shark Bay, Western Australia. In this population, unrelated male Indo-Pacific bottlenose dolphins (*Tursiops aduncus*) form three alliance levels, or "orders," in competition over access to females (Connor and Krützen 2015; see Connor and King, this volume). At the first level, pairs or trios work together to herd single estrus females in events termed "consortships" that can last hours to weeks (Connor and Krützen 2015; Connor et al. 1992). In turn, each male is a member of a larger second-order alliance (4–14 males) that cooperate in the pursuit and defense of females (Connor et al. 1992). Some males are also members of a third-order alliance, where two or more second-order alliances cooperate in the defense and theft of females (Connor and Krützen 2015; Connor et al. 2011).

At the simplest level, first-order alliances could be explained by socially influenced collaboration, where the behavior of one male attempting to herd a female influences another male to also try and herd her but with no coordination between the two (or three) males. This strategy could be successful because single males are likely unable to coerce females into consortships on their own due to a notable lack of sexual dimorphism (Connor 2010). However, there are a number of features of this behavior that argue for a more complex mechanism. In order to successfully herd females during consortships, males need to coordinate their behavior in space and time: consortships can last for several weeks and males take turns guarding the female in order to balance foraging needs with herding. Also, if a female attempts to escape, the males typically angle off to either side, effectively cutting the distance if she changes direction, rather than chasing directly behind her (Connor et al. 1992). These behaviors highlight the cooperative nature of herding in which males need to jointly and actively control female movements, suggesting they are engaging in actively coordinated collaboration.

Whether males are flexibly using vocal signals to coordinate their cooperative actions remains unknown. Vocal communication between allied male dolphins does appear to play an important role in the herding of females, with individuals using different call types that appear to elicit different responses in their male partners (King et al. 2019). For example, while males use a female-directed threat vocalization (narrowband, low-frequency pulses called "pops") to induce the female to remain close during consortships (Connor and Smolker 1996; Vollmer et al. 2015), males also increase pop rate during guard switches (King et al. 2019). Allied males have also been shown to engage in vocal matching exchanges with a shared whistle type immediately preceding a change in group behavior (King et al. 2019). Future work should closely examine male vocalization use to determine whether and how they use vocal signals to recruit cooperative partners and coordinate actions with each other.

Experimentally Testing Proximate Mechanisms of Cooperative Acts in Facilities

So far, we have seen that many of the cooperative behaviors observed in wild cetaceans can be explained by socially influenced collaboration, which does not require the use of intentional strategies to facilitate coordination with a partner. Of course, just because a lower-level mechanism could theoretically explain an observed behavior does not mean that the animals are not using a higher-level mechanism. However, these mechanisms are difficult to differentiate between based purely on observation. This is particularly true in the wild, where it can be nearly impossible to conduct controlled experiments. Instead, experimental evidence with animals in more controlled settings is required. As a result, insight into the proximate mechanisms underlying cooperative behavior in numerous species has come from experiments in which animals in human care have been presented with a task that requires cooperation in order for them to succeed (Duguid and Melis 2020; Noë 2006).

In the most common experimental paradigm, two animals must simultaneously pull two ropes or handles in the same direction in order to bring food within their reach (Hirata and Fuwa 2007; Melis et al. 2006). Then, in the critical condition, one of the animals is allowed access to the apparatus first. If that animal simply pulls without a partner, this suggests they do not understand the need for a cooperative partner to succeed at the task. But if that animal waits for a partner before pulling, this suggests that they understand the cooperative nature of the task, at least to the extent of knowing that a partner is required. Such studies have suggested that some animals, such as chimpanzees (Chalmeau 1994; Chalmeau and Gallo 1996; Hirata and Fuwa 2007; Melis et al. 2006; Suchak et al. 2014), orangutans (*Pongo pygmaeus* [Chalmeau et al. 1997]), capuchins (*Sapajus apella* [Mendres and de Waal 2000]), elephants (*Elephas maximus* [Plotnik et al. 2011]), wolves (Marshall-Pescini et al. 2017), hyenas (*Crocuta crocuta* [Drea and Carter 2009]), keas (*Nestor notabilis* [Heaney et al. 2017]), peach-fronted conures (*Eupsittula aurea* [Torres Ortiz et al. 2020]), and bottlenose dolphins (Yamamoto et al. 2019) understand the need for a partner, while others, such as otters (*Pteronura brasiliensis* and *Aonyx cinerea* [Schmelz et al. 2017]), rooks (*Corvus frugilegus* [Seed et al. 2008]), ravens (*C. corax* [Massen et al. 2015]), African grey parrots (*Psittacus erithacus* [Péron et al. 2011]), and blue-throated macaws (*Ara glaucogularis* [Tassin de Montaigu et al. 2020]) do not.

As multiple researchers have noted, however, success in the delayed partner condition could arise either from understanding the need for cooperation or by associatively learning a behavioral contingency such as "pull when another individual is present" or "pull when there is tension on the rope" (Duguid and Melis 2020; Seed and Jensen 2011; Vesper et al. 2010; Völter et al. 2017). Thus, while waiting for a partner may be considered a necessary condition for attributing understanding in this task (since an animal that does not wait clearly does not understand the necessity of that partner), it is not sufficient.

A second paradigm that has been used to test cooperative behavior involves presenting animals with a food-baited container that can be opened by pulling rope handles in opposite directions to separate a cap from the body of the container. Studies using this methodology, so far exclusively with bottlenose dolphins, have suggested that dolphins actively coordinated their behavior to solve this novel task (Kuczaj et al. 2015; Matrai et al. 2021a, 2021b, 2022), swam synchronously with the device as a form of cooperative play (Matrai et al. 2021a, 2021b, 2022), and used vocal behavior to coordinate their cooperation (Eskelinen et al. 2016). Although we agree that bottlenose dolphins do in fact use acoustic communication to actively coordinate cooperative behavior (King et al. 2021b; Sørensen et al. 2023), we must also point out that the nature of this particular methodology does not inherently support a cooperative interpretation over a competitive "tug-of-war" interpretation (King et al. 2016). In fact, if the dolphins were each trying to open the container independently, we would expect exactly the results that were found, including (i) the first dolphin to grab a handle on the apparatus immediately starting to swim away before a second dolphin grabs the opposite handle (Kuczaj et al. 2015); (ii) dolphins swimming together while each holding a handle (similar to when two dogs grab the same stick that neither wants to relinquish), either before the cap comes off or when the handles are attached to unopenable caps (Matrai et al. 2021a, 2021b, 2022); and (iii) an increase in burst-pulse vocalizations (which are commonly associated with aggressive signaling) while the dolphins are interacting with the apparatus (Eskelinen et al. 2016). The only exception to this pattern was a study in which, after learning to open the container being held by a human, a dolphin would subsequently bring the free-floating container over to a human partner for opening (Matrai et al. 2020), which suggests either an understanding or at least an associative learning that a human is required to open the container.

A third paradigm for assessing cooperative understanding and strategies in bottlenose dolphins involves a task in which two dolphins are asked to swim across a lagoon and each press their own underwater button simultaneously (operationalized as within 1 second of each other), whether sent together or with a delay between partners (Jaakkola et al. 2018; King et al. 2021b). The problematic issues of the previous two experimental paradigms are solved here by requiring tight behavioral synchronization that eliminates the possibility of succeeding through independent action, social facilitation, or responding to an environmental cue such as tension in the rope (Duguid and Melis 2020; Jaakkola et al. 2018; King et al. 2021b). Studies using this paradigm have shown that dolphins can work together with extreme precision even when they have to wait for a partner, demonstrating both their understanding of the cooperative nature of the task, as well as active behavioral coordination with that partner (Jaakkola et al. 2018). To do so, they used one of two behavioral strategies: either waiting nearby for their partner and swimming together when approaching the buttons, or swimming to other areas of the lagoon until their partner was released and then coordinating at the buttons. In both cases, the dolphins were significantly more likely to cooperate successfully when they used whistles prior to pressing their buttons (King et al. 2021b), showing that they used vocal signals to facilitate their coordinated, cooperative actions.

In an extension of this study, anthropogenic noise was played through an underwater speaker when the animals were engaging with the cooperative task in order to assess how noise disturbance would impair the dolphins' ability to work together (Sørensen et al. 2023). In this study, the dolphins performed the task while equipped with sound and movement recording DTAGs (Johnson and Tyack 2003; fig. 10.2) and were exposed to increasing levels of anthropogenic noise that covered the same frequency range as dolphin whistles. As in previous studies, the dolphins used whistles to coordinate their button presses, but as the noise got louder, the dolphins struggled to successfully solve the task and their success rate decreased. It has previously been shown that humans will use information about noise in their surroundings and how important it is to get their message across, to decide how much to actively modulate the level and duration of their vocalizations (Lane and Tranel 1971). In this study, while the dolphins were less successful in the presence of noise, they did increase the amplitude of their whistles in response to increasing noise (referred to as a Lombard response), and they also nearly doubled their whistle durations, oriented more toward their partner and spent significantly more time in close proximity as the noise got louder, suggesting that both dolphins were actively attempting to overcome the effects of noise to continue communicating with each other (Sørensen et al. 2023).

This body of work demonstrates that bottlenose dolphins are capable of actively coordinated collaboration whereby they understand their partner's role in a cooperative context (Jaakkola et al. 2018) and can flexibly modify their vocalizations and behavior to facilitate cooperative success (Sørensen et al. 2023; King et al. 2021b). Like chimpanzees (Duguid and Melis 2020), both dolphins in this study were motivated to reach a common goal that required them to work together in close temporal and spatial proximity, and they were intentionally employing active strategies by increasing the amplitude and duration of their communication signals and changing their orientation to coordinate with each other (Sørensen et al. 2023). Further, the fact that dolphins communicate to coordinate cooperative behavior (King et al. 2021b; Sørensen et al. 2023) suggests they are capable of shared intentionality. Evidence for shared intentionality includes "*communication to initiate and coordinate the intention to collaborate*" (Duguid and Melis 2020). In the numerous cooperation experiments with animals in human care that have explored the cognition underlying coordination toward a shared goal, it is only studies with great apes (Botting and Bastian 2019; Melis and Tomasello 2019) and dolphins (King et al. 2021b; Sørensen et al. 2023) that have provided evidence that nonhuman animals can use communication signals to facilitate the successful execution of coordinated cooperative actions. One study has shown that peach-fronted conures increase redundancy of their vocalizations when they cannot see their partner during a cooperative rope-pulling task, but it remains to be determined exactly how these vocalizations are being used (Torres Ortiz et al. 2020).

Long-Term Cooperative Relationships and Their Cognitive Conceptualization

We now move from discussing the cognitive mechanisms underlying individual cooperative acts, to the cognitive processes underpinning long-term cooperative relationships. As we discussed earlier in the chapter, unrelated male dolphins in Shark Bay, Western Australia, form multilevel alliances, where males engage in coordinated efforts to compete with rival alliances over access to females (Connor and Krützen 2015; Connor et al. 1992; Connor et al. 2022; see Connor and King, this volume). At the first level, males cooperate in pairs or trios to herd single estrus females during events termed consortships (Connor and Krützen 2015; Connor et al. 1992). Males also cooperate in teams of up to 14 males in second-order alliances to attack other alliances for access to females and to defend against such attacks (Connor et al. 1992; Connor and Krützen 2015). Second-order alliances can remain stable for decades and are considered the core unit of male social organization (Randić et al. 2012; King et al. 2021a). First-order allies are chosen from within a male's second-order alliance and vary in composition and stability; with some males showing clear preferences for particular individuals with whom to partner as first-order allies, while others will consort females with numerous males from within their second-order alliance (Connor et al. 2001; Connor and Krützen 2015). Male dolphins in Shark Bay also form between-group alliances, i.e., third-order alliances, where second-order alliances associate preferentially and cooperate in contests over access to females (Connor et al. 2022, 2011). These strategic between-group alliances increase male access to females, thereby increasing their reproductive success (Connor et al. 2022).

To keep track of many different individual and alliance level relationships, bottlenose dolphins exhibit a suite of cognitive skills to monitor individual behavior (King et al. 2022; King et al. 2021a). For example, they produce individual vocal labels called "signature whistles," analogous to human names (Caldwell and Caldwell 1965; Sayigh et al. 1999; Janik et al. 2006; Bruck et al. 2022; see Oswald et al., this volume), and have a social memory of these whistles that spans up to at least two decades (Bruck 2013). These signature whistles play a central role in the recognition of cooperative partners, underpinning a social concept of cooperative team membership in this population (King et al. 2021a). In order to understand the significance of this cooperation-based concept, it is important to note that relationship strength and consortship history can be highly differentiated among males, such that social bond strengths among second-order allies may be comparable to social bond strengths among some third-order allies (King et al. 2021a). This means that the strength of social bonds alone does not distinguish second-order allies from third-order allies. Which poses the question, how do male dolphins categorize these different alliance levels?

To explore the cognitive processes underpinning long-term cooperative relationships, a sound playback experiment of allied male signature whistles was conducted, where males were exposed to the playback of a signature whistle from a second-order ally and a third-order ally (King et al. 2021a), while drone-mounted video was used to record fine-scale behavioral responses to the playback (e.g., head turns, change in body orientation, approach/avoidance distance; King and Jensen 2023; King et al. 2021a). These experiments revealed two important results. The first is that males exhibited a much stronger response to second-order allies compared to third-order allies, even when relationship strength was comparable (King et al. 2021a). The second is that males responded strongly to second-order allies irrespective of relationship strength and consortship history, i.e., males responded strongly to the signature whistles of second-order allies they had never been documented herding a female with (King et al. 2021a). This is a striking result given that males have notably different signature whistles from their second-order allies (King et al. 2018), are generally unrelated (Gerber et al. 2020, 2021), and social bond strength is highly differentiated; meaning that whistle similarity, relationship strength, and kinship cannot be used as indicators of second-order alliance membership. Instead, the results are suggestive of male dolphins forming a social concept of cooperative team membership at the second-order alliance level; based on their experience with each individual's arbitrary signature whistle and previous cooperative investment in the team (King et al. 2021a). It is highly likely that dolphins can use memory of previous cooperative interactions with individuals to classify their relationships, given that they have social memory that spans decades (Bruck 2013) and are capable of episodic-like memory, i.e., the recollection of personally experienced events (Davies et al. 2022).

This work shows that in the mind of the male dolphin, second- and third-order alliances are not equivalent. While third-order alliances are known to be important in providing additional support in the capture or defense of females (Connor et al. 2022), it is the second-order alliance that provides consistent, cooperative support in the defense and theft of females from rivals, despite high risk of physical injury (King et al. 2021a). Second-order alliances are stable over decades, with lost individuals rarely replaced (Connor et al. 2022; King et al. 2021a); thus individuals have ample opportunity to invest in "team" members based on the value of future by-product benefits and reciprocity (Connor 1986; Connor 2010).

What Can We Conclude About the Cognitive Mechanisms Underlying Cetacean Cooperation?

In this chapter, we have summarized some of the key studies on cetacean cooperation and considered the proximate mechanisms that could theoretically explain the observed behavior. We have shown that in many cases, cooperative behaviors that appear complex can be achieved by simple mechanisms that do not necessarily involve intentional cooperation. However, it is important to note that demonstrating "intentional" cooperation is setting a high bar and requires experimental evidence where confounding variables can be controlled for. This is where the synergy between experiments conducted with animals in human care and observational studies with wild animals is important. Experiments in facilities have allowed us to test the cognitive mechanisms underlying cooperation, something we are not able to do in the wild. For example, cooperation experiments with bottlenose dolphins in human care have shown that the species certainly possesses the cognitive skills to actively coordinate cooperative tasks in the wild. Studies like these are helping us interpret the behavior of cetaceans in the wild and have allowed us to refine our predictions of what individuals are likely doing in the field to facilitate cooperative behavior. Finally, we have considered the cognitive skills underlying long-term cooperative relationships, with a focus on concept learning in dolphin alliances. Studying cetacean social cognition in the wild is a burgeoning area of research thanks to the significant technological advancements that have been made in recent years (King and Jensen 2023). Our hope is that the synergism between experimental work in facilities and field-based studies with wild animals will continue, allowing us to determine how cetaceans use vocal signals to coordinate cooperation in the wild, how individuals recruit allies to cooperative endeavors and how cooperative partnerships are formed and maintained within the socioecological settings in which their cognition and communication systems have evolved.

Acknowledgments

S. L. K. was supported by the Branco Weiss Fellowship—Society in Science. We would like to thank the humans and dolphins at Dolphin Research Center for their cooperation in the button-press studies discussed in this chapter. Those studies were supported by a research grant from Jim and Marjorie Sanger to Dolphin Research Center.

References

Bailey, Ida, Julia P. Myatt, and Alan M. Wilson. 2013. "Group Hunting Within the Carnivora: Physiological, Cognitive and Environmental Influences on Strategy and Cooperation." *Behavioral Ecology and Sociobiology* 67 (1): 1–17. https://doi.org/10.1007/s00265-012-1423-3.

Benoit-Bird, Kelly J., and Whitlow W. L. Au. 2009. "Phonation Behaviour of Cooperatively Foraging Spinner Dolphins." *Journal of the Acoustical Society of America* 125 (1): 539–46. https://doi.org/10.1121/1.2967477.

Boesch, Christophe. 2002. "Cooperative Hunting Roles Among Tai Chimpanzees." *Human Nature* 13 (1): 27–46. https://doi.org/10.1007/s12110-002-1013-6.

Botting, Jennifer, and Meredith Bastian. 2019. "Orangutans (*Pongo pygmaeus* and Hybrid) and Gorillas (*Gorilla gorilla gorilla*) Modify Their Visual, but Not Auditory, Communicative Behaviors, Depending on the Attentional State of a Human Experimenter." *International Journal of Primatology* 40: 244–62. https://doi.org/10.1007/s10764-019-00083-0.

Bruck, Jason N. 2013. "Decades-Long Social Memory in Bottlenose Dolphins." *Proceedings of the Royal Society B* 280: 20131726. https://doi.org/10.1098/rspb.2013.1726.

Bruck, Jason N., Sam F. Walmsley, and Vincent M. Janik. 2022. "Cross-Modal Perception of Identity by Sound and Taste in Bottlenose Dolphins." *Science Advances* 8 (20): eabm7684. https://doi.org/10.1126/sciadv.abm7684.

Caldwell, Melba C., and David K. Caldwell. 1965. "Individualised Whistle Contours in Bottlenose Dolphins (*Tursiops truncatus*)." *Nature* 207: 434–44. https://doi.org/10.1038/207434a0.

Cantor, Mauricio, Damien R. Farine, and Fábio G. Daura-Jorge. 2023. "Foraging Synchrony Drives Resilience in Human-Dolphin Mutualism." *Proceedings of the National Academy of Sciences* 120 (6): e2207739120. https://doi.org/10.1073/pnas.2207739120.

Chalmeau, Raphael. 1994. "Do Chimpanzees Cooperate in a Learning Task?" *Primates* 35 (3): 385–92. https://doi.org/10.1007/BF02382735.

Chalmeau, Raphael, and Alain Gallo. 1996. "What Chimpanzees (*Pan troglodytes*) Learn in a Cooperative Task." *Primates* 37 (1): 39–47. https://doi.org/10.1007/BF02382918.

Chalmeau, Raphael, Karine Lardeux, Pierre Brandibas, and Alain Gallo. 1997. "Cooperative Problem Solving by Orangutans (*Pongo pygmaeus*)." *International Journal of Primatology* 18 (1): 23–32. https://doi.org/10.1023/A:1026337006136.

Connor, Richard C. 1986. "Pseudo-Reciprocity: Investing in Mutualism." *Animal Behaviour* 34 (5): 1562–66. https://doi.org/10.1016/S0003-3472(86)80225-1.

Connor, Richard C. 2010. "Cooperation Beyond the Dyad: On Simple Models and a Complex Society." *Philosophical Transactions of the Royal Society B* 365 (1553): 2687–97. https://doi.org/10.1098/rstb.2010.0150.

Connor, Richard C., Michael R. Heithaus, and Lynne M. Barrè. 2001. "Complex Social Structure, Alliance Stability and Mating Access in a Bottlenose Dolphin 'Super-Alliance.'" *Proceedings of the Royal Society B* 268 (1464): 263–67. https://doi.org/10.1098/rspb.2000.1357.

Connor, Richard C., and Michael Krützen. 2015. "Male Dolphin Alliances in Shark Bay: Changing Perspectives in a 30-Year Study." *Animal Behaviour* 103 (May): 223–35. https://doi.org/10.1016/j.anbehav.2015.02.019.

Connor, Richard C., Michael Krützen, Simon J. Allen, William B. Sherwin, and Stephanie L. King. 2022. "Strategic Intergroup Alliances Increase Access to a Contested Resource in Male Bottlenose Dolphins." *Proceedings of the National Academy of Sciences* 119 (36): e212172311. https://doi.org/10.1073/pnas.2121723119.

Connor, Richard C., and Rachel A. Smolker. 1996. "'Pop' Goes the Dolphin: A Vocalization Male Bottlenose Dolphins Produce During Consortships." *Behaviour* 133: 643–62. https://doi.org/10.1163/156853996X00404.

Connor, Richard C., Rachel A. Smolker, and Andrew F. Richards. 1992. "Two Levels of Alliance Formation Among Male Bottlenose Dolphins (*Tursiops* sp.)." *Proceedings of the National Academy of Sciences* 89 (3): 987–90. https://doi.org/10.1073/pnas.89.3.987.

Connor, Richard C., Jana J. Watson-Capps, William B. Sherwin, and Michael Krützen. 2011. "A New Level of Complexity in the Male Alliance Networks of Indian Ocean Bottlenose Dolphins (*Tursiops* sp.)." *Biology Letters* 7 (4): 623–26. https://doi.org/10.1098/rsbl.2010.0852.

Davies, James R., Elias Garcia-Pelegrin, Luigi Baciadonna, Cristina Pilenga, Livio Favaro, and Nicola S. Clayton 2022. "Episodic-Like Memory in Common Bottlenose Dolphins." *Current Biology* 32: 3436–42. https://doi.org/10.1016/j.cub.2022.06.032.

Drea, Christine M., and Allisa N. Carter. 2009. "Cooperative Problem Solving in a Social Carnivore." *Animal Behaviour* 78 (4): 967–77. https://doi.org/10.1016/j.anbehav.2009.06.030.

Duffy-Echevarria, Erin E., Richard C. Connor, and David J. St. Aubin. 2008. "Observations of Strand-Feeding Behavior by Bottlenose Dolphins (*Tursiops truncatus*) in Bull Creek, South Carolina." *Marine Mammal Science* 24 (1): 202–6. https://doi.org/10.1111/j.1748-7692.2007.00151.x.

Duguid, Shona, and Alicia P. Melis. 2020. "How Animals Collaborate: Underlying Proximate Mechanisms." *Wiley Interdisciplinary Reviews: Cognitive Science* 11 (5): e1529. https://doi.org/10.1002/wcs.1529.

D'Vincent, Cynthia G., Russell M. Nilson, and Richard E. Hanna. 1985. "Vocalization and Coordinated Feeding Behavior of the Humpback Whale in Southeastern Alaska." *Scientific Reports of the Whales Research Institute* (Tokyo) 36: 41–47.

Eskelinen, Holli C., Kelley A. Winship, Brittany L. Jones, Audra E. M. Ames, and Stan A. Kuczaj. 2016. "Acoustic Behavior Associated with Cooperative Task Success in Bottlenose Dolphins (*Tursiops truncatus*)." *Animal Cognition* 19 (4): 789–97. https://doi.org/10.1007/s10071-016-0978-1.

Gazda, Stefanie K. 2016. "Driver-Barrier Feeding Behavior in Bottlenose Dolphins (*Tursiops truncatus*): New Insights from a Longitudinal Study." *Marine Mammal Science* 32 (3): 1152–60. https://doi.org/10.1111/mms.12314.

Gazda, Stefanie K., Richard C. Connor, Robert K. Edgar, and Frank Cox. 2005. "A Division of Labour with Role Specialization in Group-Hunting Bottlenose Dolphins (*Tursiops truncatus*) off Cedar Key, Florida." *Proceedings of the Royal Society B* 272 (1559): 135–40. https://doi.org/10.1098/rspb.2004.2937.

Gerber, Livia, Richard C. Connor, Stephanie L. King, et al. 2020. "Affiliation History and Age Similarity Predict Alliance Formation in Adult Male Bottlenose Dolphins." *Behavioral Ecology* 31 (2): 361–70. https://doi.org/10.1093/beheco/arz195.

Gerber, Livia, Samuel Wittwer, Simon J. Allen, et al. 2021. "Cooperative Partner Choice in Multi-Level Male Dolphin Alliances." *Scientific Reports* 11 (1): 1–10. https://doi.org/10.1038/s41598-021-85583-x.

Guinet, Christophe, and Jérome Bouvier. 1995. "Development of Intentional Stranding Hunting Techniques in Killer Whale (*Orcinus orca*) Calves at Crozet Archipelago." *Canadian Journal of Zoology* 73 (1): 27–33. https://doi.org/10.1139/z95-004.

Hamilton, Rebecca A., Stefanie K. Gazda, Stephanie L. King, Josefin Starkhammar, and Richard C. Connor. 2022. "Bottlenose Dolphin Communication during a Role-Specialized Group Foraging Task." *Behavioural Processes* 200: 104691. https://doi.org/https://doi.org/10.1016/j.beproc.2022.104691.

Hamilton, William D. 1963. "The Evolution of Altruistic Behavior." *American Naturalist* 97: 354–56. https://doi.org/10.1086/497114.

Hamilton, William D. 1964. "The Genetical Evolution of Social Behaviour. I." *Journal of Theoretical Biology* 7 (1): 1–16. https://doi.org/10.1016/0022-5193(64)90038-4.

Heaney, Megan, Russell D Gray, and Alex H Taylor. 2017. "Keas Perform Similarly to Chimpanzees and Elephants When Solving Collaborative Tasks." *PLOS ONE* 12 (2): 1–13. https://doi.org/10.1371/journal.pone.0169799.

Herbert-Read, James E., Pawel Romanczuk, Stefan Krause, et al. 2016. "Proto-Cooperation: Group Hunting Sailfish Improve Hunting Success by Alternating Attacks on Grouping Prey." *Proceedings of the Royal Society B* 283: 20161671. https://doi.org/10.1098/rspb.2016.1671.

Hirata, Satoshi, and Kohki Fuwa. 2007. "Chimpanzees (*Pan troglodytes*) Learn to Act with Other Individuals in a Cooperative Task." *Primates* 48: 13–21. https://doi.org/10.1007/s10329-006-0022-1.

Hoese, Hinton D. 1971. "Dolphin Feeding out of Water in a Salt Marsh." *Journal of Mammalogy* 52: 222–23. https://doi.org/10.2307/1378455.

Hurtado, Maria José, Renée Fénéron, and Patrick Gouat. 2013. "Specialization in Building Tasks in the Mound-Building Mouse, *Mus spicilegus*." *Animal Behaviour* 85 (6): 1153–60. https://doi.org/10.1016/j.anbehav.2013.03.020.

Jaakkola, Kelly, Emily Guarino, Katy Donegan, and Stephanie L. King. 2018. "Bottlenose Dolphins Understand Their Partner's Role in a Cooperative Task." *Proceedings of the Royal Society B* 285 (1887): 20180948. https://doi.org/https://doi.org/10.1098/rspb.2018.0948.

Janik, Vincent M., Laela S. Sayigh, and Randall S. Wells. 2006. "Signature Whistle Shape Conveys Identity Information to Bottlenose Dolphins." *Proceedings of the National Academy of Sciences* 103 (21): 8293–97. https://doi.org/10.1073/pnas.0509918103.

Johnson, Mark P., and Peter L. Tyack. 2003. "A Digital Acoustic Recording Tag for Measuring the Response of Wild Marine Mammals to Sound." *IEEE Journal of Oceanic Engineering* 28 (1): 3–12. https://doi.org/10.1109/JOE.2002.808212.

King, Stephanie L., Simon J. Allen, Richard C. Connor, and Kelly Jaakkola. 2016. "Cooperation or Dolphin 'Tug-of-War'? Comment on Kuczaj et al. and Eskelinen et al." *Animal Cognition* 19 (6): 1227–29. https://doi.org/10.1007/s10071-016-1026-x.

King, Stephanie L., Simon J. Allen, Michael Krützen, and Richard C. Connor. 2019. "Vocal Behaviour of Allied Male Dolphins During Cooperative Mate Guarding." *Animal Cognition* 22 (6): 991–1000. https://doi.org/10.1007/s10071-019-01290-1.

King, Stephanie L., Richard C. Connor, Michael Krützen, and Simon J. Allen. 2021a. "Cooperation-Based Concept Formation in Male Bottlenose Dolphins." *Nature Communications* 12 (1): 2373. https://doi.org/10.1038/s41467-021-22668-1.

King, Stephanie L., Richard C. Connor, and Stephen H. Montgomery. 2022. "Social and Vocal Complexity in Bottlenose Dolphins." *Trends in Neurosciences* 45 (12): 881–83. https://doi.org/10.1016/j.tins.2022.09.006.

King, Stephanie L., Whitney F. Friedman, Simon J. Allen, et al. 2018. "Bottlenose Dolphins Retain Individual Vocal Labels in Multi-Level Alliances." *Current Biology* 28 (12): 1993–99. https://doi.org/10.1016/j.cub.2018.05.013.

King, Stephanie L., Emily Guarino, Katy Donegan, Christina McMullen, and Kelly Jaakkola. 2021b. "Evidence That Bottlenose Dolphins Can Communicate with Vocal Signals to Solve a Cooperative Task." *Royal Society Open Science* 8: 202073. https://doi.org/10.1098/rsos.202073.

King, Stephanie L., and Frants H. Jensen. 2023. "Rise of the Machines: Integrating Technology with Playback Experiments to Study Cetacean Social Cognition in the Wild." *Methods in Ecology and Evolution* 14 (8): 1873–86. https://doi.org/10.1111/2041-210x.13935.

Kuczaj, Stan A., Kelley A. Winship, and Holli C. Eskelinen. 2015. "Can Bottlenose Dolphins (*Tursiops truncatus*) Cooperate When Solving a Novel Task?" *Animal Cognition* 18 (2): 543–50. https://doi.org/10.1007/s10071-014-0822-4.

Lane, Harlan, and Bernard Tranel. 1971. "The Lombard Sign and the Role of Hearing in Speech." *Journal of Speech and Hearing Research* 14 (4): 677–709. https://doi.org/10.1044/jshr.1404.677.

Lopez, Juan Carlos, and Diana Lopez. 1985. "Killer Whales (*Orcinus orca*) of Patagonia, and Their Behavior of Intentional Stranding While Hunting Nearshore." *Journal of Mammalogy* 66 (1): 181–83. https://doi.org/10.2307/1380981.

Marshall-Pescini, Sarah, Jonas F. L. Schwarz, Inga Kostelnik, Zsófia Virányi, and Friederike Range. 2017. "Importance of a Species' Socioecology: Wolves Outperform Dogs in a Conspecific Cooperation Task." *Proceedings of the National Academy of Sciences* 114 (4): 11793–98. https://doi.org/10.1073/pnas.1709027114.

Massen, Jorg J. M., Caroline Ritter, and Thomas Bugnyar. 2015. "Tolerance and Reward Equity Predict Cooperation in Ravens (*Corvus corax*)." *Scientific Reports* 5: 1–11. https://doi.org/10.1038/srep15021.

Matrai, Eszter, Shaw Ting Kwok, Michael Boos, and Ákos Pogány. 2021a. "Cognitive Enrichment Device Provides Evidence for Intersexual Differences in Collaborative Actions in Indo-Pacific Bottlenose Dolphins (*Tursiops aduncus*)." *Animal Cognition* 24 (6): 1215–25. https://doi.org/10.1007/s10071-021-01510-7.

Matrai, Eszter, Shaw Ting Kwok, Michael Boos, and Ákos Pogány. 2021b. "Group Size, Partner Choice and Collaborative Actions in Male Indo-Pacific Bottlenose Dolphins (*Tursiops aduncus*)." *Animal Cognition* 25 (1): 179–93. https://doi.org/10.1007/s10071-021-01541-0.

Matrai, Eszter, Shaw Ting Kwok, Michael Boos, and Ákos Pogány. 2022. "Testing Use of the First

Multi-Partner Cognitive Enrichment Devices by a Group of Male Bottlenose Dolphins." *Animal Cognition* 25 (4): 961–73. https://doi.org/10.1007/s10071-022-01605-9.

Matrai, Eszter, Angela K. W. Ng, Michelle M. H. Chan, Suzanne M. Gendron, and Kathleen M. Dudzinski. 2020. "Testing Use of a Potential Cognitive Enrichment Device by an Indo-Pacific Bottlenose Dolphin (*Tursiops aduncus*)." *Zoo Biology* 39 (3): 156–67. https://doi.org/10.1002/zoo.21536.

Melis, Alicia P., Brian Hare, and Michael Tomasello. 2006. "Chimpanzees Recruit the Best Collaborators." *Science* 311 (5765): 1297–300. https://doi.org/10.1126/science.1123007.

Melis, Alicia P, and Michael Tomasello. 2019. "Chimpanzees (*Pan troglodytes*) Coordinate by Communicating in a Collaborative Problem-Solving Task." *Proceedings of the Royal Society B* 286: 20190408. https://doi.org/10.1098/rspb.2019.0408.

Mendres, Kimberly A., and Frans B. de Waal. 2000. "Capuchins Do Cooperate: The Advantage of an Intuitive Task." *Animal Behaviour* 60 (4): 523–29. https://doi.org/10.1006/anbe.2000.1512.

Mine, Joseph G., Katie E. Slocombe, Erik P. Willems, et al. 2022. "Vocal Signals Facilitate Cooperative Hunting in Wild Chimpanzees." *Science Advances* 8 (30): eabo5553. https://doi.org/10.1126/sciadv.abo5553.

Muro, Cristina, Ramón Escobedo, Lee Spector, and Raymond P. Coppinger. 2011. "Wolf-Pack (*Canis lupus*) Hunting Strategies Emerge from Simple Rules in Computational Simulations." *Behavioural Processes* 88 (3): 192–97. https://doi.org/10.1016/j.beproc.2011.09.006.

Noë, Ronald. 2006. "Cooperation Experiments: Coordination Through Communication Versus Acting Apart Together." *Animal Behaviour* 71 (1): 1–18. https://doi.org/10.1016/j.anbehav.2005.03.037.

Péron, Franck, Lauriane Rat-Fischer, Mathilde Lalot, Laurent Nagle, and Dalila Bovet. 2011. "Cooperative Problem Solving in African Grey Parrots (*Psittacus erithacus*)." *Animal Cognition* 14 (4): 545–53. https://doi.org/10.1007/s10071-011-0389-2.

Petricig, R. O. 1995. "Bottlenose Dolphins (*Tursiops truncatus*) in Bull Creek, South Carolina." PhD diss., University of Rhode Island.

Pitman, Robert L., and John W. Durban. 2012. "Cooperative Hunting Behavior, Prey Selectivity and Prey Handling by Pack Ice Killer Whales (*Orcinus orca*), Type B, in Antarctic Peninsula Waters." *Marine Mammal Science* 28 (1): 16–36. https://doi.org/10.1111/j.1748-7692.2010.00453.x.

Plotnik, Joshua M., Richard Lair, Wirot Suphachoksahakun, and Frans B. M. de Waal. 2011. "Elephants Know When They Need a Helping Trunk in a Cooperative Task." *Proceedings of the National Academy of Sciences* 108 (12): 5116–21. https://doi.org/10.1073/pnas.1101765108.

Pryor, Karen, and Jon Lindbergh. 1990. "A Dolphin-Human Fishing Cooperative in Brazil." *Marine Mammal Science* 6: 77–82. https://doi.org/10.1111/j.1748-7692.1990.tb00228.x.

Randić, Srđan, Richard C. Connor, William B. Sherwin, and Michael Krützen. 2012. "A Novel Mammalian Social Structure in Indo-Pacific Bottlenose Dolphins (*Tursiops* sp.): Complex Male Alliances in an Open Social Network." *Proceedings of the Royal Society B* 279 (1740): 3083–90. https://doi.org/10.1098/rspb.2012.0264.

Rossbach, Kelly Aann 1999. "Cooperative Feeding among Bottlenose Dolphins (*Tursiops truncatus*) near Grand Bahama Island, Bahamas." *Aquatic Mammals* 25: 163–67.

Sayigh, Laela. S., Peter L. Tyack, Randall S. Wells, Andrew. R. Solow, Michael. D. Scott, and A. Blair Irvine. 1999. "Individual Recognition in Wild Bottlenose Dolphins: A Field Test Using Playback Experiments." *Animal Behaviour* 57: 42–50. https://doi.org/10.1006/anbe.1998.0961.

Schmelz, Martin, Shona Duguid, Manuel Bohn, and Christoph J. Völter. 2017. "Cooperative Problem Solving in Giant Otters (*Pteronura brasiliensis*) and Asian Small-Clawed Otters (*Aonyx cinerea*)." *Animal Cognition* 20 (6): 1107–14. https://doi.org/10.1007/s10071-017-1126-2.

Seed, Amanda M., Nicola S. Clayton, and Nathan J. Emery. 2008. "Cooperative Problem Solving in Rooks (*Corvus frugilegus*)." *Proceedings of the Royal Society B* 275 (1641): 1421–29. https://doi.org/10.1098/rspb.2008.0111.

Seed, Amanda M., and Keith Jensen. 2011. "Animal Behaviour: Large-Scale Cooperation." *Nature* 472: 424–25. https://doi.org/10.1038/472424a.

Similä, Tiu, and Fernando Ugarte. 1993. "Surface and Underwater Observations of Cooperatively Feeding Killer Whales in Northern Norway." *Canadian Journal of Zoology* 71 (8): 1494–99. https://doi.org/10.1139/z93-210.

Simões-Lopes, Paulo C. 1991. "Interaction of Coastal Populations of *Tursiops truncatus* (Cetacea, Delphinidae) with the Mullet Artisanal Fisheries in Southern Brazil." *Biotemas* 4 (2): 83–94.

Simões-Lopes, Paulo C., Marta E. Fabián, and João O. Menegheti. 1998. "Dolphin Interactions with the Mullet Artisanal Fishing on Southern Brazil: A Qualitative and Quantitative Approach." *Revista Brasileira de Zoologia* 15 (3): 709–26. https://doi.org/10.1590/s0101-8175 1998000300016.

Sørensen, Pernille M., Abigail Haddock, Emily Guarino, et al. 2023. "Anthropogenic Noise Impairs Cooperation in Bottlenose Dolphins." *Current Biology* 33: 1–6. https://doi.org/10.1016/j.cub.2022.12.063.

Spottiswoode, Claire N., Keith S. Begg, and Colleen M. Begg. 2016. "Reciprocal Signaling in Honeyguide-Human Mutualism." *Science* 353 (6297): 387–89. https://doi.org/10.1126/science.aaf4885.

Stander, P. E. 1992. "Cooperative Hunting in Lions: The Role of the Individual." *Behavioral Ecology and Sociobiology* 29 (6): 445–54. https://doi.org/10.1007/BF00170175.

Suchak, Malini, Timothy M. Eppley, Matthew W. Campbell, and Frans B. M. de Waal. 2014. "Ape Duos and Trios: Spontaneous Cooperation with Free Partner Choice in Chimpanzees." *PeerJ* 2: e417. https://doi.org/10.7717/peerj.417.

Tassin de Montaigu, Cannelle, Kresimir Durdevic, Désirée Brucks, Anastasia Krasheninnikova, and Auguste von Bayern. 2020. "Blue-Throated Macaws (*Ara glaucogularis*) Succeed in a Cooperative Task Without Coordinating Their Actions." *Ethology* 126 (2): 267–77. https://doi.org/10.1111/eth.12973.

Tomasello, Michael. 2022. "The Coordination of Attention and Action in Great Apes and Humans." *Philosophical Transactions of the Royal Society B* 377: 20210093. https://doi.org/10.1098/rstb.2021.0093.

Torres Ortiz, Sara, Alejandro C. Castro, Thorsten J. S. Balsby, and Ole N. Larsen. 2020. "Problem-Solving in a Cooperative Task in Peach-Fronted Conures (*Eupsittula aurea*)." *Animal Cognition* 23: 265–75. https://doi.org/10.1007/s10071-019-01331-9.

Torres Ortiz, Sara, Johanna Stedt, Henrik Skov Midtiby, Henrik Dyrberg Egemose, and Magnus Wahlberg. 2021. "Group Hunting in Harbour Porpoises (*Phocoena phocoena*)." *Canadian Journal of Zoology* 99 (6): 511–20. https://doi.org/10.1139/cjz-2020-0289.

Trivers, Robert L. 1971. "The Evolution of Reciprocal Altruism." *Quarterly Review of Biology* 46: 35–57. https://doi.org/10.1086/406755.

Vesper, Cordula, Stephen Butterfill, Günther Knoblich, and Natalie Sebanz. 2010. "A Minimal Architecture for Joint Action." *Neural Networks* 23 (8–9): 998–1003. https://doi.org/10.1016/j.neunet.2010.06.002.

Vollmer, Nicole L., Lee-Ann C. Hayek, Michael R. Heithaus, and Richard C. Connor. 2015. "Further Evidence of a Context-Specific Agonistic Signal in Bottlenose Dolphins: The Influence of Consortships and Group Size on the Pop Vocalization." *Behaviour* 152: 1979–2000. https://doi.org/10.1163/1568539X-00003311.

Völter, Christoph J., Federico Rossani, and Josep Call. 2017. "Social Manipulation in Nonhuman Primates: Cognitive and Motivational Determinants." *Neuroscience and Biobehavioral Reviews* 82: 76–94. https://doi.org/10.1016/j.neubiorev.2016.09.008.

West, Stuart A., Ashleigh S. Griffin, and Andy Gardner. 2007. "Evolutionary Explanations for Cooperation." *Current Biology* 17 (16): 661–72. https://doi.org/10.1016/j.cub.2007.06.004.

Wiley, David, Colin Ware, Alessandro Bocconcelli, et al. 2011. "Underwater Components of Humpback Whale Bubble-Net Feeding Behaviour." *Behaviour* 148: 575–602. https://doi.org/10.1163/000579511X570893.

Yamamoto, Chisato, Nobuyuki Kashiwagi, Mika Otsuka, Mai Sakai, and Masaki Tomonaga. 2019. "Cooperation in Bottlenose Dolphins: Bidirectional Coordination in a Rope-Pulling Task." *PeerJ* 2019 (10): 1–16. https://doi.org/10.7717/peerj.7826.

PART IV

The Cultures of Cetacean Societies

11

Cultural Transmission of Foraging Behavior in Cetaceans

SONJA WILD AND LUKE RENDELL

Abstract

Cetaceans have rich cultural lives and the past two decades have seen a steady increase in studies on their foraging cultures. Increased availability of detailed long-term data on individuals' behavior, and statistical advances within the social network analysis framework that make it possible to identify and quantify the importance of social learning in the spread of behavior in wild animal populations, have both contributed to this. Cetacean sociality, and capacity for behavioral innovations and social learning, has led to the emergence of a large diversity of foraging techniques that target different foraging niches, ranging from intentional stranding to catch onshore prey and the depredation of fishing lines, through to tool use and cooperative hunting with humans. Increasingly, studies on foraging cultures now go beyond merely identifying socially learned behaviors to consider the far-reaching consequences culture can have on societies, its interaction with the environment, its fitness consequences and its implications for conservation efforts. Yet, there is more to learn. Integrative approaches combining observational, environmental, and genomic data provide a promising avenue to better understand the mechanisms leading to the emergence and maintenance of cultures in cetaceans, and to predict the resilience of those cultures in a changing world.

Introduction

Foraging is a key life process for any animal, and often the focus for cooperative activity in social animals. This is certainly true for cetaceans, and the decades since the publication of *Cetacean Societies* have seen a flowering of research with one overarching conclusion—the diverse social structures of cetaceans are often the conduit for the cultural transmission of foraging behavior. Cetaceans are known for engaging in a diverse variety of foraging behaviors, ranging from killer whales (*Orcinus orca*) intentionally stranding themselves to catch onshore prey (Lopez and Lopez 1985), to sperm whales (*Physeter macrocephalus*) picking fish off longline hooks (Sigler et al. 2008), and cooperative hunting efforts between dolphins (*Tursiops truncatus*) and human fishers (Simões-Lopes et al. 1998). When Rendell and Whitehead (2001) first proposed the notion that some of these behavioral differences could be cultural, i.e., information or behavior that is socially learned from conspecifics (Boyd and Richerson 1996), reactions were mixed. While some were not opposed to the idea that cetaceans could have cultures (Connor 2001), others argued that calling behavioral differences in cetaceans "cultural" was premature (Mann 2001), unconvincing (Galef 2001) or even purely speculative (Deputte 2001). Nevertheless, these early claims of culture in cetaceans (Rendell and Whitehead 2001) as well as primates (Whiten et al. 1999) fueled fundamental discussion about how to define culture, what constitutes evidence for culture in animals, and appropriate methods to identify cultural traditions in wild animal populations, for which laboratory experiments are not feasible. Now, more than two decades later, the field of animal culture has become firmly established. New experimental approaches as well as analytical methods have been developed to identify and quantify social learning in the wild (Franz and Nunn 2009; Harrison and van de Waal 2022), and the number of studies on social learning, traditions, and animal culture across a broad range of taxa has steadily increased (Whiten 2019, 2021).

In the field of cetacean culture, much of this progress is owed to the increased availability of long-term data, and with it the increased knowledge about the behavior that different individuals engage in. Studying cetaceans is logistically challenging. Given their mobile lifestyle in an aquatic environment with virtually no physical boundaries, rates of recurrent sightings of the same individuals might be rare, and collecting data on individual behavioral repertoires is therefore a time-intensive endeavor. Increased availability of long-term data has enabled us to get a better picture of the behavioral similarities and differences among individuals. Photographic and video-graphic tools with ever increasing resolutions allow for individual identification at a distance based on dorsal fins, scars, and color patterns (Würsig and Jefferson 1990); the mapping of social structure (Weiss et al. 2021); and the attribution of observed behaviors to particular individuals. Thus advances in technology have played a major role in improving our ability to track behaviors at the level of the individual.

However, collecting high-resolution data on individuals' behaviors is not sufficient to detect cultural phenomena. Generating such data also creates the need for suitable analytical approaches to discover relevant patterns for identifying cultural transmission. Therefore, the development of novel statistical methods for identifying and quantifying culture in wild animal populations, such as network-based diffusion analysis (NBDA; Franz and Nunn 2009; Hoppitt et al. 2010) has also played an important role in the advancement of the field of cetacean culture.

In this chapter, we will discuss the novel insights we have gained into cetacean culture over the past two decades. Thereby, we focus on foraging cultures and refer to Hersh et al. (this volume) for a thorough discussion of vocal cultures and Brakes and Whitehead (this volume) for a full treatment of the conservation implications of culture. After a brief introduction to the topic of animal culture, we will provide an overview of the methods that have been used to identify foraging cultures in wild populations of cetaceans. We will then review the latest developments in methods and discoveries in foraging cultures in toothed whales (*Odontocetes*) and baleen whales (*Mysticetes*), and briefly discuss the consequences of cultural behavior, before providing a conclusion and outlook on promising avenues that could take the field further.

What Is Culture?

Not all animals behave the same way. Behavioral differences can be found between populations, social units, or even among individuals within the same unit. A recent widely reported example is the interactions between killer whales and small vessels off the Atlantic coast of the Iberian Peninsula (Esteban et al. 2022). This behavior arose in a specific community and then spread to create a significant wildlife interaction issue, but that has not been seen in any other killer whale community even though they are relatively well studied (for cetaceans) wherever they are accessible (Cantor et al., this volume).

How does such behavioral diversity emerge? There are three main drivers that are responsible for shaping behavioral development and variation (Boyd and Richerson 1996). First, environmental factors can influence what behaviors individuals acquire and express, for example through differences in resource availability. Individuals living in similar habitats are then expected to behave in similar ways because they are exposed to the same environmental and learning conditions. Second, individuals can be genetically predisposed to behave in certain ways. Finally, individuals can acquire particular behaviors through social learning of innovations that might differ among social units. Importantly, these drivers do not act in isolation but can interact and together shape the behavior that individuals acquire and express (Bateson and Martin 2000).

Where does culture come into this? Culture is (broadly) defined as shared behavior within a community that is acquired through some form of social learning (Boyd and Richerson 1996; Whitehead and Rendell 2014). Individuals may be able to innovate novel behavior, for example by exploiting a novel resource, but the important part for the behavior to become cultural, is that it is then transmitted socially to other individuals in its social unit.

A Historical Perspective: The Ethnographic Method

How can we attribute observed patterns of behavioral variation to culture? How do we know whether a behavior is transmitted socially, rather than driven by environmental or genetic factors? Can we attempt to quantify the relative contributions of these factors? Various analytical methods for identifying culture in animal populations have been proposed, discussed, refined, or dismissed over the past 20 years. While they differ in their underlying mechanisms, their common goal is to identify whether observed behavioral differences are—at least in part—due to social learning. Here, we provide a historical perspective on different methods that have been used to identify cetacean foraging cultures, discussing their strengths and weaknesses.

Early claims of culture were often based on the "ethnographic" approach, also called the "method of exclusion." The ethnographic method identifies patterns of variation in the behavioral repertoires between social units and infers social transmission as at least partially responsible by excluding genetic and ecological factors as sufficient explanations for generating behavioral differences among social units. While the roots of the ethnographic method originally lie in studies investigating cultural variation in humans, the method soon became popular among primatologists (Van Schaik et al. 2003; Whiten et al. 1999), before being adopted for studying cultural behavior in other taxa. In the cetacean world, the ethnographic method was first used to assess patterns of transmission of "sponging," a foraging behavior involving tool use observed in a population of bottlenose dolphins (*Tursiops aduncus*) in Shark Bay, Western Australia. During sponging, dolphins carry a conically shaped marine sponge on their rostrum, apparently to serve as protection from abrasions while probing the substrate for buried prey (Mann et al. 2008; fig. 11.1, left). Sponging is a female-biased activity, and most sponging dolphins belong to the same maternal lineage, i.e., they share the same mitochondrial haplotype (Mann et al. 2012; Kopps et al. 2014). The behavior is also spatially restricted, occurring almost exclusively in deep-water channels where sponges are more abundant (Tyne et al. 2012). Using the ethnographic approach Krützen et al. (2005) considered 10 different pathways of autoso-

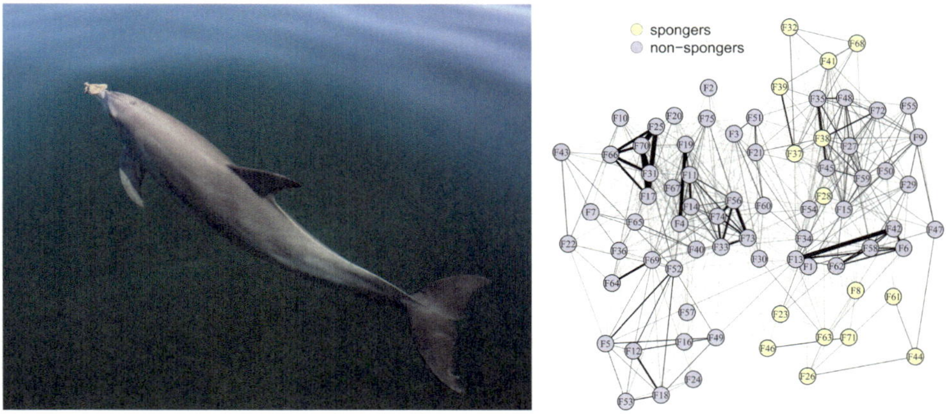

FIGURE 11.1. (*Left*) A bottlenose dolphin carrying a sponge. (*Right*) Social network color coded by foraging strategy (spongers in yellow versus non-spongers in light purple). Each node represents an individual, with edge thickness depicting dyadic association strengths.
Sources: (*Left*) Photo courtesy Sonja Wild, Shark Bay Dolphin Research. (*Right*) Marfurt et al. 2022.

mal, X-linked, or mtDNA-linked genetic inheritance of sponging, which were all inconsistent with the observed patterns of sponging. Furthermore, they argued that environmental factors were unlikely to be solely responsible for the transmission of sponging, since both spongers and non-spongers foraged in the deep-water channels, and that therefore, cultural transmission between mother and primarily female offspring was the only logical explanation for generating the observed patterns. However, around this time the ethnographic method was increasingly criticized on two broad fronts (Laland and Janik 2006, 2007). First, it was argued that it is almost impossible to consider all potential genetic and environmental factors and that this could lead to false claims of culture. Second, while the exclusion of genetic and environmental factors as the sole drivers of behavioral variation can point toward social learning as being causal, these three drivers also can simultaneously shape the observed behavioral variation. If this leads to correlations with genetic and environmental factors, then important social components can be missed.

Methodological Advances in Identifying Cultural Transmission in Wild Cetaceans

Following debates on the utility of the ethnographic approach (Laland and Janik 2007; Krützen et al. 2007), more quantitative methods were developed that allowed a more nuanced approach in disentangling relative contributions of the three drivers of behavioral development. While the underlying mechanics differ, these approaches rest on the assumption that if behaviors are socially transmitted, then more closely associated individuals should share similar behavioral repertoires or be more likely to pass on a specific behavior (Coussi-Korbel and Fragaszy 1995).

Repertoire-based methods—often used when only snapshots of data on behavioral repertoires are available—compare scores of behavioral similarity among individuals with measures of association strengths, genetic relatedness, or similarities in habitat use. For example, in a study investigating the transmission mechanisms of bottlenose dolphins learning to accept food handouts from recreational fishers in Western Australia, Donaldson et al. (2012) used a generalized linear model framework to demonstrate that dolphins spending more time in areas of high boat density and in association with previously conditioned individuals were more likely to start accepting food handouts themselves. These findings indicate that both social learning and environmental factors—here habitat use—were partially responsible for the spread of this behavior. Perhaps the biggest drawback of such an approach is that by averaging dyadic association across multiple individuals, data on individual connections is lost and the structural autocorrelation that is inherent to social networks is not taken into account. Matrix correlations involving permutations such as multiple regression quadratic assignment procedure (MRQAP; Dekker et al. 2007) can represent a suitable alternative in that case. They allow the inclusion of multiple matrices while accounting for structural autocorrelation in the social network. In the context of cetacean foraging cultures, matrix correlations were used to examine whether sponging dolphins in Shark Bay were culturally distinct from non-sponging individuals, by determining social preferences among dolphins while controlling for location, sex, and matrilineal relatedness (Mann et al. 2012). Female spongers associated more strongly with other female spongers belonging to the same maternal lineage and with those using similar areas, consistent with homophilic tendencies of preferably associating with those engaging in similar behavior (Marfurt et al. 2022; fig. 11.1, right). One challenge that remains with repertoire-based approaches is that it can be difficult to disentangle whether associated individuals behave similarly because they learned the behavior from one another, or because they chose to associate with individuals that behave in similar ways due to homophily.

The increased availability of long-term data has enabled us to take a different approach for identifying culture by tracking the spread of novel behavior through space and time. This has fueled the development of novel statistical methods such as NBDA that infers social learning if the diffusion is predicted by the connections within the social network and estimates the

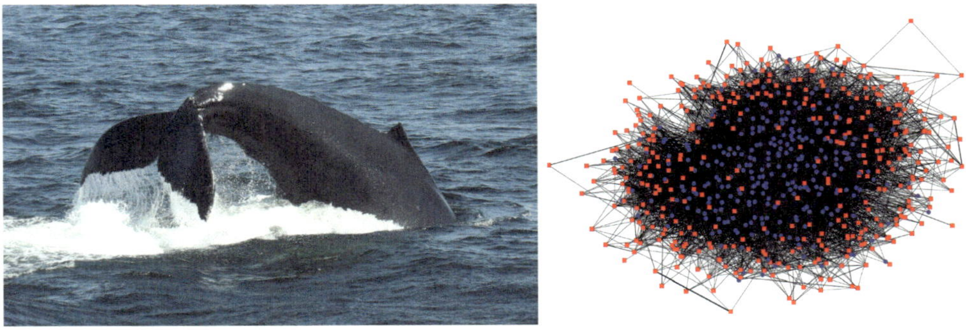

FIGURE 11.2. (*Left*) Lobtail-feeding humpback whale. (*Right*) Social network of whales sighted at least 20 times. Blue circular nodes are individuals observed lobtail feeding; red square nodes are those never observed lobtail feeding. The network was laid out by spring embedding using NetDraw software, which draws well-connected nodes into the center of the network. Note how well-connected individuals are more likely to exhibit the lobtail-feeding behavior.
Sources: (*Left*) Photo courtesy Jennifer Allen, Ocean Alliance. (*Right*) Allen et al. 2013.

strength of social transmission based on how closely the spread follows the social network (Franz and Nunn 2009; Hoppitt et al. 2010). NBDA additionally allows the inclusion of individual-level variables that can influence the social or asocial learning rate, such as sex, age, or rank (Hasenjager et al. 2020b). In the field of cetacean foraging cultures, NBDA was used for the first time to investigate transmission patterns of "lobtail feeding" in a population of humpback whales (*Megaptera novaeangliae*) in the Gulf of Maine, USA, a feeding technique used to target sand lance (Allen et al. 2013). During lobtail feeding, whales strike the water's surface with the ventral side of their flukes one to four times, before engaging in so-called bubble feeding, which consists of producing bubbles underneath and around prey followed by a lunge through those bubbles (Wiley et al. 2011; fig. 11.2, left). In their study, Allen et al. (2013) tracked the spread of lobtail feeding through the social network over a period of 27 years, with NBDA models indicating that up to 87% of whales that acquired lobtail feeding did so by social transmission from their associates (fig. 11.2, right). Environmental factors also seemed to play an important role in the acquisition of the behavior. The biomass of sand lance had a positive effect on the whales' learning rates, and whales that spent proportionally more time within the Stellwagen sanctuary area—where sand lance gather for spawning—were more likely to learn the behavior (Allen et al. 2013). In contrast, having a lobtail-feeding mother had no influence on a whale's learning rate, which is inconsistent with a strong role for genetic factors or vertical learning. The authors therefore concluded that lobtail feeding was a horizontally transmitted foraging culture.

The use of NBDA to identify and quantify cultural transmission in various animal taxa has increased dramatically over the past decade (e.g., honeybees, *Apis mellifera* [Hasenjager et al. 2020a]), and several extensions of the method have been developed that allow the inclusion of dynamic networks (Hobaiter et al. 2014), of spatial locations as diffusion data (Nightingale et al. 2015; Klump et al. 2021), or multiple networks that allow the disentangling of different pathways of transmission (Farine et al. 2015). The latter approach was used to assess pathways of transmission of two tool-using behaviors in the Shark Bay bottlenose dolphin population in Western Australia. We have previously encountered the use of sponges as foraging tools as an example of a foraging culture identified by the ethnographic method. Using a more quantitative approach, Wild et al. (2019a) tracked the spread of sponging over 12 years and included four different networks in a multi-network NBDA to assess pathways of transmission of sponging: a network with home range overlaps as a proxy for the influence of environmental factors, a network with dyadic relatedness to assess the importance of a genetic predisposition, a mother-offspring network to test for vertical social learning, and finally, a network reflecting dyadic association strength to test for horizontal learning among associates. The spread of sponging was best predicted by the mother-offspring network, indicative of a strong role of vertical social transmission (Wild et al. 2019a), which corroborates previous results (Krützen et al. 2005). A similar approach using multi-network NBDA was used to assess transmission mechanisms of "shelling," a second foraging strategy performed by a subset of individuals within the Shark Bay dolphin population. During shelling, dolphins chase prey into empty shells of giant gastropods, bring the shell to the water's surface and shake it above the surface to drain the water—and with it the prey—out of the shell (Allen et al. 2011). Multi-network NBDA revealed that shelling spreads through non-vertical learning among associates with no apparent influence of environmental or genetic factors (Wild et al. 2020), which stands in contrast to the usually strong reliance on vertical social learning between mother and offspring to acquire foraging behavior in this population (Wild et al. 2019a; Mann and Sargeant 2003; Sargeant et al. 2005), and, in fact, in toothed whales in general (Guinet and Bouvier 1995; Whitehead 2017; Simões-Lopes et al. 2016; Whitehead and Rendell 2014).

In summary, statistical methods for identifying cultural factors in foraging behavior have evolved considerably over the past decades (and continue to do so), allowing for a more refined picture of the importance of foraging cultures in cetaceans and the different pathways of transmission. There is no single best method for identifying cultural behavior, and researchers should carefully weigh the pros and cons of different methods based on their questions, their available data, and the logistic and ethical considerations that also shape research activity.

Foraging Cultures in Toothed and Baleen Whales—an Overview

In this section, we provide an overview of nearly two decades of research on foraging cultures in cetaceans, virtually all of which have appeared since the publication of the original *Cetacean Societies*. Despite the methodological advances, it remains the case that the best-studied species are those four that the original book focused on, the killer, sperm, and humpback whales, and the bottlenose dolphin. We therefore organize our overview of foraging cultures in cetacean societies along those same lines. We then look to the future with a survey of recent behavioral observations that are also indicative of areas where further research could reveal significant social learning components to cetacean foraging.

Bottlenose dolphins show a broad diversity of foraging strategies that reflect the diversity of their prey, ranging from fish and squid to crustaceans and more. These include both individual hunting behaviors, such as "kerplunking," which involves fluke-slaps when foraging in shallow water (Connor et al. 2000), preparing cuttlefish (Finn et al. 2009), or beach hunting (Sargeant et al. 2005), as well as strategies that require cooperation with conspecifics (Ramos et al. 2022) or humans (Cantor et al. 2023; see King et al., this volume). Many of these behaviors are most likely cultural, based on their distribution in time and space in relation to social and genetic population structure (Krützen et al. 2005), as well as well-established social learning and cooperation abilities (King et al. 2021; Jaakkola et al. 2013, 2018). Yet we are a long way away from grasping the actual breadth of different strategies and the role of culture in their transmission, since most of our information stems from a few coastal communities of bottlenose dolphins, while information on the foraging behavior of communities inhabiting rivers and offshore waters around the globe remains scarce.

What we do know from the coastal populations of bottlenose dolphins is that they often live in well delineated communities that are separated from others through geographic barriers, and have become adapted to their particular foraging niches, with cultural transmission having an important role in this adaptation. The innovation and cultural transmission of feeding tactics that allow the occupation of these coastal niches have had apparent coevolutionary effects in defining coastal ecotypes as they separate from ancestral oceanic populations (Louis et al. 2021, 2023). Long-term studies such as that on the Shark Bay dolphin population suggest the existence of several distinct cultural foraging lineages within the same population, with different maternal lineages typically engaging in a specific set of foraging behaviors that target different foraging niches (Mann and Sargeant 2003). How do such matrilineal foraging cultures emerge? Calves being exposed to all of their mothers' activities during infancy is likely a crucial factor in their acquisition of specific foraging tactics through vertical social learning. We also know, however, that horizontal transmission can occur in dolphin populations (Bossley et al. 2018; Wild et al. 2020), and the interplay between learning in infancy and adulthood remains to be fully understood.

Like bottlenose dolphins, killer whales show a diversity of foraging strategies globally, ranging from intentional stranding (Lopez and Lopez 1985), creating waves to wash prey off ice floes (Pitman and Durban 2012) and depredating longlines (Visser 2000). These specializations have led to the emergence of "ecotypes" of killer whales that differ in appearance, habitat, prey choice, and hunting strategies (Ford 2009). For example, the fish-eating communities off the northwest coast of North America, termed "residents" consume salmon, with a prey base that can be dominated by a single species, while mammal eaters consume marine mammals (Ford et al. 1998). So reliable are these distinctions that mammal prey species have learned to recognize the non-threatening fish-eaters and moderate their flight response as a result (Deecke et al. 2002). How do such profoundly different diets and hunting behaviors evolve? First, killer whales can be picky eaters. Individuals that are brought into captivity can show great unwillingness to try unfamiliar food (Ford and Ellis 1999; Riesch et al. 2012). Second, they are culturally conservative. While demonstratively capable of behavioral innovations and social transmission of behaviors (Esteban et al. 2022), and even imitation (Abramson et al. 2013), young individuals rely almost exclusively on their mothers as sources for social information, and behaviors are shared within their social unit of closely (maternally) related individuals. So profoundly do these cultural dietary specializations affect the animals' lives that genetic selection pressures themselves can be changed, for example, with respect to dietary metabolism (Foote et al. 2016).

Sperm whale foraging takes place typically at hundreds of meters of depth and so it is unsurprisingly the least known, with behavior largely inferred from biologging techniques using tags that log depth, movement, acoustics, and a range of other environmental features. Unfortunately, these attachments typically do not last more than 48 hours, so unless researchers can reliably tag the same individuals, they are not sufficient for building up a robust long-term picture of an individual's behavior over time. Invasive satellite tags can produce longer-term position data, but little more currently. Intermediate in duration between these two are focal follow techniques where groups can be followed at sea for periods of a week or more, collecting position, acoustic, genetic, and fine-scale association data (Whitehead 2003). During such follows, foraging success can be indexed from surface defecation. This approach has found that different sperm whale vocal clans (Hersh et al. 2022) show variation in their foraging success that can be linked to differences in movement patterns during foraging. These findings suggest that clans also have specific "explore and exploit" strategies at intermediate scales (movements on the scale of days to weeks). We also have well-documented evidence of cultural diffusion from the increase in spatial extent over time of a specific learned behavior in which sperm whales take fish from longlines (Hanselman et al. 2018; Schakner et al. 2014, 2016).

The role of culture in baleen whale foraging is less well understood than in odontocetes, with the notable exception of lobtail feeding. While migration behavior to foraging grounds is thought to be influenced by vertical transmission of routes from mother to calf (Baker et al. 2013), the specific tactics on feeding grounds often involve individual foraging, with little behavioral specialization required to gather the abundant small prey that are the main foraging target. Nonetheless, there are a number of recent studies that suggest an important role for social transmission in large-scale behavioral decision-making (e.g., Iwata et al. 2017). It is here also worth surveying recent behavioral reports on cetaceans with a view to the future of this area of research. As we shall see, there are several examples of behavioral observations suggestive of local adaptation, and sometimes sequence learning, and others that are so social in their exhibition that it is hard to see how some form of social learning would not be involved.

For example, bottlenose dolphins gather during cuttlefish spawning season in the Australian Gulf of Spencer and use a sequence of processing tactics to optimize their nutrition (Finn et al. 2009). The prey cuttlefish is first shaken violently to kill it and to open its ink sac, since the ink contains compounds that are noxious to digestion; it is then inverted and rubbed on the seafloor to strip off the thin skin covering the dorsal side of the cuttlebone. This is done until the cuttlebone, of little nutritional value and a physical irritant, separates from the flesh, which is then consumed. The tactic shows adaptation in movement to arrive at the right time, and a relatively complex processing sequence. Furthermore, it is one of three processing methods that have now been documented for this prey in three different populations, with dolphins in the Sado estuary of Portugal just taking the cuttlefish head, and those off Bunbury, southwest Australia, using a similar sequence as those in the Gulf of

Spencer, but with robustly population-specific sequence idiosyncrasies such as method of carrying (Smith and Sprogis 2016). This is comparable to the variation in foraging tactics this dolphin genus uses to catch various species of mullet (family Mugilidae)—off Laguna in Brazil they cooperate with fishers (Cantor et al. 2023), in other areas they use cooperative mud-ring feeding (Ramos et al. 2022; Engleby and Powell 2019), and in yet others they use less cooperative variants of intentional beaching tactics (Sargeant et al. 2005; Fox and Young 2012; McGovern et al. 2020; Samuelson et al. 2020; Silber and Fertl 1995; Jiménez and Alava 2015). The causal factors behind these variations remain unknown, but the variation within and between communities is highly suggestive of a role for cultural transmission.

These behaviors are ripe for further research, and we suspect they will continue to reveal significant social learning components. New observations of previously undescribed foraging behaviors continue to accumulate—for example, in the northern Gulf of Mexico, bottlenose dolphins have been observed attempting to take large catfish, and it is notable that the outcome is still sometimes in doubt since they are not always successful (Ronje et al. 2017). There are associated risks of getting pieces of large prey stuck in the throat or esophagus, which suggests that this behavior is not fully optimized, compared to, for example, foraging on cuttlefish. Eating poorly processed or unsuitable food when exploring new resources can be fatal (IJsseldijk et al. 2015). Thus, we see the double-edged sword of culture—it allows for rapid adaptation to new resources, especially those involving specific tactics to capture and process, but it also carries risks in exploration and specialization, which we shall return to later.

In two baleen whale species, there have been recent reports of a technique referred to as "trap feeding" (also referred to as "tread-water feeding"), involving a single animal suspending vertically in the water with the jaw open 90 degrees and just at the surface of the water (fig. 11.3). The approach of the drifting whale panics prey fish into leaping, often into the agape jaw of the whale. This has been reported in Bryde's whales (*Balaenoptera brydei*) in the Gulf of Thailand (Iwata et al. 2017) and, through apparently independent inventions, in humpback whales off northeastern Vancouver Island, Canada (McMillan et al. 2019). Compared to the high-energy lunge-feeding seen in rorquals this is a relaxed pose, held for up to 30 seconds, which is much cheaper in energy terms. It is possible that in the Gulf of Thailand the low oxygen nature of the waters this behavior is seen in concentrates prey species at the surface, making this a behavioral adaptation to local ecology (Iwata et al. 2017). They also note that mothers with calves regularly use this tactic, presenting social learning opportunities, and suggest it is a cultural pattern.

FIGURE 11.3. Trap or tread-water feeding Bryde's whale. (A) The whale hangs vertically at the surface and opens its mouth so the lower jaw drops to the surface and also forms channels into the mouth. (B) Once sufficient prey have entered the "trap," the whale closes its mouth to engulf them.
Source: Photos courtesy of Takashi Iwata.

McMillan et al. (2019) described the tactic being used to feed on herring and suggest trap feeding by humpbacks is a local, learned, feeding technique because the researchers observed its innovation. Two animals were seen in the focal population doing this in the first year of study and by 2015, 16 out of a known local population of 55 individuals were known to have performed trap feeding. The foraging tactic showed a strong growth in use during the study on a trajectory measured through the social network in a similar way to the lobtail-feeding study previously described.

In a study that illustrates the revolutionary role that combining new technologies is playing in increasing our understanding of cetacean foraging, Cade et al. (2021) reported on the presence and dynamics of aggregations of whales they termed "supergroups" because of their unusually large nature. A supergroup of blue whales (*B. musculus*) consists of up to 40 individuals visible from a boat at sea level, so all within about 5 km. By combining biologging with passive acoustics and remote sensing satellite data, the researchers found that the establishment of these supergroups was related to the discovery of large but ephemeral prey patches and also associated with the whales making a lot of a specific call type, the "D" call, not a song element. They produced models informed by these data that showed the pace of formation of whale aggregations could not simply be explained by the whales encountering a given prey patch—instead, they suggest the calls cue whales in the immediate vicinity to the presence of the patch. The prey patches were also sufficiently large that there was no cost to attracting others (and indeed, the presence of multiple whales foraging on the same patch may be advantageous if it confuses prey escape responses). In this case the calls themselves might or might not be culturally acquired, but their use acts to rapidly and efficiently spread ephemeral social information about prey patches; often these patches were associated with known topographic features such as canyons, so over time, knowledge of these locations is also spread. A further example of how horizontal transmission of social information might work in interesting ways is emerging in blue whales. Recent research has shown that the timing of a measurable shift in blue whale singing behavior—they start singing more in the day as well as at night—reflects seasonal changes in the ecology of feeding grounds off the Californian coast (Oestreich et al. 2020). Each year, the later these waters were productive the later the shift to singing in the day occurred, and the later the whales left. The change in singing was always a clear predictor of the whales' commencing migration to breeding grounds each year, so possibly this is a mechanism for sharing ecologically relevant information on a large spatial scale.

Cooperation, in the sense of individuals working together for the common benefit, is a feature of many cetacean foraging tactics that are good foci for ongoing research on the role of culture. These include such intricate specializations as killer whales moving prey-bearing ice floes into convenient positions to then cooperate to generate waves that spill the pinniped prey into the water (Pitman and Durban 2012b). The most famous example of this is the sequence shown in episode 4 of the BBC documentary series *Frozen Planet II*, "Frozen South," which first aired in 2022. Alternative cooperative tactics are shown by herring-feeding killer whales in the waters off Norway and Iceland, where the specific tactics involved have been shown to be tuned to the specific life stages of the herring prey as they progress over the season (Samarra and Miller 2015). This "carousel feeding," which involves cooperatively herding prey and then using tail slaps to stun them occurs at depth or at the surface depending on the prey life stage. The whales must decide whether to herd a prey school to the surface before feeding, and the smaller school sizes of feeding and spawning herring, as opposed to over-wintering fish, likely makes it less energetically efficient to do so, so they forage at depth. Bottlenose dolphins also engage in cooperative foraging, for example in the form of mud-ring feeding where groups cooperate to herd fish into mud plume "nets" formed by a single animal beating the seabed with their tail, showing clear division of labor (Gazda et al. 2005; fig. 11.4). The most recent observations describe calf exposure to, and apparent juvenile practicing of, the mud-ring creation behavior, suggesting again a role for social learning (Engleby and Powell 2019). The relative experimental tractability of this genus has led to studies aimed at revealing the cognitive abilities underlying these behaviors in the wild (see King et al., this volume)—in one study, Jaakkola et al. (2018) showed how pairs of dolphins in human care could achieve

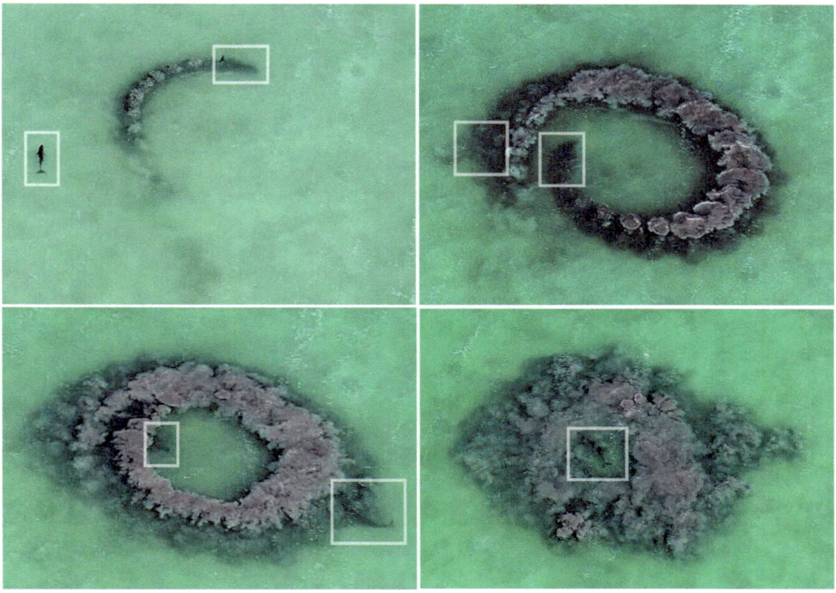

FIGURE 11.4. Bottlenose dolphin mud-ring feeding. The sequence of panels shows the progression of the behavior as a single dolphin produces a mud ring by beating their tail on the seabed, which eventually comes to encircle a shoal of fish, whereupon other dolphins in the group join to take the trapped fish. Gray boxes highlight individual dolphins.
Source: Photos courtesy of Eric Angel Ramos (see Ramos et al. 2021).

synchrony within 350 ms while undertaking a novel coordination task. It is difficult to conceive how some form of social learning could not be involved in such coordination when faced with novel behavioral challenges.

Another example is the long-known but little-studied cooperative group foraging shown by humpbacks in the waters off Alaska and northern British Columbia each summer (Wray et al. 2021). While humpbacks can often use bubble nets to encircle prey schools by themselves, they also work together in certain areas. The forces pushing this social version of bubble-net feeding are not fully known but may relate to prey patchiness and abundance. Nonetheless, it has been shown that these whales repeatedly find each other and forage together, with some relationships documented over a 12-year period and the strongest associations observed between bubble-net feeders rather than, say, females in similar reproductive phases. The social nature of this foraging technique strongly suggests social learning will play some role in its development.

All these behaviors are clearly ripe for further investigation. How are they organized? How are they distributed within the regional population—is it just a few individuals, groups, or an entire community or ecotype? How do the behaviors affect the fitness of the individuals involved? Are there cooperative rules to the behaviors? How do young animals begin to be involved in the behavior? All these outstanding questions are pertinent to understanding the role of social learning in these cooperative foraging tactics.

The last aspect of social foraging we briefly touch on are local cooperative foraging traditions known in bottlenose dolphins and killer whales. These human interactions are covered in more detail in Cantor et al. (this volume) but are still relevant in any consideration of cultural foraging tactics. In killer whales, we do not know of any such extant behavior, but it is known by historical account that one community did cooperate with human whalers in Twofold Bay, Australia, in attacking, tiring, and killing humpbacks as they migrated past the bay (Clode 2002; Reeves et al. 2023). In bottlenose dolphins however, members of a population that cooperate with fishers in Laguna, Brazil, are becoming increasingly famous. This interaction has been comprehensively shown to benefit both fishers and the dolphins and to be based on a complex system of mutual signaling (Cantor et al. 2023). The localized nature of this interaction, as well as the social and communicative aspects of its performance, which are unlikely to be genetically evolved responses, undoubtedly persists through cultural transmission on the human side, and has to be considered a prime candidate for a significant cultural component on the dolphin side also.

Finally, we can also hope that this research theme will continue to expand beyond the original "fab four" species from *Cetacean Societies*. We can highlight recently reported variation in foraging tactics between narwhal (*Monodon monoceros*) populations as a prime example of the kind of research that will achieve this going forward. Three North Atlantic populations from Baffin Bay, east Greenland, and northern Hudson Bay apparently exploit different depth layers when foraging, and it would be an interesting study to understand whether this is local ecological adaptation and whether it is manifested by cultural means or not (Watt et al. 2015).

Culture and the Interplay Between Human Activity and Cetacean Foraging

While there are more detailed treatments of human-cetacean interactions elsewhere in this volume (Cantor et al.), we briefly discuss here the way that these interactions can be impacted by the kinds of cultural transmission we have been focused on. Human activity can and has created novel foraging opportunities for several cetacean species, with multiple delphinid species learning how to remove fish efficiently from various forms of fishing gear

(Bearzi and Reeves 2022). In many cases, the spread in space and time of these tactics is entirely consistent with a cultural transmission account, with identified sources and documented spread (Weinpress-Galipeau et al. 2021; Amelot et al. 2022; Schakner et al. 2014). Other cases show idiosyncratic group-specific behaviors, including cooperating with humans in specific contexts (e.g., Reeves et al. 2023) consistent with local cultures. In some cases, this is known to have persisted across multiple generations of humans and non-humans, almost certainly by cultural transmission in both species involved (Cantor et al. 2023). The relevance of innovation and cultural transmission in promoting the discovery and passing on of these opportunities within cetacean societies is what we note here—the interactions can be positive or negative (Cantor et al., this volume), but their invention and rapid spread is often down to culture.

Consequences of Foraging Culture

The study of animal culture is increasingly reaching beyond merely identifying socially learned behavioral differences among groups to include investigations into the far-reaching consequences of cultural behavior on a range of ecological, biological, and evolutionary processes (Cantor et al., this volume) and its implications for biodiversity conservation (Brakes and Whitehead, this volume). Examples of how culture impacts these processes are manifold, but here we focus on foraging in particular.

First, foraging cultures are often related to behavioral specialization, which in the foraging domain refers to the narrowing of a foraging niche and a loss of capacity to use alternative food resources, a process that is often driven and maintained through social transmission within a group (Whitehead and Ford 2018). Naturally, foraging specializations are tightly linked to predator-prey dynamics, with prey availability shaping selection pressures on different culturally learned behaviors. Thereby, social units differing in their repertoires of culturally transmitted behaviors may experience different selection pressures (Brakes et al. 2021; Brakes and Whitehead, this volume). For example, following an unprecedented marine heatwave that caused a mass die-off of the habitat-forming seagrass in Shark Bay, the resident bottlenose dolphins experienced significant long-term decline in both reproductive rates and survival (Wild et al. 2019b). However, the survival of sponging dolphins was less severely affected by the heatwave compared to non-sponging dolphins, suggesting that the cultural knowledge of sponge tool use allowed them access to a foraging niche that escaped the adverse effects of the heatwave, at least to an extent (Wild, et al. 2019b). But specializations are risky, since they are vulnerable to fluctuations in a single resource (Whitehead and Ford 2018) and culturally driven foraging specialization can lead to so-called evolutionary traps if a foraging niche experiences significant reductions in yield or even depletion, and the formerly well-adapted behavioral trait targeting the niche becomes obsolete or even maladaptive (Schlaepfer et al. 2002).

Second, culture is tightly linked to sociality and can shape and be shaped by demographic processes (Cantor and Whitehead 2013). For example, a population of bottlenose dolphins in Moreton Bay, Australia, were found to have formed two distinct communities based on whether individuals engaged in following commercial prawn trawlers for foraging or not (Chilvers and Corkeron 2001). After trawling efforts had been substantially reduced in the area, the partitioning into the two communities disappeared and "trawler" and "non-trawler" dolphins were found to associate and interact with each other (Ansmann et al. 2012). In the extreme case, culturally driven foraging specializations combined with strong homophilic tendencies can even lead to the formation of so-called ecotypes and reproductive isolation between culturally different social units and could ultimately potentially result in speciation. The best-studied example of cetacean ecotypes with overlapping habitat—but which differ in appearance, diet, behavior, and largely no longer interbreed with each other—are the killer whale ecotypes (Riesch et al. 2012).

The interplay between culturally driven foraging specializations and predator-prey dynamics is not a one-way street. Just like prey availability can tip the balance toward preservation or extirpation of a foraging specialization, cultural behavior in the predator can shape biological and evolutionary processes in prey. For example, some groups of fish-eating killer whales in the northeast Pacific Ocean (northern residents) who specifically target Chinook salmon (*Oncorhynchus tshawytscha*) as their preferred prey species, have increased in abundance in the past few decades. At the same time, Chinook salmon experienced a shift toward faster growth and earlier maturation. Simulations on the predator-prey dynamics suggested that the selective targeting of large fish by the killer whales may be one of the causes of this evolutionary shift in life history in their prey, eventually resulting in smaller overall body sizes (Ohlberger et al. 2019).

Finally, cultural processes are important determinants of a species' capacity to respond to environmental changes, as learning is one of the key components of behavioral plasticity (Whitehead 2010). For example, reliance on vertical learning typically found in the matrilineal societies of sperm whales and killer whales (Whitehead and Rendell 2014), leads to behavioral conservatism with slow-to-change behaviors, which may impede behavioral adaptations (Whitehead et al. 2004). Meanwhile, a capacity for horizontal learning allows for a much more rapid spread of novel adaptive behavior (Whitehead et al. 2004). In recognition of the consequences of cultural knowledge on survival, and the importance of learning processes for a species' capacity to adapt to environmental changes, efforts are being made to include animal culture in biodiversity conservation, and—in an attempt to improve resilience—to not only conserve genetic but also cultural diversity (Brakes et al. 2019, 2021; Brakes and Whitehead, this volume).

Conclusions and Outlook

There have been significant advancements in the study of cetacean foraging cultures in the past two decades. Since the early cautious claims of putative cultural behavior, it has now been firmly established that many nonhuman animals—including cetaceans—possess culture. Beyond merely identifying socially transmitted behavior, studies now increasingly include investigations into the multitude of consequences cultural processes have for societies, their environment, and conservation, which has also contributed to broader debates about animal rights (Marino and White 2022). Yet, there is still work to do. As field sampling methods evolve and higher-resolution data on individuals' behaviors, genomics, and health conditions become available, more integrative approaches can provide insights into the mechanisms underlying the emergence and maintenance of cultural traits, as well as the fitness consequences for their bearers. In combination with models of forecasted environmental changes such as climate-driven shifts in resource availability, we may be able to better predict the resilience or vulnerability of specific cultural units. Higher-resolution and more multifaceted data also require the development of appropriate analytical approaches. For example, we envisage analytical tools for better integration of behavioral, genomic, and environmental data into a social network analysis framework, e.g., through multilayer network approaches (Silk et al. 2018; Finn et al. 2019), or more readily available tools for dealing with missing data, which—given the logistics of studying cetaceans—will always remain a challenge.

Acknowledgments

We thank Jenny Allen, Eric Angel Ramos and Takashi Iwata for providing photographs for the figures. We also thank Ellen Garland and Andy Foote for helpful feedback on earlier drafts.

References

Abramson, José Z., Victoria Hernández-Lloreda, Josep Call, and Fernando Colmenares. 2013. "Experimental Evidence for Action Imitation in Killer Whales (*Orcinus orca*)." *Animal Cognition* 16 (January): 11–22. https://doi.org/10.1007/s10071-012-0546-2.

Allen, Jenny A., Mason T. Weinrich, William Hoppitt, and Luke Rendell. 2013. "Network-Based Diffusion Analysis Reveals Cultural Transmission of Lobtail Feeding in Humpback Whales." *Science* 340 (6131): 485–88. https://doi.org/10.1126/science.1231976.

Allen, Simon J., Lars Bejder, and Michael Krützen. 2011. "Why Do Indo-Pacific Bottlenose Dolphins (*Tursiops* sp.) Carry Conch Shells (*Turbinella* sp.) in Shark Bay, Western Australia?" *Marine Mammal Science* 27 (2): 449–54. https://doi.org/10.1111/j.1748-7692.2010.00409.x.

Amelot, Morgane, Floriane Plard, Christophe Guinet, John P. Y. Arnould, Nicolas Gasco, and Paul Tixier. 2022. "Increasing Numbers of Killer Whale Individuals Use Fisheries as Feeding Opportunities Within Subantarctic Populations." *Biology Letters* 18 (2). https://doi.org/10.1098/rsbl.2021.0328.

Ansmann, Ina C., Guido J. Parra, B. Louise Chilvers, and Janet M. Lanyon. 2012. "Dolphins Restructure Social System After Reduction of Commercial Fisheries." *Animal Behaviour* 84 (3): 575–81. https://doi.org/10.1016/j.anbehav.2012.06.009.

Baker, C. Scott, Debbie Steel, John Calambokidis, et al. 2013. "Strong Maternal Fidelity and Natal Philopatry Shape Genetic Structure in North Pacific Humpback Whales." *Marine Ecology Progress Series* 494 (December): 291–306. https://doi.org/10.3354/meps10508.

Bateson, Paul P. G., and Paul R. Martin. 2000. *Design for a Life: How Behavior and Personality Develop*. Simon & Schuster.

Bearzi, Giovanni, and Randall R. Reeves. 2022. "Marine Mammals Foraging Around Fishing Gear or Preying upon Fishing Catch and Bait: It May Not Be 'Depredation.'" *ICES Journal of Marine Science* 79 (8): 2178–83. https://doi.org/10.1093/icesjms/fsac173.

Bossley, Mike, Aude Steiner, Philippa Brakes, Juliet Shrimpton, Charlotte Foster, and Luke Rendell. 2018. "Tail Walking in a Bottlenose Dolphin Community: The Rise and Fall of an Arbitrary Cultural 'Fad.'" *Biology Letters* 14 (20180314). https://doi.org/10.1098/rsbl.2018.0314.

Boyd, Robert, and Peter J. Richerson. 1996. "Why Culture Is Common, but Cultural Evolution Is Rare." *Proceedings of the British Academy* 88: 77–93.

Brakes, Philippa, Emma L. Carroll, Sasha R. X. Dall, et al. 2021. "A Deepening Understanding of Animal Culture Suggests Lessons for Conservation." *Proceedings of the Royal Society B* 288 (1949): 20202718. https://doi.org/10.1098/rspb.2020.2718.

Brakes, Philippa, Sasha R. X. Dall, Lucy M. Aplin, et al. 2019. "Animal Cultures Matter for Conservation." *Science* 363 (6431): 1032–34. https://doi.org/10.1126/science.aaw3557.

Cade, David E., James A. Fahlbusch, William K. Oestreich, et al. 2021. "Social Exploitation of Extensive, Ephemeral, Environmentally Controlled Prey Patches by Supergroups of Rorqual Whales." *Animal Behaviour* 182 (December): 251–66. https://doi.org/10.1016/j.anbehav.2021.09.013.

Cantor, Mauricio, Damien. R. Farine, and Fábio. G. Daura-Jorge. 2023. "Foraging Synchrony Drives Resilience in Human-Dolphin Mutualism." *Proceedings of the National Academy of Sciences* 120 (6): e2207739120. https://doi.org/10.1073/pnas.2207739120.

Cantor, Mauricio, and Hal Whitehead. 2013. "The Interplay Between Social Networks and Culture: Theoretically and Among Whales and Dolphins." *Philosophical Transactions of the Royal Society B* 368 (20120340). https://doi.org/10.1098/rstb.2012.0340.

Chilvers, B. Louise, and Peter J. Corkeron. 2001. "Trawling and Bottlenose Dolphins' Social Structure." *Proceedings of the Royal Society B* 268: 1901–5. https://doi.org/10.1098/rspb.2001.1732.

Clode, Danielle. 2002. *Killers in Eden: The True Story of Killer Whales and Their Remarkable Partnership with the Whalers of Twofold Bay*. Allen and Unwin.

Connor, Richard. 2001. "Individual Foraging Specializations in Marine Mammals: Culture and Ecology." *Behavioural and Brain Sciences* 24 (2): 329–30. https://doi.org/10.1017/S0140525X01283964.

Connor, Richard C., Michael R. Heithaus, Per Berggren, and Jennifer L. Miksis. 2000. "'Kerplunking': Surface Fluke-Splashes During Shallow-Water Bottom Foraging by Bottlenose Dolphins." *Marine Mammal Science* 16 (3): 646–53. https://doi.org/10.1111/j.1748-7692.2000.tb00959.x.

Coussi-Korbel, Sabine, and Dorothy M. Fragaszy. 1995. "On the Relation Between Social Dynamics and Social Learning." *Animal Behaviour* 50: 1441–53. https://doi.org/10.1016/0003-3472(95)80001-8.

Deecke, Volker B., Peter J. B. Slater, and John K. B. Ford. 2002. "Selective Habituation Shapes Acoustic Predator Recognition in Harbour Seals." *Nature* 420 (6912): 171–73. https://doi.org/10.1038/nature01030.

Dekker, David, David Krackhardt, and Tom A. B. Snijders. 2007. "Sensitivity of MRQAP Tests to Collinearity and Autocorrelation Conditions." *Psychometrika* 72 (4): 563–81. https://doi.org/10.1007/s11336-007-9016-1.

Deputte, Bertrand L. 2001. "Culture in Cetaceans: Why Put the Cart Before the Horse?" *Behavioural and Brain Sciences* 24 (2). https://doi.org/10.1017/S0140525X01303965.

Donaldson, Rebecca, Hugh Finn, Lars Bejder, David Lusseau, and Michael Calver. 2012. "The Social Side of Human-Wildlife Interaction: Wildlife Can Learn Harmful Behaviours from Each Other." *Animal Conservation* 15 (October): 427–35. https://doi.org/10.1111/j.1469-1795.2012.00548.x.

Engleby, Laura K., and Jessica R. Powell. 2019. "Detailed Observations and Mechanisms of Mud Ring Feeding by Common Bottlenose Dolphins (*Tursiops truncatus truncatus*) in Florida Bay, Florida, U.S.A." *Marine Mammal Science* 35 (3): 1162–72. https://doi.org/10.1111/mms.12583.

Esteban, Ruth, Alfredo López, Álvaro Garcia de los Rios, et al. 2022. "Killer Whales of the Strait of Gibraltar, an Endangered Subpopulation Showing a Disruptive Behavior." *Marine Mammal Science* 38 (4): 1699–709. https://doi.org/10.1111/mms.12947.

Farine, Damien R., Lucy M. Aplin, Ben C. Sheldon, and William Hoppitt. 2015. "Interspecific Social Networks Promote Information Transmission in Wild Songbirds." *Proceedings of the Royal Society B* 282 (1803): 20142804. https://doi.org/10.1098/rspb.2014.2804.

Finn, Julian, Tom Tregenza, and Mark Norman. 2009. "Preparing the Perfect Cuttlefish Meal: Complex Prey Handling by Dolphins." *PLOS ONE* 4 (1). https://doi.org/10.1371/journal.pone.0004217.

Finn, Kelly R., Matthew J. Silk, Mason A. Porter, and Noa Pinter-Wollman. 2019. "The Use of Multilayer Network Analysis in Animal Behaviour." *Animal Behaviour* 149: 7–22. https://doi.org/10.1016/j.anbehav.2018.12.016.

Foote, Andrew D., Nagarjun Vijay, María C. Ávila-Arcos, et al. 2016. "Genome-Culture Coevolution Promotes Rapid Divergence of Killer Whale Ecotypes." *Nature Communications* 7 (11693). https://doi.org/10.1038/ncomms11693.

Ford, John K. B. 2009. "Killer Whale: *Orcinus orca*." In *Encyclopedia of Marine Mammals*, edited by William F. Perrin, Bernd Würsig, and Johannes Gerardus Marie Thewissen, 650–57. Academic Press.

Ford, John K. B., and Graeme M. Ellis. 1999. *Transients: Mammal-Hunting Killer Whales of British Columbia, Washington, and Southeastern Alaska*. UBC Press.

Ford, John K. B., Graeme M. Ellis, Lance G. Barrett-Lennard, Alexandra B. Morton, Rod S. Palm, and Kenneth C. Balcomb III. 1998. "Dietary Specialization in Two Sympatric Populations of Killer Whales (*Orcinus orca*) in Coastal British Columbia and Adjacent Waters." *Canadian Journal of Zoology* 76 (8): 1456–71. https://doi.org/10.1139/z98-089.

Fox, Adam G., and Robert F. Young. 2012. "Foraging Interactions Between Wading Birds and Strand-Feeding Bottlenose Dolphins (*Tursiops truncatus*) in a Coastal Salt Marsh." *Canadian Journal of Zoology* 90 (6): 744–52. https://doi.org/10.1139/Z2012-043.

Franz, Mathias, and Charles L. Nunn. 2009. "Network-Based Diffusion Analysis: A New Method for Detecting Social Learning." *Proceedings of the Royal Society B* 276 (1663): 1829–36. https://doi.org/10.1098/rspb.2008.1824.

Galef, Bennet G. 2001. "Where's the Beef? Evidence of Culture, Imitation, and Teaching, in Cetaceans?" *Behavioural and Brain Sciences* 24 (2). https://doi.org/10.1017/S0140525X01343960.

Gazda, Stefanie K., Richard C. Connor, Robert K. Edgar, and Frank Cox. 2005. "A Division of Labour with Role Specialization in Group-Hunting Bottlenose Dolphins (*Tursiops truncatus*) off Cedar Key, Florida." *Proceedings of the Royal Society B* 272 (1559): 135–40. https://doi.org/10.1098/rspb.2004.2937.

Guinet, Christophe, and Jérome Bouvier. 1995. "Development of Intentional Stranding Hunting Techniques in Killer Whale (*Orcinus orca*) Calves at Crozet Archipelago." *Canadian Journal of Zoology* 73: 27–33. https://doi.org/10.1139/z95-004.

Hanselman, Dana H., Brian J. Pyper, and Megan J. Peterson. 2018. "Sperm Whale Depredation on Longline Surveys and Implications for the Assessment of Alaska Sablefish." *Fisheries Research* 200 (April): 75–83. https://doi.org/10.1016/j.fishres.2017.12.017.

Harrison, Rachel A., and Erica van de Waal. 2022. "The Unique Potential of Field Research to Understand Primate Social Learning and Cognition." *Current Opinion in Behavioral Sciences* 45 (101132). https://doi.org/10.1016/j.cobeha.2022.101132.

Hasenjager, Matthew J., William Hoppitt, and Ellouise Leadbeater. 2020a. "Network-Based Diffusion Analysis Reveals Context-Specific Dominance of Dance Communication in Foraging Honeybees." *Nature Communications* 11 (1). https://doi.org/10.1038/s41467-020-14410-0.

Hasenjager, Matthew J., Ellouise Leadbeater, and William Hoppitt. 2020b. "Detecting and Quantifying Social Transmission Using Network-Based Diffusion Analysis." *Journal of Animal Ecology* 90 (1): 8–26. https://doi.org/10.1111/1365-2656.13307.

Hersh, Taylor A., Shane Gero, Luke Rendell, et al. 2022. "Evidence from Sperm Whale Clans of Symbolic Marking in Non-Human Cultures." *Proceedings of the National Academy of Sciences* 119 (37): e2201692119. https://doi.org/10.1073/pnas.2201692119.

Hobaiter, Catherine, Timothée Poisot, Klaus Zuberbühler, William Hoppitt, and Thibaud Gruber. 2014. "Social Network Analysis Shows Direct Evidence for Social Transmission of Tool Use in Wild Chimpanzees." *PLOS Biology* 12 (9): e1001960. https://doi.org/10.1371/journal.pbio.1001960.

Hoppitt, William, Neeltje Boogert, and Kevin Laland. 2010. "Detecting Social Transmission in Networks." *Journal of Theoretical Biology* 263 (4): 544–55. https://doi.org/10.1016/j.jtbi.2010.01.004.

IJsseldijk, Lonneke L., Mardik F. Leopold, Elisa L. Bravo Rebolledo, et al. 2015. "Fatal Asphyxiation in Two Long-Finned Pilot Whales (*Globicephala melas*) Caused by Common Soles (*Solea solea*)." *PLOS ONE* 10 (11). https://doi.org/10.1371/journal.pone.0141951.

Iwata, Takashi, Tomonari Akamatsu, Surasak Thongsukdee, Phaothep Cherdsukjai, Kanjana Adulyanukosol, and Katsufumi Sato. 2017. "Tread-Water Feeding of Bryde's Whales." *Current Biology* 27 (21): 1154–55. https://doi.org/10.1016/j.cub.2017.09.045.

Jaakkola, Kelly, Emily Guarino, Katy Donegan, and Stephanie L. King. 2018. "Bottlenose Dolphins Can Understand Their Partner's Role in a Cooperative Task." *Proceedings of the Royal Society B* 285 (1887): 20180948. https://doi.org/10.1098/rspb.2018.0948.

Jaakkola, Kelly, Emily Guarino, Mandy Rodriguez, and Jane Hecksher. 2013. "Switching Strategies: A Dolphin's Use of Passive and Active Acoustics to Imitate Motor Actions." *Animal Cognition* 16 (5): 701–9. https://doi.org/10.1007/s10071-013-0605-3.

Jiménez, Pedro Jose, and Juan José Alava. 2015. "Strand-Feeding by Coastal Bottlenose Dolphins (*Tursiops truncatus*) in the Gulf of Guayaquil, Ecuador." *Latin American Journal of Aquatic Mammals* 10 (1): 33–37. https://doi.org/10.5597/lajam00191.

King, Stephanie L., Emily Guarino, Katy Donegan, Christina McMullen, and Kelly Jaakkola. 2021. "Evidence That Bottlenose Dolphins Can Communicate with Vocal Signals to Solve a Cooperative Task." *Royal Society Open Science* 8 (3). https://doi.org/10.1098/rsos.202073.

Klump, Barbara C., John M. Martin, Sonja Wild, Jana K. Hörsch, Richard E. Major, and Lucy M. Aplin. 2021. "Innovation and Geographic Spread of a Complex Foraging Culture in an Urban Parrot." *Science* 373: 456–60. https://doi.org/10.1126/science.abe7808.

Kopps, Anna M., Corinne Y. Ackermann, William B. Sherwin, Simon J. Allen, Lars Bejder, and Michael Krützen. 2014. "Cultural Transmission of Tool Use Combined with Habitat Specializations Leads to Fine-Scale Genetic Structure in Bottlenose Dolphins." *Proceedings of the Royal Society B* 281: 20133245. https://doi.org/10.1098/rspb.2013.3245.

Krützen, Michael, Janet Mann, Michael R Heithaus, Richard C Connor, Lars Bejder, and William B Sherwin. 2005. "Cultural Transmission of Tool Use in Bottlenose Dolphins." *Proceedings of the National Academy of Sciences* 102 (25): 8939–43. https://doi.org/10.1073/pnas.0500232102.

Krützen, Michael, Carel van Schaik, and Andrew Whiten. 2007. "The Animal Cultures Debate: Response to Laland and Janik." *Trends in Ecology and Evolution* 22 (1): 6. https://doi.org/10.1016/j.tree.2006.10.011.

Laland, Kevin N., and Vincent M. Janik. 2006. "The Animal Cultures Debate." *Trends in Ecology and Evolution* 21 (10): 542–47. https://doi.org/10.1016/j.tree.2006.06.005.

Laland, Kevin N., and Vincent M. Janik. 2007. "Response to Krützen et al.: Further Problems with the 'Method of Exclusion.'" *Trends in Ecology and Evolution* 22 (1): 7. https://doi.org/10.1016/j.tree.2006.10.010.

Lopez, Juan C., and Diana Lopez. 1985. "Killer Whales (*Orcinus orca*) of Patagonia, and Their Behavior of Intentional Stranding While Hunting Nearshore." *Journal of Mammalogy* 66 (1): 181–83. https://doi.org/10.2307/1380981.

Louis, Marie, Marco Galimberti, Frederick Archer, et al. 2021. "Selection on Ancestral Genetic Variation Fuels Repeated Ecotype Formation in Bottlenose Dolphins." *Science Advances* 7 (44): eabg1245. https://doi.org/10.1126/sciadv.abg1245.

Louis, Marie, Petra Korlević, Milaja Nykänen, et al. 2023. "Ancient Dolphin Genomes Reveal Rapid Repeated Adaptation to Coastal Waters." *Nature Communications* 14 (1): 4020. https://doi.org/10.1038/s41467-023-39532-z.

Mann, Janet. 2001. "Cetacean Culture: Definitions and Evidence." *Behavioral and Brain Sciences* 24 (2): 343. https://doi.org/10.1017/S0140525X01433966.

Mann, Janet, and Brooke Sargeant. 2003. "Like Mother, like Calf: The Ontogeny of Foraging

Traditions in Wild Indian Ocean Bottlenose Dolphins (*Tursiops* sp.)." In *The Biology of Traditions: Models and Evidence*, edited by Dorothy M. Fragaszy and Susan Perry, 236–66. Cambridge University Press.

Mann, Janet, Brooke L. Sargeant, Jana J. Watson-Capps, et al. 2008. "Why Do Dolphins Carry Sponges?" *PLOS ONE* 3 (12): e3868. https://doi.org/10.1371/journal.pone.0003868.

Mann, Janet, Margaret A. Stanton, Eric M. Patterson, Elisa J. Bienenstock, and Lisa O. Singh. 2012. "Social Networks Reveal Cultural Behaviour in Tool-Using Dolphins." *Nature Communications* 3 (January): 980. https://doi.org/10.1038/ncomms1983.

Marfurt, Svenja M., Simon J. Allen, Manuela R. Bizzozzero, et al. 2022. "Association Patterns and Community Structure Among Female Bottlenose Dolphins: Environmental, Genetic and Cultural Factors." *Mammalian Biology* 102 (4): 1373–87. https://doi.org/10.1007/s42991-022-00259-x.

Marino, Lori, and Thomas I. White. 2022. "Cetacean Personhood, Rights, and Flourishing." In *Marine Mammals: The Evolving Human Factor*, edited by Giuseppe Notarbartolo di Sciara and Bernd Würsig. Ethology and Behavioral Ecology of Marine Mammals. Springer. https://doi.org/10.1007/978-3-030-98100-6.

McGovern, Barry, Tess Gridley, Bridget S. James, and Simon Elwen. 2020. "Risky Business? A Note on Repeated Live Strandings of Common Bottlenose Dolphins (*Tursiops truncatus*) While Foraging in a Shallow Water Environment." *Marine Mammal Science* 36 (1): 305–14. https://doi.org/10.1111/mms.12634.

McMillan, Christie J., Jared R. Towers, and Jackie Hildering. 2019. "The Innovation and Diffusion of 'Trap-Feeding,' a Novel Humpback Whale Foraging Strategy." *Marine Mammal Science* 35 (3): 779–96. https://doi.org/10.1111/mms.12557.

Nightingale, Glenna F., Kevin N. Laland, William Hoppitt, and Peter Nightingale. 2015. "Bayesian Spatial NBDA for Diffusion Data with Home-Base Coordinates." *PLOS ONE* 10 (7): 1–19. https://doi.org/10.1371/journal.pone.0130326.

Oestreich, William K., James A. Fahlbusch, David E. Cade, et al. 2020. "Animal-Borne Metrics Enable Acoustic Detection of Blue Whale Migration." *Current Biology* 30 (23): 4773–79. https://doi.org/10.1016/j.cub.2020.08.105.

Ohlberger, Jan, Daniel E. Schindler, Eric J. Ward, Timothy E. Walsworth, and Timothy E. Essington. 2019. "Resurgence of an Apex Predator and the Decline in Prey Body Size." *Proceedings of the National Academy of Sciences* 116 (52). https://doi.org/10.1073/pnas.191093011.

Pitman, Robert L., and John W. Durban. 2012. "Cooperative Hunting Behavior, Prey Selectivity and Prey Handling by Pack Ice Killer Whales (*Orcinus orca*), Type B, in Antarctic Peninsula Waters." *Marine Mammal Science* 28 (1): 16–36. https://doi.org/10.1111/j.1748-7692.2010.00453.x.

Ramos, Eric A., Leomir Santoya, Joel Verde, et al. 2022. "Lords of the Rings: Mud Ring Feeding by Bottlenose Dolphins in a Caribbean Estuary Revealed from Sea, Air, and Space." *Marine Mammal Science* 38 (1): 364–73. https://doi.org/10.1111/mms.12854.

Reeves, Isabella M., John A. Totterdell, Emma L. Betty, et al. 2023. "Ancestry Testing of 'Old Tom,' a Killer Whale Central to Mutualistic Interactions with Human Whalers." *Journal of Heredity* 114 (6): 598–611. https://doi.org/10.1093/jhered/esad058.

Rendell, Luke, and Hal Whitehead. 2001. "Culture in Whales and Dolphins." *Behavioral and Brain Sciences* 24 (2): 309–24. https://doi.org/10.1017/S0140525X0100396X.

Riesch, Rüdiger, Lance G. Barrett-Lennard, Graeme M. Ellis, John K. B. Ford, and Volker B. Deecke. 2012. "Cultural Traditions and the Evolution of Reproductive Isolation: Ecological

Speciation in Killer Whales?" *Biological Journal of the Linnean Society* 106 (1): 1–17. https://doi.org/10.1111/j.1095-8312.2012.01872.x.

Ronje, Errol I., Kevin P. Barry, Carrie Sinclair, et al. 2017. "A Common Bottlenose Dolphin (*Tursiops truncatus*) Prey Handling Technique for Marine Catfish (Ariidae) in the Northern Gulf of Mexico." *PLOS ONE* 12 (7). https://doi.org/10.1371/journal.pone.0181179.

Samarra, Filipa I. P., and Patrick J. O. Miller. 2015. "Prey-Induced Behavioural Plasticity of Herring-Eating Killer Whales." *Marine Biology* 162 (4): 809–21. https://doi.org/10.1007/s00227-015-2626-8.

Samuelson, Mystera M., Ashley Millan Ambert, Cristina Díaz Clark, and Moby Solangi. 2020. "Description of a Unique Solitary Feeding Strategy Observed in Bottlenose Dolphins (*Tursiops truncatus*) in the Biloxi Marsh, Louisiana." *Animal Behavior and Cognition* 7 (4): 505–13. https://doi.org/10.26451/abc.07.04.03.2020.

Sargeant, Brooke L., Janet Mann, Per Berggren, and Michael Krützen. 2005. "Specialization and Development of Beach Hunting, a Rare Foraging Behavior, by Wild Bottlenose Dolphins (*Tursiops* sp.)." *Canadian Journal of Zoology* 83: 1400–1410. https://doi.org/10.1139/Z05-136.

Schakner, Zachary A., Michael G. Buhnerkempe, Mathew J. Tennis, et al. 2016. "Epidemiological Models to Control the Spread of Information in Marine Mammals." *Proceedings of the Royal Society B* 283 (1844): 20162037. https://doi.org/http://dx.doi.org/10.1098/rspb.2016.2037.

Schakner, Zachary A., Chris Lunsford, Janice Straley, Tomoharu Eguchi, and Sarah L. Mesnick. 2014. "Using Models of Social Transmission to Examine the Spread of Longline Depredation Behavior Among Sperm Whales in the Gulf of Alaska." *PLOS ONE* 9 (10). https://doi.org/10.1371/journal.pone.0109079.

Schlaepfer, Martin A., Michael C. Runge, and Paul W. Sherman. 2002. "Ecological and Evolutionary Traps." *Trends in Ecology and Evolution* 17 (10). https://doi.org/10.1016/S0169-5347(02)02580-6.

Sigler, Michael F., Chris R. Lunsford, Janice M. Straley, and Joseph B. Liddle. 2008. "Sperm Whale Depredation of Sablefish Longline Gear in the Northeast Pacific Ocean." *Marine Mammal Science* 24 (1): 16–27. https://doi.org/10.1111/j.1748-7692.2007.00149.x.

Silber, Gregory K., and Dagmar Fertl. 1995. "Intentional Beaching by Bottlenose Dolphins (*Tursiops truncatus*) in the Colorado River Delta, Mexico." *Aquatic Mammals* 21 (3): 183–86. https://www.researchgate.net/publication/230729883.

Silk, Matthew J., Kelly R. Finn, Mason A. Porter, and Noa Pinter-Wollman. 2018. "Can Multilayer Networks Advance Animal Behavior Research?" *Trends in Ecology and Evolution* 33 (6): 376–78. https://doi.org/10.1016/j.tree.2018.03.008.

Simões-Lopes, Paulo C., Fábio G. Daura-Jorge, and Mauricio Cantor. 2016. "Clues of Cultural Transmission in Cooperative Foraging Between Artisanal Fishermen and Bottlenose Dolphins, *Tursiops truncatus* (Cetacea: Delphinidae)." *Zoologia* 33 (6): 4–7. https://doi.org/10.1590/S1984-4689zool-20160107.

Simões-Lopes, Paulo C., Marta E. Fabián, and João O. Menegheti. 1998. "Dolphin Interactions with the Mullet Artisanal Fishing on Southern Brazil: A Qualitative and Quantitative Approach." *Revista Brasileira de Zoologia* 15 (3): 709–26. https://doi.org/10.1590/S0101-8175 1998000300016.

Smith, Holly C., and Kate R. Sprogis. 2016. "Seasonal Feeding on Giant Cuttlefish (*Sepia apama*) by Indo-Pacific Bottlenose Dolphins (*Tursiops aduncus*) in South-Western Australia." *Australian Journal of Zoology* 64 (1): 8–13. https://doi.org/10.1071/ZO15075.

Tyne, Julian A., Neil R. Loneragan, Anna M. Kopps, Simon J. Allen, Michael Krützen, and Lars Bejder. 2012. "Ecological Characteristics Contribute to Sponge Distribution and Tool Use in Bottlenose Dolphins *Tursiops* sp." *Marine Ecology Progress Series* 444: 143–53. https://doi.org/10.3354/meps09410.

Van Schaik, Carel P., Marc Ancrenaz, Gwendolyn Borgen, et al. 2003. "Orangutan Cultures and the Evolution of Material Culture." *Science* 299 (5603): 102–5. https://doi.org/10.1126/science.1078004.

Visser, Ingrid N. 2000. "Killer Whale (*Orcinus orca*) Interactions with Longline Fisheries in New Zealand Waters." *Aquatic Mammals* 26 (3): 241–52.

Watt, Cortney A., Jack R. Orr, Mads P. Heide-Jørgensen, Nynne H. Nielsen, and Steven H. Ferguson. 2015. "Differences in Dive Behaviour among the World's Three Narwhal *Monodon Monoceros* Populations Correspond with Dietary Differences." *Marine Ecology Progress Series* 525 (April): 273–85. https://doi.org/10.3354/meps11202.

Weinpress-Galipeau, Meghan, Hannah Baker, Bethany Wolf, Bill Roumillat, and Patricia A. Fair. 2021. "An Adaptive Bottlenose Dolphin Foraging Tactic, 'Shipside Feeding,' Using Container Ships in an Urban Estuarine Environment." *Marine Mammal Science* 37 (3): 1159–65. https://doi.org/10.1111/mms.12806.

Weiss, Michael N., Daniel W. Franks, Deborah A. Giles, et al. 2021. "Age and Sex Influence Social Interactions, but Not Associations, Within a Killer Whale Pod." *Proceedings of the Royal Society B* 288 (1953): 20210617. https://doi.org/10.1098/rspb.2021.0617.

Whitehead, Hal. 2003. *Sperm Whales: Social Evolution in the Ocean*. University of Chicago Press.

Whitehead, Hal. 2010. "Conserving and Managing Animals That Learn Socially and Share Cultures." *Learning and Behavior* 38 (3): 329–36. https://doi.org/10.3758/LB.38.3.329.

Whitehead, Hal. 2017. "Gene-Culture Coevolution in Whales and Dolphins." *Proceedings of the National Academy of Sciences* 114 (30): 7814–21. https://doi.org/10.1073/pnas.1620736114.

Whitehead, Hal, and John K. B. Ford. 2018. "Consequences of Culturally-Driven Ecological Specialization: Killer Whales and Beyond." *Journal of Theoretical Biology* 456: 279–94. https://doi.org/10.1016/j.jtbi.2018.08.015.

Whitehead, Hal, and Luke Rendell. 2014. *The Cultural Lives of Whales and Dolphins*. University of Chicago Press.

Whitehead, Hal, Luke Rendell, Richard W. Osborne, and Bernd Würsig. 2004. "Culture and Conservation of Non-Humans with Reference to Whales and Dolphins: Review and New Directions." *Biological Conservation* 120 (3): 427–37. https://doi.org/10.1016/j.biocon.2004.03.017.

Whiten, Andrew. 2019. "Cultural Evolution in Animals." *Annual Review of Ecology, Evolution, and Systematics* 50: 27–48. https://doi.org/10.1146/annurev-ecolsys-110218-025040.

Whiten, Andrew. 2021. "The Burgeoning Reach of Animal Culture." *Science* 372 (6537): eabe6514. https://doi.org/10.1126/science.abe6514.

Whiten, Andrew, Jane Goodall, William C. McGrew, et al. 1999. "Cultures in Chimpanzees." *Nature* 399 (June): 682–85. https://doi.org/10.1038/21415.

Wild, Sonja, Simon J. Allen, Michael Krützen, Stephanie L. King, Livia Gerber, and William J. E. Hoppitt. 2019a. "Multi-Network-Based Diffusion Analysis Reveals Vertical Cultural Transmission of Sponge Tool Use Within Dolphin Matrilines." *Biology Letters* 15 (7): 20190227. https://doi.org/10.1098/rsbl.2019.0227.

Wild, Sonja, William J. E. Hoppitt, Simon J. Allen, and Michael Krützen. 2020. "Integrating Genetic, Environmental, and Social Networks to Reveal Transmission Pathways of a Dolphin Foraging Innovation." *Current Biology* 30 (15): 3024–30. https://doi.org/10.1016/j.cub.2020.05.069.

Wild, Sonja, Michael Krützen, Robert W. Rankin, William J. E. Hoppitt, Livia Gerber, and Simon J. Allen. 2019b. "Long-Term Decline in Survival and Reproduction of Dolphins Following a Marine Heatwave." *Current Biology* 29 (7): R239–40. https://doi.org/10.1016/j.cub.2019.02.047.

Wiley, David, Colin Ware, Alessandro Bocconcelli, et al. 2011. "Underwater Components of Humpback Whale Bubble-Net Feeding Behaviour." *Behaviour* 148 (5–6): 575–602. https://doi.org/10.1163/000579511X570893.

Wray, Janie, Eric M. Keen, and Éadin N. O'Mahony. 2021. "Social Survival: Humpback Whales (*Megaptera novaeangliae*) Use Social Structure to Partition Ecological Niches Within Proposed Critical Habitat." *PLOS ONE* 16 (6): e0245409. https://doi.org/10.1371/journal.pone.0245409.

Würsig, Bernd, and Thomas A. Jefferson. 1990. "Methods of Photo-Identification for Small Cetaceans." *Reports of the International Whaling Commission* 12.

12

Vocal Cultures in Cetaceans

TAYLOR A. HERSH, OLGA A. FILATOVA, AND ELLEN C. GARLAND

Abstract

Over the last two decades our understanding of the role that culture plays in structuring animal societies and modulating animal interactions has rapidly expanded. For animals that primarily communicate acoustically, like cetaceans, cultural processes can impact how certain vocalizations are learned and shared. This chapter explores three of the best-known examples of vocal culture in cetaceans: killer whale (*Orcinus orca*) vocal dialects, sperm whale (*Physeter macrocephalus*) coda dialects, and humpback whale (*Megaptera novaeangliae*) songs. For each, we evaluate and synthesize the evidence for vocal culture by describing the vocal repertoire and discussing how repertoires are learned, modified, and shared among individuals. These examples illustrate how social learning and vocal culture can operate in free-ranging animals at ocean-basin scales. We have only started to scratch the surface of vocal culture in the ocean; the coming decade will undoubtedly provide many more examples and a rich foundation for comparative studies to understand complex communication.

Introduction

In the original *Cetacean Societies*, cetacean culture was rarely mentioned (Mann et al. 2000). Exceptions include a brief discussion of vertically transmitted cultural knowledge in sperm whales (*Physeter macrocephalus*) and a half-page section on matrilineal cetacean cultures (Whitehead and Weilgart 2000; Whitehead and Mann 2000). In the intervening decades, it has become clear that culture—"information or behaviors shared within a group and acquired from conspecifics through some form of social learning"—can manifest across all aspects of behavior for human *and* nonhuman animals (Whitehead and Rendell 2014; Fragaszy and Perry 2003). As our understanding of animal culture has evolved, so has our thinking about what it means to "have culture," and how best to study and conserve cultural species (Brakes et al. 2021, 2019; Whitehead et al. 2023; Brakes and Whitehead, this volume). Often, this means breaking down cultural traits or traditions into components and studying how those parts are socially transmitted, learned, and changed.

Social learning is any learning process that is facilitated by observing or interacting with another animal or the products of its behaviors (Hoppitt and Laland 2008). Such socially learned information can flow vertically (from parent to offspring), horizontally (from peer to peer), or obliquely (from elders to young), and it can occur across sensory domains. For animals living in marine environments, like cetaceans, sound is a key signaling modality, as it typically travels much farther and faster than chemical, visual, or tactile cues (Mann et al. 2000). Accordingly, this chapter focuses on the acoustic domain of cetacean communication, specifically investigating culturally transmitted vocalizations (i.e., vocal culture).

The following sections introduce the "poster children" of cetacean vocal culture: killer whale (*Orcinus orca*) vocal dialects, sperm whale coda dialects, and humpback whale (*Megaptera novaeangliae*) songs. For each species, we describe the vocal repertoire, discuss transmission mechanisms and directions, and summarize the evidence for vocal culture. These three species have notable differences and similarities in terms of life history, social structure, sound production anatomy, and behavioral transmission mechanisms, allowing us to investigate links between vocal complexity, social complexity, and cultural transmission in wild animals. Killer whales and sperm whales have matrilineal social structures, and their dialects are transmitted through maternal lines; in contrast, humpback whales lack stable social groups, and their songs are produced only by males, who learn from other males within a population (although songs can also be shared between populations). In terms of acoustic structure, killer whale calls and humpback whale songs are represented by a wide variety of sounds, while sperm whales generally only produce clicks, with subsequent complexity arising from temporal combinations of these simple components. The adaptive benefits of the calls and songs also differ between the species, with songs in humpback whales likely produced as part of a courtship display to attract females. In contrast, the group-specific calls of killer whales and sperm whales are hypothesized to function in aiding the recognition of in- versus out-group members, which may help promote cooperation and avoid inbreeding. Collectively, these three case studies give us complementary insights and perspectives on cetacean communication, and provide robust, parsimonious evidence for social learning and vocal culture over various scales.

Killer Whales

Killer whales occur in all oceans, but local populations can have specific feeding preferences, hunting techniques, social organizations, and many other population-specific traits, forming more or less discrete ecotypes (Ford et al. 1998; Pitman and Ensor 2003). The main reason for this diversity is that a major portion of the killer whale behavioral repertoire is not innate but rather transmitted through social learning (Ford and Ellis 2006; Baird 2000; Wild and Rendell, this volume). The vocal repertoire is no exception: calves learn vocal dialects from their mothers and other maternal relatives (Deecke et al. 2000; Crance et al. 2014).

Killer whales produce a variety of sounds that have traditionally been divided into three categories: clicks, pulsed calls, and whistles (Ford 1989). Clicks are short, broadband, impulsive sounds mostly used for echolocation (Au et al. 2004). Clicks are produced individually or in series, and click series can vary in length and repetition rate. Fast click trains are perceived by the human ear as continuous buzzes or pulsed calls with the fundamental frequency corresponding to the pulse repetition rate (Murray et al. 1998). Narrowband tonal sounds are termed "whistles" and often have fundamental frequencies exceeding those of pulsed calls. Whistles are sometimes reported to be discernible from pulsed calls by timbre (Thomsen et al. 2001), but this distinction is not well defined, and it is often impossible to discern whistles from high-frequency pulsed calls. Some authors suggest that clicks and whistles are two extremes of a perceptual continuum with typical pulsed calls in the middle (Murray et al. 1998).

Killer whale sounds vary not only in structure, but also in their degree of stereotypy, which ranges from highly stereotyped to completely variable (Ford 1989). This pattern applies to both pulsed calls and whistles (Ford 1989; Riesch et al. 2006), though stereotyped pulsed calls are more common than stereotyped whistles. Killer whales also produce aberrant sounds in which the structure of a stereotyped sound is altered in a systematic fashion (Ford 1989).

Sets of stereotyped sounds form vocal dialects specific to closely related and frequently associating killer whale groups. The social structure of killer whales is shaped by maternal kinship. In the best-studied ecotype, the so-called residents from the North Pacific, the basic unit of social structure is the matriline, which consists of a mature female and several generations of her offspring. Both male and female offspring stay in their natal matriline for life. When a matriarch dies, her daughters (along with their direct maternal descendants) begin to spend less time together, gradually separating into their own matrilines (Bigg et al. 1990). In some cases, a female with her offspring can separate from her natal matriline while her mother is still alive (Stredulinsky et al. 2021). Sets of closely associating matrilines form "pods," initially defined by Bigg et al. (1990) as matrilines that spend at least 50% of their time together. In other ecotypes, social structure can be more fluid, with some offspring normally leaving their natal group and joining other groups, but maternal kinship still plays a major role in social bond formation (Baird and Dill 1996; Esteban et al. 2016; Nielsen et al. 2023).

In the North Pacific "resident" killer whale ecotype, different pods have distinct dialects comprising a mix of unique and shared stereotyped call types (Ford 1991). This led to a transformation of the definition of a "pod" over time, from "matrilines that spend at least 50% of their time together" to "matrilines that share the same dialect," because dialects proved to be more stable over time than social associations (Ford and Ellis 2002; Ivkovich et al. 2010). A set of pods that share at least one call type is referred to as an "acoustic clan" (Ford 1991).

In general, animal acoustic repertoires are transmitted genetically and not learned. However, eight animal groups—songbirds, hummingbirds, parrots, cetaceans, pinnipeds, bats, elephants, and humans—are capable of vocal production learning: the ability to modify the form of vocalizations after experience with the vocalizations of others (Janik and Slater 2000; Vernes et al. 2021). Current evidence suggests that only these eight groups have the capacity for vocal culture. Killer whale sounds are transmitted from generation to generation culturally rather than genetically, with calves adopting the call repertoire of their matrilineal unit. Their fathers can belong to another matrilineal unit, a different pod, or even a different clan (Barrett-Lennard 2000; but see Ford et al. 2018, 2011) and therefore have a different call repertoire than their offspring. If dialects were transmitted genetically, offspring repertoires should be intermediate between the mother's and father's dialects, but in reality, they are almost identical to their mother's and do not exhibit paternal input (Bowles et al. 1988). Furthermore, a large amount of evidence in favor of a cultural instead of a genetic transmission of dialects has accumulated over the last two decades. Deecke et al. (2000) showed that structural modifications to stereotyped call types were transmitted horizontally between two matrilines that did not interbreed, confirming that killer whales are capable of vocal learning. Foote et al. (2006) provided additional evidence of vocal learning: they reported that an abandoned juvenile killer whale in waters off Vancouver Island imitated sea lion barks (that individual's call usage also differed from its natal pod). Additionally, some studies on captive animals have demonstrated that killer whales are capable of imitating calls of both conspecifics and humans (Abramson et al. 2018). For example, Bain (1986) described a 10-year-old Icelandic female imitating calls from a 13-year-old Canadian female after sharing a tank for several years. Crance et al. (2014) reported that two young captive-born males learned new calls and altered their repertoires to match that of an adult male in the same tank.

There is currently little doubt that killer whale dialects are learned rather than innate and are generally stable over decadal timescales (Ford 1991; Deecke et al. 2000; Foote et al. 2006; Crance et al. 2014; Abramson et al. 2018). The recording datasets of resident killer whales from the area around Vancouver Island span multiple decades, but no substantial change in call usage has been reported so far (Ford 1991; Foote et al. 2008). It is not yet clear how divergent dialects evolve and how they change with time. Ford (1991) suggested that dialects evolve through the accumulation of random learning errors; in this case, higher call similarity should occur in recently diverged groupings. Indeed, shared call types are generally more similar within than between matrilines (Nousek et al. 2006) and between matrilines of the same pod than between pods (Ford 1991; Miller and Bain 2000). However, other studies suggest that vocal change can occur with different speeds in different call types or even syllable types (Deecke et al. 2000; Filatova et al. 2013, 2017) and can be driven by processes beyond random error accumulation (Filatova et al. 2013, 2015, 2017). Specifically, it is possible that killer whales can borrow some call features from other matrilines through horizontal transmission or, conversely, alter their calls to diverge from similar calls of related matrilines (fig. 12.1; Filatova et al. 2013, 2017); besides, call change may be limited by some physical constraints in the call structure (Filatova et al. 2016). Nevertheless, most inferences on cultural evolution of killer whale dialects were made through the comparison of contemporary repertoires, and careful analysis of long-term time series data is needed for better understanding of these processes.

Sperm Whales

Sperm whales are the largest toothed whale, with mature females and males ranging from 11 to 16 m in length (Whitehead 2018). Up to one-third of an individual's body length is taken up by their boxy head and extremely modified "nose" (Whitehead 2018). Made up of specialized sound production

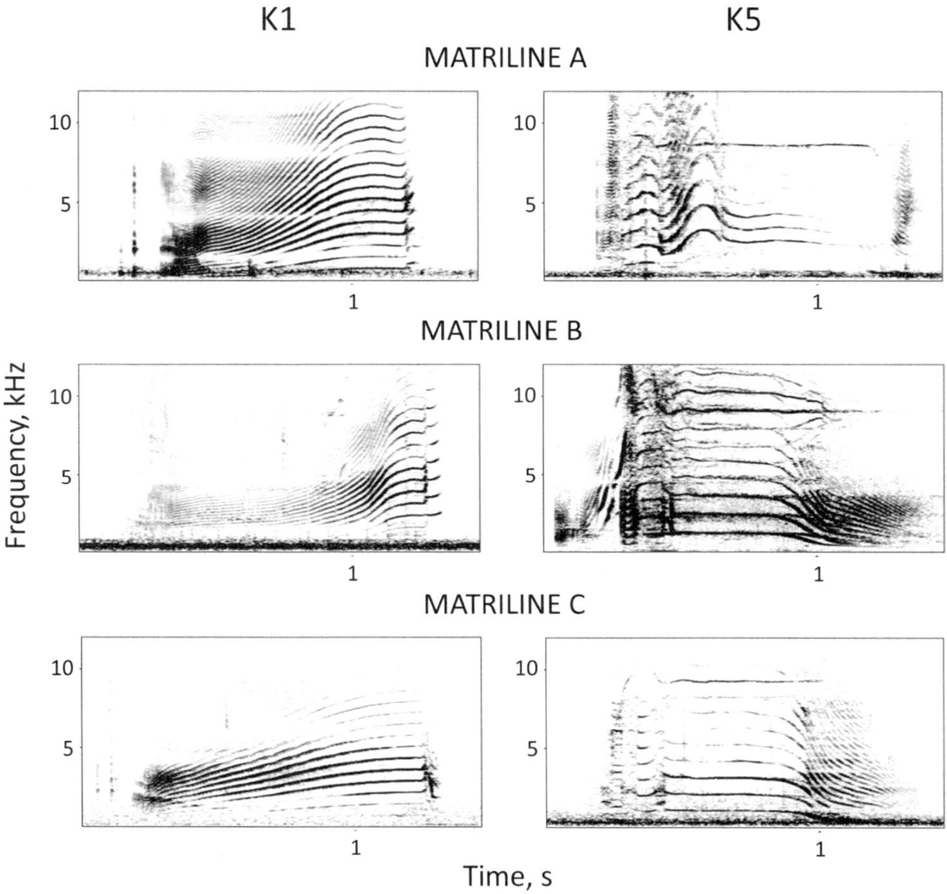

FIGURE 12.1. Similarity patterns of killer whale calls are not always consistent across dialects. Here, the call type on the left, K1, is more similar in matrilines A and B, while the call type on the right, K5, is more similar in matrilines B and C. This pattern is unlikely to arise through the steady accumulation of random errors and suggests more complex evolutionary processes, for example, horizontal transmission.

organs, this nose allows sperm whales to generate clicks (Møhl et al. 2003; Huggenberger et al. 2016). These clicks can be produced and packaged in different ways to yield different types of vocalizations, including echolocation, creaks, clangs, and codas. Of relevance to any discussion of cetacean vocal culture are codas—stereotyped, Morse code-like patterns of clicks made by sperm whales in social situations (Watkins and Schevill 1977). Sperm whales produce different "coda types" by varying the number of clicks, rhythm, and tempo, and individuals only associate if they use a similar "dialect" of coda types (Rendell and Whitehead 2003).

Sperm whale societies are hierarchically and matrilineally structured, with trends in coda type usage mapping onto social structure. Related and unrelated female and immature whales live in temporally stable "social units" (Christal et al. 1998). While daughters typically remain in their mother's social unit for life, males disperse prior to the onset of sexual maturity and live largely solitary lives, only returning to the warmer waters favored by social units to breed (Best 1979). Different social units will associate for hours, days, or weeks at a time in "groups," but group formation is limited to social units that use similar dialects. This gives rise to a cultural level of population structure, known as "vocal clans" (Rendell and Whitehead 2003).

Clans were first named and described in the South Pacific in 2003. Rendell and Whitehead (2003) co-opted the term "clan" from the killer whale literature, citing similarities between killer whale acoustic clans and sperm whale vocal clans. They also note some key differences, the most important being that killer whale acoustic clans are not socially isolated: pods from different acoustic clans have different dialects but socialize in the same way as pods from the same acoustic clan, whereas sperm whale vocal clans are socially segregated. In their work, Rendell and Whitehead (2003) showed that the coda repertoires of social units and groups could be reliably assigned to one of six clans. Several clans exhibited distinctive rhythmic "motifs" in their codas, reminiscent of rhythmic preferences seen in different human musical cultures (Jacoby and McDermott 2017). Sympatric clans occupying the same waters were readily documented along the South American coastline, and photo-identification records provided the first evidence that whales exclusively interact with same-clan individuals.

In the intervening decades, sympatric clans have been documented in all major oceans and are likely the norm rather than the exception (Gero et al. 2016a; Huijser et al. 2020; Amorim et al. 2020; Hersh et al. 2022). Originally defined based on acoustic differences, there is now ample evidence that clans vary along other behavioral dimensions as well, including how they move, forage, associate, distribute, and dive (Marcoux et al. 2007; Whitehead and Rendell 2004; Cantor and Whitehead 2015; Vachon et al. 2022; Eguiguren et al. 2019; Hersh et al. 2022). Culture is thus pervasive and paramount in this species, but clans are still distinguished (and defined) based on acoustic differences.

For multiple tiers of sperm whale social structure—vocal clan, social unit, and individual—there is evidence of identity information being encoded in codas. At the vocal clan level, "identity codas" are coda types used (almost) exclusively by one clan, whereas "non-identity codas" are coda types used by multiple clans (Hersh et al. 2021; fig. 12.2). In addition to allowing researchers to tell different clans apart, there is quantitative evidence that identity codas act as symbolic markers of cultural identity among Pacific Ocean sperm

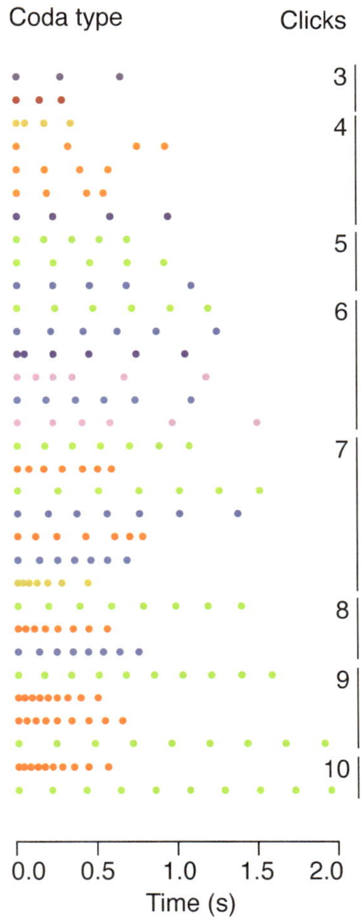

FIGURE 12.2. Examples of identity coda types produced by different Pacific Ocean sperm whale vocal clans. Each row represents one identity coda type, and each dot represents a single click. Note that the click spacing, total duration, and number of clicks can vary. Colors correspond to different clans: Slow Increasing (purple), Short (red), Rapid Increasing (yellow), Palindrome (orange), Regular (green), Plus-One (blue), and Four-Plus (pink).

Source: Reproduced with modification and permission from Hersh et al. (2022).

whales (Hersh et al. 2022). At a finer scale, research in the Caribbean Sea and North Atlantic Ocean suggests that different coda types convey social unit and individual identity (Gero et al. 2016b; Oliveira et al. 2016).

As in killer whales, the available empirical and theoretical evidence suggests dialects are culturally rather than genetically transmitted. For one, clans do not differ genetically when compared using biparentally inherited DNA markers, implying that males will mate with females from multiple clans (Konrad et al. 2018; Lyrholm et al. 1999). In contrast, clans do show some variation in mitochondrial haplotype distributions, which likely reflects stable clan membership within matrilines—calves inherit both their mitochondrial DNA and their dialect from their mothers (Rendell et al. 2012). However, the most common mitochondrial haplotypes are shared by multiple clans, implying that variation in the mitochondrial control region cannot be the primary driver of the observed clan structure (Rendell et al. 2012; Alexander et al. 2016). Agent-based models grounded in empirical data show that clans can arise from cultural transmission via biased social learning of codas, but not from genetic inheritance or individual learning (Cantor et al. 2015).

While not as concrete as that for killer whales or humpback whales, there is evidence that sperm whales are vocal learners. Sperm whale calves produce a greater diversity of coda types than adults, reminiscent of characteristic babbling seen in other taxa with socially learned vocalizations (Gero et al. 2016b; Schulz et al. 2011; Ter Haar et al. 2021; Fernandez et al. 2021). Sperm whales may also meet a core criterion of the "vocal learning and rhythmic pattern perception hypothesis"—namely, that vocal learners can recognize a given rhythmic pattern across a range of tempi, as evidenced by their ability to *produce* the same rhythmic pattern across a range of tempi (Rouse et al. 2021; Hersh et al. 2022). Regardless of whether this is vocal usage learning (i.e., existing signals are produced in a new context or sequence) or vocal production learning (Janik and Slater 2000; Vernes et al. 2021), it is a clear example of social learning and cultural transmission. Finally, while coda output by social units generally appears to be stable over days to years (Rendell and Whitehead 2005), fine-scale coda structural parameters (such as total duration) have significantly changed over decadal spans in certain clans (Hersh 2021).

The disparate life histories of male and female sperm whales mean that most of what we know about culture, codas, and dialects is limited to females and their young. While mature males occasionally make codas when socializing with adult females, codas are rarely recorded in male-only waters at higher latitudes (Marcoux et al. 2006). However, two coda repertoires recorded from male sperm whales off Kaikōura, New Zealand were assigned to the "Short" clan (Hersh et al. 2022). This may indicate that the Short clan dialect is evolutionarily basal, with males reverting to a more ancestral dialect on the rare occasions in which they make codas, but additional research on the evolutionary dynamics of dialects is requisite to tease this and other possibilities apart.

Humpback Whales

Humpback whales are highly migratory baleen whales that live in all oceans (Clapham 2018; Dunlop and Parks, this volume). They annually migrate from summer feeding grounds located in polar regions to low-latitude winter breeding and calving grounds (Chittleborough 1965). Humpback whales display strong, maternally inherited site fidelity (Baker et al. 1986), or migratory culture, to breeding and feeding grounds, allowing distinct populations to be identified worldwide (Jackson et al. 2015).

Unlike killer whales and sperm whales, humpback whales do not form matrilineal societies. Instead, humpback whales form temporary, unrepeated associations, suggesting they possess a simple and individually based social structure (Dunlop 2022). The exceptions to this rule are the one-year associations between mother and calf, and a number of summer feeding groups that may exhibit stable, long-term associations for cooperative feeding (Ramp et al. 2010; Weinrich et al. 2006; Wild and Rendell, this volume). Given that complex social structures and complex vocal repertoires tend to go together (Freeberg 2006), the humpback whale may be an enigma with its relatively simple social structure but complex (in terms of number of sound types and multilevel organization) acoustic repertoire (Clapham 1996; Dunlop 2022). As social groups are typically classified based on animals within a few body lengths, this ignores that these highly vocal animals may be in acoustic groupings over a few kilometers, forming "heards" not "herds" (Tyack 2022), and thus creating a more complex network of changing social groups on the breeding grounds (Dunlop 2022; Dunlop and Parks, this volume).

All age classes and both sexes produce social calls (Dunlop et al. 2007; Fournet et al. 2018; Indeck et al. 2021). The large (over 30 sound types) and varied repertoire can be used in feeding contexts and during social interactions. Researchers are only just starting to unravel the function of the multitude of social calls heard in a variety of social contexts, including a presumed mother-calf "contact call" (Dunlop et al. 2008; Fournet et al. 2018; Indeck et al. 2021; Rekdahl et al. 2013). In contrast, the best-studied acoustic display in baleen whales is the highly elaborate song of the humpback whale.

Male humpback whales sing a long, stereotyped, repetitive, sexually selected, and culturally transmitted vocal display—termed "song"—primarily on the winter breeding grounds and when migrating to and from these areas (Clapham and Mattila 1990; Glockner 1983; Herman 2017; Noad et al. 2000; Owen et al. 2019; Payne and Payne 1985; Payne and McVay 1971). Song is structured in a nested hierarchy with the shortest sound (to our ears) termed a "unit"; a few units arranged into a stereotyped "phrase"; phrases repeated multiple times to make a "theme"; and a few different themes typically sung in a stereotyped order to make a "song" (Payne and McVay 1971). At the upper level, different versions of the song are termed "song types," which are composed of different themes (Garland et al. 2011).

Within a population, there is strong cultural conformity to the current song norm and most males will sing the same song type at any point in time (Payne and Payne 1985; Payne et al. 1983). The song constantly evolves each year at all levels of the song hierarchy (Payne and Payne 1985), and males must incorporate these changes into their own song to maintain the observed cultural conformity. Song is therefore horizontally culturally transmitted among peers within acoustic contact to allow constant updating to each individual singer's song. This is in conjunction with the constant drive for novelty, for change, where it is hypothesized that novel songs are more "attractive" to females, potentially creating runaway evolution in the complexity of the display (Cerchio et al. 2001; Garland et al. 2022; Noad et al. 2000).

At a wider geographic scale, songs within an ocean basin are similar but this depends on the distance between populations (Garland et al. 2011; Darling et al. 2019; Payne and Guinee 1983), while populations located in different ocean basins typically sing different song types (Payne and Guinee 1983). Transmission of song between genetically distinct populations is suggested to occur due to three mechanisms (Payne and Guinee 1983): (i) song exchange on shared or partially shared feeding grounds or migratory routes; (ii) individual whales switching breeding grounds between seasons; or (iii) whales visiting two breeding grounds within a season. All mechanisms have been documented to occur (Calambokidis et al. 2001; Clapham and Mattila 1990; Noad et al. 2000; Schall et al. 2022; Stimpert et al. 2012; Garland et al. 2013; Garrigue et al. 2007), and none are mutually exclusive. For example, song exchange between multiple populations on partially shared migratory routes has been documented in the South Pacific during the southern migration past the Kermadec Islands (Owen et al. 2019) and northern migration through Cook Strait, New Zealand (Warren et al. 2020). Such song exchange may lead to a single, shared song type across an entire ocean basin, as is documented in the North Pacific (see Darling et al. 2019).

Alongside the progressive song evolution documented worldwide, Southern Hemisphere humpback whale populations also display a more rapid and dramatic form of song change, termed a "song revolution" (Garland et al. 2011; Noad et al. 2000). Song from the western Australian population (located in the Indian Ocean basin) appeared in the eastern Australian population (South Pacific Ocean basin) and replaced the existing song over the course of two years (Noad et al. 2000). This drastic and complete change to the cultural trait of the eastern Australian population over two years subsequently spread across the South Pacific (Garland et al. 2011; fig. 12.3). Two decades of song revolutions have now been traced spreading eastward across the South Pacific (Allen et al. 2018; Garland et al. 2017, 2013, 2011; Owen

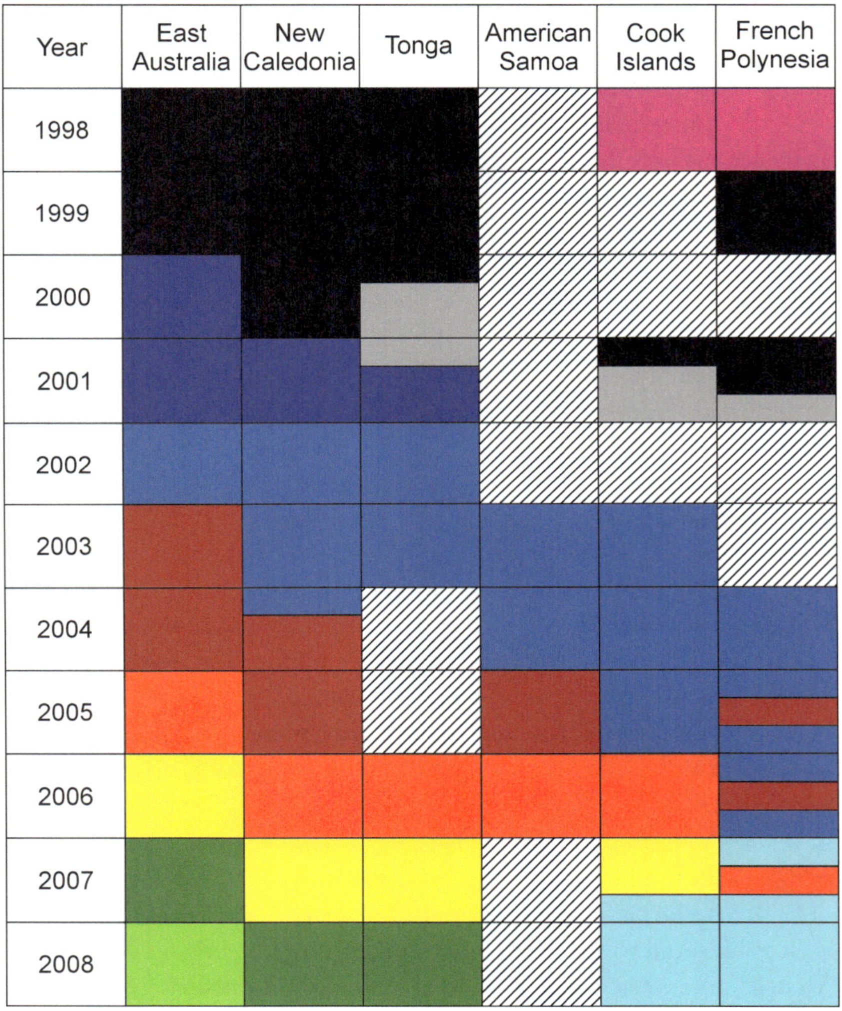

FIGURE 12.3. Song types identified in the South Pacific region from 1998 to 2008. Populations are listed from west to east across the region. Each color represents a distinct song type; song type colors are as follows: black, gray, pink, dark blue, blue, light blue, dark red, light red, yellow, dark green, and light green. Two colors within the same year and location indicate that both song types were present. In these cases, the seasons are broken into three periods (early, middle, and late) to indicate when a new song type was recorded. Crosshatching indicates no data available.
Source: Reprinted from Garland et al. (2011) with permission from Elsevier.

et al. 2019; Warren et al. 2020), including into the eastern South Pacific (Ecuador; Schulze et al. 2022). Global agent-based models support the hypothesis that song revolutions spread from larger to smaller populations, creating the west-east song spread from the large western Australian population eastward to ever smaller populations (Jackson et al. 2015; Zandberg et al. 2021). Such revolutionary songs appear to be less complex (in terms of unit repertoire and length) in comparison to the existing song (Allen et al. 2018; Garland et al. 2022), potentially suggesting an upper limit of learning.

Little is known, however, about how humpback whales learn their song. Evidence suggests that new songs are learned in segments, as revolutionary song themes have been documented spliced into existing songs (Garland et al. 2017). A "switch when similar" rule was also discovered, suggesting whales insert new themes at the location in the song that is the most similar in arrangement and sound types to create a smooth transition from old to new songs (Garland et al. 2017). The coming decade will advance many of our unanswered questions regarding the function of this cultural trait that showcases cultural transmission at a vast scale (Garland et al. 2011).

Conclusion

As these case studies illustrate, there is evidence from both mysticetes and odontocetes of socially transmitted vocalizations. In killer whales and sperm whales, young individuals learn their dialects from associated adults; in humpback whales, males horizontally transmit modifications to songs, giving rise to ocean basin–spanning song revolutions. In all three species, vocal cultural transmission plays a role in structuring populations and modulating conspecific interactions. Perhaps unsurprisingly, these "poster children" of vocal cultural transmission are also three of the best-studied cetacean species in the world, having been the subject of intensive research efforts since the mid- and late twentieth century (Payne and McVay 1971; Watkins and Schevill 1977; Schevill and Watkins 1966). As research on other cetacean species—such as bowhead whales, short-finned pilot whales, and fin whales—has expanded, so too has evidence of vocal cultural transmission (Stafford et al. 2018; Van Cise et al. 2018; Širović et al. 2017).

It is worth reiterating that culture is a multifaceted phenomenon. To fully understand the role culture plays in cetacean lives and societies, we must integrate acoustic insights with different types (e.g., hormonal, epigenetic, and life history) and scales (e.g., individual, group, and population) of data. Technological advances, including the development of autonomous recording devices and the refinement of multisensor animal-borne tags, are making such holistic research increasingly feasible (Croft et al., this volume).

When *Cetacean Societies* was published in 2000, culture was but a footnote; in *The Evolution of Cetacean Societies*, an entire part with multiple chapters has been dedicated to the topic. This expansion is a direct reflection of our growing understanding of the role that culture broadly, and vocal culture specifically, plays in structuring cetacean societies (Whitehead and Rendell 2014; Brakes et al. 2021, 2019). We look forward to what the next 20 years in cetacean acoustics research will bring, as researchers continue to collect, analyze, and interpret acoustic data within the broader ecological, behavioral, and cultural contexts in which it is produced.

Acknowledgments

T. A. H. was supported by the Oregon Gray Whale License Plate Program at the Marine Mammal Institute (Oregon State University) and Max Planck Independent Research Group Leader funding to Dr. Andrea Ravignani. E. C. G. is funded by a Royal Society University Research Fellowship (UF160081 and URF\R\221020). O. A. F. was supported by the Human Frontier Science Program (grant nr. RGP0045/2022).

References

Abramson, José Z., Ma Victoria Hernández-Lloreda, Lino García, Fernando Colmenares, Francisco Aboitiz, and Josep Call. 2018. "Imitation of Novel Conspecific and Human Speech Sounds in the Killer Whale (*Orcinus orca*)." *Proceedings of the Royal Society B* 285 (1871): 20172171. https://doi.org/10.1098/rspb.2017.2171.

Alexander, Alana, Debbie Steel, Kendra Hoekzema, et al. 2016. "What Influences the Worldwide Genetic Structure of Sperm Whales (*Physeter macrocephalus*)?" *Molecular Ecology* 25 (12): 2754–72. https://doi.org/10.1111/mec.13638.

Allen, Jenny A., Ellen C. Garland, Rebecca A. Dunlop, and Michael J. Noad. 2018. "Cultural Revolutions Reduce Complexity in the Songs of Humpback Whales." *Proceedings of the Royal Society B* 285 (1891): 20182088. https://doi.org/10.1098/rspb.2018.2088.

Amorim, Thiago Orion Simões, Luke Rendell, Juliana Di Tullio, Eduardo R Secchi, Franciele R Castro, and Artur Andriolo. 2020. "Coda Repertoire and Vocal Clans of Sperm Whales in the Western Atlantic Ocean." *Deep-Sea Research Part I: Oceanographic Research Papers* 160: 103254. https://doi.org/10.1016/j.dsr.2020.103254.

Au, Whitlow W. L., John K. B. Ford, John K. Horne, and Kelly A. Newman Allman. 2004. "Echolocation Signals of Free-Ranging Killer Whales (*Orcinus orca*) and Modeling of Foraging for Chinook Salmon (*Oncorhynchus tshawytscha*)." *Journal of the Acoustical Society of America* 115 (2): 901–9. https://doi.org/10.1121/1.1642628.

Bain, David E. 1986. "Acoustic Behavior of Orcinus: Sequences, Periodicity, Behavioral Correlates and an Automated Technique for Call Classification." In *Behavioral Biology of Killer Whales*, edited by B. C. Kirkevold and J. S. Lockard, 335–71. Alan R. Liss.

Baird, Robin W. 2000. "The Killer Whale." *Cetacean Societies: Field Studies of Dolphins and Whales*, edited by Janet Mann, Richard C. Connor, Peter L. Tyack, and Hal Whitehead, 127–53. University of Chicago Press.

Baird, Robin W., and Lawrence M. Dill. 1996. "Ecological and Social Determinants of Group Size in Transient Killer Whales." *Behavioral Ecology* 7 (4): 408–16. https://doi.org/10.1093/beheco/7.4.408.

Baker, C. Scott, Louis M. Herman, Anjanette Perry, et al. 1986. "Migratory Movement and Population Structure of Humpback Whales (*Megaptera novaeangliae*) in the Central and Eastern North Pacific." *Marine Ecology Progress Series* 31 (2): 105–19. https://www.jstor.org/stable/24817567.

Barrett-Lennard, Lance G. 2000. "Population Structure and Mating Patterns of Killer Whales (*Orcinus orca*) as Revealed by DNA Analysis." PhD diss., University of British Columbia.

Best, Peter B. 1979. "Social Organization in Sperm Whales, *Physeter macrocephalus*." In *Behavior of Marine Animals*, edited by Howard E. Winn and Bori L. Olla, 227–89. Springer. https://doi.org/10.1007/978-1-4684-2985-5_7.

Bigg, Michael A., Peter F. Olesiuk, Graeme M. Ellis, John K. B. Ford, and Kenneth C. Balcomb. 1990. "Social Organization and Genealogy of Resident Killer Whales (*Orcinus orca*) in the Coastal Waters of British Columbia and Washington State." *Report of the International Whaling Commission* 12: 383–405.

Bowles, Ann E., W. Glenn Young, and Edward D. Asper. 1988. Ontogeny of Stereotyped Calling of a Killer Whale Calf (*Orcinus orca*) During Her First Year. *Rit Fiskideildar* 11: 225–50.

Brakes, Philippa, Emma L. Carroll, Sasha R. X. Dall, et al. 2021. "A Deepening Understanding of Animal Culture Suggests Lessons for Conservation." *Proceedings of the Royal Society B* 288 (1949): 20202718. https://doi.org/10.1098/rspb.2020.2718.

Brakes, Philippa, Sasha R. X. Dall, Lucy M. Aplin, et al. 2019. "Animal Cultures Matter for Conservation." *Science* 363 (6431): 1032–34. https://doi.org/10.1126/science.aaw3557.

Calambokidis, John, Gretchen H. Steiger, Janice M. Straley, et al. 2001. "Movements and Population Structure of Humpback Whales in the North Pacific." *Marine Mammal Science* 17 (4): 769–94. https://doi.org/10.1111/j.1748-7692.2001.tb01298.x.

Cantor, Mauricio, Lauren G. Shoemaker, Reniel B. Cabral, César O. Flores, Melinda Varga, and Hal Whitehead. 2015. "Multilevel Animal Societies Can Emerge from Cultural Transmission." *Nature Communications* 6 (1): 8091. https://doi.org/10.1038/ncomms9091.

Cantor, Mauricio, and Hal Whitehead. 2015. "How Does Social Behavior Differ Among Sperm Whale Clans?" *Marine Mammal Science* 31 (4): 1275–90. https://doi.org/10.1111/mms.12218.

Cerchio, Salvatore, Jeff K. Jacobsen, and Thomas F. Norris. 2001. "Temporal and Geographical Variation in Songs of Humpback Whales, *Megaptera novaeangliae*: Synchronous Change in Hawaiian and Mexican Breeding Assemblages." *Animal Behaviour* 62 (2): 313–29. https://doi.org/10.1006/anbe.2001.1747.

Chittleborough, R. Graham. 1965. "Dynamics of Two Populations of the Humpback Whale, *Megaptera novaeangliae* (Borowski)." *Marine and Freshwater Research* 16 (1): 33–128. https://doi.org/10.1071/MF9650033.

Christal, Jenny, Hal Whitehead, and Erland Lettevall. 1998. "Sperm Whale Social Units: Variation and Change." *Canadian Journal of Zoology* 76 (8): 1431–40. https://doi.org/10.1139/cjz-76-8-1431.

Clapham, Phillip J. 1996. "The Social and Reproductive Biology of Humpback Whales: An Ecological Perspective." *Mammal Review* 26 (1): 27–49. https://doi.org/10.1111/j.1365-2907.1996.tb00145.x.

Clapham, Phillip J. 2018. "Humpback Whale: *Megaptera novaeangliae*." In *Encyclopedia of Marine Mammals*, 3rd ed., edited by Bernd Würsig, Johannes G. M. Thewissen, and Kit M. Kovacs, 489–92. Academic Press. https://doi.org/10.1016/B978-0-12-804327-1.00154-0.

Clapham, Phillip J., and David K. Mattila. 1990. "Humpback Whale Songs as Indicators of

Migration Routes." *Marine Mammal Science* 6 (2): 155–60. https://doi.org/10.1111/j.1748-7692.1990.tb00238.x.

Crance, Jessica L., Ann E. Bowles, and Alan Garver. 2014. "Evidence for Vocal Learning in Juvenile Male Killer Whales, *Orcinus orca*, from an Adventitious Cross-Socializing Experiment." *Journal of Experimental Biology* 217 (8): 1229–37. https://doi.org/10.1242/jeb.094300.

Darling, James D., Jo Marie V. Acebes, Oscar Frey, Jorge Urbán, and Manami Yamaguchi. 2019. "Convergence and Divergence of Songs Suggests Ongoing, but Annually Variable, Mixing of Humpback Whale Populations throughout the North Pacific." *Scientific Reports* 9 (1): 1–14. https://doi.org/10.1038/s41598-019-42233-7.

Deecke, Volker B., John K. B. Ford, and Paul Spong. 2000. "Dialect Change in Resident Killer Whales: Implications for Vocal Learning and Cultural Transmission." *Animal Behaviour* 60: 629–38. https://doi.org/10.1006/anbe.2000.1454.

Dunlop, Rebecca A. 2022. "Humpback Whales: A Seemingly Socially Simple Whale with Communicative Complexity." In *Ethology and Behavioral Ecology of Mysticetes*, edited by Christopher W. Clark and Ellen C. Garland, 223–46. Springer. https://doi.org/10.1007/978-3-030-98449-6_10.

Dunlop, Rebecca A., Douglas H. Cato, and Michael J. Noad. 2008. "Non-Song Acoustic Communication in Migrating Humpback Whales (*Megaptera novaeangliae*)." *Marine Mammal Science* 24 (3): 613–29. https://doi.org/10.1111/j.1748-7692.2008.00208.x.

Dunlop, Rebecca A., Michael J. Noad, Douglas H. Cato, and Dale Stokes. 2007. "The Social Vocalization Repertoire of East Australian Migrating Humpback Whales (*Megaptera novaeangliae*)." *Journal of the Acoustical Society of America* 122 (5): 2893–905. https://doi.org/10.1121/1.2783115.

Eguiguren, Ana, Enrico Pirotta, Mauricio Cantor, Luke Rendell, and Hal Whitehead. 2019. "Habitat Use of Culturally Distinct Galápagos Sperm Whale *Physeter macrocephalus* Clans." *Marine Ecology Progress Series* 609: 257–70. https://doi.org/10.3354/meps12822.

Esteban, Ruth, Philippe Verborgh, Pauline Gauffier, Joan Giménez, Andrew D. Foote, and Renaud de Stephanis. 2016. "Maternal Kinship and Fisheries Interaction Influence Killer Whale Social Structure." *Behavioral Ecology and Sociobiology* 70 (1): 111–22. https://doi.org/10.1007/s00265-015-2029-3.

Fernandez, Ahana A., Lara S. Burchardt, Martina Nagy, and Mirjam Knörnschild. 2021. "Babbling in a Vocal Learning Bat Resembles Human Infant Babbling." *Science* 373 (6557): 923–26. https://doi.org/10.1126/science.abf92.

Filatova, Olga A., Alexandr M. Burdin, and Erich Hoyt. 2013. "Is Killer Whale Dialect Evolution Random?" *Behavioural Processes* 99: 34–41. https://doi.org/10.1016/j.beproc.2013.06.008.

Filatova, Olga A., Tatiana V. Ivkovich, Mikhail A. Guzeev, Alexandr M. Burdin, and Erich Hoyt. 2017. "Social Complexity and Cultural Transmission of Dialects in Killer Whales." *Behaviour* 154 (2): 171–94. https://doi.org/10.1163/1568539X-00003417.

Filatova, Olga A., Filipa I. P. Samarra, Lance G. Barrett-Lennard, et al. 2016. "Physical Constraints of Cultural Evolution of Dialects in Killer Whales." *Journal of the Acoustical Society of America* 140 (5): 3755–64. https://doi.org/10.1121/1.4967369.

Filatova, Olga A., Filipa I. P. Samarra, Volker B. Deecke, John K. B. Ford, Patrick J. O. Miller, and Harald Yurk. 2015. "Cultural Evolution of Killer Whale Calls: Background, Mechanisms and Consequences." *Behaviour* 152 (15): 2001–38. https://doi.org/10.1163/1568539X-00003317.

Foote, Andrew D., Rachael M. Griffin, David Howitt, Lisa Larsson, Patrick J. O. Miller, and

A. Rus Hoelzel. 2006. "Killer Whales Are Capable of Vocal Learning." *Biology Letters* 2 (4): 509–12. https://doi.org/10.1098/rsbl.2006.0525.

Foote, Andrew D., Richard W. Osborne, and A. Rus Hoelzel. 2008. "Temporal and Contextual Patterns of Killer Whale (*Orcinus orca*) Call Type Production." *Ethology* 114 (6): 599–606. https://doi.org/10.1111/j.1439-0310.2008.01496.x.

Ford, John K. B. 1989. "Acoustic Behaviour of Resident Killer Whales (*Orcinus orca*) off Vancouver Island, British Columbia." *Canadian Journal of Zoology* 67 (3): 727–45. https://doi.org/10.1139/z89-105.

Ford, John K. B. 1991. "Vocal Traditions Among Resident Killer Whales (*Orcinus orca*) in Coastal Waters of British Columbia." *Canadian Journal of Zoology* 69 (6): 1454–83. https://doi.org/10.1139/z91-206.

Ford, John K. B., and Graeme M. Ellis. 2002. "Reassessing the Social Organization of Resident Killer Whales in British Columbia." In *Fourth International Orca Symposium and Workshop*, 23–28. Centre d'Etudes Biologiques de Chizé / Centre national de la recherche scientifique.

Ford, John K. B., and Graeme M. Ellis. 2006. "Selective Foraging by Fish-Eating Killer Whales *Orcinus orca* in British Columbia." *Marine Ecology Progress Series* 316: 185–99. https://doi.org/10.3354/meps316185.

Ford, John K. B., Graeme M. Ellis, Lance G. Barrett-Lennard, Alexandra B. Morton, Rod S. Palm, and Kenneth C. Balcomb III. 1998. "Dietary Specialization in Two Sympatric Populations of Killer Whales (*Orcinus orca*) in Coastal British Columbia and Adjacent Waters." *Canadian Journal of Zoology* 76 (8): 1456–71. https://doi.org/10.1139/z98-089.

Ford, Michael J., M. Bradley Hanson, Jennifer A. Hempelmann, et al. 2011. "Inferred Paternity and Male Reproductive Success in a Killer Whale (*Orcinus orca*) Population." *Journal of Heredity* 102 (5): 537–53. https://doi.org/10.1093/jhered/esr067.

Ford, Michael J., Kim M. Parsons, Eric J. Ward, et al. 2018. "Inbreeding in an Endangered Killer Whale Population." *Animal Conservation* 21 (5): 423–32. https://doi.org/10.1111/acv.12413.

Fournet, Michelle E. H., Christine M. Gabriele, David C. Culp, Fred Sharpe, David K. Mellinger, and Holger Klinck. 2018. "Some Things Never Change: Multi-Decadal Stability in Humpback Whale Calling Repertoire on Southeast Alaskan Foraging Grounds." *Scientific Reports* 8 (1): 1–13. https://doi.org/10.1038/s41598-018-31527-x.

Fragaszy, Dorothy M., and Susan Perry, eds. 2003. *The Biology of Traditions: Models and Evidence*. Cambridge University Press.

Freeberg, Todd M. 2006. "Social Complexity Can Drive Vocal Complexity: Group Size Influences Vocal Information in Carolina Chickadees." *Psychological Science* 17 (7): 557–61. https://doi.org/10.1111/j.1467-9280.2006.01743.x.

Garland, Ellen C., Claire Garrigue, and Michael J. Noad. 2022. "When Does Cultural Evolution Become Cumulative Culture? A Case Study of Humpback Whale Song." *Philosophical Transactions of the Royal Society B* 377 (1843): 20200313. https://doi.org/10.1098/rstb.2020.0313.

Garland, Ellen C., Jason Gedamke, Melinda L. Rekdahl, Michael J. Noad, Claire Garrigue, and Nick Gales. 2013. "Humpback Whale Song on the Southern Ocean Feeding Grounds: Implications for Cultural Transmission." *PLOS ONE* 8 (11): e79422. https://doi.org/10.1371/journal.pone.0079422.

Garland, Ellen C., Anne W. Goldizen, Melinda L. Rekdahl, et al. 2011. "Dynamic Horizontal Cultural Transmission of Humpback Whale Song at the Ocean Basin Scale." *Current Biology* 21 (8): 687–91. https://doi.org/10.1016/j.cub.2011.03.019.

Garland, Ellen C., Luke Rendell, Luca Lamoni, M. Michael Poole, and Michael J. Noad. 2017.

"Song Hybridization Events During Revolutionary Song Change Provide Insights into Cultural Transmission in Humpback Whales." *Proceedings of the National Academy of Sciences* 114 (30): 7822–29. https://doi.org/10.1073/pnas.1621072114.

Garrigue, Claire, Trish Franklin, Rochelle Constantine, et al. 2007. "First Assessment of Interchange of Humpback Whales Between Oceania and the East Coast of Australia." *Journal of Cetacean Research and Management*, special issue 3: 269–74. https://doi.org/10.47536/jcrm.vi.314.

Gero, Shane, Anne Bøttcher, Hal Whitehead, and Peter T. Madsen. 2016a. "Socially Segregated, Sympatric Sperm Whale Clans in the Atlantic Ocean." *Royal Society Open Science* 3 (6): 160061. https://doi.org/10.1098/rsos.160061.

Gero, Shane, Hal Whitehead, and Luke Rendell. 2016b. "Individual, Unit and Vocal Clan Level Identity Cues in Sperm Whale Codas." *Royal Society Open Science* 3: 150372. https://doi.org/10.1098/rsos.150372.

Glockner, Deborah A. 1983. "Determining the Sex of Humpback Whales (*Megaptera novaeangliae*) in Their Natural Environment." In *Communication and Behavior of Whales*, edited by Roger Payne, 447–64. Westview Press.

Haar, Sita M. Ter, Ahana A. Fernandez, Maya Gratier, et al. 2021. "Cross-Species Parallels in Babbling: Animals and Algorithms." *Philosophical Transactions of the Royal Society B* 376 (1836): 20200239. https://doi.org/10.1098/rstb.2020.0239.

Herman, Louis M. 2017. "The Multiple Functions of Male Song Within the Humpback Whale (*Megaptera novaeangliae*) Mating System: Review, Evaluation, and Synthesis." *Biological Reviews* 92 (3): 1795–818. https://doi.org/10.1111/brv.12309.

Hersh, Taylor A. 2021. "Dialects over Space and Time: Cultural Identity and Evolution in Sperm Whale Codas." PhD diss., Dalhousie University.

Hersh, Taylor A., Shane Gero, Luke Rendell, and Hal Whitehead. 2021. "Using Identity Calls to Detect Structure in Acoustic Datasets." *Methods in Ecology and Evolution* 12 (9): 1668–78. https://doi.org/10.1111/2041-210x.13644.

Hersh, Taylor A., Shane Gero, Luke Rendell, et al. 2022. "Evidence from Sperm Whale Clans of Symbolic Marking in Non-Human Cultures." *Proceedings of the National Academy of Sciences* 119 (37): e2201692119. https://doi.org/10.1073/pnas.2201692119.

Hoppitt, Will, and Kevin N. Laland. 2008. "Social Processes Influencing Learning in Animals: A Review of the Evidence." *Advances in the Study of Behavior* 38: 105–65. https://doi.org/10.1016/S0065-3454(08)00003-X.

Huggenberger, Stefan, Michel André, and Helmut H. A. Oelschlaeger. 2016. "The Nose of the Sperm Whale: Overviews of Functional Design, Structural Homologies and Evolution." *Journal of the Marine Biological Association of the United Kingdom* 96 (4): 783–806. https://doi.org/10.1017/S0025315414001118.

Huijser, Leonie A. E., Vanessa Estrade, Imogen Webster, Laurent Mouysset, Adele Cadinouche, and Violaine Dulau-Drouot. 2020. "Vocal Repertoires and Insights into Social Structure of Sperm Whales (*Physeter macrocephalus*) in Mauritius, Southwestern Indian Ocean." *Marine Mammal Science* 36 (2): 638–57. https://doi.org/10.1111/mms.12673.

Indeck, Katherine L., Elisa Girola, Maëlle Torterotot, Michael J. Noad, and Rebecca A. Dunlop. 2021. "Adult Female-Calf Acoustic Communication Signals in Migrating East Australian Humpback Whales." *Bioacoustics* 30 (3): 341–65. https://doi.org/10.1080/09524622.2020.1742204.

Ivkovich, Tatiana, Olga A Filatova, Alexandr M Burdin, Hal Sato, and Erich Hoyt. 2010. "The

Social Organization of Resident-Type Killer Whales (*Orcinus orca*) in Avacha Gulf, Northwest Pacific, as Revealed Through Association Patterns and Acoustic Similarity." *Mammalian Biology* 75 (3): 198–210. https://doi.org/10.1016/j.mambio.2009.03.006.

Jackson, Jennifer A., Andrea Ross-Gillespie, Doug Butterworth, et al. 2015. "Southern Hemisphere Humpback Whale Comprehensive Assessment: A Synthesis and Summary: 2005–2015." Report No. SC/66a/SH3 to the Scientific Committee of the International Whaling Commission, 1–38.

Jacoby, Nori, and Josh H. McDermott. 2017. "Integer Ratio Priors on Musical Rhythm Revealed Cross-Culturally by Iterated Reproduction." *Current Biology* 27 (3): 359–70. https://doi.org/10.1016/j.cub.2016.12.031.

Janik, Vincent M., and Peter J. B. Slater. 2000. "The Different Roles of Social Learning in Vocal Communication." *Animal Behaviour* 60 (1): 1–11. https://doi.org/10.1006/anbe.2000.1410.

Konrad, Christine M., Timothy R. Frasier, Luke Rendell, Hal Whitehead, and Shane Gero. 2018. "Kinship and Association Do Not Explain Vocal Repertoire Variation Among Individual Sperm Whales or Social Units." *Animal Behaviour* 145: 131–40. https://doi.org/10.1016/j.anbehav.2018.09.011.

Lyrholm, Thomas, Olof Leimar, Bo Johanneson, and Ulf Gyllensten. 1999. "Sex-Biased Dispersal in Sperm Whales: Contrasting Mitochondrial and Nuclear Genetic Structure of Global Populations." *Proceedings of the Royal Society B* 266 (1417): 347–54. https://doi.org/https://doi.org/10.1098/rspb.1999.0644.

Mann, Janet, Richard C. Connor, Peter L. Tyack, and Hal Whitehead, eds. 2000. *Cetacean Societies: Field Studies of Dolphins and Whales*. University of Chicago Press.

Marcoux, Marianne, Hal Whitehead, and Luke Rendell. 2006. "Coda Vocalizations Recorded in Breeding Areas Are Almost Entirely Produced by Mature Female Sperm Whales (*Physeter macrocephalus*)." *Canadian Journal of Zoology* 84 (4): 609–14. https://doi.org/10.1139/z06-035.

Marcoux, Marianne, Hal Whitehead, and Luke Rendell. 2007. "Sperm Whale Feeding Variation by Location, Year, Social Group and Clan: Evidence from Stable Isotopes." *Marine Ecology Progress Series* 333: 309–14. https://doi.org/10.3354/meps333309.

Miller, Patrick J. O., and David E Bain. 2000. "Within-Pod Variation in the Sound Production of a Pod of Killer Whales, *Orcinus orca*." *Animal Behaviour* 60 (5): 617–28. https://doi.org/10.1006/anbe.2000.1503.

Møhl, Bertel, Magnus Wahlberg, Peter T. Madsen, Anders Heerfordt, and Anders Lund. 2003. "The Monopulsed Nature of Sperm Whale Clicks." *Journal of the Acoustical Society of America* 114 (2): 1143–54. https://doi.org/10.1121/1.1586258.

Murray, Scott O., Eduardo Mercado, and Herbert L. Roitblat. 1998. "Characterizing the Graded Structure of False Killer Whale (*Pseudorca crassidens*) Vocalizations." *Journal of the Acoustical Society of America* 104 (3): 1679–88. https://doi.org/10.1121/1.424380.

Nielsen, Mia L. K., Samuel Ellis, Michael N. Weiss, et al. 2023. "Temporal Dynamics of Mother-Offspring Relationships in Bigg's Killer Whales: Opportunities for Kin-Directed Help by Post-Reproductive Females." *Proceedings of the Royal Society B* 290 (2000): 20230139. https://doi.org/10.1098/rspb.2023.0139.

Noad, Michael J., Douglas H. Cato, M. M. Bryden, Micheline-N. Jenner, and K. Curt S. Jenner. 2000. "Cultural Revolution in Whale Songs." *Nature* 408 (6812): 537. https://doi.org/10.1038/35046199.

Nousek, Anna E., Peter J. B. Slater, Chao Wang, and Patrick J. O. Miller. 2006. "The Influence of

Social Affiliation on Individual Vocal Signatures of Northern Resident Killer Whales (*Orcinus orca*)." *Biology Letters* 2 (4): 481–84. https://doi.org/10.1098/rsbl.2006.0517.

Oliveira, Cláudia, Magnus Wahlberg, Mónica A. Silva, et al. 2016. "Sperm Whale Codas May Encode Individuality as Well as Clan Identity." *Journal of the Acoustical Society of America* 139 (5): 2860–69. https://doi.org/10.1121/1.4949478.

Owen, Clare, Luke Rendell, Rochelle Constantine, et al. 2019. "Migratory Convergence Facilitates Cultural Transmission of Humpback Whale Song." *Royal Society Open Science* 6 (9): 190337. https://doi.org/10.1098/rsos.190337.

Payne, Katharine, and Roger Payne. 1985. "Large Scale Changes over 19 Years in Songs of Humpback Whales in Bermuda." *Zeitschrift für Tierpsychologie* 68 (2): 89–114. https://doi.org/10.1111/j.1439-0310.1985.tb00118.x.

Payne, Katharine, Peter Tyack, and Roger Payne. 1983. "Progressive Changes in the Songs of Humpback Whales *Megaptera novaeangliae*: A Detailed Analysis of Two Seasons in Hawaii." In *Communication and Behavior of Whales*, edited by Roger Payne, 9–57. Westview Press.

Payne, Roger, and Linda N. Guinee. 1983. "Humpback Whales (*Megaptera novaeangliae*) Songs as an Indicator of 'Stocks.'" In *Communication and Behavior of Whales*, edited by Roger Payne, 333–58. Westview Press.

Payne, Roger S., and Scott McVay. 1971. "Songs of Humpback Whales: Humpbacks Emit Sounds in Long, Predictable Patterns Ranging over Frequencies Audible to Humans." *Science* 173 (3997): 585–97. https://doi.org/10.1126/science.173.3997.585.

Pitman, Robert L, and Paul Ensor. 2003. "Three Forms of Killer Whales (*Orcinus orca*) in Antarctic Waters." *Journal of Cetacean Research and Management* 5 (2): 131–40. https://doi.org/10.47536/jcrm.v5i2.813.

Ramp, Christian, Wilhelm Hagen, Per Palsbøll, Martine Bérubé, and Richard Sears. 2010. "Age-Related Multi-Year Associations in Female Humpback Whales (*Megaptera novaeangliae*)." *Behavioral Ecology and Sociobiology* 64 (10): 1563–76. https://doi.org/10.1007/s00265-010-0970-8.

Rekdahl, Melinda L., Rebecca A. Dunlop, Michael J. Noad, and Anne W. Goldizen. 2013. "Temporal Stability and Change in the Social Call Repertoire of Migrating Humpback Whales." *Journal of the Acoustical Society of America* 133 (3): 1785–95. https://doi.org/10.1121/1.4789941.

Rendell, Luke, Sarah L. Mesnick, Merel L. Dalebout, Jessica Burtenshaw, and Hal Whitehead. 2012. "Can Genetic Differences Explain Vocal Dialect Variation in Sperm Whales, *Physeter macrocephalus*?" *Behavior Genetics* 42: 332–43. https://doi.org/10.1007/s10519-011-9513-y.

Rendell, Luke, and Hal Whitehead. 2003. "Vocal Clans in Sperm Whales (*Physeter macrocephalus*)." *Proceedings of the Royal Society B* 270 (1512): 225–31. https://doi.org/10.1098/rspb.2002.2239.

Rendell, Luke, and Hal Whitehead. 2005. "Spatial and Temporal Variation in Sperm Whale Coda Vocalizations: Stable Usage and Local Dialects." *Animal Behaviour* 70 (1): 191–98. https://doi.org/10.1016/j.anbehav.2005.03.001.

Riesch, Rüdiger, John K. B. Ford, and Frank Thomsen. 2006. "Stability and Group Specificity of Stereotyped Whistles in Resident Killer Whales, *Orcinus orca*, off British Columbia." *Animal Behaviour* 71 (1): 79–91. https://doi.org/10.1016/j.anbehav.2005.03.026.

Rouse, Andrew A., Aniruddh D. Patel, and Mimi H. Kao. 2021. "Vocal Learning and Flexible Rhythm Pattern Perception Are Linked: Evidence from Songbirds." *Proceedings of the National Academy of Sciences* 118 (29): e2026130118. https://doi.org/10.1073/pnas.2026130118.

Schall, Elena, Divna Djokic, Erin C Ross-Marsh, et al. 2022. "Song Recordings Suggest Feeding Ground Sharing in Southern Hemisphere Humpback Whales." *Scientific Reports* 12 (1): 1–11. https://doi.org/10.1038/s41598-022-17999-y.

Schevill, William E., and William A. Watkins. 1966. "Sound Structure and Directionality in *Orcinus* (Killer Whale)." *Zoologica: New York Zoological Society* 51: 71–76.

Schulz, Tyler M., Hal Whitehead, Shane Gero, and Luke Rendell. 2011. "Individual Vocal Production in a Sperm Whale (*Physeter macrocephalus*) Social Unit." *Marine Mammal Science* 27 (1): 149–66. https://doi.org/10.1111/j.1748-7692.2010.00399.x.

Schulze, Josephine N., Judith Denkinger, Javier Oña, M. Michael Poole, and Ellen C. Garland. 2022. "Humpback Whale Song Revolutions Continue to Spread from the Central into the Eastern South Pacific." *Royal Society Open Science* 9 (8): 220158. https://doi.org/10.1098/rsos.220158.

Širović, Ana, Erin M Oleson, Jasmine Buccowich, Ally Rice, and Alexandra R Bayless. 2017. "Fin Whale Song Variability in Southern California and the Gulf of California." *Scientific Reports* 7 (1): 1–11. https://doi.org/10.1038/s41598-017-09979-4.

Stafford, Kathleen M., Christian Lydersen, Øystein Wiig, and Kit M. Kovacs. 2018. "Extreme Diversity in the Songs of Spitsbergen's Bowhead Whales." *Biology Letters* 14 (4): 20180056. https://doi.org/10.1098/rsbl.2018.0056.

Stimpert, Alison K., Lindsey E. Peavey, Ari S. Friedlaender, and Douglas P. Nowacek. 2012. "Humpback Whale Song and Foraging Behavior on an Antarctic Feeding Ground." *PLOS ONE* 7 (12): e51214. https://doi.org/10.1371/journal.pone.0051214.

Stredulinsky, Eva H., Chris T. Darimont, Lance Barrett-Lennard, Graeme M. Ellis, and John K. B. Ford. 2021. "Family Feud: Permanent Group Splitting in a Highly Philopatric Mammal, the Killer Whale (*Orcinus orca*)." *Behavioral Ecology and Sociobiology* 75 (3): 1–17. https://doi.org/10.1007/s00265-021-02992-8.

Thomsen, Frank, D Franck, and John K. B. Ford. 2001. "Characteristics of Whistles from the Acoustic Repertoire of Resident Killer Whales (*Orcinus orca*) off Vancouver Island, British Columbia." *Journal of the Acoustical Society of America* 109 (3): 1240–46. https://doi.org/10.1121/1.1349537.

Tyack, Peter L. 2022. "Social Organization of Baleen Whales." In *Ethology and Behavioral Ecology of Mysticetes*, edited by Christopher W. Clark and Ellen C. Garland, 147–75. Springer. https://doi.org/10.1007/978-3-030-98449-6_7.

Vachon, Felicia, Taylor A. Hersh, Luke Rendell, Shane Gero, and Hal Whitehead. 2022. "Ocean Nomads or Island Specialists? Culturally Driven Habitat Partitioning Contrasts in Scale Between Geographically Isolated Sperm Whale Populations." *Royal Society Open Science* 9 (5): 211737. https://doi.org/10.1098/rsos.211737.

Van Cise, Amy M., Sabre D. Mahaffy, Robin W. Baird, T. Aran Mooney, and Jay Barlow. 2018. "Song of My People: Dialect Differences Among Sympatric Social Groups of Short-Finned Pilot Whales in Hawai'i." *Behavioral Ecology and Sociobiology* 72 (12): 193. https://doi.org/10.1007/s00265-018-2596-1.

Vernes, Sonja C., Buddhamas Pralle Kriengwatana, Veronika C. Beeck, et al. 2021. "The Multi-Dimensional Nature of Vocal Learning." *Philosophical Transactions of the Royal Society B* 376 (1836): 20200236. https://doi.org/10.1098/rstb.2020.0236.

Warren, Victoria E., Rochelle Constantine, Michael Noad, Claire Garrigue, and Ellen C. Garland. 2020. "Migratory Insights from Singing Humpback Whales Recorded Around Central New Zealand." *Royal Society Open Science* 7 (11): 201084. https://doi.org/10.1098/rsos.201084.

Watkins, William A., and William E. Schevill. 1977. "Sperm Whale Codas." *Journal of the Acoustical Society of America* 62 (6): 1485–90. https://doi.org/10.1121/1.381678.

Weinrich, Mason T., Howard Rosenbaum, C. Scott Baker, Alexis L. Blackmer, and Hal Whitehead. 2006. "The Influence of Maternal Lineages on Social Affiliations Among Humpback Whales (*Megaptera novaeangliae*) on Their Feeding Grounds in the Southern Gulf of Maine." *Journal of Heredity* 97 (3): 226–34. https://doi.org/10.1093/jhered/esj018.

Whitehead, Hal. 2018. "Sperm Whale: *Physeter macrocephalus*." In *Encyclopedia of Marine Mammals*, edited by Bernd Würsig, Johannes G. M. Thewissen, and Kit M. Kovacs, 919–25. Academic Press. https://doi.org/10.1016/C2015-0-00820-6.

Whitehead, Hal, John K. B. Ford, and Andrew G. Horn. 2023. "Using Culturally Transmitted Behavior to Help Delineate Conservation Units for Species at Risk." *Biological Conservation* 285: 110239. https://doi.org/10.1016/j.biocon.2023.110239.

Whitehead, Hal, and Janet Mann. 2000. "Female Reproductive Strategies of Cetaceans: Life Histories and Calf Care." In *Cetacean Societies: Field Studies of Dolphins and Whales*, edited by Janet Mann, Richard C. Connor, Peter L. Tyack, and Hal Whitehead, 219–46. University of Chicago Press.

Whitehead, Hal, and Luke Rendell. 2004. "Movements, Habitat Use and Feeding Success of Cultural Clans of South Pacific Sperm Whales." *Journal of Animal Ecology* 73 (1): 190–96. https://doi.org/10.1111/j.1365-2656.2004.00798.x.

Whitehead, Hal, and Luke Rendell. 2014. *The Cultural Lives of Whales and Dolphins*. University of Chicago Press.

Whitehead, Hal, and Linda Weilgart. 2000. "The Sperm Whale: Social Females and Roving Males." In *Cetacean Societies: Field Studies of Dolphins and Whales*, edited by Janet Mann, Richard C. Connor, Peter Tyack, and Hal Whitehead, 154–72. University of Chicago Press.

Zandberg, Lies, Robert F Lachlan, Luca Lamoni, and Ellen C Garland. 2021. "Global Cultural Evolutionary Model of Humpback Whale Song." *Philosophical Transactions of the Royal Society B* 376 (1836): 20200242. https://doi.org/10.1098/rstb.2020.0242.

13

Culture and Conservation

PHILIPPA BRAKES AND HAL WHITEHEAD

Abstract

Culture is important to cetaceans, likely all cetaceans, even though our knowledge is currently greatly skewed to just a few species. Evolving methods will provide new insights into the processes of cultural transmission as well as how culture interacts with conservation issues. Most fundamentally, the diversity of behavior driven by culture is part of the global biodiversity that the conservation impetus is trying to preserve. Culture can increase resilience to environmental change through the spread of adaptive innovations. But culture can also create vulnerability when conformism inhibits adaptation or when traits that cause conflict with humans, such as depredation, spread quickly through a population. Behavioral patterning created by social learning can be important when delineating conservation units. We suggest how conservation managers should integrate scientific knowledge on cultural processes into policy and practice and consider how culture can be incorporated into wider conservation biodiversity frameworks. Knowledge of cetacean cultures can change the conservation landscape in significant ways, and we urge conservation managers to work to conserve cultural capacity.

Introduction

Much of cetacean behavior is—or at least seems to be—culturally determined (i.e., transmitted by social learning; see box 13.1 for definition of culture), and culture is important across a broad range of behavioral domains (Whitehead and Rendell 2014). From foraging strategies learned from mothers (Kopps and Sherwin 2012), or in collaboration with humans (Cantor et al. 2023), to migration (Carroll et al. 2014) and song (Payne and McVay 1971). While we have only recently developed the tools to identify and document nonhuman

Box 13.1. Definitions of Culture and Social Learning

These definitions (or variants) are generally used by biologists studying cultural behavior in nonhumans.

Culture. Information or behavior—shared within a community—that is acquired from conspecifics through some form of social learning (Whitehead and Rendell 2014).

Social learning. Learning that is influenced by observation of, or interaction with, another animal or its products (Heyes 1994).

cultures, some of these cultures are likely ancient. Studying the cultural histories of cetaceans is challenging. Nevertheless, since the advent of scientific monitoring, we can demonstrate that the foraging strategies of some killer whale cultural groups are at least 50 years, or two generations, old (Ford and Ellis 2014; see the section "Conservation of Biodiversity" in this chapter). These cultures may have both contemporary relevance, as well as potential evolutionary significance (Whitehead et al. 2019).

The advent of scientific methods to identify and measure cultural behavior in cetaceans brings an important responsibility for wildlife managers (Whitehead 2010; Brakes et al. 2019). Culture is important because it affects behavior, in particular, how behavior is transmitted among individuals. Thus, as began to be recognized in about 2005 (Whitehead et al. 2004; Ryan 2006), conservation should include a consideration of culture. (The significance of cultural behavior for conservation is not mentioned in *Cetacean Societies* [2000]. One of us was the first author of the chapter on conservation.)

Especially when coupled with innovations, social learning can change population behavior fast, for good (if the new behavior helps ameliorate anthropogenic effects, e.g., hook avoidance in fish; Lovén Wallerius et al. 2020) or ill (if the new behavior increases conflict with humans, e.g., crop raiding in elephants). Additionally, cultural conformism (Henrich and Boyd 1998) can inhibit behavioral plasticity and adaptation to anthropogenic change (Whitehead 2010). Culture can also constrain behavior within social boundaries, setting up potential evolutionarily significant units (Brakes et al. 2019). A good understanding of these factors will help the management of human behavior to the benefit of species and ecosystems. For some species, social learning can be manipulated to achieve conservation outcomes (e.g., using social learning to give reintroduced or translocated animals appropriate behavior for their new habitat, or showing naive animals appropriate migratory paths; Brakes et al. 2021). We suspect that this will rarely be the case for cetaceans. But we can use evolving knowledge in this field to help understand the patterning created by social learning and culture and consider what this means for their conservation.

The key question from a management perspective may be whether there is exchange of heritable information that may influence the survival, reproduction, or dispersal of animals; their interactions with humans; or their responses to management initiatives or conservation intervention. Social learning and culture present several challenging questions for policymakers:

a. What does the variation and patterning of behavior created by social learning mean for the delineation of social units for conservation purposes?
b. How can the processes of social learning be utilized to increase the efficacy and efficiency of conservation interventions?
c. How should conservation managers integrate the emergent scientific knowledge in this field into policy and practice?

On first inspection, these may seem reasonably straightforward. In practice, distilling salient, generic management advice in this field is challenging. Social systems are, by nature, multifaceted. When these systems come under pressure from human activities the outcomes related to the flow of information across social networks, within and between populations, can be complex. Traditional conservation has focused on conserving genetic diversity across generational timescales. Genetic adaptation, shaped by the processes of natural selection, requires time for adaptive variants to rise to sufficiently high frequency; the serendipity of an advantageous de novo mutation followed by the processes of selection; or genetic rescue, which is challenging in marine mammals. The processes of social learning can operate both within and between generations, elevating nimble decision-making, but from a conservation perspective, the speed of transmission is perhaps both the most advantageous and also the most precarious aspect of these processes. For example, social learning between peers can facilitate the spread of adaptive foraging strategies in more behaviorally plastic species such as bottlenose dolphins (Ansmann et al. 2012). In contrast, the spread of maladaptive information, within a generation (or shorter), has the potential to significantly exacerbate conservation issues.

Typically, then, we need information—the more the better—on how communities of animals are structured and how they behave. The behavioral data should be as extensive as possible, spatially and temporally, as this spatial and temporal variation has proved key to understanding cetacean culture. We need to collaborate across and between ocean basins (McDonald et al. 2006; Hersh et al. 2022), and we need to rescue those old recordings, scan data and focal follows from their archaic recording formats and disintegrating data sheets (e.g., Whitehead et al. 2021).

In the following six sections we consider some of the issues facing cetacean scientists, conservationists, and managers. In each section we will start with the scientific perspective, and then, where appropriate, discuss implications for policy and management.

Which Species Have Culture?

Our knowledge of the distribution of culture across nonhuman species is skewed toward some of the most studied species. We have good knowledge of bird song and chimpanzee culture, a fair understanding of parts of culture of orangutans (Krützen et al. 2011) and New Caledonian crows (Hunt and Gray 2003), but for the great majority of vertebrates, while there is some evidence for social learning, there is less evidence of temporally stable cultures (Whiten 2017). What about cetaceans?

Study of cetacean behavior (Wild and Rendell, this volume; Hersh et al., this volume), and consequent knowledge of culture, has been overwhelmingly focused on four genera: bottlenose dolphins, killer whales, sperm whales, and humpback whales. For each genus we have accumulated a substantial body of evidence that social learning, and culture, has a role in a very wide range of behavior, in movements and migrations, foraging behavior, social behavior, vocalizations, and even play (Whitehead and Rendell 2014; see also Wild and Rendell, this volume; Hersh et al., this volume). We have much more limited, but persuasive, evidence for culture in a growing number of other cetacean species from blue whales (McDonald et al. 2006) to Irrawaddy dolphins (Ambie et al. 2023).

The conservation of cultural species and the ecosystems in which they live must be based on knowledge of those species and their cultures. But, for most cetaceans, that knowledge is not easily gained. Some insist on high evidentiary standards before attributing culture to nonhumans (Laland and Janik 2006). However, conservation decisions are typically based upon the "best available evidence" (Cook et al. 2016) and use the "precautionary approach" (Cooney 2012), both of which guide us in the use of less than perfect information of which we have plenty on cetacean culture (Whitehead and Rendell 2014).

From these perspectives we believe that, despite the very skewed levels of evidence, we should at least consider the role of culture in conservation deliberations for all species of Cetacea. Where we know little, it will often make sense to extrapolate from phylogenetically related species, for instance assuming that baleen whale migration routes are at least partially socially learned given the evidence particularly for humpback whales, or, from all the data on killer whales that social units of pilot whales (and related species) may have distinctive cultural behavior. If culture itself, or correlates of the incidence of cultural behavior, influences the intensity and nature of field research on a species, then the great variation in evidence for culture among species of Cetacea might reflect real differences in the importance of culture. However, there is no evidence that such a correlation exists and so it is precautionary to assume that culture is significant for all cetacean species (Whitehead and Rendell 2014). Evidence for culture in cetaceans generally requires either extremely detailed long-term studies at the level of the individual (e.g., bottlenose dolphins in Shark Bay); well-documented changes in behavior over time (e.g., humpback song and lobtail feeding); or stable modular social systems (e.g., sperm whales and killer whales). However, if behavior does not change clearly over time and the social system is not modular, this does not necessarily mean there is no culture, it may just mean we have not detected it. Bottlenose whales have been studied for over 30 years using individual identification but have yet to yield evidence for culture. Nevertheless, there could be a great deal of cultural transmission in this species, but because these conditions do not hold, we cannot detect it.

Documenting Culture

We need to document cetacean culture. The most direct methods of establishing the role of culture in nonhumans—involving fostering, translocation, or direct observation of social learning and behavior change among individuals (Laland and Hoppitt 2003; Laland and Janik 2006)—are generally unavailable for cetaceans outside the captive environment, and there is very restricted scope within it. Instead, we typically use data on how, in a population of a target species, behavior varies with space, time, and across the social network. A major challenge is to disentangle the effects of social learning, individual learning, environmental variation, and genetic causation. All of these can cause behavioral patterning within a population. An important initial perspective on identifying nonhuman culture was that genetic and environmental causation needed to be ruled out before culture could be accepted (Laland and Hoppitt 2003; Laland and Janik 2006), a procedure often known as the ethnographic method (see Wild and Rendell, this volume). This was, and is, difficult, and is unduly restrictive as kinship, social learning, individual learning, and environmental variation often operate together both for humans and nonhumans (Whitehead et al. 2019).

However, three types of evidence, available for some cetaceans, can help identify the significance of cultural transmission:

1. When behavior changes synchronously among individuals within a lifetime (e.g., the songs of humpback and blue whales; Payne 2000; McDonald et al. 2009), genetic determination is not supported.
2. If communities with distinct behavior are sympatric (i.e., use the same environment) as with killer whales in the northeastern Pacific and Antarctic (Riesch et al. 2012), environmental causation can usually be ruled out.
3. Variation in social signals, such as the greeting ceremonies of the "resident" killer whale communities (Riesch et al. 2012), the codas of sperm whales (Hersh et al. 2022) and the songs of baleen whales, probably have little environmental determination.

If we show or assume that culture is important for a species, then the bottom line is, how does culture manifest for this species? This is a fundamental question. It includes considerations such as: How is behavioral information transmitted through the social network (vertically from mothers, horizontally from peers, or . . .), and are there biases in the transmission (such as conformism, leadership, content biases)? These are often difficult questions for wild cetaceans, but techniques such as network-based diffusion analysis can provide remarkable illumination of transmission mechanisms, as well as the roles of environment and genetics in well-studied populations (e.g., Allen et al. 2013).

The nature, extent, and processes of cetacean cultural behavior are of interest as natural phenomena, but they also have important implications for conservation. From this perspective, there needs to be emphasis on documenting cultural behavior that relates to our conservation concerns, such as data on migratory behavior when thinking about climate change (Keith and Bull 2017), or foraging behavior when the focus is interactions with fisheries (Straley 2012).

The timeframes of cultural processes are key to evaluating significance for conservation. An ephemeral fad-like culture (such as tail walking in bottlenose dolphins; Bossley et al. 2018) is unlikely to have the same significance for conservation as a temporally stable foraging culture passed through generations, as in killer whales (Ford et al. 2010). Additionally, cultural content that is critical to survival or reproduction, such as foraging behavior, is more likely to be important than play-like behavior. Nevertheless, if a fad such as tail walking were to reduce fertility or survival within a social group, then conservationists should keep a close watch on its propagation.

Social learning processes driving cultural evolution can be important to conservation in both the more urgent near term (e.g., in relation to provisioning or depredation; Mann and Kemps 2003; Sigler et al. 2008), but also in the longer term in relation to issues such as hybridization or speciation. Conservation policy needs to respond to within-generation shifts in population behavior, while continuing to monitor longer-term trends. This may be particularly the case when cultural transmission impacts vital rates (survival, reproduction, and dispersal).

Conservation of Biodiversity

The primary goal of conservation can be seen as the preservation of biodiversity, the "full variety of life on Earth" (Takacs 1996). Social learning drives behavioral diversity, and thus a part of overall biodiversity is cultural.

How to preserve cultural biodiversity? First, we need to know it is there, so the systematic, broad, and consistent documentation of behavior described in the "Documenting Culture" section in this chapter is fundamental. Second, we need to know that the behavioral patterns are stable, as naturally labile behavior patterns may not specifically need to be preserved as part of biodiversity. However, conserving the capacity for cultural variation within populations may be valuable, to increase the probability of useful variants arising (Brakes et al. 2021). Showing that behavior patterns are stable is not trivial. Ideally, we have records of behavior over generations (McDonald et al. 2009), but such time series are very rare for cetaceans. However, strong correlations between behavioral patterns and genetic variation (e.g., mitochondrial haplotypes with matrilineally transmitted cultural behavior) can provide strong indications of stability (Yurk et al. 2002; Rendell et al. 2012; Kopps et al. 2014).

Then, given stable patterns of behavioral variability, we can think about protection. Part of this comes down to protecting the behavioral substrate. For instance, if a type of behavior includes eating, playing with, or interacting with another type of organism or environmental feature, then these organisms and features need to be conserved. So, both the preferred prey of southern resident killer whales, Chinook salmon (Ford et al. 2010), and the beaches on which the northern resident killer whales rub (Riesch et al. 2012), need to be protected. Second, the social network needs to support the social learning of the behavior. We need to prevent communities becoming too small or too partitioned to allow the behavior, or the knowledge of the behavior, to persist. The merging of previously distinct social communities—observed in some cetacean societies when the social elements get small (Konrad et al. 2018)—will tend to reduce cultural biodiversity. So, conservation policy and action should be aimed at maintaining complex social networks, as well as the external resources essential for behavioral diversity.

For such actions to have priority, there needs to be an acceptance by those organizations and individuals who structure conservation actions that cultural information and cultural processes are important elements of biodiversity. This acceptance has been slow coming, but there has been some progress. The Committee on the Status of Endangered Wildlife in Canada (COSEWIC) assesses the conservation status of Canadian species. COSEWIC delineates population units to be assessed and protected as "discrete and evolutionarily significant, where discrete means that there is currently very little transmission of heritable (cultural or genetic) information from other such units, and evolutionarily significant means that the unit harbors heritable adaptive traits or an evolutionary history not found elsewhere in Canada" (COSEWIC, Operations and Procedures Manual, app. F5). Thus, in this respect, cultural information is set on a level footing with genetic information.

More broadly, the UN Environment Programme's Convention on the Conservation of Migratory Species of Wild Animals (CMS), which has been exploring the significance of social learning and culture for the conservation of a wide range of vertebrate species, developed a "Concerted Action" for sperm whale social units in the eastern tropical Pacific (Brakes et al. 2019). These social units can be identified by their coda vocalizations (Rendell and Whitehead 2003) and may utilize their habitat differently, as indicated by movement patterns and foraging success (Whitehead and Rendell 2004). Building capacity and fostering collaboration between researchers across the CMS range states of these social units may provide important insights on habitat use to guide the management of these sperm whale clans (Eguiguren et al. 2019).

Resilience

As the Earth is changed by humans, most comprehensively through climate change, but also in many other ways, we need other organisms to persist. Resilience—the ability to maintain function in the face of exogenous disturbance (Hodgson et al. 2015)—becomes particularly valuable. Social learning and culture can be important generators of resilience. Animals can learn new and adaptive ways quickly from one another, without each having to make the change independently. The models for social learning can be an inventor of a new behavior, an individual that learned it elsewhere, or animals that learned at one or more steps from these original demonstrators. New and adaptive behavior can spread fast (e.g., Whitehead et al. 2021) and promote resilience in the face of change, without being constrained (at least in the short term) by the intergenerational timescale of the processes of natural selection of genes. The enhanced behavioral plasticity of cultural transmission may provide enhanced opportunities for adapting to ecological changes as they arise, as shown by the trawler-feeding bottlenose dolphins in Moreton Bay, Australia, who switched their behavior and modified their social network when the fishery they exploited was terminated (Ansmann et al. 2012).

As scientists, we would like to know how this works, so that we can document cultural transmission, and assess the capacity for resilience. We need information on the spread of new behavior through a social network. And this, in turn, needs much observation and data collection, both of changes in behavior and the social network, typically over a long time span. A particularly good example is Allen et al.'s (2013) study of the spread of a new feeding behavior through a social network of humpback whales over 27 years (see Wild and Rendell, this volume).

While we generally think of social learning and culture as enhancing adaptation to environmental change, cultural conformism can constrain behavior in ways that are not adaptive in the face of a changing environment—individuals are driven to conform to the behavior of the majority, even if that is no longer the best option. They may continue to select a scarce food type, or a polluted habitat, even though more available prey types or more pristine environments are available. The southern resident community of killer whales seems to show both these types of cultural maladaptation (Ford et al. 2010). Once again, to understand how culture may inhibit resilience we need to understand how the incorporation of new behavior may be resisted across a social network, using detailed long-term data on both behavior and network structure.

Spread of Harmful Traits

The spread of new behavior through a population can be harmful to its members, setting up new conservation problems. Such behaviors typically involve the animals using human activities to their short-term benefit but long-term cost. They may depredate fishing gear or beg from recreational vessels (Lauriano et al. 2009; Sigler et al. 2008; Noke and Odell 2002). Such behaviors can spread fast through a population via social learning. A nice meal is easily

FIGURE 13.1. (*Left*) An underwater camera catches a sperm whale carefully removing a sablefish from a longline off southeast Alaska, a seemingly socially learned behavior. (*Right*) A small crowd of tourists watch as a volunteer feeds a resident adult female Indo-Pacific bottlenose dolphin at Monkey Mia, Western Australia.
Sources: (*Left*) Photo courtesy of SEASWAP (see Schakner et al. 2014). (*Right*) Photo courtesy of Simon J. Allen, Shark Bay Dolphin Research.

obtained, but the animals may get caught in nets, rammed by angry fishers, or have a beer bottle thrust into their blowhole by an inebriated boater.

To try to ameliorate such situations, we need to understand the behavior itself and make changes to human behavior. An example is the detailed study of the depredation of fishers' longlines off southeast Alaska by sperm whales (the SEASWAP project; fig. 13.1, left). The study showed that the behavior is likely socially learned (Schakner et al. 2014) and SEASWAP is searching for changes in the practice of the fishery that might reduce depredation (Straley 2012). Knowledge of the social learning process can also be important. For instance, in Shark Bay, Western Australia, bottlenose dolphins were provisioned at a tourist beach (fig. 13.1, right). Detailed research showed that calves learn foraging methods from their mothers (Sargeant and Mann 2009) and that calves born to provisioned females had twice the mortality of calves born to non-provisioned females (Mann and Kemps 2003). These results led to changes in feeding policies, which reduced calf experience of provisioning and led to a dramatic increase in the survival of the calves of provisioned females (Mann and Kemps 2003; Foroughirad and Mann 2013).

Good knowledge of the behavior (including the role of social learning) of humans and cetaceans can lead to robust management. Then, as with the provisioned dolphins of Shark Bay, an adverse interaction may be resolved or mitigated. Often however, especially with depredation of fisheries, too much is at stake for both parties (livelihoods and sustenance, respectively) for a simple solution. These examples illustrate that, as in the case of the dedicated SEASWAP consortium, interpreting how the patterning created by social learning and culture can interface with human activities and conservation is truly a wicked problem (Head 2008; Rittel 1973).

Population Delineation

There has been a long debate about how best to divide a population for conservation purposes, or whether to divide at all. Theoretically, this has focused on the concept of the "evolutionarily significant unit" (ESU; Waples 1991; Fraser and Bernatchez 2001; Ryder 1986; Casacci et al. 2014), with almost all of the emphasis on using genetic data to delineate such units (Casacci et al. 2014). Genetic data can be very informative about phylogenetic patterns and evolutionary processes, but the assumption has often been that biodiversity equals genetic diversity (see Fraser and Bernatchez 2001). We (and others) disagree, seeing biodiversity as more equivalent to phenotypic diversity, with phenotypic diversity itself being the result of genetic, cultural, ecological, and other evolutionary processes (Laland et al. 2015).

To use culturally transmitted behavior to delineate conservation units, we need to know about the behavior, and, in particular, about its patterning in space, time, and with the social network. The most promising, and used, types of behavior for delineating conservation units are songs and calls, migrations, foraging behavior including diet, and social signals. Songs and calls are relatively easily documented, especially for cetaceans, and have often been used to indicate discrete population elements (e.g., blue whales [McDonald et al. 2006], sperm whales [Hersh et al. 2022]). Usually, song or call dialects are not evolutionarily significant, but if they proscribe mating opportunities (as in killer whales; Barrett-Lennard 2000) or social relationships (e.g., the xenophobia of killer whale ecotypes and communities; Barrett-Lennard 2011; Barrett-Lennard et al. 1996) they may be considered evolutionarily significant.

Migrations are especially useful for delineating conservation units, as distinct stable migratory patterns indicate both that population elements are discrete and that the distinction is evolutionarily significant, as the members of the units are using different habitats in a consistent fashion. Examples of using migrations to define conservation units include beluga whales in Canada, and the Distinct Population Segments of humpback whales identified by the US National Marine Fisheries Service. Differences in foraging behavior and diet are more challenging as evidence for distinct conservation units as they are more parsimoniously ascribed to ecological differences and the behaviors may be plastic. When the different putative conservation units are sympatric and the foraging/diet distinctions are striking, as with some pairs of killer whale ecotypes (Riesch et al. 2012), distinct conservation units are more readily supported. Social signals are largely free from the ecologically determined counterargument but have been little used for the delineation of conservation units (an exception is the greeting ceremonies shown by southern resident killer whales, but not northern residents; Ford et al. 2000).

Showing stability is challenging. The behavioral distinctions between putative conservation units need to be stable, not necessarily the details of the behavior itself. Humpback whale songs change with time, but at any time songs in different oceans are distinct so this is good evidence for discreteness. At best we only have observations spanning two to three cetacean generations when looking for the stability of behavioral traditions in cetaceans, but we can navigate this using methods mentioned in the "Documenting Culture" section in this chapter such as genetic correlations indicating stability.

The identities of the putative conservation units being considered need to originate somewhere. They could be named or suspected subspecies, or be based upon previous delineations including those made for a different fundamental purpose (e.g., "management units"), or use habitat distinctions (e.g., ocean areas for pelagic species, or estuaries for inshore dolphins). However, it is much better if the data themselves suggest putative conservation units, either subjectively (e.g., McDonald et al. 2006 for blue whales) or, even better, objectively using clustering techniques (e.g., Hersh et al. 2022 for sperm whales). This is especially the case where the potential conservation units are to some degree sympatric, as with sperm whale clans.

Having come up with putative conservation units, we need to assess the evidence. This is perhaps the hardest and least rigorous part of the process. It is possible to use measures of population division such as the cultural fixation index (CF_{ST}), which can be used to quantify cultural differences between populations (Boesch et al. 2020), following the very widely used fixation index (F_{ST}) of molecular genetics. But it is not clear what values of CF_{ST}, or F_{ST}, indicate discreteness. To our knowledge, CF_{ST} has not been used in the delineation of cetacean population units. In well-studied populations, such as humpback whales in the North Pacific or North Atlantic, we may have estimated rates of interchange between alternative migratory endpoints of seasonal migrations, and these can guide delineation decisions (Cheeseman et al. 2022). Furthermore, evidence from culturally transmitted behavior has to be used in conjunction with genetic data, morphological data, information from traditional knowledge and other information types, to make these decisions. This is hard!

These are some of the scientific issues when using cultural data to delineate conservation units within species—which can receive different listings (e.g., "critically endangered," "threatened," or "least concern"), and thence different levels of protection. There are "subpopulations" (IUCN), "distinct population segments" (USA), "designatable units" (Canada), and so on. The definitions and practices for delineating conservation units vary across agencies, but all require some measure of discreteness between the different units, and sometimes there is an additional requirement for the units to each be evolutionarily significant (Brakes et al. 2021, fig. 13.1). Discreteness implies that if the unit were removed, repopulation from other units would not happen or take a very long time, i.e., demographic isolation. Evolutionary significance requires that each unit contains information that is unique and that cannot be reconstituted (at least in the short term). Traditionally, this information was assumed to be genetic, but there is now some awareness (e.g., in the COSEWIC and CMS processes; see "Conservation of Biodiversity" section in this chapter) that culturally transmitted information can be evolutionarily significant. As an example, if migratory routes are socially learned, and the animals making a particular migration are lost, then the presence, and ecological role, of the species in particular habitats is gone and the resilience of the species to environmental change may be reduced. This has been argued for baleen whales whose profound population collapses caused by industrial whaling may have led to the loss of culturally transmitted knowledge of important habitat (Clapham et al. 2008). Such reasoning suggests the listing of groups of animals with distinct socially learned migratory routes that are stable over generations as separate conservation units. This was the essence of the reasoning behind COSEWIC's listing of the belugas that summer in eastern and western Hudson Bay as distinct units, even though the units mix and likely mate across units in the winter and spring (COSEWIC 2020).

The integration of cultural data into the COSEWIC assessments provides an important benchmark for other management authorities. But there remain many important questions related to how behavioral granulation created by cultural processes should influence thresholds in management systems, such as for the calculation of sustainable bycatch thresholds (e.g., "potential biological removals" [PBR]) or in the strike or catch-limit algorithm used by the International Whaling Commission (IWC) to calculate whaling quotas.

Conclusion

Emergent understanding of the processes of cultural transmission—and their consequences—gives rise to some fundamental questions for conservation scientists and managers, the answers to which may change the conservation landscape in significant ways.

- Should we conserve all cultures, irrespective of content, in the interest of biocultural diversity?

- Should we focus effort only on cultures that appear to create resilience or that contain potentially adaptive traits or variants?
- Should we focus on cultures in endangered species, or small isolated populations? Or in small culturally discrete units within wider populations?
- Or should we primarily focus efforts on managing populations where social learning is creating vulnerability, for example through isolation, or specialization creating overdependence on one prey type?
- How should we manage apparently maladaptive cultures when they arise? Should we assume that they are self-limiting in the same way that genetic maladaptation ultimately is limited by the processes of natural selection?
- Should we focus on specific content, e.g., migration or foraging, which may be critical to survival or reproduction?
- How do social network connections influence the flow of information and what does this mean for conservation?
- How does social learning interface with genetic evolution to create a complex of gene-culture coevolution?

This is a long list of questions. Only some of which we can begin to address.

We need decisions that are as well informed as possible. To achieve this, it may be useful to look toward specific indicators, such as: specialization, rapid spread of innovation, specific social structures, life history characteristics such as post-reproductive lifespan, behavioral plasticity, or conservatism.

Science has described only some of the cultural processes that may be in play in shaping cultural cetacean populations. Some of these processes may further interact or generate as-yet undescribed feedback loops. Amid all this uncertainty, conservationists and managers need some clarity on when to act. We have provided embryonic ideas for a new generation of conservation models and assessments that may capture some of the important phenotypic diversity generated by cultural processes. Nevertheless, the evidence is now unequivocal that cultural processes in nonhumans, including cetaceans, have far-reaching consequences for conservation. From shaping domestic policy through national forums (e.g., COSEWIC), to collaboration among range states in international policy instruments, such as CMS, a new conservation paradigm, beyond genes alone, is being ushered in (Brakes et al. 2019, 2021).

Table 13.1 provides examples of specific instances where cultural process and conservation interface, either in generating resilience or creating vulnerability. As a stimulus for future assessments, we also provide a list of examples where cultural processes have been, or could be, incorporated in policy settings (table 13.2).

Conservation managers should use cultural information to augment genetic assessments but also work to conserve cultural capacity within and

TABLE 13.1. Implications of social learning and culture for conservation efforts

Conservation considerations	*Process*	*Examples*
Delineating conservation units	Social learning generates putative conservation units	Social learning of acceptable mates sets up distinct conservation units in killer whales (Barrett-Lennard 2000). Social learning of migration routes sets up conservation units in beluga whales (O'Corry-Crowe et al. 2018). Social learning of communication vocalizations sets up social barriers in sperm whales (Cantor et al. 2015).
Resilience and "buffering"	Social learning provides access to novel, abundant or additional resources	Indo-Pacific bottlenose dolphins in Western Australia were buffered against the cascading effects of a marine heatwave on the food web through socially learned foraging strategies. Survival of "spongers"—users of a complex foraging technique reliant on social learning—declined less during the heatwave (5.9%) compared with non-spongers (12.2%; Wild et al. 2019).
Potential resilience *or* potential vulnerability and ecological traps	Social learning provides access to resources/habitat, novel behavior to protect from threat, or constrains behavior so that threat is not avoided	Population projections from surveys of critically endangered vaquita indicated that more vaquitas have survived than expected. One reason may be that surviving adult vaquitas have learned to avoid entanglement in gillnets (Rojas-Bracho et al. 2022). It remains to be seen whether this behavior is transmitted to offspring in this critically endangered population. Migration in southern right whales, which is maternally led (Carroll et al. 2015; Valenzuela et al. 2009), could create an "ecological trap" if climate change results in spatial shifts in productivity away from areas traditionally used by these whales (Keith and Bull 2017). It is unknown whether the species has sufficient behavioral plasticity to exploit new areas or prey resources (Seyboth et al. 2016; Leaper et al. 2006).
Vulnerability	Social learning creates vulnerability through specialization on vulnerable or declining resource(s)/habitat	The calving rate of southern resident killer whales, which have a socially learned specialization on a declining chinook salmon population, differed by 50% between years of low and high salmon abundance (Ward et al. 2009), suggesting a strong link between salmon abundance and fertility. Mutualism between bottlenose dolphins and Brazilian fishers, the result of a rare example of inter-species cultural transmission between species, has existed for over a century and may be vulnerable to a reduction in abundance in the target mullet population, which may increase the exposure of dolphins to bycatch in alternative fisheries (Cantor et al. 2023).
Social learning as conservation tool	Knowledge of processes of social learning used to improve conservation outcome	Depredation from longlines in southeast Alaska by sperm whales is likely socially learned (Schakner et al. 2014) and changes in fisheries practice may reduce depredation (Straley 2012).

TABLE 13.2. Cultural processes and policy settings

Forum/mechanism	*Species*	*Details*
Culture has been incorporated into domestic or international management instruments		
COSEWIC (Canada)	Beluga	Beluga that summer in eastern and western Hudson Bay are listed as distinct units. Although the units mix and likely mate across units in winter and spring, they have distinct socially learned migratory routes that are stable over generations, which warrant management as separate conservation units, with separate status (Eastern Hudson Bay—Threatened; Western Hudson Bay—Not at Risk; COSEWIC 2020).
EPBC (Australia)	Southern right whale	Southern right whale fidelity to wintering grounds is vertically transmitted and led to recognition of demographically independent populations (including under the IWC stock assessments and Australian domestic legislation), as the result of socially learned migratory destinations (Carroll et al. 2015; Brakes et al. 2021).
Department of Conservation (New Zealand)	Southern right whale	Similarly, the New Zealand population is recognized as a distinct stock by the IWC and the New Zealand Department of Conservation (Carroll et al. 2012).
ESA (USA)	Killer whale	Southern residents recognized as "Distinct Population Segment" in 2004 and listed as endangered in 2005, based on "differences in cultural traditions" (https://www.federalregister.gov/d/05-22859).
	False killer whale	The insular population of false killer whales of the main Hawaiian Islands was assessed under the ESA (Baird 2009; Oleson et al. 2010), and it was concluded that "Hawaiian insular false killer whales represent a significant cultural unit independent of other false killer whale populations" (Oleson et al. 2010). The population was listed as endangered under the ESA in 2012, and a subsequent review recommended action to recover the population, including addressing threats from fisheries, as social clusters have different injury rates (NOAA 2022).
CMS	Sperm whale	Sperm whale social units in the eastern tropical Pacific form clans that can be identified by their coda vocalizations (Rendell and Whitehead 2003), but they also differ in their movement patterns, feeding success, and other attributes (e.g., Whitehead and Rendell 2004). A "Concerted Action" under CMS is fostering regional capacity-building and collaboration between academic institutions and Range States—sperm whales in the area move between several CMS Range States—using integrated research methods to better elucidate the importance of clan structure for conservation (Brakes et al. 2019), for example, elucidating habitat use by the different clans (Eguiguren et al. 2019).

(*continued*)

TABLE 13.2. *(continued)*

Forum/mechanism	*Species*	*Details*
Culture has potential for incorporation into management		
IMMAs		The IMMA concept, developed by the IUCN Joint SSC/WCPA Marine Mammal Protected Areas Task Force ("MMPA Task Force" or "Task Force"), is modeled on the successful example of "Important Bird and Biodiversity Areas" (IBAs). Criteria are intended to capture critical aspects of biology, ecology, and population structure, and they encompass vulnerability, distribution, abundance, special attributes, and key life cycle activities (IUCN Marine Mammal Protected Areas Task Force 2022). We argue that there is scope under the following sub-criterion for incorporating important aspects of cetacean culture: **Criterion D—Special Attributes** Sub-criterion D1—Distinctiveness *Areas that sustain populations with important genetic, behavioral or ecologically distinctive characteristics.*
IWC		Data on social learning and culture across foraging, migration, and communication may be informative for stock assessments (see Southern right whale example). Specifically, the strike limit algorithm (SLA) for a population is dependent on two key parameters, maximum sustainable yield (MSY) and survival. We argue that both parameters may be informed by data on cultural transmission and the delineation of cultural units.
CITES		CITES non-detriment findings (NDFs) provide an assessment of whether any trade will have a detrimental effect on the survival of a species. This determines whether trade would allow a species to be maintained "throughout its range and at a level consistent with its role in the ecosystems in which it occurs" (CITES Article IV, paragraph 3). We argue these assessments could benefit from data on cultural granulation across populations, particularly where cultural transmission is relevant to survival, reproduction, or dispersal of population segments.
World Heritage	Irrawaddy and bottlenose dolphins and human fishers	A proposal to recognize the Irrawaddy River Corridor being considered as a potential World Natural Heritage Site includes cooperation between Irrawaddy dolphins and fishers (UNESCO 2014). Similarly, domestic legislation in Brazil already recognizes cultural mutualism between bottlenose dolphins and fishers as Intangible Cultural Heritage. We argue there is scope to recognize this cross-species, cooperative cultural behavior internationally under the UNESCO Convention for the Safeguarding of the Intangible Cultural Heritage 2003, to safeguard this unique interspecies cultural transmission. But such efforts must be individually tailored and "logistically and politically feasible, ethically engaged, locally appropriate, and highly collaborative" (Van der Wal et al. 2022).

Notes: Acronyms included in the table are as follows: CITES—Convention on International Trade in Endangered Species of Wild Fauna and Flora; CMS—Convention on the Conservation of Migratory Species of Wild Animals (also known as the Bonn Convention); COSEWIC—Committee on the Status of Endangered Wildlife in Canada; EPBC—Environment Protection and Biodiversity Act 1999 (Australia); ESA—Endangered Species Act 1973 (USA); IMMA—Important Marine Mammal Areas; IUCN—The International Union for the Conservation of Nature; IWC—International Whaling Commission; SSC—Species Survival Commission (IUCN); WCPA—World Commission on Protected Areas (IUCN).

between populations. This may be achieved by maintaining complex social networks, as well as conserving the ecological and other resources necessary for behavioral diversity to thrive.

Finally, beyond the functional value of cetacean culture to conservation, cetacean cultures may also have aesthetic value to some human populations, particularly Indigenous populations. Indeed, we believe that these cultures have intrinsic value of their own.

Acknowledgments

The authors would like to thank the many researchers whose painstaking work—often over decades and in challenging field conditions—have given us an understanding of cetacean cultures. We would also like to acknowledge the members of the UNEP Convention on the Conservation of Migratory Species (CMS), Expert Group on Animal Culture and Social Complexity and the CMS Secretariat for their efforts in this area. P. B. would like to acknowledge the vision and support of Whale and Dolphin Conservation, and in particular Chris Butler-Stroud, and H. W. the vital collaborations with Luke Rendell.

References

Allen, Jenny A., Mason T. Weinrich, William Hoppitt, and Luke Rendell. 2013. "Network-Based Diffusion Analysis Reveals Cultural Transmission of Lobtail Feeding in Humpback Whales." *Science* 340 (6131): 485–88. https://doi.org/10.1126/science.1231976.

Ambie, Samantha, Cindy Peter, Gianna Minton, Jenny Ngeian, Anna N. Zulkifli Poh, Aazani Mujahid, and Andrew A. Tuen. 2023. "Utilizing Interview-Based Data to Measure Interactions of Artisanal Fishing Communities and Cetacean Populations in Kuching, Bay, Sarawak, East Malaysia." *Ocean and Coastal Management* 239 (106592). https://doi.org/10.2139/ssrn.4330069.

Ansmann, Ina C., Guido J. Parra, B. Louise Chilvers, and Janet M. Lanyon. 2012. "Dolphins Restructure Social System After Reduction of Commercial Fisheries." *Animal Behaviour* 84 (3): 575–81. https://doi.org/10.1016/j.anbehav.2012.06.009.

Baird, Robin W. 2009. "A Review of False Orcas in Hawaiian Waters: Biology, Status, and Risk Factors." Report prepared for the US Marine Mammal Commission under Order No. E40475499.

Barrett-Lennard, Lance Godfrey. 2000. "Population Structure and Mating Patterns of Killer Whales (*Orcinus orca*) as Revealed by DNA Analysis." PhD diss., University of British Columbia.

Barrett-Lennard, Lance Godfrey. 2011. "Killer Whale Evolution: Populations, Ecotypes, Species, Oh My!" *Journal of the American Cetacean Society* 40 (1): 48–53.

Barrett-Lennard, Lance G., John K. B. Ford, and Kathy A. Heise. 1996. "The Mixed Blessing of Echolocation: Differences in Sonar Use by Fish-Eating and Mammal-Eating Killer Whales." *Animal Behaviour* 51: 553–65. https://doi.org/10.1006/anbe.1996.0059.

Boesch, Christophe, Ammie K. Kalan, Roger Mundry, et al. 2020. "Chimpanzee Ethnography Reveals Unexpected Cultural Diversity." *Nature Human Behaviour* 4 (9): 910–16. https://dx.doi.org/10.1038/s41562-020-0890-1.

Bossley, Mike, Aude Steiner, Philippa Brakes, Juliet Shrimpton, Charlotte Foster, and Luke Rendell. 2018. "Tail Walking in a Bottlenose Dolphin Community: The Rise and Fall of an Arbitrary Cultural 'Fad.'" *Biology Letters* 14: 20180314. https://doi.org/10.1098/rsbl.2018.0314.

Brakes, Philippa, Emma L. Carroll, Sasha R. X. Dall, et al. 2021. "A Deepening Understanding of Animal Culture Suggests Lessons for Conservation." *Proceedings of the Royal Society B* 288 (1949): 20202718. https://doi.org/10.1098/rspb.2020.2718.

Brakes, Philippa, Sasha R. X. Dall, L. M. Aplin, et al. 2019. "Animal Cultures Matter for Conservation." *Science* 363 (6431): 1032–34. https://doi.org/10.1126/science.aaw3557.

Cantor, Mauricio, Damien R. Farine, and Fábio G. Daura-Jorge. 2023. "Foraging Synchrony Drives Resilience in Human-Dolphin Mutualism." *Proceedings of the National Academy of Sciences* 120 (6): e2207739120. https://doi.org/10.1073/pnas.2207739120.

Cantor, Mauricio, Lauren G. Shoemaker, Reniel B. Cabral, César O. Flores, Melinda Varga, and Hal Whitehead. 2015. "Multilevel Animal Societies Can Emerge from Cultural Transmission." *Nature Communications* 6 (1): 8091. https://doi.org/10.1038/ncomms9091.

Carroll, Emma L., C. Scott Baker, M. Watson, et al. 2015. "Cultural Traditions Across a Migratory Network Shape the Genetic Structure of Southern Right Whales Around Australia and New Zealand." *Scientific Reports* 5: 16182. https://doi.org/doi.org/10.1038/srep16182.

Carroll, Emma L., Simon J. Childerhouse, Mark Christie, et al. 2012. "Paternity Assignment and Demographic Closure in the New Zealand Southern Right Whale." *Molecular Ecology* 21 (16): 3960–73. https://doi.org/10.1111/j.1365-294X.2012.05676.x.

Carroll, Emma L., William J. Rayment, Alana M. Alexander, et al. 2014. "Reestablishment of Former Wintering Grounds by New Zealand Southern Right Whales." *Marine Mammal Science* 30 (1): 206–20. https://doi.org/10.1111/mms.12031.

Casacci, Luca P., Francesca Barbero, and Emilio Balletto. 2014. "The 'Evolutionarily Significant Unit' Concept and Its Applicability in Biological Conservation." *Italian Journal of Zoology* 81 (2): 182–93. https://doi.org/10.1080/11250003.2013.870240.

Cheeseman, Ted, Katherina Audley, Astrid Frisch, et al. 2022. "A Happy Collaboration: Basin-Scale, High-Throughput, AI-Assisted Photo-ID Matching for North Pacific Humpback Whales." In *Abstracts of the 24th Biennial Conference on the Biology of Marine Mammals*, Palm Beach, Florida, USA, August 1–5, 122.

Clapham, Phillip J., Alex Aguilar, and Leila T. Hatch. 2008. "Determining Spatial and Temporal Scales for Management: Lessons from Whaling." *Marine Mammal Science* 24 (1): 183–201. https://doi.org/10.1111/j.1748-7692.2007.00175.x.

Cook, Carly N., Kelly de Bie, David A. Keith, and Prue F. E. Addison. 2016. "Decision Triggers Are a Critical Part of Evidence-Based Conservation." *Biological Conservation* 195: 46–51. https://doi.org/https://doi.org/10.1016/j.biocon.2015.12.024.

Cooney, Rosie. 2012. "From Promise to Practicalities: The Precautionary Principle in Biodiversity Conservation and Sustainable Use." In *Biodiversity and the Precautionary Principle: Risk, Uncertainty and Practice in Conservation and Sustainable Use*, edited by R. Cooney and B. Dickson, 25–33. Routledge. https://doi.org/10.4324/9781849770583.

COSEWIC. 2020. "COSEWIC Assessment and Status Report on the Beluga Whale *Delphinapterus leucas*, Eastern High Arctic—Baffin Bay Population, Cumberland Sound Population, Ungava Bay Population, Western Hudson Bay Population, Eastern Hudson Bay Population and James Bay Pop." https://www.canada.ca/en/environment-climate-change/services/species-risk-public-registry.html.

Eguiguren, Ana, Enrico Pirotta, Mauricio Cantor, Luke Rendell, and Hal Whitehead. 2019.

"Habitat Use of Culturally Distinct Galápagos Sperm Whale *Physeter macrocephalus* Clans." *Marine Ecology Progress Series* 609: 257–70. https://doi.org/10.3354/meps12822.

Ford, John K. B., and Graeme M. Ellis. 2014. "You Are What You Eat: Ecological Specializations and Their Influence on the Social Organization and Behaviour of Killer Whales." In *Primates and Cetaceans: Field Research and Conservation of Complex Mammalian Societies*, edited by J. Yamagiwa and L. Karczmarski, 75–98. Springer. https://doi.org/10.1007/978-4-431-54523-1_4.

Ford, John K. B., Graeme M. Ellis, and Kenneth C. Balcomb. 2000. *Killer Whales*. UBC Press.

Ford, John K. B., Graeme M. Ellis, Peter F. Olesiuk, and Kenneth C. Balcomb. 2010. "Linking Killer Whale Survival and Prey Abundance: Food Limitation in the Oceans' Apex Predator?" *Biology Letters* 6 (1): 139–42. https://doi.org/10.1098/rsbl.2009.0468.

Foroughirad, Vivienne, and Janet Mann. 2013. "Long-Term Impacts of Fish Provisioning on the Behavior and Survival of Wild Bottlenose Dolphins." *Biological Conservation* 160: 242–49. https://doi.org/10.1016/j.biocon.2013.01.001.

Fraser, Dylan J., and Louis Bernatchez. 2001. "Adaptive Evolutionary Conservation: Towards a Unified Concept for Defining Conservation Units." *Molecular Ecology* 10: 2741–52. https://dx.doi.org/10.1046/j.1365-294X.2001.t01-1-01411.x.

Head, Brian W. 2008. "Wicked Problems in Public Policy." *Public Policy* 3: 101. https://doi.org/10.1007/978-3-030-94580-0.

Henrich, Joe, and Robert Boyd. 1998. "The Evolution of Conformist Transmission and the Emergence of Between-Group Differences." *Evolution and Human Behavior* 19: 215–41. https://doi.org/10.1016/S1090-5138(98)00018-X.

Hersh, Taylor A., Shane Gero, Luke Rendell, et al. 2022. "Evidence from Sperm Whale Clans of Symbolic Marking in Non-Human Cultures." *Proceedings of the National Academy of Sciences* 119 (37): e2201692119. https://dx.doi.org/10.1073/pnas.2201692119.

Heyes, Ceclia M. 1994. "Social Learning in Animals: Categories and Mechanisms." *Biological Reviews of the Cambridge Philosophical Society* 69 (2): 207–31. https://www.ncbi.nlm.nih.gov/pubmed/8054445.

Hodgson, Dave, Jenni L. McDonald, and David J. Hosken. 2015. "What Do You Mean, 'Resilient'?" *Trends in Ecology and Evolution* 30 (9): 503–6. https://doi.org/10.1016/j.tree.2015.06.010.

Hunt, Gavin R., and Russell D. Gray. 2003. "Diversification and Cumulative Evolution in New Caledonian Crow Tool Manufacture." *Proceedings of the Royal Society B* 270: 867–74. https://doi.org/10.1098/rspb.2002.2302.

IUCN Marine Mammal Protected Areas Task Force. 2022. "Guidance on the Use of Selection Criteria for the Identification of Important Marine Mammal Areas (IMMAs)—Marine Mammal Protected Areas Task Force." https://www.marinemammalhabitat.org/download/guidance-on-the-use-of-selection-criteria-for-the-identification-of-important-marine-mammal-areas-immas/.

Keith, Sally A., and Joseph W. Bull. 2017. "Animal Culture Impacts Species' Capacity to Realise Climate-Driven Range Shifts." *Ecography* 40 (2): 296–304. https://doi.org/10.1111/ecog.02481.

Konrad, Christine M., Shane Gero, Timothy R. Frasier, and Hal Whitehead. 2018. "Kinship Influences Sperm Whale Social Organization Within, but Generally Not Among, Social Units." *Royal Society Open Science* 5 (8): 180914. https://doi.org/10.1098/rsos.180914.

Kopps, Anna M., Corinne Y. Ackermann, Willima B. Sherwin, Simon J. Allen, Lars Bejder, and Michael Krützen. 2014. "Cultural Transmission of Tool Use Combined with Habitat

Specializations Leads to Fine-Scale Genetic Structure in Bottlenose Dolphins." *Proceedings of the Royal Society B* 281 (1782): 20133245. https://doi.org/10.1098/rspb.2013.3245.

Kopps, Anna M., and William B. Sherwin. 2012. "Modelling the Emergence and Stability of a Vertically Transmitted Cultural Trait in Bottlenose Dolphins." *Animal Behaviour* 84 (6): 1347–62. https://doi.org/10.1016/j.anbehav.2012.08.029.

Krützen, Michael, Erik P. Willems, and Carel P. van Schaik. 2011. "Culture and Geographic Variation in Orangutan Behavior." *Current Biology* 21 (21): 1808–12. https://doi.org/10.1016/j.cub.2011.09.017.

Laland, Kevin N., and W. Hoppitt. 2003. "Do Animals Have Culture?" *Evolutionary Anthropology* 12: 150–59. https://dx.doi.org/10.1002/evan.10111.

Laland, Kevin N., and Vincent M. Janik. 2006. "The Animal Cultures Debate." *Trends in Ecology and Evolution* 21: 542–47. https://doi.org/10.1016/j.tree.2006.06.005.

Laland, Kevin N., Tobias Uller, Marcus W. Feldman, et al. 2015. "The Extended Evolutionary Synthesis: Its Structure, Assumptions and Predictions." *Proceedings of the Royal Society B* 282 (1813): 20151019. https://doi.org/10.1098/rspb.2015.1019.

Lauriano, Giancarlo, L. Caramanna, Marco Scarnò, and Franco Andaloro. 2009. "An Overview of Dolphin Depredation in Italian Artisanal Fisheries." *Journal of the Marine Biological Association of the United Kingdom* 89 (August): 921–29. https://doi.org/10.1017/S0025315409000393.

Leaper, Russell, Justin Cooke, Phil Trathan, Keith Reid, Victoria Rowntree, and Roger S. Payne. 2006. "Global Climate Drives Southern Right Whale (*Eubalaena australis*) Population Dynamics." *Biology Letters* 2 (2): 289–92. https://doi.org/10.1098/rsbl.2005.0431.

Lovén Wallerius, Magnus, Jörgen I. Johnsson, Steven J. Cooke, and Robert Arlinghaus. 2020. "Hook Avoidance Induced by Private and Social Learning in Common Carp." *Transactions of the American Fisheries Society* 149: 498–511. https://doi.org/10.1002/tafs.10246.

Mann, Janet, and Courtney Kemps. 2003. "The Effects of Provisioning on Maternal Care in Wild Bottlenose Dolphins, Shark Bay, Australia." In *Marine Mammals and Humans: Towards a Sustainable Balance*, edited by R. Kirkwood. Melbourne University Press.

McDonald, Mark A., John A. Hildebrand, and Sarah L. Mesnick. 2009. "Worldwide Decline in Tonal Frequencies of Blue Whale Songs." *Endangered Species Research* 9 (1): 13–21. https://dx.doi.org/10.3354/esr00217.

McDonald, Mark A., Sarah L. Mesnick, and John A. Hildebrand. 2006. "Biogeographic Characterization of Blue Whale Song Worldwide: Using Song to Identify Populations." *Journal of Cetacean Research and Management* 8: 55–65. https://dx.doi.org/10.47536/jcrm.v8i1.702.

NOAA. 2022. "Main Hawaiian Islands Insular False Killer Whale (*Pseudorca crassidens*) Distinct Population Segment 5-Year Review: Summary and Evaluation." https://media.fisheries.noaa.gov/2022-04/main-hawaiian-islands-insular-false-killer-whale-5-year-review-final-2022-signed.pdf.

Noke, Wendy. D., and Daniel. K. Odell. 2002. "Interactions Between the Indian River Lagoon Blue Crab Fishery and the Bottlenose Dolphin, *Tursiops truncatus*." *Marine Mammal Science* 18: 819–32. https://dx.doi.org/10.1111/j.1748-7692.2002.tb01075.x.

O'Corry-Crowe, Greg, Robert Suydam, Lori Quakenbush, et al. 2018. "Migratory Culture, Population Structure and Stock Identity in North Pacific Beluga Whales (*Delphinapterus leucas*)." *PLOS ONE* 13 (3): e0194201. https://dx.doi.org/10.1371/journal.pone.0194201.

Oleson, Erin M., Christofer H. Boggs, Karin A. Forney, et al. 2010. *Status Review of Hawaiian*

Insular False Killer Whales (Pseudorca crassidens) Under the Endangered Species Act. NOAA Technical Memorandum NMFS-PIFSC-22.

Payne, Katharine 2000. "The Progressively Changing Songs of Humpback Whales: A Window on the Creative Process in a Wild Animal." In *The Origins of Music*, edited by N. L. Wallin, B Merker, and S Brown, 135–50. MIT Press.

Payne, Roger S., and Scott McVay. 1971. "Songs of Humpback Whales: Humpbacks Emit Sounds in Long, Predictable Patterns Ranging over Frequencies Audible to Humans." *Science* 173 (3997): 585–97. https://doi.org/10.1126/science.173.3997.585.

Rendell, Luke, Sarah L. Mesnick, Merel L. Dalebout, Jessica Burtenshaw, and Hal Whitehead. 2012. "Can Genetic Differences Explain Vocal Dialect Variation in Sperm Whales, *Physeter macrocephalus*?" *Behavior Genetics* 42: 332–43. https://doi.org/10.1007/s10519-011-9513-y.

Rendell, Luke E., and Hal Whitehead. 2003. "Vocal Clans in Sperm Whales (*Physeter macrocephalus*)." *Proceedings of the Royal Society B* 270 (1512): 225–31. https://doi.org/10.1098/rspb.2002.2239.

Riesch, Rüdiger, Lance G. Barrett-Lennard, Graeme M. Ellis, John K. B. Ford, and Volker B. Deecke. 2012. "Cultural Traditions and the Evolution of Reproductive Isolation: Ecological Speciation in Killer Whales?" *Biological Journal of the Linnean Society* 106 (1): 1–17. https://doi.org/10.1111/j.1095-8312.2012.01872.x.

Rittel, Horst W. J. and Melvin M. Webber 1973. "Planning Problems Are Wicked." *Polity* 4: 155–69.

Rojas-Bracho, Lorenzo, Barbara Taylor, Cormac Booth, et al. 2022. "More Vaquita Porpoises Survive Than Expected." *Endangered Species Research* 48. https://doi.org/10.3354/esr01197.

Ryan, Sadie J. 2006. "The Role of Culture in Conservation Planning for Small or Endangered Populations." *Conservation Biology* 20, 1321–24. https://doi.org/10.1111/j.1523-1739.2006.00347.x.

Ryder, Oliver A. 1986. "Species Conservation and Systematics: The Dilemma of Subspecies." *Trends in Ecology and Evolution* 1: 9–10.

Sargeant, Brooke L., and Janet Mann. 2009. "Developmental Evidence for Foraging Traditions in Wild Bottlenose Dolphins." *Animal Behaviour* 78 (3): 715–21. https://doi.org/10.1016/j.anbehav.2009.05.037.

Schakner, Zachary A., Chris Lunsford, Janice Straley, Tomoharu Eguchi, and Sarah L. Mesnick. 2014. "Using Models of Social Transmission to Examine the Spread of Longline Depredation Behavior Among Sperm Whales in the Gulf of Alaska." *PLOS ONE* 9 (10). https://doi.org/10.1371/JOURNAL.PONE.0109079.

Seyboth, Elisa, Karina R. Groch, Luciano D. Rosa, Keith Reid, Paulo. A. C. Flores, and Eduardo. R. Secchi. 2016. "Southern Right Whale (*Eubalaena australis*) Reproductive Success Is Influenced by Krill (*Euphausia superba*) Density and Climate." *Scientific Reports* 6 (February): 1–8. https://doi.org/10.1038/srep28205.

Sigler, Michael F., Chris R. Lunsford, Janice M. Straley, and Joseph B. Liddle. 2008. "Sperm Whale Depredation of Sablefish Longline Gear in the Northeast Pacific Ocean." *Marine Mammal Science* 24 (1): 16–27. https://dx.doi.org/10.1111/j.1748-7692.2007.00149.x.

Straley, Jan. 2012. "Sperm Whales and Fisheries: An Alaskan Perspective of a Global Problem." *Journal of the American Cetacean Society* 41 (1): 38–41.

Takacs, David. 1996. *The Idea of Biodiversity: Philosophies of Paradise.* Johns Hopkins University Press.

UNESCO World Heritage Centre. 2014. "Ayeyawady River Corridor." https://whc.unesco.org/en/tentativelists/5870/.

Valenzuela, Luciano O., Mariano Sironi, Victoria J. Rowntree, and Jon Seger. 2009. "Isotopic and Genetic Evidence for Culturally Inherited Site Fidelity to Feeding Grounds in Southern Right Whales (*Eubalaena australis*)." *Molecular Ecology* 18 (5): 782–91. https://doi.org/10.1111/j.1365-294X.2008.04069.x.

Van der Wal, Jessica E. M., Claire N. Spottiswoode, Natalie T. Uomini, et al. 2022. "Safeguarding Human-Wildlife Cooperation." *Conservation Letters* 15 (4). https://doi.org/10.1111/CONL.12886.

Waples, Robin S. 1991. "Pacific Salmon, *Oncorhynchus* spp. and the Definition of 'Species' Under the Endangered Species Act." *Marine Fisheries Review* 53: 11–22.

Ward, Eric J., Elizabeth E. Holmes, and Ken C. Balcomb. 2009. "Quantifying the Effects of Prey Abundance on Killer Whale Reproduction." *Journal of Applied Ecology* 46 (3): 632–40. https://doi.org/10.1111/j.1365-2664.2009.01647.x.

Whitehead, Hal. 2010. "Conserving and Managing Animals That Learn Socially and Share Cultures." *Learning and Behavior* 38: 329–36. https://doi.org/10.3758/LB.38.3.329.

Whitehead, Hal, Kevin N. Laland, Luke Rendell, Rose Thorogood, and Andrew Whiten. 2019. "The Reach of Gene-Culture Coevolution in Animals." *Nature Communications* 10 (1): 2405. https://doi.org/10.1038/s41467-019-10293-y.

Whitehead, Hal, and Luke Rendell. 2004. "Movements, Habitat Use and Feeding Success of Cultural Clans of South Pacific Sperm Whales." *Journal of Animal Ecology* 73 (1): 190–96. https://doi.org/10.1111/j.1365-2656.2004.00798.x.

Whitehead, Hal, and Luke Rendell. 2014. *The Cultural Lives of Whales and Dolphins*. University of Chicago Press.

Whitehead, Hal, Luke Rendell, Richard W. Osborne, and Bernd Würsig. 2004. "Culture and Conservation of Non-Humans with Reference to Whales and Dolphins: Review and New Directions." *Biological Conservation* 120 (3): 431–41. https://doi.org/10.1016/j.biocon.2004.03.017.

Whitehead, Hal, Tim D. Smith, and Luke Rendell. 2021. "Adaptation of Sperm Whales to Open-Boat Whalers: Rapid Social Learning on a Large Scale?" *Biology Letters* 17: 20210030. https://dx.doi.org/10.1098/rsbl.2021.0030.

Whiten, Andrew. 2017. "A Second Inheritance System: The Extension of Biology Through Culture." *Interface Focus* 7 (5): 20160142. https://doi.org/10.1098/rsfs.2016.0142.

Wild, Sonja, Michael Krützen, Robert W. Rankin, William. J. E. Hoppitt, Livia Gerber, and Simon J. Allen. 2019. "Long-Term Decline in Survival and Reproduction of Dolphins Following a Marine Heatwave." *Current Biology* 29 (7): R239–40. https://doi.org/10.1016/j.cub.2019.02.047

Yurk, Harald, Lance Barrett-Lennard, John K. B. Ford, Craig O. Matkin. 2002. "Cultural Transmission Within Maternal Lineages: Vocal Clans in Resident Killer Whales in Southern Alaska." *Animal Behaviour* 63 (6): 1103–19. https://doi.org/10.1006/anbe.2002.3012.

PART V

Human and Cetacean Societies

14

Intersections Between Human and Cetacean Societies: From Collision to Coexistence

MAURICIO CANTOR, DIPANI SUTARIA, SHANAN ATKINS, AND SIMON ALLEN

Abstract

Intersections between human and cetacean societies span time, space, and nature. From distant through modern history, our interactions have had negative, neutral, and positive outcomes for either or both cetaceans and humans. Despite sweeping attitudinal shifts in the twentieth century and advancements in understanding cetacean cognition, culture, and their pivotal roles in ecosystem function, human-cetacean interactions still span ambivalence to reverence, and conflict to cooperation. In the twenty-first century, targeted cetacean hunting continues, and their incidental capture in fisheries has several species on an extinction trajectory. Negative outcomes for cetaceans where our societies overlap also stem from habitat modification, ship strikes, chemical and noise pollution and, perhaps humans' greatest imposition, climate breakdown. This chapter focuses on intentional human-cetacean interactions—hunting and capture, fisheries conflicts, cetacean-based tourism, and cooperative foraging between nonmechanized fisheries and dolphins. Cetacean societies and humans collide, contest, and cooperate, so students of cetaceans must generate and share knowledge collaboratively, strive for responsible innovation, and use positive connections with cetaceans to leverage decisions that benefit marine ecosystems broadly. Expanding protected areas and reducing destructive fisheries are starting points, and our society has a critical role in turning the tide from collision and contest to coexistence with cetacean societies.

Introduction

Human-cetacean interactions have been occurring since the Old Stone Age (Stringer et al. 2008). Whales, dolphins, and porpoises inspired and intimidated ancient societies and they have been key figures in modern human history, being part of our material and intangible cultures. Cetaceans have been deified as they have been crucified. From stranded animals to traditional whaling and cooperative fisheries, cetaceans have provided food and comfort to Aboriginal communities; via modern whaling, cetaceans have lubricated the machinery of the Industrial Revolution that embodied our global ecological success and propelled the subsequent global ecological crisis (Ceballos and Ehrlich 2018). Cetaceans have helped compose myth and legend, and they continue to inspire art, literature, scientific endeavors, and revolutionary social movements (Samuels and Tyack 2000). From the opportunism and exploration of thousands of years past, humans and cetacean societies now meet in rivers, along coastlines, and on the high seas, with interactions that span the broadest possible spectra, from ambivalence to reverence and conflict to cooperation (fig. 14.1; Whitehead et al. 2000; Allen 2014). Human society is now tasked with the responsibility of recognizing how and where interactions with cetacean societies (and with the natural world as a whole) cause more harm than good, and being strategic about how to reduce conflicts such that our terrestrial and their aquatic societies can coexist and persist.

So, how far have we come since the turn of the century? By way of example: When *Cetacean Societies* was being written, there was a mass stranding of Cuvier's beaked whales (*Ziphius cavirostris*) in Greece that "probably resulted from an unusually strong behavioral reaction to a loud military sonar being tested in the area" (as reported in Frantzis 1998, cited in Whitehead et al. 2000). When this chapter was being written, another mass stranding of the species occurred on Cyprian shores immediately after military exercises, and preliminary results point to acoustic trauma as the cause (Tonay et al. 2024). When *Cetacean Societies* was published, there had been no modern-day cetacean extinctions. In 2002, the last confirmed sighting of a free-ranging Yangtze River dolphin (*Lipotes vexillifer*) occurred, and the species was declared functionally extinct in 2006 (Turvey et al. 2007). It is highly likely more will follow before the next two decades of the century are done. In *Cetacean Societies*, Whitehead et al. (2000) stated, "There are also great difficulties in assessing the effects of any climate change on cetaceans, . . . so the implications for whale, dolphin, and porpoise populations of the systematic changes in our environment that are taking place right now will remain almost entirely conjectural." In 2023, a vast body of literature now exists, reporting striking correlations between climatic conditions and cetacean (and other taxa) habitat use, reproductive success, and survivorship from multidecadal datasets, as well as sophisticated modeling and reviews (e.g., Wild et al. 2019; Agrelo et al. 2021; Nelms et al. 2021; Williamson et al. 2021). The picture painted

FIGURE 14.1. Interactions between human and cetacean societies span time, space, and nature. (*Top left*) Dolphin sculptures on Rhodes, Greece, depict evidence of thousands of years of coexistence between human and cetacean societies. (*Top right*) Socializing dolphins with an industrial tanker ship passing are indicative of modern times, in which there is increasing overlap between human activities and cetacean habitats globally. (*Middle left*) The resulting interactions can be, at least for one party, unintentionally negative, such as this case of humpback dolphin mortality through bycatch in shark nets set to create the perception of protection for bathers from sharks in South Africa; (*middle right*) unintentionally benign or positive, like this close encounter between recreational water users and Hawaiian spinner dolphins; (*bottom left*) intentionally negative, as in commercial whaling; or (*bottom right*) intentionally positive, such as artisanal cooperative fishing in Brazil.
Sources: Photos by (*top left*) John Boulton, (*top right*) Brett Atkins, (*middle left*) Dave Savides, (*middle right*) Simon J. Allen, (*bottom left*) public domain, Museum of New Zealand, (*bottom right*) Mauricio Cantor.

is uncommonly a positive one. On the other hand, several populations of baleen whales have rebounded since the cessation of commercial whaling to the extent that they are no longer considered threatened. There is growing awareness of the multiple ecosystem services provided by cetaceans, many including cultural and nonmaterial values (Cook et al. 2020), such that human communities worldwide hold them in such high regard that injuring or hunting whales is not part of some cultures and is considered offensive, even criminal. There are also more environmental NGOs, community groups, and citizen scientists devoted to cetacean monitoring and advocacy than existed in the prior century, reflecting an expanding circle of beings whose interests we are willing to value as we do our own.

This chapter takes a deep dive into the breadth of the intentional human-cetacean interactions, symbolic of the larger relationship between humans and nature. Scientific evidence is synthesized about the instances in which humans and cetaceans are in conflict (hunting and capture, fisheries), where they both conflict and coexist (tourism), and the few where they align and even collaborate (cooperative foraging), with discussion of their implications for social structures and social processes in cetaceans. The first two decades of the twenty-first century are examined as to whether there has been progress toward more effective conservation and coexistence since the publication of *Cetacean Societies*. After covering the history of interactions between humans and cetaceans, the future role our society must play in cetacean conservation is pondered upon. It seems high time to face conflicts as opportunities for change, and use the lessons learned to help flip the nature of human-cetacean interactions from negative to neutral or positive.

Where Human and Cetacean Societies Collide

Although modern-day cetacean-human interactions remain diverse, they could reasonably be characterized as predominantly beneficial to humans and neutral or negative for cetaceans. Despite the high esteem granted to cetaceans by some segments of human society, anthropogenic activities have marked impacts on cetacean societies. The life history characteristics of cetaceans, being long-lived (from decades to centuries), slow to reach sexual maturity, with low reproductive rates (single offspring produced at a time, one to multiyear interbirth intervals), along with their cognitive and cooperative capabilities (e.g., Marino et al. 2007), allow them to cope with seasonal or annual environmental and demographic stochasticity. However, these same traits also render cetaceans vulnerable to threatening processes that originate from anthropogenic sources of disturbance or mortality. The list is long, including disturbance and vessel strike from increasing shipping and boating activity (fig. 14.2, top left); habitat loss and degradation through coastal and offshore development; proliferation of underwater noise from anthropogenic sources; depletion of food resources through commercial and recreational fishing; catchment runoff (including chemical contaminants); and climate

FIGURE 14.2. Collisions between human and cetacean societies. (*Top left*) Industrial shipping traffic is now prevalent in all oceans, exposing many cetacean societies to increased noise pollution and risk of vessel strike. (*Top right*) A crew paddles their *umiaq* (skin boat) off Utqiagvik, Alaska, where Aboriginal subsistence whaling continues today. (*Middle left*) The structure of cetacean societies can be influenced by their proclivity to interact with fisheries, such as when dolphins scavenge prey disturbed by or discarded from trawlers. (*Middle right*) Cetacean foraging budgets and habitat use can be impacted by the proliferation of aquaculture facilities, such as this fish farm in Western Australia. Cetacean-based tourism, such as (*bottom left*) the global whale-watching industry and (*bottom right*) food-provisioning programs to promote close encounters, can have mixed effects on cetacean societies but also provide platforms for education, outreach, and research.
Sources: Photos by (*top left*) Brett Atkins, (*top right*) Kate Stafford, (*others*) Simon J. Allen.

breakdown, including both gradual ocean warming and acidification, as well as more frequent and intense extreme weather events. These human-induced threats originate largely from unintentional human-cetacean interactions, where the interests of human development intersect with the lives of wild cetaceans.

COLLISIONS THROUGH HUNTING AND CAPTURE

The intentional capture and removal of cetaceans from wild populations is motivated by numerous factors and can take many forms, including both industrial and Indigenous whaling; hunting for consumption, bait, or cultural practices; and live capture for displays and oceanariums. In practice, their population and social consequences can be similar. Indiscriminate or targeted removals can reduce population sizes, remove reproducing individuals, cause breakdowns in sex ratios, dismantle social structures and disrupt the social transmission of critical behaviors (e.g., Williams and Lusseau 2006).

The discourse on the history of whaling and hunting covers millennia. Human societies have explored and celebrated cetaceans as resources for nutrition and fuel, for ornaments and ceremonies, from ancient to current times (e.g., Clapham and Baker 2018; Ellis 2018; Huntington et al. 2021). Indeed, the exploitation of cetaceans in Europe, for example, predates the arrival of anatomically modern humans, with evidence of subsistence hunting by archaic humans (Neanderthals) in the Paleolithic Age from Gibraltar (Stringer et al. 2008). The exploitation of cetaceans and other marine mammals played a significant role in the spread and settling of human societies, and their economic and social development in many parts of the world (e.g., Estes et al. 2006; Huntington et al. 2021). Leaping ahead to the late eighteenth and early nineteenth centuries, these practices expanded well beyond subsistence and artisanal shore-based whaling; chase boats and factory ships were introduced, rendering these pursuits genuinely commercial endeavors. The United States had the largest whaling fleet, followed by Great Britain, France, Germany, and Portugal. Right (*Eubalaena* spp.) and sperm whales (*Physeter macrocephalus*) were most sought after by the commercial fleet, and they declined precipitously well before the international moratorium on whaling was established in 1986. Some populations of these species have not recovered (Vighi et al. 2021; Whitehead and Shin 2022). Other examples of the substantial impact of modern whaling include the near extinction of fin whales (*Balaenoptera physalus quoyi*) of the Southern Hemisphere by the twentieth century, which were extirpated from their feeding grounds in Antarctic waters; and the vast reduction of right, sperm, and humpback whales (*Megaptera novaeangliae*) around southern Australia and New Zealand by the late 1920s (Bannister 2020; Dawbin 1986). In addition, commercial whaling so depleted the numbers of bowhead whales (*Balaena mysticetus*) in the Pacific Arctic that, in 1977, the International Whaling Commission (IWC) imposed a moratorium on subsistence hunting. Had it continued, the communities that have relied on bowhead whales for nutritional, cultural, and spiritual subsistence for centuries would have been irreparably harmed. The international moratorium on modern commercial whaling remains in place; yet it is still carried out by Norway, Iceland, and Japan.

Aboriginal subsistence whaling also continues today—now recognized by the IWC as Indigenous practices quite distinct from commercial whaling. Subsistence whaling does not seek to maximize profit or catches but instead to enable local communities to catch enough cetaceans to satisfy their cultural and nutritional needs while safeguarding cetacean populations from the risk of extirpation. The practice is deeply rooted in Indigenous Arctic cultures—Denmark's Greenland, Russia's Chukotka, Canada's Artic, and the US's Alaska (fig. 14.2, top right)—and continues to contribute to the physical and material well-being of people in challenging environments, as well as to their social, spiritual, and traditional values (e.g., Huntington et al. 2021). Outside of polar regions, traditional whaling continues in some Indonesian communities (e.g., Alvard and Nolin 2002) and in St. Vincent and the Grenadines' Bequia (e.g., Fielding and Kiszka 2021), under other cultural practices and motives, such as dolphin drive hunting (e.g., Ellis 2018). Not all whaling practices by today's human societies are similar in intent and scale, nor in the magnitude of their impacts on cetacean societies.

A recent assessment of the artisanal and Aboriginal hunts of cetaceans from St. Vincent and the Grenadines revealed records of a single whaler taking thousands of odontocetes, including short-finned pilot whales (*Globicephala macrorhynchus*), killer whales (*Orcinus orca*), and other small cetaceans, and large whales including humpback and Bryde's whales (*Balaenoptera brydei* [Fielding and Kiszka 2021]). Dolphin hunting drives have ceased in Newfoundland, Canada, but continue in the Faroe Islands. Traditionally called Grindadráp, this cultural event is community driven, where long-finned pilot whales (*G. melas*) and Atlantic white-sided dolphins (*Lagenorhynchus acutus*) are killed. In the Solomon Islands, dolphin drives (primarily *Stenella longirostris*) were aimed at collecting teeth that are used as traditional currency, for dowry, adornment, and direct selling, although meat is also consumed. Interestingly, the capture and export of cetaceans for oceanariums that took place in the Solomon Islands in 2003 caught only Indo-Pacific bottlenose dolphins

(*Tursiops aduncus*), as the islanders were unable to drive other cetaceans into the bay for hunting with the traditional techniques (Oremus et al. 2013). Beyond subsistence takes, there is considerable hunting pressure on small- and midsized cetaceans worldwide for fishing bait, as well as for medicinal, aphrodisiac, and other cultural uses. For instance, Robards and Reeves (2011) documented 87 species of marine mammals being consumed in at least 54 countries, and Avila et al. (2018) found at least 89 marine mammal species being targeted in directed hunts. Approximately 100,000 small cetaceans across 56 species were still being intentionally killed each year worldwide (Altherr and Hodgins 2018).

The intentional pursuit and capture of cetaceans for display in oceanariums and human entertainment first took place in the 1930s and grew over the ensuing 90 years (Corkeron 2022). Although this industry contributed markedly to raising awareness and understanding of cetaceans, at least some sectors of society have questioned the ethics and aesthetics of the practice for many decades. The number of individuals removed for the oceanarium/dolphinarium industry is relatively trivial by comparison to that of intentional hunting or incidental takes in fisheries, but live captures function, in practice, as sources of additive mortality (i.e., additional to existing mortality rates). The population-level consequences from where they were taken are therefore akin to intentional killing.

The hunting or removal of several individuals not only contributes to declining populations but also perpetrates further social consequences. For instance, whaling is one plausible explanation of the formation of different types of multilevel societies in sperm whales in different ocean basins (Whitehead et al. 2012). Early open-boat whaling of the eighteenth century took place primarily in the Northwest Atlantic, taking thousands of sperm whales per year before declining in the late nineteenth century. Modern mechanized whaling, on the other hand, killed tens of thousands of sperm whales per year between 1920 and 1980 in all areas but the Northwest Atlantic, particularly affecting populations in the Pacific. This marked difference in takes led to the destruction of social units and loss of matrilines in the Pacific, which were likely followed by merging and increasing group size, and social homogeneity within units. In response, sperm whales may have learned to avoid whaling boats, and the cultural transmission of that knowledge may have helped them to avoid hunts in the nineteenth century: North Pacific whalers' logbooks indicate that the success rate of harpooning a sperm whale fell by 58% over the first few years of hunts (Whitehead et al. 2021).

The hunting or removal of key individuals in a society, such as the repositories of ecological and cultural knowledge, can be particularly problematic. The fitness of individuals in some cetacean societies seems positively influenced by both their social bond strengths as juveniles and adults and, particularly, the early stages of social learning and bond formation (e.g., Stanton and Mann 2012; Gerber et al. 2022). Older resident killer whales, for example, often retain key knowledge of salmon (*Salmo* spp.) movements (Brent et al. 2015); matriarch killer whales and sperm whale calves have strong social connections within and between social entities. Removing socially central individuals can cause social rearrangement or social disruption and decrease the chances of long-term survival and fitness in others (e.g., Williams and Lusseau 2006). In highly social cetaceans, sociality makes the population resilient to demographic stress, and the mortality of key individuals weakens associations between surviving individuals and decreases the fitness of the social group (Busson et al. 2019).

For the most part, human societies are at cultural crossroads with cetaceans. Cetacean societies with cultures that humans have not yet completely grasped are at risk of being lost to perhaps the most prolonged and destructive human-cetacean interaction: hunting (Brakes and Whitehead, this volume). In sharp contrast, there are still modern-day cultural practices in which cetaceans are openly celebrated. In addition to the Indigenous Arctic cultures that jointly exploit and worship cetaceans (e.g., Huntington et al. 2021), baleen whales are traditionally revered in some parts of south Asia. In India, Pakistan, and Bangladesh, for example, they are never approached by fishers and have not been intentionally targeted or hunted (Sutaria et al. 2017). Fishers perceive baleen whales as a sign of goodwill and luck, and they are locally referred to in different west Indian languages as *maccha raja*, *maccha babu*, *dev masa*, and *kadal acchan*, meaning "king of fish," "father of fish," "god of fish," and "father of the sea," respectively. Conversely, large odontocetes such as sperm and killer whales are perceived as a source of danger and, hence, avoided. Upon observing any of the large whales, fishers usually carry out a small ceremony by lighting incense on the fishing vessel before proceeding—an intriguing, different perspective from the historical view of cetaceans in other parts of the world. The ancient Greeks and Australian Aborigines, for example, held dolphins in such high regard that injuring or killing one was viewed in the same light as killing a human. Dolphins were seen as sleek, graceful, and a good omen for sailors (or as pests to fishers), while the great whales were monstrous, mysterious, and fearful (Allen 2014). Taken together, these studies point to variation across cetacean and human societies in how they interact and manage their intentional interactions.

COLLISIONS THROUGH FISHERIES AND AQUACULTURE

Intentional human-cetacean interactions also occur when we share or compete for common interests, such as the same prey. Fisheries can alter the distribution, availability, or predictability of prey targeted by cetaceans, making them scarcer (e.g., through overfishing), or more concentrated in space or time (e.g., through long lining, purse seining, trawling, or aquaculture; Allen et al. 2017; fig. 14.2, middle left and right). Many cetacean-fisheries interactions are unintentional on the part of humans but detrimental to cetaceans nonetheless, with entanglement and bycatch being the result. Human fishers deploying fishing gear in areas where cetaceans are present can lead to bycatch (e.g., Ballance et al. 2021; Nelms et al. 2021), and gillnets, for instance, remain a persistent mortality threat worldwide (e.g., Reeves et al. 2013). Even entanglements that are not immediately lethal, such as those in the high tensile strength ropes of newer lobster pots and traps, can result in nutritional stress and sublethal injuries that reduce fecundity rates and compromise the reproductive performance of endangered species, like North Atlantic right whales (*Eubalaena glacialis*; Johnson et al. 2005; Moore 2021; Frasier et al. 2023).

Interactions with fisheries can be intentional, having been initiated not only by humans, as illustrated in the previous section, but also by cetaceans. When cetaceans are attracted to fishing gear to extract prey, they can cause negative outcomes for humans through damage or loss of bait, catch, or gear (e.g., Tixier et al. 2021). These fisheries interactions can rearrange the spatial distribution of cetaceans, influencing their movements, behavioral budgets, foraging tactics, and even social structure (e.g., Bonizzoni et al. 2022).

Cetacean behavior upon approach to fishing activity is generally characterized by patrolling of, and scavenging around, fishing gear and operations. These interactions are often perceived as having no negative consequences for cetaceans or humans. Humpback (*Sousa* spp.) and bottlenose (*Tursiops* spp.) dolphins, among others, scavenge on prey disturbed by trawling or consume the discards, a shift in foraging behavior that could change group social dynamics and fitness (e.g., Ansmann et al. 2012). Bottlenose dolphins also spend a considerable amount of time foraging around fishing gear (e.g., Díaz López 2006). This type of fidelity to foraging around fisheries could have resulted from there being low prey availability in the area, or by observational learning from their mothers or peers that finding prey around fishing operations is an easy (if risky) endeavor (e.g., Jaiteh et al. 2013).

There are many examples of human-centered foraging tactics in dolphins that likely emerged from a few innovators (via individual learning) and then spread (via social learning) to others in the population. Some of these foraging behaviors are sources of conflict and can be maladaptive, in the sense that they may be profitable in the short term for the cetaceans involved but are risky or carry long-term consequences. Examples of cetaceans approaching gear to extract prey include humpback dolphins in India (Jog et al. 2018), tucuxi (*Sotalia fluviatilis*) and Amazon River dolphins (*Inia geoffrensis*) in Brazil (Hallwass et al. 2024), killer whales off the Crozet Islands (Tixier et al. 2017), and sperm whales in Alaska (Hanselman et al. 2018). Cetaceans gain relatively easy access to prey, and participants can sometimes even experience higher survival. Some killer whales in the Strait of Gibraltar and off the Crozet Islands, for instance, learned how to exploit longline fisheries and divided into distinct groups; these individuals tended to have higher survival and, sometimes, higher reproductive success than those that did not have access to the high-quality trapped prey (Esteban et al. 2016; Tixier et al. 2017). These types of interactions can, nevertheless, be lethal. First, mortality can occur if there is ingestion of gear. Nets, hooks, lures, and other gear that damage the digestive or respiratory tracts have been detected in postmortem assessment of stranded animals (e.g., Wells et al. 2008). Second, the interaction can be costly for humans. Bottlenose dolphins that interact with gill nets in Italian waters tend to damage gear (over two-thirds of fishing days involve net damage), resulting in negative fisher perceptions toward them (Díaz López 2006). The economic loss of gear damage and catch loss to fishers are sometimes offset by a more positive perception that dolphins help to maintain fish productivity, such as among artisanal fishers and river dolphins in the Amazon (Hallwass et al. 2024). More often, however, economic losses can lead to dangerous deterrent methods and retaliatory killing (e.g., Tixier et al. 2017). Globally, 20 odontocete species were reported extracting fish and damaging gear in 44 countries across 214 fisheries in all sectors (commercial, artisanal, and recreational) between 1979 and 2019 (70% post-2000; Tixier et al. 2021).

Foraging in association with fisheries can also alter the pattern, quality, or nature of social relationships among cetaceans (e.g., Harnish et al. 2023); alter the types of behaviors in which they engage (e.g., Díaz López 2020); and, ultimately, shape the structure of their societies (e.g., Chilvers and Corkeron 2001). By adopting a human-centered foraging tactic, individuals can become more peripheral or central in their social networks or become part of a socially segregated community of individuals who also engage in the same foraging tactic. Examples come from resident bottlenose dolphins that are exposed to fisheries or aquaculture that increase the availability of prey at specific locations, thereby altering the cetaceans' space use and then social behavior (e.g., Harnish et al. 2023). A striking example comes from bottlenose dolphins in

Moreton Bay, Australia that had split into social communities of individuals that foraged on discarded bycatch associated with prawn trawlers, and those that did not (Chilvers and Corkeron 2001). These communities were largely socially segregated and did not interact with each other despite home range overlap, showing how distinct foraging tactics can influence social cohesion. Interestingly, after a reduction in prawn trawling due to more strict fishing regulations, that social segregation diminished, dismantling the social communities, and increasing the association strength among dolphins into a more well-mixed society (Ansmann et al. 2012). Similar effects of human-centered foraging tactics on dolphin social patterns are found around fish farms. In Italy, a long-term study showed that the high-density food patches around aquaculture have changed the group composition and social structure of dolphins by also socially separating the population into those that forage around fish farms and those that do not (Díaz López and Shirai 2008).

These studies demonstrate the diverse impacts that fisheries can have on cetacean societies, but also highlight the plasticity of some species, making them resilient to human-induced physical changes in their environment. When individual cetaceans learn these novel tactics socially (e.g., Wild and Rendell, this volume), then changes in social structure promoted by using these tactics imply changes in the pathways by which such novel behaviors can spread. For example, if bolder individuals that explore novel foraging tactics also occupy more central positions in their social networks (e.g., Díaz López 2020), then they likely influence not only social stability but may also act as key "influencers" on the behavioral spreading dynamics. Since cetaceans are generally avid imitators and learners, any maladaptive human-centered foraging tactics can potentially spread through the population (e.g., Donaldson et al. 2012), scaling the negative effects that individuals experience up to the social group, community, or population level. Nevertheless, not all intentional interactions between humans and cetaceans are negative, and some targeted close encounters between the two can lead to neutral or even positive outcomes.

COLLISION AND COEXISTENCE THROUGH TOURISM

Targeted close encounters between humans and cetaceans are driven by the desire of some sections of human society to interact with charismatic megafauna and, for some, by the socioeconomic benefits these encounters can yield. Cetacean-based tourism is now a vast industry involving hundreds, if not thousands, of both human and cetacean communities, and includes nation states that once exploited cetaceans for food and other resources, as well as some that still do (Hoyt 2001; Higham et al. 2016). Globally, a variety of resident and migratory cetaceans are now regularly exposed to commercial tourism operations, most frequently in the form of boat-based "whale watching," as well as swim-with programs and, in a few locations, food provisioning. Benefits to our society, aside from the obvious economic gains, include employment opportunities, expanded infrastructure, exposure of visitors to wildlife and natural areas, support for scientific endeavors, and marine environmental education. Benefits to cetaceans through tourism, aside from some animals appearing to seek out interaction, include the indirect conservation value to cetacean societies and ecosystems through increased public awareness and activism (Hoyt 2001; Allen 2014).

Human-cetacean interactions through whale-watching tourism were formerly viewed uncritically as benign or indeed "good for whales," but there have been calls for the industry to be recognized as consumptive in nature, particularly where intensive operations exist (Higham et al. 2016). Some individuals, communities, and even species of cetacean seem more amenable to interaction with tourism operations, but the nature of this subset of human-cetacean interactions is, again, highly variable. Short-term changes in cetacean behavior and movements in response to vessel approach have been widely documented (e.g., Heckel et al. 2001; Constantine et al. 2004), and how best to manage this industry has been the subject of considerable investigation since the turn of the century (e.g., Lusseau 2003; Allen et al. 2007). At least some of these close encounters continue to provide alternate livelihoods, deliver regional economic and infrastructure gains, and can help to sustain research and conservation action (fig. 14.2, bottom left).

Swim-with-cetacean tourism can also elicit immediate changes to cetacean behavioral budgets. In Kaikōura, New Zealand, for example, dusky dolphins (*Lagenorhynchus obscurus*) spent less time resting and socializing and more time milling when swim-with-dolphin operations were present, and group dynamics were influenced (Lundquist et al. 2012). In Port Phillip Bay, Australia, the presence of swim-with-dolphins tour vessels significantly reduced the time dolphins spent foraging, and there was a fourfold increase in the probability of transitioning from foraging to milling (Filby et al. 2017). Swim-with-dolphin tourism can have lingering behavioral consequences. In Mozambique, for example, dolphins are more likely to travel than socialize, rest, or forage even up to 20 minutes after a swim-with event is over or a tour vessel has passed (Rocha et al. 2023).

Human food provisioning of dolphins for tourism purposes happens under both regulated and unregulated conditions in the Americas and Australia (e.g., Mann et al. 2021; fig. 14.2, bottom right). The social consequences are mixed, depending upon many variables. In Bunbury, Western Australia, for example, provisioned females experience a more limited social environment than their non-provisioned counterparts, with social associations among the provisioned females occurring within relatively small home ranges around the provisioning site (Senigaglia et al. 2022). Provisioned females also had reduced reproductive success via lower calf survival (Senigaglia et al. 2019). By contrast, provisioned female dolphins in Shark Bay, where they have been hand-fed since the 1960s, had significantly lower calf mortality than their non-provisioned counterparts in recent decades (Mann et al. 2021), implying some benefit to provisioning, while the broader population suffered the deleterious effects of a marine heatwave (Wild et al. 2019).

Although importance lies in carefully managing the animal welfare issues that arise from close encounters and provisioning wild animals, as well as addressing the potential impacts of intensive cetacean-based wildlife tourism, it should be noted that few studies have been able to draw causative links between exposure to tourism and long-term negative impacts on cetacean vital rates. There is certainly a need for long-term research of human-cetacean interactions through tourism. It will be important to include other drivers of fluctuations in dolphin abundance, movements, and fitness (i.e., prey abundance/distribution, predator prevalence, environmental conditions, and degradation) into future empirical research and modeling efforts. Despite the array of short-term impacts of such close interaction between humans and cetaceans, this industry's potential to foster admiration and conservation awareness should be noted. This can play an important part in flipping the nature of human-cetacean interactions from negative to neutral or positive. In Zanzibar, Tanzania, for instance, bottlenose and humpback dolphins were hunted for consumption and bait until 1996. After a local boat began to take tourists to see or swim with the dolphins in 1992, however, this activity grew as an economic alternative to the direct take of dolphins and, in some cases, as an alternative livelihood to other fishing activities that resulted in bycatch (Stensland et al. 2006). Further examples suggest that, when conducted responsibly and ethically, these close encounters can help cetaceans gain more space in the conservation debate through a growing crowd of supporters of better regulations and protection through legal instruments (Hoyt 2001).

Where Human and Cetacean Societies Collaborate

Beyond inadvertently beneficial interactions between cetacean and human societies through shared ecosystem services, including predation regulating marine food webs (fig. 14.3, top left), cycling nutrients between deep and surface waters, and carbon storage (Cook et al. 2020; Kiszka et al. 2022), there are also intentional interactions that provide mutual benefits for cetaceans and humans. When cetaceans and humans work together to find food, teaming up seems to provide greater benefits than the sum of the parts. Some humans and cetaceans are avid social predators, with complementary hunting skills: cetaceans being adept in prey pursuit and herding; humans being unparalleled tool users. Historical and recent accounts of human-cetacean cooperation indicate that, throughout human history, some of our societies coevolved strategies to achieve the common goal of catching prey (Cram et al. 2022). Such cooperative hunting can provide both material (e.g., food and income) and nonmaterial benefits (e.g., recreation and cultural belongingness) for both parties (Cantor et al. 2023). Since cultural practices like these tend not to leave fossilized evidence, it is plausible that other mutually beneficial human-wildlife interactions were once more widespread and ecologically important than we appreciate today (Fairholme 1856; van der Wal et al. 2022). Nevertheless, human-cetacean cooperation is a comparatively rare phenomenon. Among many putative cases (Cram et al. 2022), those with sufficient evidence of mutual benefits for both humans and cetaceans include cooperative hunting between whalers and killer whales, and between artisanal fishers and a few small dolphin species.

HUMAN–KILLER WHALE COOPERATIVE HUNTING

Whalers have cooperated with killer whales in at least two locations to hunt large marine mammals. In Twofold Bay, Australia, and Chukotka, Russia, killer whales would herd and attempt to trap their quarry against the water surface or shoreline, thereby increasing their accessibility to humans (Wellings 1944; Holzlehner 2015). Between the late 1800s and about 1930, approximately 30 killer whales cooperated with Australian colonial whalers (fig. 14.3, top right); some harassed humpback whales, while others would alert the shore-based whalers to the presence of the whales by performing surface-active behaviors in the bay, then leading the whalers to the hunt (Allen 2014; Cram et al. 2022). A successful cooperative hunt meant that humans would harpoon the humpback more efficiently than they could without the killer whales' help, then leave the killer whales to feast on the tongue, before returning later to retrieve the whale's carcass and tow it to shore. Importantly, recent research included interviews with an Aboriginal elder, who recounted the extensive history of the mutualism between local Indigenous people and killer whales when hunting baleen whales prior to European colonization (Reeves

FIGURE 14.3. Collaborations between human and cetacean societies. (*Top left*) The top-down regulation of marine food webs is an example of how cetacean societies inadvertently provide ecosystem services to human societies. Historically, there have also been intentional positive interactions between human and cetacean societies, such as (*top right*) the (now extinct) artisanal cooperative whaling between killer whales and whalers in Australia, and (*middle left*) the contemporary artisanal cooperative fishing between bottlenose dolphins and fishers in Brazil. (*Middle right*) This collaboration provides both material (e.g., prey capture) and nonmaterial (e.g., social connectedness) benefits for both parties. Students of cetaceans are well placed to facilitate broader society's connection to cetaceans by, for instance, (*bottom left*) working with Indigenous participants to merge Western science and traditional ecological knowledge, and (*bottom right*) using innovative, noninvasive field methods, such as drones and hydrophone arrays.
Sources: Photos by (*top left*) Dipani Sutaria, (*top right*) C. E. Wellings, public domain, (*middle left and right*) Fábio G. Daura-Jorge, (*bottom left and right*) Simon J. Allen.

et al. 2023). The relationship thereby appears considerably older than previously reported and, in fact, more in line with reports of Aboriginal tribesmen working with coastal dolphins in southeast Queensland, which was suggested to have been occurring for as long as they could remember (Fairholme 1856; Reeves et al. 2023).

Although cooperative hunting was not reported with the same details in Russia, it apparently involved similar hunting and communication techniques between humans and killer whales to capture a more diverse range of whales and pinnipeds (Holzlehner 2015). Despite persisting for many decades, considerably longer in the Australian case, the whaler-killer whale cooperation struggled to persist in the conditions that eventually crashed the global whaling industry—declining baleen whale populations and declining demand for their oil, as petroleum fueled the rapid industrialization of human societies, including the efficiency of modern whaling (Clapham and Baker 2018). In Australia, the cooperation ceased in the late 1920s; perhaps its deathblow being the deliberate killing of two knowledgeable killer whales by settlers, corresponding with the disruption and departure of the cooperative pod, or perhaps their extirpation (Reeves et al. 2023). This highlights that cultural practices involving human-wildlife cooperation are vulnerable to declines in prey availability, in motivation in at least one of the parties, and in their compatible knowledge (van der Wal et al. 2022).

HUMAN-DOLPHIN COOPERATIVE FISHING

Historically, artisanal human fishers have cooperated with wild dolphins to catch fast-moving fish schools in coastal, estuarine, and riverine habitats. Well-documented cases involve Lahille's bottlenose dolphin (*Tursiops truncatus gephyreus*) along the coast of southern Brazil (Cantor et al. 2023); Irrawaddy dolphins (*Orcaella brevirostris*) in the Ayeyarwady river in Myanmar (Smith et al. 2009); Indian humpback dolphins (*Sousa plumbea*) in Ashtamudi Estuary in southwestern India (Kumar et al. 2012); Indo-Pacific bottlenose dolphins in Moreton Bay, eastern Australia (Fairholme 1856); and bottlenose and Atlantic humpback dolphins (*S. teuszii*) in El-Mamghar on the coast of Mauritania (Busnel 1973). Although circumstances differ, there are common themes around all such mutually beneficial interactions.

All these interactions occur in shallow coastal habitats where small-scale artisanal fisheries operate and involve small species of dolphins foraging alone or in small groups for schooling fish, often migratory mullet (Mugilidae). The primary role performed by dolphins is to herd fish schools from deeper waters to the shoreline, making them available to fishers using short-range gear, such as casting and hand nets (or spears in the case of Australian Aboriginal people). Fishers who have learned to interpret the dolphins' foraging behavior wait in specific shallow spots along the shoreline (standing or in moored canoes) for cues by the dolphins that disclose where the fish school is and the time to act (fig. 14.3, middle left). By immediately deploying their fishing gear on the concentrated fish, fishers gain easy access and experience high fishing success (e.g., Simões-Lopes et al. 1998). Dolphins can benefit from fishers disrupting the antipredator defense of fish and targeting the fish either trapped or evading the nets (Cantor et al. 2023). In most cases, gains for dolphins remain inconclusive, although a recent study has shown that dolphins can not only benefit from the trapped fish, but also consistently adjust their foraging behavior to match the fishers' reactions to their cues. In Laguna, southern Brazil, dolphins dive for longer and time the onset of their echolocation clicks to match the time it takes for the nets to sink and close over the fish (Cantor et al. 2023). Importantly, dolphins only engage in echolocation clicks that are indicative of prey capture attempt when fishers cast their nets in the right time and place following the dolphins' foraging cues, further demonstrating the importance of their synchronized foraging for the success of both societies (King et al., this volume).

Through cooperation, both humans and dolphins gain increased access to food resources; but benefits can go beyond immediate foraging success. In Brazil, dolphins that regularly cooperate with net-casting fishers experience, in the long-term, increased survival and reduced home ranges, which reduce the risk of bycatch in other fisheries (Cantor et al. 2023). Cooperative fishing also provides nonmaterial benefits. The dolphin social network is shaped into communities of those that engage in cooperation with humans, and those that do not, within which members have distinct vocal communication and strong social relationships (Romeu et al. 2017; Machado et al. 2019a). Fishers who regularly cooperate with dolphins reap similar long-term, nonmaterial benefits. In addition to increased catch and monetary return (Simões-Lopes et al. 1998), cooperative fishers experience a range of ecosystem services, from increased well-being and recreation to strong social relationships (fig. 14.3, middle right), cultural pride, and a sense of belonging (Machado et al. 2019b). Human-dolphin cooperation is culturally transmitted among artisanal fishers (Peterson et al. 2008; Smith et al. 2009), and likely among dolphins as well (Simões-Lopes et al. 2016). Small-scale fisheries, such as these involving mutually beneficial interactions with wild dolphins, are more equitable and sustainable, perpetrating lower impacts on fish stocks than the more destructive and less selective fisheries that also take dolphins as bycatch (Cantor et al. 2023).

Unfortunately, most cases of dolphin-fisher cooperation are already extinct or in decline. Like the whaler-killer whale cooperation, declines in prey availability (due to whaling), reduced motivation or the loss of compatible interspecific knowledge (due to changes in the livelihoods of humans and dolphins) are the likely culprits (van der Wal et al. 2022; Cantor et al. 2023). Mutually beneficial human-cetacean interactions remain important at local and regional scales, and should be safeguarded for their intrinsic ecological, cultural, and intangible nonmaterial value for both human and wildlife parties (van der Wal et al. 2022). While the ecological importance of coexisting with cetaceans is undeniable for human society (Cook et al. 2020; Kiszka et al. 2022), the global reality is such that fewer local human societies are likely to continue to depend economically on mutually beneficial foraging interactions with cetaceans. Nevertheless, this should not halt efforts to preserve such intangible cultural heritage, nor stop humanity from being inspired by the lesson these interactions carry: through cooperation, coexistence is possible.

Evolving Toward Coexistence Between Human and Cetacean Societies

Human-cetacean interactions will likely increase in frequency, intensity, and diversity as our societies overlap in space and needs (Moore 2021). While a few cetacean populations are recovering from unsustainable exploitation, the human population has surpassed eight billion and, when combined with the mammals grown for human consumption, vastly dominates the global biomass of wild terrestrial and marine mammals combined (Greenspoon et al. 2023). Humanity has thus transformed the balance of biodiversity. What steps can be taken to increase the chances that tomorrow's societies of humans and cetaceans can coexist? How can students and chroniclers of human-cetacean interactions turn the tide from collision to coexistence?

LEARNING FROM FAILURE AND SUCCESS

Some advocate learning from success, others from failure. Identifying and trying to replicate success can guide and motivate others, particularly in the face of overwhelming doom and gloom (Knowlton 2021). Although our interactions with cetaceans are predominantly negative in outcome, successes from which we can learn exist and are worth celebrating. When compared with other taxa, progress for cetaceans has been promising. For example, 12 of 16 cetacean populations increased after being listed under the United States Endangered Species Act (Valdivia et al. 2019). Identifying and replicating success helps to sustain and amplify existing efforts by defining pathways toward positive human-wildlife interactions, like the Seeds of a Good Anthropocene initiative, for example (Bennett et al. 2016). "Seeds" can be social initiatives, new technologies, economic tools, social-ecological projects, organizations, movements, or new ways of acting. One such seed is Marine Spatial Planning, which has improved the balance between the needs of humans and marine biodiversity (Merrie and Olson 2014).

The benefits of learning from success are real, yet there are limitations. First, over-sampling success and undersampling failure leads to biased interpretations of outcomes, inaccurate causal attributions, and underestimation of the role of luck; and second, it lowers a culture of tolerance of experimentation and reduces the motivation to search for alternatives (Catalano et al. 2018). Two critical steps in learning from failure include acknowledging that failure is inevitable in complex systems and rarely has a single cause; and using direct, useful, fact-based feedback to identify and explore underlying causes (Catalano et al. 2018). Biologists working with the vaquita (*Phocoena sinus*) and northern right whale (*Eubalaena borealis*) examined conservation failures and provided useful advice: interventions are needed before populations reach critically low levels; and conservation strategies must be as dynamic and flexible as a cetacean's changing distribution (Frasier and Reeves 2017; Rojas-Bracho et al. 2019). It seems most prudent to use both success and failure in finding evidence to guide future interactions (e.g., Sutherland et al. 2019; Ballance et al. 2021).

LEVERAGE POINTS FOR HUMAN-CETACEAN COEXISTENCE

How might change be implemented to reduce the negative impacts of our interactions with cetaceans? The direct drivers of negative impacts on cetaceans and biodiversity loss, such as fishing, habitat degradation, and pollution, are underpinned by indirect drivers, demographic (e.g., human population dynamics), sociocultural (e.g., human consumption patterns), economic (e.g., trade), technological, and others (Díaz et al. 2019). There is value in considering both social and ecological elements interactively, as a social-ecological system. Systems thinking provides a framework to address intractable problems in other sectors and is useful for conservation (Knight et al. 2019). A valuable systems-thinking concept uses leverage points: key points in a system where interventions can result in disproportionately large changes (Chan et al. 2020). The leverage points concept was used by the Intergovernmental Science-Policy Platform on Biodiversity and Ecosystem Services to identify

TABLE 14.1. The eight key leverage points of intervention

(i)	Enabling visions of a good quality of life that do not entail ever-increasing material consumption.
(ii)	Lowering total consumption and waste by addressing both population growth and per capita consumption differently in different contexts.
(iii)	Unleashing existing, widely held values of responsibility to effect new social norms for sustainability.
(iv)	Addressing inequalities, especially regarding income and gender, that undermine the capacity for sustainability.
(v)	Ensuring inclusive decision-making and the fair and equitable sharing of benefits.
(vi)	Accounting for nature's deterioration from both local economic activities and tele-couplings (e.g., international trade).
(vii)	Ensuring environmentally friendly technological and social innovation, considering rebound effects and investment regimes.
(viii)	Promoting education, knowledge generation, and the maintenance of different knowledge systems.

the most important elements of transformative change from current trends toward more sustainable ones (Chan et al. 2020; table 14.1).

These actions apply to all members of society, yet some have greater applicability to students of human-cetacean interactions, most obviously the leverage point focused on knowledge. The work of knowledge creators in the field of human-cetacean interactions should be salient, credible, and legitimate (Game et al. 2015). The format, timing, and resolution of our science must be appropriate; such knowledge must be the result of a robust scientific process; and the research process must account for the concerns and needs of stakeholders germane to policy and conservation practice. Regarding knowledge sharing, collaborations must be forged with others: policy makers who might regulate human-cetacean interactions; implementers who may action scientists' recommendations; and the public. Experts recommend shifting the view of knowledge as a "thing" that is transferred to viewing knowledge as a "process of relating" that involves negotiating meaning among partners (Roux et al. 2006).

To improve cetacean science's impact on policy and management, scientists should work to understand their audience; to clarify the need and types of evidence required; to gather "just enough" evidence; and then share and discuss these findings together (Fisher et al. 2020). Such transdisciplinary knowledge production, involving people with backgrounds across scientific and societal (e.g., policy, practice) bodies of knowledge, allows people to coevolve their understanding of human-cetacean interactions, reconcile their diverse perspectives, and co-produce appropriate knowledge to serve the common purpose of mitigating human impacts on cetaceans (Roux et al. 2017). Note that "knowledge" means more than Western science and includes local, traditional, and Indigenous knowledge (fig. 14.3, bottom left). Engaging with multiple perspectives entails acknowledging the existence and validity of differing views, values, and types of knowledge (Knight et al. 2019). Integrating different types of knowledge can improve management of human-wildlife conflicts. For example, Indigenous, holistic, intergenerational resource-nurturing principles and practices are actively shaping marine futures for narwhal management in the Canadian Arctic and marine management in New Zealand (Armitage et al. 2011; Le Heron et al. 2019). These are linked to the justice and equality leverage points.

Students of human-cetacean interactions are also likely to be active at the "environmentally friendly innovation and investment" leverage point, placing a high premium on identifying and developing innovations to reduce human's impact on cetaceans. Cetacean researchers generally strive to innovate and use noninvasive research methods (fig. 14.3, bottom right), and recent examples include the use of drones to reduce, among other risks, disturbance to cetaceans, and passive acoustic monitoring to estimate abundance in mark-recapture models using identity whistles (Longden et al. 2020). Innovative research techniques can improve our understanding of human-cetacean interactions specifically, e.g., combining GIS and local knowledge to identify high-risk bycatch areas (Verutes et al. 2020). Technological and behavioral innovations can help to reduce negative human-cetacean interactions, such as reducing bycatch, including the current drive to develop "ropeless" fishing gear to reduce whale entanglement (Myers et al. 2019).

Students of human-cetacean interactions can be active at the "unleashing values of responsibility" leverage point. Society is increasingly disconnected from nature, yet many people have connections to cetaceans, which evoke strong feelings of care and protection. As large, mobile, and charismatic predators, they can also be considered keystone, flagship, umbrella, or indicator species within the marine ecosystem (Williamson et al. 2021). Students of cetaceans are in a unique position to facilitate society's connection to cetaceans and should harness this connection to leverage responsible behavior and choices that benefit the marine ecosystem generally.

"Reducing total consumption and waste" is another relevant leverage point. While directed hunts continue, consider how total consumption of cetaceans has declined compared to just decades ago. This important success, the moratorium on whaling, was achieved by widespread international collaboration between cetacean scientists and policymakers. Sadly, "waste" through bycatch continues globally, even in well-studied circumstances and wealthy states where its lack of sustainability has been demonstrated (e.g., Manlik et al. 2022), and remains the greatest threat to many populations and species. Yet, even in bycatch, some success can be celebrated, as interventions implemented across at least some jurisdictions have seen substantial reductions therein (e.g., dolphin catches in tuna fisheries are a fraction of historical numbers; Ballance et al. 2021; but see Dolman et al. 2022).

Many marine scientists believe that drastically improving the situation for marine biodiversity is an achievable "Grand Challenge" for humanity that is an ethical obligation and a smart economic objective to achieve a sustainable future (Duarte et al. 2020). Many marine scientists, engineers, and others are invested and already dedicating their expertise and time to this end.

Concluding Remarks

Our synthesis highlights that the intersections between human and cetacean societies have spanned broad spectra for centuries, and they continue to do so in the twenty-first century. From brief, opportunistic encounters to enduring intentional relationships, the nature of human-cetacean interactions still varies, from negative to neutral or positive. Students and advocates of cetaceans can play an important role in promoting coexistence between human and cetacean societies. Safeguarding the future of cetaceans requires consciousness of how to generate and share knowledge, innovation, and solutions—especially when these directly change how the marine environment is explored, by replacing destructive fishing methods, promoting genuine transparency in fisheries, and putting a real end to commercial whaling. Human society's inherent connection to cetaceans must be harnessed to unleash a sense of responsibility to protect them and, by extension, aquatic habitats. Learning from both success and failure, the working party of marine scientists that is finding and sowing seeds for a future of improved interactions between human and cetacean societies must be allowed to grow.

Acknowledgments

We thank the editors for the invitation to contribute and the reviewers for their insightful comments and suggestions. M. C. received support from the Marine Mammal Research Program Fund at Oregon State University and the National Geographic Society (NGS-101549R-23). S. A. received support from the Gesellschaft zur Rettung der Delphine.

References

Agrelo, Macarena, Fábio G. Daura-Jorge, Victoria J. Rowntree, et al. 2021. "Ocean Warming Threatens Southern Right Whale Population Recovery." *Science Advances* 7 (42): eabh2823. https://doi.org/10.1126/sciadv.abh2823.

Allen, Simon J. 2014. “From Exploitation to Adoration: The Historical and Contemporary Contexts of Human-Cetacean Interactions.” In *Whale-Watching: Sustainable Tourism and Ecological Management*, edited by James E. S. Higham Lars Bejder, and Rob Williams. Cambridge University Press.

Allen, Simon J., Kenneth H. Pollock, Phil J. Bouchet, et al. 2017. “Preliminary Estimates of the Abundance and Fidelity of Dolphins Associating with a Demersal Trawl Fishery.” *Scientific Reports* 7 (1): 4995. https://doi.org/10.1038/s41598-017-05189-0.

Allen, Simon J., Holly Smith, Kelly Waples, and Robert Harcourt. 2007. “The Voluntary Code of Conduct for Dolphin Watching in Port Stephens, Australia: Is Self-Regulation an Effective Management Tool?” *Journal of Cetacean Research and Management* 9 (2): 159–66. https://doi.org/10.47536/jcrm.v9i2.684.

Altherr, Sandra, and Nicola Hodgins. 2018. *Small Cetaceans, Big Problems: A Global Review of the Impacts of Hunting on Small Whales, Dolphins and Porpoises*. Wiltshire: Whale and Dolphin Conservation.

Alvard, Michael S., and David A. Nolin. 2002. “Rousseau’s Whale Hunt? Coordination Among Big-Game Hunters.” *Current Anthropology* 43 (4): 533–59. https://doi.org/10.1086/341653.

Ansmann, Ina C., Guido J. Parra, B. Louise Chilvers, and Janet M. Lanyon. 2012. “Dolphins Restructure Social System After Reduction of Commercial Fisheries.” *Animal Behaviour* 84 (3): 575–81. https://doi.org/10.1016/j.anbehav.2012.06.009.

Armitage, Derek, Fikret Berkes, Aaron Dale, Erik Kocho-Schellenberg, and Eva Patton. 2011. “Co-Management and the Co-Production of Knowledge: Learning to Adapt in Canada’s Arctic.” *Global Environmental Change* 21 (3): 995–1004. https://doi.org/10.1016/j.gloenvcha.2011.04.006.

Avila, Isabel C., Kristin Kaschner, and Carsten F. Dormann. 2018. “Current Global Risks to Marine Mammals: Taking Stock of the Threats.” *Biological Conservation* 221: 44–58. https://doi.org/10.1016/j.biocon.2018.02.021.

Ballance, Lisa T., Tim Gerrodette, Cleridy E. Lennert-Cody, Robert L. Pitman, and Dale Squires. 2021. “A History of the Tuna-Dolphin Problem: Successes, Failures, and Lessons Learned.” *Frontiers in Marine Science* 8: 754755. https://doi.org/10.3389/fmars.2021.754755.

Bannister, John. 2020. “Status of Southern Right Whales (*Eubalaena australis*) off Australia.” *Journal of Cetacean Research and Management* 2: 103–10. https://doi.org/10.47536/jcrm.vi.273.

Bennett, Elena M., Martin Solan, Reinette Biggs, et al. 2016. “Bright Spots: Seeds of a Good Anthropocene.” *Frontiers in Ecology and the Environment* 14 (8): 441–48. https://doi.org/10.1002/fee.1309.

Bonizzoni, Silvia, Sheryl Hamilton, Randall R. Reeves, Tilen Genov, and Giovanni Bearzi. 2022. “Odontocete Cetaceans Foraging Behind Trawlers, Worldwide.” *Reviews in Fish Biology and Fisheries* 32 (3): 827–77. https://doi.org/10.1007/s11160-022-09712-z.

Brent, Lauren J. N., Daniel W. Franks, Emma A. Foster, Kenneth C. Balcomb, Michael A. Cant, and Darren P. Croft. 2015. “Ecological Knowledge, Leadership, and the Evolution of Menopause in Killer Whales.” *Current Biology* 25 (6): 746–50. https://doi.org/10.1016/j.cub.2015.01.037.

Busnel, R.-G. 1973. “Symbiotic Relationship Between Man and Dolphins.” *Transactions of the New York Academy of Sciences* 35 (2) Series II: 112–31. https://doi.org/10.1111/j.2164-0947.1973.tb01511.x.

Busson, Marine, Matthieu Authier, Christophe Barbraud, et al. 2019. "Role of Sociality in the Response of Killer Whales to an Additive Mortality Event." *Proceedings of the National Academy of Sciences* 116 (24): 11812–17. https://doi.org/10.1073/pnas.1817174116.

Cantor, Mauricio, Damien. R. Farine, and Fábio. G. Daura-Jorge. 2023. "Foraging Synchrony Drives Resilience in Human-Dolphin Mutualism." *Proceedings of the National Academy of Sciences* 120 (6): e2207739120. https://doi.org/10.1073/pnas.2207739120.

Catalano, Allison S., Kent Redford, Richard Margoluis, and Andrew T. Knight. 2018. "Black Swans, Cognition, and the Power of Learning from Failure." *Conservation Biology* 32 (3): 584–96. https://doi.org/10.1111/cobi.13045.

Ceballos, Gerardo, and Paul Ehrlich. 2018. "The Misunderstood Sixth Mass Extinction." *Science* 360: 1080–81. https://doi.org/10.1126/science.aau0191.

Chan, Kai M. A., David R. Boyd, Rachelle K. Gould, et al. 2020. "Levers and Leverage Points for Pathways to Sustainability." *People and Nature* 2 (3): 693–717. https://doi.org/10.1002/pan3.10124.

Chilvers, Louise B., and Peter J. Corkeron. 2001. "Trawling and Bottlenose Dolphins' Social Structure." *Proceedings of the Royal Society B* 268 (1479): 1901–5. https://doi.org/10.1098/rspb.2001.1732.

Clapham, Phillip J., and C. Scott Baker. 2018. "Whaling, Modern." In *Encyclopedia of Marine Mammals*, edited by William F. Perrin, Bernd Würsig, and Johannes Gerardus Marie Thewissen. Academic Press.

Constantine, Rochelle, Dianne H. Brunton, and Todd Dennis. 2004. "Dolphin-Watching Tour Boats Change Bottlenose Dolphin (*Tursiops truncatus*) Behaviour." *Biological Conservation* 117 (3): 299–307. https://doi.org/10.1016/j.biocon.2003.12.009.

Cook, David, Laura Malinauskaite, Brynhildur Davíðsdóttir, Helga Ögmundardóttir, and Joe Roman. 2020. "Reflections on the Ecosystem Services of Whales and Valuing Their Contribution to Human Well-Being." *Ocean and Coastal Management* 186: 105100. https://doi.org/10.1016/j.ocecoaman.2020.105100.

Corkeron, Peter. 2022. "Marine Mammal Captivity, an Evolving Issue." In *Marine Mammals: The Evolving Human Factor*, edited by Giuseppe Notarbartolo di Sciara and Bernd Würsig. Springer.

Cram, Dominic L., Jessica E. M. van der Wal, Natalie Uomini, et al. 2022. "The Ecology and Evolution of Human-Wildlife Cooperation." *People and Nature* 4 (4): 841–55. https://doi.org/10.1002/pan3.10369.

Dawbin, William H. 1986. "Right Whales Caught in Waters Around South Eastern Australia and New Zealand During the Nineteenth and Early Twentieth Centuries." *Report of the International Whaling Commission* 10: 261–67.

Díaz, Sandra, Josef Settele, Eduardo S. Brondízio, et al. 2019. "Pervasive Human-Driven Decline of Life on Earth Points to the Need for Transformative Change." *Science* 366 (6471): eaax3100. https://doi.org/10.1126/science.aax3100.

Díaz López, Bruno. 2006. "Interactions Between Mediterranean Bottlenose Dolphins (*Tursiops truncatus*) and Gillnets off Sardinia, Italy." *ICES Journal of Marine Science* 63 (5): 946–51. https://doi.org/10.1016/j.icesjms.2005.06.012.

Díaz López, Bruno. 2020. "When Personality Matters: Personality and Social Structure in Wild Bottlenose Dolphins, *Tursiops truncatus*." *Animal Behaviour* 163: 73–84. https://doi.org/10.1016/j.anbehav.2020.03.001.

Díaz López, Bruno, and Julia Andrea B. Shirai. 2008. "Marine Aquaculture and Bottlenose Dolphins' (*Tursiops truncatus*) Social Structure." *Behavioral Ecology and Sociobiology* 62 (6): 887–94.

Dolman, Sarah J., Charlotte N. Breen, Philippa Brakes, Andy Butterworth, and Simon J. Allen. 2022. "The Individual Welfare Concerns for Small Cetaceans from Two Bycatch Mitigation Techniques." *Marine Policy* 143: 105126. https://doi.org/10.1016/j.marpol.2022.105126.

Donaldson, Rebecca, Hugh Finn, Lars Bejder, David Lusseau, and Michael Calver. 2012. "The Social Side of Human-Wildlife Interaction: Wildlife Can Learn Harmful Behaviours from Each Other." *Animal Conservation* 15 (5): 427–35. https://doi.org/10.1111/j.1469-1795.2012.00548.x.

Duarte, Carlos M., Susana Agusti, Edward Barbier, et al. 2020. "Rebuilding Marine Life." *Nature* 580 (7801): 39–51. https://doi.org/10.1038/s41586-020-2146-7.

Ellis, Richard. 2018. "Whaling, Aboriginal and Western Traditional." In *Encyclopedia of Marine Mammals*, edited by Bernd Würsig, Johannes Gerardus Marie Thewissen and Kit M. Kovacs. Academic Press.

Esteban, Ruth, Philippe Verborgh, Pauline Gauffier, Joan Giménez, Christophe Guinet, and Renaud de Stephanis. 2016. "Dynamics of Killer Whale, Bluefin Tuna and Human Fisheries in the Strait of Gibraltar." *Biological Conservation* 194: 31–38. https://doi.org/10.1016/j.biocon.2015.11.031.

Estes, James A., Douglas R. DeMaster, Daniel F. Doak, Terri M. Williams, and Robert L. Brownell Jr., eds. 2006. *Whales, Whaling, and Ocean Ecosystems*. University of California Press.

Fairholme, J. K. E. 1856. "The Blacks of Moreton Bay and the Porpoises." *Proceedings of the Zoological Society of London* 24: 353–54.

Fielding, Russell, and Jeremy J. Kiszka. 2021. "Artisanal and Aboriginal Subsistence Whaling in Saint Vincent and the Grenadines (Eastern Caribbean): History, Catch Characteristics, and Needs for Research and Management." *Frontiers in Marine Science* 8: 668597. https://doi.org/10.3389/fmars.2021.668597.

Filby, Nicole E., Fredrik Christiansen, Carol Scarpaci, and Karen A. Stockin. 2017. "Effects of Swim-with-Dolphin Tourism on the Behaviour of a Threatened Species, the Burrunan Dolphin (*Tursiops australis*)." *Endangered Species Research* 32: 479–90. https://doi.org/10.3354/esr00826.

Fisher, Jonathan R. B., Stephen A. Wood, Mark A. Bradford, and Thomas Rodd Kelsey. 2020. "Improving Scientific Impact: How to Practice Science That Influences Environmental Policy and Management." *Conservation Science and Practice* 2 (7): e210. https://doi.org/10.1111/csp2.210.

Frantzis, A. 1998. "Does Acoustic Testing Strand Whales?" *Nature* 392 (6671): 29. https://doi.org/10.1038/32068.

Frasier, Timothy R., Philip Hamilton, and Richard Pace III. 2023. "How Compromised Is Reproductive Performance in the Endangered North Atlantic Right Whale? A Proposed Method for Quantification and Monitoring." Preprint. *bioRxiv*. https://doi.org/10.1101/2023.11.21.568115.

Frasier, Timothy R., and Randall R. Reeves. 2017. "Lessons Learned from Cetacean Tragedies." *Science* 358 (6369): 1396–96. https://doi.org/10.1126/science.aar5529.

Game, Edward T., Mark W. Schwartz, and Andrew T. Knight. 2015. "Policy Relevant Conservation Science." *Conservation Letters* 8 (5): 309–11. https://doi.org/10.1111/conl.12207.

Gerber, Livia, Richard C. Connor, Simon J. Allen, et al. 2022. "Social Integration Influences Fitness in Allied Male Dolphins." *Current Biology* 32 (7): 1664–69. https://doi.org/10.1016/j.cub.2022.03.027.

Greenspoon, Lior, Eyal Krieger, Ron Sender, et al. 2023. "The Global Biomass of Wild Mammals." *Proceedings of the National Academy of Sciences* 120 (10): e2204892120. https://doi.org/10.1073/pnas.2204892120.

Hallwass, Gustavo, Paula E. R. Pereyra, Kaluan C. Vieira, et al. 2024. "Fishers' Knowledge Indicates That Collective Benefits Outweigh the Individual Costs of Coexisting with Dolphins." *Journal for Nature Conservation* 81: 126691. https://doi.org/10.1016/j.jnc.2024.126691.

Hanselman, Dana H., Brian J. Pyper, and Megan J. Peterson. 2018. "Sperm Whale Depredation on Longline Surveys and Implications for the Assessment of Alaska Sablefish." *Fisheries Research* 200: 75–83. https://doi.org/10.1016/j.fishres.2017.12.017.

Harnish, Annette E., Robin W. Baird, Enrico Corsi, et al. 2023. "Long-Term Associations of Common Bottlenose Dolphins with a Fish Farm in Hawai'i and Impacts on Other Protected Species." *Marine Mammal Science* 39 (3): 794–810. https://doi.org/10.1111/mms.13010.

Heckel, Gisela, Stephen B. Reilly, James L. Sumich, and Ileana Espejel. 2001. "The Influence of Whale Watching on the Behaviour of Migrating Gray Whales (*Eschrichtius robustus*) in Todos Santos Bay and Surrounding Waters, Baja California, Mexico." *Journal of Cetacean Research and Management* 3 (3): 227–37. https://doi.org/10.47536/jcrm.v3i3.875.

Higham, James E. S., Lars Bejder, Simon J. Allen, Peter J. Corkeron, and David Lusseau. 2016. "Managing Whale-Watching as a Non-Lethal Consumptive Activity." *Journal of Sustainable Tourism* 24 (1): 73–90. https://doi.org/10.1080/09669582.2015.1062020.

Holzlehner, Tobias. 2015. "Werewolves of the Sea: Human-Killer Whale (*Orcinus orca*) Relationship in the North Pacific." Paper presented at the 11th Conference on Hunting and Gathering Societies, Vienna, Austria.

Hoyt, Erich. 2001. *Whale Watching Worldwide: Tourism Numbers, Expenditures, and Expanding Socioeconomic Benefits*. Yarmouth Port, MA: International Fund for Animal Welfare.

Huntington, H. P., Chie Sakakibara, George Noongwook, et al. 2021. "Whale Hunting in Indigenous Arctic Cultures." In *The Bowhead Whale*, edited by John C. George and J. G. M. Thewissen. Academic Press.

Jaiteh, Vanessa F., Simon J. Allen, Jessica J. Meeuwig, and Neil R. Loneragan. 2013. "Subsurface Behavior of Bottlenose Dolphins (*Tursiops truncatus*) Interacting with Fish Trawl Nets in Northwestern Australia: Implications for Bycatch Mitigation." *Marine Mammal Science* 29 (3): E266–81. https://doi.org/10.1111/j.1748-7692.2012.00620.x.

Jog, Ketki, Mihir Sule, Isha Bopardikar, Vardhan Patankar, and Dipani Sutaria. 2018. "Living with Dolphins: Local Ecological Knowledge and Perceptions of Small Cetaceans along the Sindhudurg Coastline of Maharashtra, India." *Marine Mammal Science* 34 (2): 488–98. https://doi.org/10.1111/mms.12466.

Johnson, Amanda, Glenn Salvador, John Kenney, Jooke Robbins, Scott Kraus, Scott Landry, and Phil Clapham. 2005. "Fishing Gear Involved in Entanglements of Right and Humpback Whales." *Marine Mammal Science* 21 (4): 635–45. https://doi.org/10.1111/j.1748-7692.2005.tb01256.x.

Kiszka, Jeremy J., Matthew S. Woodstock, and Michael R. Heithaus. 2022. "Functional Roles and Ecological Importance of Small Cetaceans in Aquatic Ecosystems." *Frontiers in Marine Science* 9: 803173. https://doi.org/10.3389/fmars.2022.803173.

Knight, Andrew T., Carly N. Cook, Kent H. Redford, et al. 2019. "Improving Conservation Practice with Principles and Tools from Systems Thinking and Evaluation." *Sustainability Science* 14: 1531–48. https://doi.org/10.1007/s11625-019-00676-x.

Knowlton, Nancy. 2021. "Ocean Optimism: Moving Beyond the Obituaries in Marine Conservation."

Annual Review of Marine Science 13: 479–99. https://doi.org/10.1146/annurev-marine-040220-101608.

Kumar, A. Biju, Raj Smrithy, and Kumaran Sathasivam. 2012. "Dolphin-Assisted Cast Net Fishery in the Ashtamudi Estuary, South-West Coast of India." *Indian Journal of Fisheries* 59 (3): 143–48. https://epubs.icar.org.in/index.php/IJF/article/view/14038.

Le Heron, Erena, June Logie, Will Allen, et al. 2019. "Diversity, Contestation, Participation in Aotearoa New Zealand's Multi-Use/User Marine Spaces." *Marine Policy* 106: 103536. https://doi.org/10.1016/j.marpol.2019.103536.

Longden, Emma G., Simon H. Elwen, Barry McGovern, Bridget S. James, Clare B. Embling, and Tess Gridley. 2020. "Mark-Recapture of Individually Distinctive Calls—A Case Study with Signature Whistles of Bottlenose Dolphins (*Tursiops truncatus*)." *Journal of Mammalogy* 101 (5): 1289–301. https://doi.org/10.1093/jmammal/gyaa081.

Lundquist, David, Neil J. Gemmell, and Bernd Würsig. 2012. "Behavioural Responses of Dusky Dolphin Groups (*Lagenorhynchus obscurus*) to Tour Vessels off Kaikoura, New Zealand." *PLOS ONE* 7 (7): e41969. https://doi.org/10.1371/journal.pone.0041969.

Lusseau, David. 2003. "Effects of Tour Boats on the Behavior of Bottlenose Dolphins: Using Markov Chains to Model Anthropogenic Impacts." *Conservation Biology* 17 (6): 1785–93. https://doi.org/10.1111/j.1523-1739.2003.00054.x.

Machado, Alexandre M. S., Mauricio Cantor, Ana P. B. Costa, et al. 2019a. "Homophily Around Specialized Foraging Underlies Dolphin Social Preferences." *Biology Letters* 15 (4): 20180909. https://doi.org/10.1098/rsbl.2018.0909.

Machado, Alexandre M. S., Fábio G. Daura-Jorge, Dannieli F. Herbst, et al. 2019b. "Artisanal Fishers' Perceptions of the Ecosystem Services Derived from a Dolphin-Human Cooperative Fishing Interaction in Southern Brazil." *Ocean and Coastal Management* 173: 148–56. https://doi.org/10.1016/j.ocecoaman.2019.03.003.

Manlik, Oliver, Robert C. Lacy, William B. Sherwin, Hugh Finn, Neil R. Loneragan, and Simon J. Allen. 2022. "A Stochastic Model for Estimating Sustainable Limits to Wildlife Mortality in a Changing World." *Conservation Biology* 36 (4): e13897. https://doi.org/10.1111/cobi.13897.

Mann, Janet, Vivienne Foroughirad, Molly H. F. McEntee, et al. 2021. "Elevated Calf Mortality and Long-Term Responses of Wild Bottlenose Dolphins to Extreme Climate Events: Impacts of Foraging Specialization and Provisioning." *Frontiers in Marine Science* 8: 617550. https://doi.org/10.3389/fmars.2021.617550.

Marino, Lori, Richard C. Connor, R. Ewan Fordyce, et al. 2007. "Cetaceans Have Complex Brains for Complex Cognition." *PLOS Biology* 5 (5): e139. https://doi.org/10.1371/journal.pbio.0050139.

Merrie, Andrew, and Per Olsson. 2014. "An Innovation and Agency Perspective on the Emergence and Spread of Marine Spatial Planning." *Marine Policy* 44: 366–74. https://doi.org/10.1016/j.marpol.2013.10.006.

Moore, Michael J. 2021. *We Are All Whalers: The Plight of Whales and Our Responsibility*. University of Chicago Press.

Myers, Hannah J., Michael J. Moore, Mark F. Baumgartner, et al. 2019. "Ropeless Fishing to Prevent Large Whale Entanglements: Ropeless Consortium Report." *Marine Policy* 107: 103587. https://doi.org/10.1016/j.marpol.2019.103587.

Nelms, Sarah E., Joanna Alfaro-Shigueto, John P. Y. Arnould, et al. 2021. "Marine Mammal Conservation: Over the Horizon." *Endangered Species Research* 44: 291–325. https://doi.org/10.3354/esr01115.

Oremus, Marc, John Leqata, Joe Hurutarau, Sue Taei, Michael Donoghue, and C. Scott Baker. 2013. "Population Status of Indo-Pacific Bottlenose Dolphins, *Tursiops aduncus*, in the Solomon Islands and Assessment of Live-Capture Sustainability." Report No. SC/65a/Forinfo32 to the Scientific Committee of the International Whaling Commission.

Peterson, Débora, Natalia Hanazaki, and Paulo C. Simões-Lopes. 2008. "Natural Resource Appropriation in Cooperative Artisanal Fishing Between Fishermen and Dolphins (*Tursiops truncatus*) in Laguna, Brazil." *Ocean and Coastal Management* 51 (6): 469–75. https://doi.org/10.1016/j.ocecoaman.2008.04.003.

Reeves, Isabella M., John A. Totterdell, Emma L. Betty, et al. 2023 "Ancestry Testing of 'Old Tom,' a Killer Whale Central to Mutualistic Interactions with Human Whalers." *Journal of Heredity* 114 (6): 598–611. https://doi.org/10.1093/jhered/esad058.

Reeves, Randall R., Kate McClellan, and Timothy B. Werner. 2013. "Marine Mammal Bycatch in Gillnet and Other Entangling Net Fisheries, 1990 to 2011." *Endangered Species Research* 20 (1): 71–97. https://doi.org/10.3354/esr00481.

Robards, Martin D., and Randall R. Reeves. 2011. "The Global Extent and Character of Marine Mammal Consumption by Humans: 1970–2009." *Biological Conservation* 144 (12): 2770–86. https://doi.org/10.1016/j.biocon.2011.07.034.

Rocha, Diana, Sarah A. Marley, Benjamin Drakeford, Jonathan Potts, and Angie Gullan. 2023. "Effects of Dolphin-Swim Activities on the Behaviour of an Indo-Pacific Bottlenose Dolphin Population off the South Coast of Mozambique." *Biological Conservation* 279: 109949. https://doi.org/10.1016/j.biocon.2023.109949.

Rojas-Bracho, Lorenzo, Frances M. D. Gulland, Cynthis R. Smith, et al. 2019. "A Field Effort to Capture Critically Endangered Vaquitas (*Phocoena sinus*) for Protection from Entanglement in Illegal Gillnets." *Endangered Species Research* 38: 11–27. https://doi.org/10.3354/esr00931.

Romeu, Bianca, Mauricio Cantor, Carolina Bezamat, Paulo C. Simões-Lopes, and Fábio G. Daura-Jorge. 2017. "Bottlenose Dolphins That Forage with Artisanal Fishermen Whistle Differently." *Ethology* 123 (12): 906–15. https://doi.org/10.1111/eth.12665.

Roux, Dirk J., Jeanne L. Nel, Georgina Cundill, Patrick O'Farrell, and Christo Fabricius. 2017. "Transdisciplinary Research for Systemic Change: Who to Learn with, What to Learn About and How to Learn." *Sustainability Science* 12: 711–26. https://doi.org/10.1007/s11625-017-0446-0.

Roux, Dirk J., Kevin H. Rogers, Harry C. Biggs, Peter J. Ashton, and Anne Sergeant. 2006. "Bridging the Science-Management Divide: Moving from Unidirectional Knowledge Transfer to Knowledge Interfacing and Sharing." *Ecology and Society* 11 (1): 4. https://www.jstor.org/stable/26267817.

Samuels, Amy, and Peter L. Tyack. 2000. "Flukeprints: A History of Studying Cetacean Societies." In *Cetacean Societies: Field Studies of Dolphins and Whales*, edited by Janet Mann, Richard C. Connor, Peter L. Tyack, and Hal Whitehead. University of Chicago Press.

Senigaglia, Valeria, Fredrik Christiansen, Lars Bejder, Kate R. Sprogis, and Mauricio Cantor. 2022. "Human Food Provisioning Impacts the Social Environment, Home Range and Fitness of a Marine Top Predator." *Animal Behaviour* 187: 291–304. https://doi.org/10.1016/j.anbehav.2022.02.005.

Senigaglia, Valeria, Fredrik Christiansen, Kate R. Sprogis, J. Symons, and Lars Bejder. 2019. "Food-Provisioning Negatively Affects Calf Survival and Female Reproductive Success in Bottlenose Dolphins." *Scientific Reports* 9 (1): 8981. https://doi.org/10.1038/s41598-019-45395-6.

Simões-Lopes, Paulo C., Fábio G. Daura-Jorge, and Mauricio Cantor. 2016. "Clues of Cultural Transmission in Cooperative Foraging Between Artisanal Fishermen and Bottlenose Dolphins, *Tursiops truncatus* (Cetacea: Delphinidae)." *Zoologia* 33 (6): e20160107. https://doi.org/10.1590/S1984-4689zool-20160107.

Simões-Lopes, Paulo C., Marta E. Fabián, and João O. Menegheti. 1998. "Dolphin Interactions with the Mullet Artisanal Fishing on Southern Brazil: A Qualitative and Quantitative Approach." *Revista Brasileira de Zoologia* 15 (3): 709–26. https://doi.org/10.1590/S0101-81751998000300016.

Smith, Brian D., Mya Than Tun, Aung Myo Chit, Han Win, and Thida Moe. 2009. "Catch Composition and Conservation Management of a Human-Dolphin Cooperative Cast-Net Fishery in the Ayeyarwady River, Myanmar." *Biological Conservation* 142 (5): 1042–49. https://doi.org/10.1016/j.biocon.2009.01.015.

Stanton, Margaret A., and Janet Mann. 2012. "Early Social Networks Predict Survival in Wild Bottlenose Dolphins." *PLOS ONE* 7 (10): e47508. https://doi.org/10.1371/journal.pone.0047508.

Stensland, Eva, Ida Carlén, Anna Särnblad, Anders Bignert, and Per Berggren. 2006. "Population Size, Distribution, and Behavior of Indo-Pacific Bottlenose (*Tursiops aduncus*) and Humpback (*Sousa chinensis*) Dolphins off the South Coast of Zanzibar." *Marine Mammal Science* 22 (3): 667–82. https://doi.org/10.1111/j.1748-7692.2006.00051.x.

Stringer, Christopher B., J. Clive Finlayson, R. Nick E. Barton, et al. 2008. "Neanderthal Exploitation of Marine Mammals in Gibraltar." *Proceedings of the National Academy of Sciences* 105 (38): 14319–24. https://doi.org/10.1073/pnas.0805474105.

Sutaria, Dipani, Mihir Sule, Ketki Jog, Isha Bopardikar, Abhishek Jamalabad, and Divya Panicker. 2017. "Baleen Whale Records from India." Paper SC/67a/CMP03rev1 presented to the International Whaling Commission Scientific Committee.

Sutherland, William J., Nigel G. Taylor, Douglas MacFarlane, et al. 2019. "Building a Tool to Overcome Barriers in Research-Implementation Spaces: The Conservation Evidence Database." *Biological Conservation* 238: 108199. https://doi.org/10.1016/j.biocon.2019.108199.

Tixier, Paul, Christophe Barbraud, Deborah Pardo, Nicolas Gasco, Guy Duhamel, and Christophe Guinet. 2017. "Demographic Consequences of Fisheries Interaction Within a Killer Whale (*Orcinus orca*) Population." *Marine Biology* 164: 1–16. https://doi.org/10.1007/s00227-017-3195-9.

Tixier, Paul, Mary-Anne Lea, Mark A. Hindell, Dirk Welsford, Camille Mazé, Sophie Gourguet, and John P. Y. Arnould. 2021. "When Large Marine Predators Feed on Fisheries Catches: Global Patterns of the Depredation Conflict and Directions for Coexistence." *Fish and Fisheries* 22 (1): 31–53. https://doi.org/10.1111/faf.12504.

Tonay, Arda M., Işıl Aytemiz Danyer, İlayda Taşkaya, et al. 2024. "Preliminary Findings on Cuvier's Beaked Whale Mass Stranding in Northern Cyprus." *Journal of the Black Sea/Mediterranean Environment* 30 (1): 1–19

Turvey, Samuel T., Robert L. Pitman, Barbara L. Taylor, et al. 2007. "First Human-Caused Extinction of a Cetacean Species?" *Biology Letters* 3 (5): 537–40. https://doi.org/10.1098/rsbl.2007.0292.

Valdivia, Abel, Shaye Wolf, and Kieran Suckling. 2019. "Marine Mammals and Sea Turtles Listed Under the US Endangered Species Act Are Recovering." *PLOS ONE* 14 (1): e0210164. https://doi.org/10.1371/journal.pone.0210164.

Van der Wal, Jessica E. M., Claire N. Spottiswoode, Natalie T. Uomini, et al. 2022. "Safeguarding

Human-Wildlife Cooperation." *Conservation Letters* 15 (4): e12886. https://doi.org/10.1111/conl.12886.

Verutes, Gregory M., Andrew F. Johnson, Marjolaine Caillat, et al. 2020. "Using GIS and Stakeholder Involvement to Innovate Marine Mammal Bycatch Risk Assessment in Data-Limited Fisheries." *PLOS ONE* 15 (8): e0237835. https://doi.org/10.1371/journal.pone.0237835.

Vighi, Morgana, Asunción Borrell, Jennifer A. Jackson, Emma L. Carroll, Maria Grazia Pennino, and Alex Aguilar. 2021. "The Missing Whales: Relevance of 'Struck and Lost' Rates for the Impact Assessment of Historical Whaling in the Southwestern Atlantic Ocean." *ICES Journal of Marine Science* 78 (1): 14–24. https://doi.org/10.1093/icesjms/fsaa205.

Wellings, C. E. 1944. "The Killer Whales of Twofold Bay, NSW, Australia, *Grampus orca*." *Australian Zoologist* 10: 291–93.

Wells, Randall S., Jason B. Allen, Suzanne Hofmann, et al. 2008. "Consequences of Injuries on Survival and Reproduction of Common Bottlenose Dolphins (*Tursiops truncatus*) Along the West Coast of Florida." *Marine Mammal Science* 24 (4): 774–94. https://doi.org/10.1111/j.1748-7692.2008.00212.x.

Whitehead, Hal, Ricardo Antunes, Shane Gero, Sarah N. P. Wong, Daniel Engelhaupt, and Luke Rendell. 2012. "Multilevel Societies of Female Sperm Whales (*Physeter macrocephalus*) in the Atlantic and Pacific: Why Are They So Different?" *International Journal of Primatology* 33: 1142–64. https://doi.org/10.1007/s10764-012-9598-z.

Whitehead, Hal, Randall R. Reeves, and Peter L. Tyack. 2000. "Science and the Conservation, Protection, and Management of Wild Cetaceans." In *Cetacean Societies: Field Studies of Dolphins and Whales*, edited by Janet Mann, Richard C. Connor, Peter L. Tyack and Hal Whitehead. University of Chicago Press.

Whitehead, Hal, and Megan Shin. 2022. "Current Global Population Size, Post-Whaling Trend and Historical Trajectory of Sperm Whales." *Scientific Reports* 12 (1): 19468. https://doi.org/10.1038/s41598-022-24107-7.

Whitehead, Hal, Tim D. Smith, and Luke Rendell. 2021. "Adaptation of Sperm Whales to Open-Boat Whalers: Rapid Social Learning on a Large Scale?" *Biology Letters* 17 (3): 20210030. https://doi.org/10.1098/rsbl.2021.0030.

Wild, Sonja, Michael Krützen, Robert W. Rankin, William J. E. Hoppitt, Livia Gerber, and Simon J. Allen. 2019. "Long-Term Decline in Survival and Reproduction of Dolphins Following a Marine Heatwave." *Current Biology* 29 (7): R239–40. https://doi.org/10.1016/j.cub.2019.02.047.

Williams, Rob, and David Lusseau. 2006. "A Killer Whale Social Network Is Vulnerable to Targeted Removals." *Biology Letters* 2 (4): 497–500. https://doi.org/10.1098/rsbl.2006.0510.

Williamson, Michael J., Mariel T. I. ten Doeschate, Rob Deaville, Andrew C. Brownlow, and Nicola L. Taylor. 2021. "Cetaceans as Sentinels for Informing Climate Change Policy in UK Waters." *Marine Policy* 131: 104634. https://doi.org/10.1016/j.marpol.2021.104634.

15

Who Are We? Assessing Diversity and Enhancing Inclusivity in Marine Mammalogy

FREDERICK ARCHER, DANTE FREEMAN, AND TARA COX

Abstract

Just as diversity in marine mammal societies is critical for ecosystem function, diversity of thought is critical for marine mammal science to move forward. However, there are both clear and hidden impediments to many people with aspirations to enter and remain in the field of marine mammal science. A survey of 879 marine mammal scientists suggested that marine mammalogists were not very diverse overall, with the majority of respondents being white, heterosexual females from the US. Although the field may be unique among STEM fields because it is female dominated (at least at early career stages), there was a notable lack of diversity across all other areas analyzed, primarily sexual orientation and identity, accessibility, and ethnicity and global representation. It is critical to recognize the disparities between the current makeup of the field and the population at large and enact practices that foster the inclusion of researchers that are more representative of a wider array of communities. Future generations of scientists will be able to bring their complete selves to their careers and work together in a more diverse and inclusive community. By bringing novel thoughts, ideas, and experiences with them, these generations will be better suited to solve the increasingly complex problems faced in the field of marine mammal science.

Introduction

Across the animal kingdom, societies have evolved for the support, protection, and welfare of their individual members. For similar reasons, the people who study them also exist in their own societies. Every researcher is the product, in part, of a unique set of identities, interactions, and group affiliations. These identities and relationships in turn directly influence their scientific interests, the questions they pose, how they address them, and the insights they uncover.

It is well understood that biodiversity in its many forms is critical to global ecosystem functioning (Tilman et al. 2014). For example, genetic diversity is the underlying engine for adaptation to changing conditions, while ecological diversity helps energy and nutrients move between trophic levels (Ekschmitt et al. 2001). Meanwhile, species diversity leads to resilient ecosystems and communities within them (Ekschmitt et al. 2001).

For much the same reasons, diversity in the workforce is critical to innovation and scientific discovery (Guterl 2014). Diversity in ethnicity, culture, gender, and experiences creates a community that is well-rounded in ideas and solutions and prepared to meet evolving challenges (Guterl 2014). Diversity of thought, brought on not only by research expertise, but also by lived experiences, can lead to great strides in science and conservation of marine mammals, strong research networks, and effective disseminations and application of science outcomes in management, policy, and public spheres. In short, forming and maintaining groups of diverse scientists is the basis for creating welcoming, effective, and successful research teams and ecosystems (Guterl 2014).

Despite its importance, there have been few improvements in diversity in the science, technology, engineering, and mathematics (STEM) workforce over the last 30 years (Allen-Ramdial and Campbell 2014; Fry et al. 2021). There are more efforts to understand the problems faced by individuals from underrepresented groups (Allen-Ramdial and Campbell 2014) and to implement and evaluate programs focused on mitigating some of these factors (Sellami et al. 2021).

Although there are serious issues faced by individuals around the globe, most research has focused on communities within the United States. Some studies have suggested that negative socioeconomic factors are the main reason minorities are underrepresented in STEM fields (Allen-Ramdial and Campbell 2014; Hanson et al. 2004). Others suggest that young people are just not aware of the career options available in STEM fields (Hall et al. 2011). Johnson (2011) emphasized that the combination of ethnicity, gender, and personal background creates a unique set of obstacles for each individual. Overall, there is a complex and poorly understood web of issues that makes it challenging for minorities to pursue STEM fields.

Marine mammal science is especially prone to, and egregious in terms of, disparities in representation. The low diversity in this field is certainly related to the earlier-mentioned issues affecting STEM fields more generally. However, it is also compounded by special challenges inherent in studying marine mammals including the limited exposure and difficulties in access to marine mammals or opportunities in marine mammalogy facing people living in non-coastal or lower socioeconomic regions. In addition, there are relatively few programs and staff in academic institutions with marine mammal expertise. Furthermore, the allure of marine mammals in the general public leads to extreme competition for an ever-shrinking number of opportunities at each step in the career ladder, creating severe restrictions, where people already at a disadvantage are quickly excluded.

Although understanding the multitude of barriers to the field is clearly a complex problem, it is critical to quantify the current composition of the people involved to begin to enact meaningful changes to improve diversity. Recognizing this need, in December 2018, the Committee on Diversity and Inclusion of the Society for Marine Mammalogy (SMM) fielded an 18-question survey to better understand the demographic composition of the field of marine mammalogy at the global level and to serve as a baseline for future surveys aimed at monitoring changes in the field (see the appendixes to this chapter). The survey was intentionally kept relatively simple and short to encourage participation, delving deeper into details of many of the topics. However, the last question was a free-text comment, where participants were encouraged to provide "suggestions for what an international marine mammal society can do to enhance diversity and inclusion." The entire marine mammal science community was encouraged to participate in the survey, which was sent out to present and past members of SMM, posted on social media sites such as Facebook and Twitter, as well as the marine mammal (MARMAM) mailing list. Over a two-week period, a total of 879 people representing 79 countries responded.

This chapter presents results from the survey as an initial effort to better understand the composition of our field. We do not attempt to comprehensively cover all factors related to demographic inequities in the field nor the full scope of their effects. Rather, four major themes were addressed in the survey: gender, sexual orientation and identity, accessibility, and ethnicity and global representation. The goal is to highlight some of the more relevant issues in each of these themes and suggest potential strategies to diversify the field. For details on these themes outside the scope of this chapter, there is a wealth of research and resources available within the references provided.

Although the survey was publicized as widely as possible, it was only available in English, which may have reduced responses from those in the field with limited English skills. In addition, survey respondents may be biased toward those who are more interested in issues of diversity and inclusion within the field and thus may not be completely representative. Finally, by virtue of the authors' being US citizens and residents, and the majority of respondents being from the US, many of our findings and recommendations may appear to be US-centric. However, many of the recommended actions can and should be more broadly applied to the entire international field of marine mammal science (see box 15.1 at the end of the chapter). The results of this survey make it clear that there is still much work to do. Thus, this chapter is part of the initial process of moving the field of marine mammal science toward being more diverse and inclusive.

Gender

The disparities between men and women in the sciences and the struggles of women to achieve equality are well-known (Hooker et al. 2017; Koch et al. 2022; Ortiz-Martínez et al. 2023). Within the US, although women earn approximately 57% of all undergraduate degrees, men who declared an undergraduate major in a STEM field were 1.31 times more likely than women to continue to graduate in a STEM field (Snyder et al. 2019; Koch et al. 2022). In 2008, women in the US, earned 38% of bachelor's degrees and made up 34.4% of graduate students, but made up less than 1% of tenure-track faculty in the top 50 research universities (Johnson 2011).

Over the past several decades, marine mammalogy, which has historically been a male-dominated field, has witnessed a significant increase in the participation of women. In the only other study to examine the demographics of marine mammalogists, Hooker et al. (2017) found that approximately 72% of the 478 respondents in their survey identified as women. We found a similar gender skew in our survey with two-thirds of the respondents (66%) identifying as women.

However, this gender skew was not uniform across all ages (fig. 15.1). The proportion of female respondents under 45 years old was about four times greater than male respondents in each age group. That difference decreased in the 45–54 age group, and the gender response rate was relatively equal in the 55–64 age group. The oldest group (>64) was composed of more men than women. Almost half of all respondents (46%) had a doctoral degree, while about a third (34%) had their master's. However, a higher proportion of women than men across age groups had their master's, while a higher proportion of men than women had their doctorate in all age groups except 55–64, in which the rates were relatively equal (fig. 15.2).

These differences support the previous findings that women have been entering the field at a greater rate than men since at least the late 1990s (Hooker et al. 2017). However, it is not clear that women currently have the same

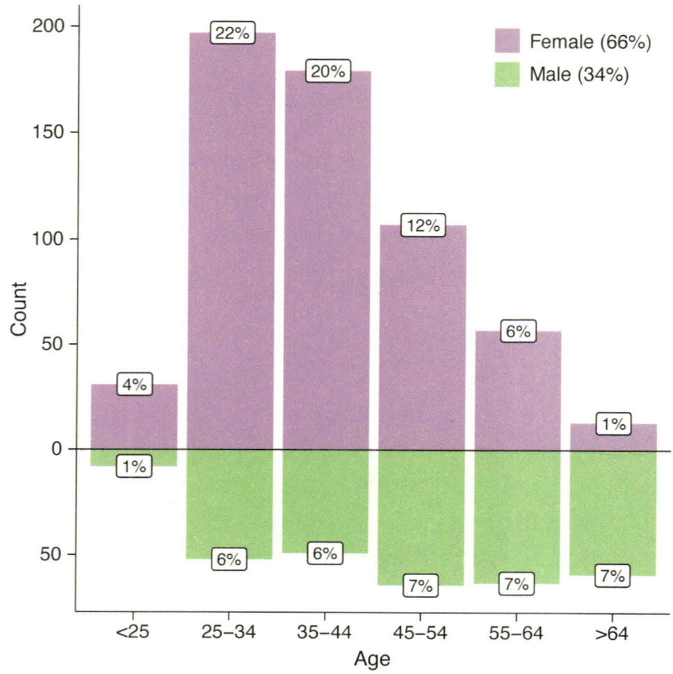

FIGURE 15.1. Number of survey respondents by age (x-axis) and gender (colors). Values in white boxes represent percentage of total respondents represented by each age class and gender combination.

opportunities as their male colleagues or will have similar career trajectories in the coming years. Studies across other STEM fields indicate that significant barriers remain. As discussed in more detail later in this chapter, there is also evidence that women in marine mammal science are more geographically mobile than their male counterparts.

Because marine mammal science often requires research at remote locations, safety is of utmost importance and a relatively unique challenge. Other fields have grappled with the challenges and have made recommendations to ensure safety for not only women but also minorities and transgender individuals, who are statistically at higher risk of harassment (reviewed in Nash 2021 and Matsuda 2023). At the most basic, all field research programs should have a clear safety protocol that explicitly addresses definitions of harassment and bullying and provides clear pathways for reporting.

It is well-known that the responsibilities of caregiving disproportionately fall to women in most societies. This places a considerable burden on women in the field as they must balance the demands of their career with their roles outside of the workplace (Hanson et al. 2004). Most saliently, in particular, it brings a large set of challenges for professional activities critical to their career such as attending and participating in conferences. A group of mothers working in science have published a list of suggestions titled "Childcare, Accommodate families, Resources, Establish social networks" (CARE) that would allow for more inclusion of parents at meetings and conferences (Calisi 2018). The first overall suggestion is financial support for either individual or on-site childcare options (Calisi 2018). The group suggests travel grants that allow scientists to bring a caregiver, such as a spouse, relative, or nanny, from home (Calisi 2018). Some examples of societies that already include these grants include the Society for Molecular Biology and Evolution, the American Society of Plant Biologists, and the European Society for Evolutionary Biology (Calisi 2018). The group also says that babywearing should

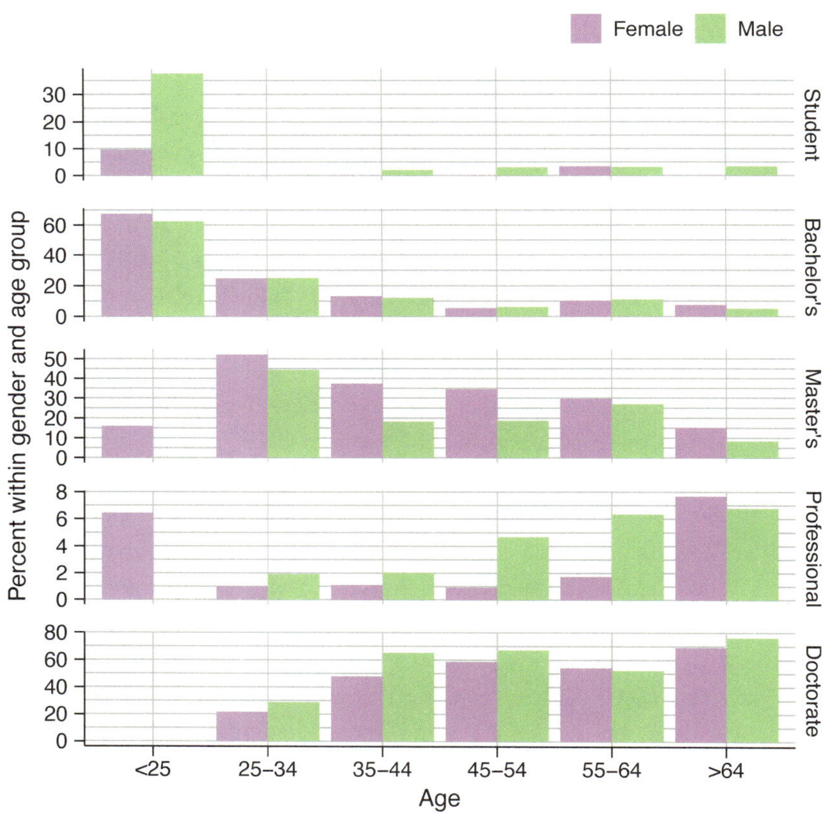

FIGURE 15.2. Percentage of respondents within gender and age group (y-axis) for each age group (x-axis) and highest degree earned (rows). Here, "Student" includes all respondents with school attendance below achieving a bachelor's degree. "Master's" includes respondents with a master's as well as those enrolled in, but who have not yet completed, a doctorate program. "Doctorate" includes those with a doctorate and those in a postdoctoral program.

be allowed at conferences, stating that the minor disturbances caused by infants are no different from any other family-friendly function and that they are a small price to pay to improve inclusion (Calisi 2018).

Sexual Orientation and Identity

Within the United States, approximately 5% of adults openly identify as lesbian, gay, bisexual, transgender, or queer (LGBTQ), but globally, this value drops to approximately 3% (Badgett et al. 2021; Flores 2021; Flores and Conron 2023). Although we did not directly ask if respondents identified as members of the LGBTQ community, we combined responses to the questions of transgender status and sexual orientation to create a proxy for LGBTQ affiliation. Under this definition, approximately 10% of all respondents identified as LGBTQ, with most in the 25–34 age group. There was little difference in overall rates of LGBTQ affiliation between respondents who identified as being male versus those who identified as being female.

The need for greater recognition of and support for those in the LGBTQ community is not unique to the field of marine mammalogy. In fact, STEM professionals who are LGBTQ experienced higher rates of harassment, professional devaluation, and mental health struggles, and less overall civic participation than their non-LGBTQ peers (Cech and Waidzunas 2021; Flores and Conron 2023). Several survey respondents pointed to the LGBTQ mixers at the SMM biennial conferences as a step forward. Continuing these mixers, but not at the cost of attendees missing scientific symposia or other opportunities for building relationships with peers, should be paramount (Burnett et al. 2020).

The fraction of respondents identifying as LGBTQ varied significantly by region (fig. 15.3), likely reflecting differences in global legal and social climates. While LGBTQ acceptance and protections are growing in many countries, there are still deep-seated social taboos, or even outright persecution in others (Wyatt-Nicol and Naylor 2013; Poushter and Kent 2020; Flores 2021; UN Women 2022; Flores and Conron 2023; Herre and Arriagada 2024). As of this writing in 2024, the International Lesbian, Gay, Bisexual, Trans, and Intersex Association (ILGA, https://database.ilga.org) has identified 57 countries where consensual homosexual acts could be punished by up to 8 or more years in prison, and 11 other countries where a death sentence could be imposed.

Although there are concerted efforts to change these laws and increase protection for the LGBTQ community across the world (United Nations 2016), the fact that the social and political landscape can represent a real threat to members of the LGBTQ community must be kept in mind when planning events in those countries. In particular, we strongly recommend

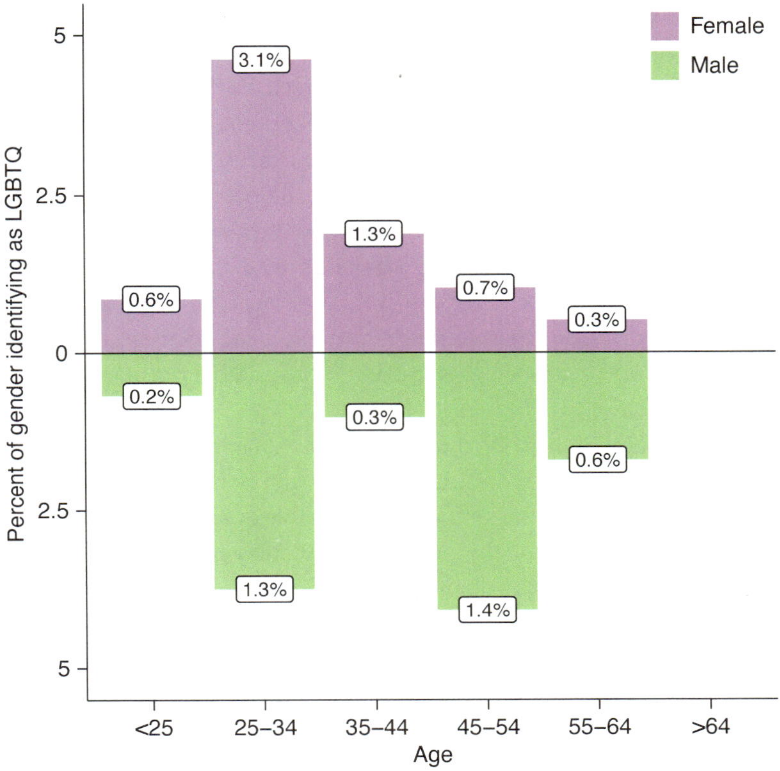

FIGURE 15.3. Percentage of each gender identifying as LGBTQ (*y*-axis) by age group (*x*-axis).

consulting reports and databases with organizations like the United Nations Human Rights Office of the High Commissioner (www.ohchr.org), ILGA, and Human Rights Watch (www.hrw.org) prior to choosing countries for conferences, workshops, or other meetings. Additionally, if research must be conducted in these countries, we recommend that participants become thoroughly familiar with the relevant local laws and customs and have medical and legal contingencies in place.

Disabilities

There is a wide array of physical and mental disabilities that pose challenges for a number of members of the marine mammal community. People with disabilities made up only 7% of faculty at four-year institutions in the United States in 2006 (Hill et al. 2010; Hartmann 2019). Likewise, approximately 9% of all respondents to the marine mammalogist survey reported living with either a physical, sensory, or learning disability. Although half of these were in the 25–34 age group, after correcting for differences in the overall number of responses in each age group, the percentage reporting a disability was relatively consistent across ages, only varying between 7% and 11%.

To increase the accessibility of educational and research opportunities, it is important to recognize the needs of individuals with disabilities and make concerted efforts to accommodate them in conferences, academic settings, and fieldwork (Peterson 2021; Rudzki and Kohl 2023). This can include providing flexible work hours, arranging lab spaces to assist people with mobility issues, searching for texts and papers in Braille or text-to-voice readers, and ensuring that work spaces have emergency equipment tailored to those with special needs, such as strobe lights in addition to audible fire alarms (Brown 2016; Sarju 2021).

The universal design (UD) framework is a valuable resource to assist in evaluating and designing spaces. UD provides comprehensive principles for designing "products and environments to be usable by all people, to the greatest extent possible, without the need for adaptation or specialized design" (Burgstahler 2021; North Carolina State University Center for UD 2024). By focusing on seven core principles of design, UD can be applied to a wide array of spaces such as labs, classrooms, and fieldwork, to address barriers to accessibility (Hartman 2019; Burgstahler 2021). Additionally, Langley-Turnbaugh et al. (2009) provide a suite of suggestions for formatting protocols, data sheets, equipment instructions, equipment, and safety measures with accessibility in mind.

Improving accessibility at conferences was mentioned by several respondents as a recommendation for the SMM. The Association for Computing Machinery's Special Interest Group on Accessible Computing has attempted to address some of these concerns by writing a guide to accessible conferences (Trewin et al. 2019). This guide mainly focuses on accommodations for people with disabilities and sensory or neurological differences (Trewin et al. 2019). It goes through the entire conference process from planning and setup to the conference proceedings and side events associated with the conference (Trewin et al. 2019). Some of the suggestions include confirming if the venue staff has had disability awareness training, locating dining options within and around the venue for dietary restrictions, and designated quiet rooms (Trewin et al. 2019). It is also important to ensure that people with disabilities are involved where possible in the planning of meetings and spaces (Peterson 2021; Sarju 2021). Disabilities can take many forms and may be hidden to a casual observer (Sarju 2021). Without the consultation and input of disabled people, opportunities to increase their inclusion in many research spaces are likely to be missed.

Ethnicity and Global Representation

The global nature of marine mammal research underscores the importance of ensuring equitable participation in the field from people across the world, including improving representation of a wide array of ethnic identities. The concept of ethnicity, which combines multiple components of identity, such as culture, ancestry, language, and religion, is notoriously difficult to define, especially in an international context (Chandra 2006; Brubaker 2009). Yet, for many, it is a defining feature of who we are. In many arenas it is difficult to fully exempt our ethnic background or how it is perceived from how we experience and interact with the rest of the world. This naturally extends into our careers and how we conduct our work. Our ethnic identities bring unique perspectives to how we approach questions and find solutions to challenges.

One of the central features of most people's identity is the country or region of the world where they grew up, live, and work. To better understand the geographic distribution of marine mammalogists, the survey asked respondents to list their countries of birth, citizenship, and residence. As expected, a large proportion of respondents (about 55%) were from the United States (table 15.1). Because approximately 60% of the other 78 countries were represented by three or fewer respondents across these three questions, we grouped all countries into five large geographical regions (table 15.1). About a quarter to a fifth of the respondents represented Europe and central Asia. The least represented regions, the Middle East and Africa, had at most 21 respondents (about 2%).

In the Hooker et al. (2017) survey of 478 members of the SMM conducted in the early 2010s, approximately 38 countries were represented, but 65% of the respondents were from the United States. The increase in the more recent survey, both in the fraction of respondents from outside the US, and the number of countries represented suggests either an increase in participation in the field outside of the US within the decade, or better dissemination, access, and buy-in to the survey due to differences in global connectivity by email and social media. Regardless, there is still a clear need to increase representation of marine mammal scientists across most of the world.

We also examined rates and patterns of movement between regions. Overall, 83% of the respondents were citizens of, and also resided in, a country within the region in which they were born (fig. 15.4). However, the rates of staying within a region varied by region: 93% of those from Canada or the United States stayed, while only approximately 72% of those from Asia, Europe, and Oceania stayed. The rates of staying within the Caribbean and Latin America, and Africa and the Middle East were even lower at 67% and

TABLE 15.1. Countries within each region and number (percent) of respondents born in, citizens of, and residing in each region

Region	*Countries*	*Birth*	*Citizen*	*Residence*
Africa and Middle East	Angola, Cameroon, Iran, Israel, Kuwait, Madagascar, Namibia, Nigeria, Republic of Cabo Verde (West Africa), Senegal, South Africa, United Arab Emirates, Zimbabwe	9 17 (2)	7 14 (2)	10 21 (2)
Canada and USA	Canada, United States	2 493 (56)	2 526 (60)	2 529 (60)
Caribbean and Latin America	Argentina, Bahamas, Brazil, Chile, Colombia, Costa Rica, Ecuador, Guatemala, Jamaica, Mexico, Peru, Puerto Rico, Uruguay, Venezuela	14 111 (13)	13 96 (11)	13 84 (10)
Central Asia and Europe	Andorra, Austria, Belgium, Britain, Croatia, Denmark, England, Estonia, Finland, France, Germany, Greece, Holland, Hungary, Iceland, Ireland, Italy, Latvia, Luxembourg, Norway, Poland, Portugal, Russia, Scotland, Serbia, Slovenia, Spain, Sweden, Switzerland, Turkey, Ukraine, United Kingdom, Wales	30 204 (23)	31 187 (21)	27 175 (20)
East and South Asia and Oceania	Australia, Bangladesh, Cambodia, China, Hong Kong, India, Indonesia, Japan, Korea, Malaysia, New Zealand, Philippines, South Korea, Sri Lanka, Taiwan, Thailand, Vietnam	16 54 (6)	16 56 (6)	17 70 (8)

Note: For Birth, Citizen, and Residence columns, top value is number of countries in region, and bottom value is number of respondents from each region with percentage in parentheses.

65%, respectively. Additionally, 44% of the respondents that stayed in their country of birth had doctorates compared with 52% of those who moved.

Across all respondents, 18% of women and 13% of men were not residing in their region of birth. However, this difference also varied by age and region, with a much greater difference between genders in age groups between 25 and 54 in most regions (fig. 15.5). Women between 35 and 54 years of age had the highest probability of moving from their region of birth in the Caribbean and Latin America, and Africa and the Middle East. Although we can only speculate about this difference in mobility between genders, an obvious possibility is that it reflects the increased difficulty of obtaining meaningful employment for women in the field in their native countries in these regions, or different expectations of women to be more flexible and move based on the careers of their partners.

Marine mammals face serious threats in many of the regions that were poorly represented in the survey. The best way to address these issues is to ensure that more researchers familiar with a particular region are involved in their conservation. Geographical and the associated economic imbalances are one of the underlying causes of the problem of "helicopter" or "parachute" science (Stefanoudis et al. 2021)—the term given to the phenomenon of relatively higher income researchers collecting data from typically lower income regions, and then returning back to their country of origin without either sufficient acknowledgment or substantive participation of local researchers in the analytical or publication process. The process also excludes local environmental knowledge from being fully incorporated into the interpretation of the data.

Thus, it is critical to ensure that projects are undertaken, from design,

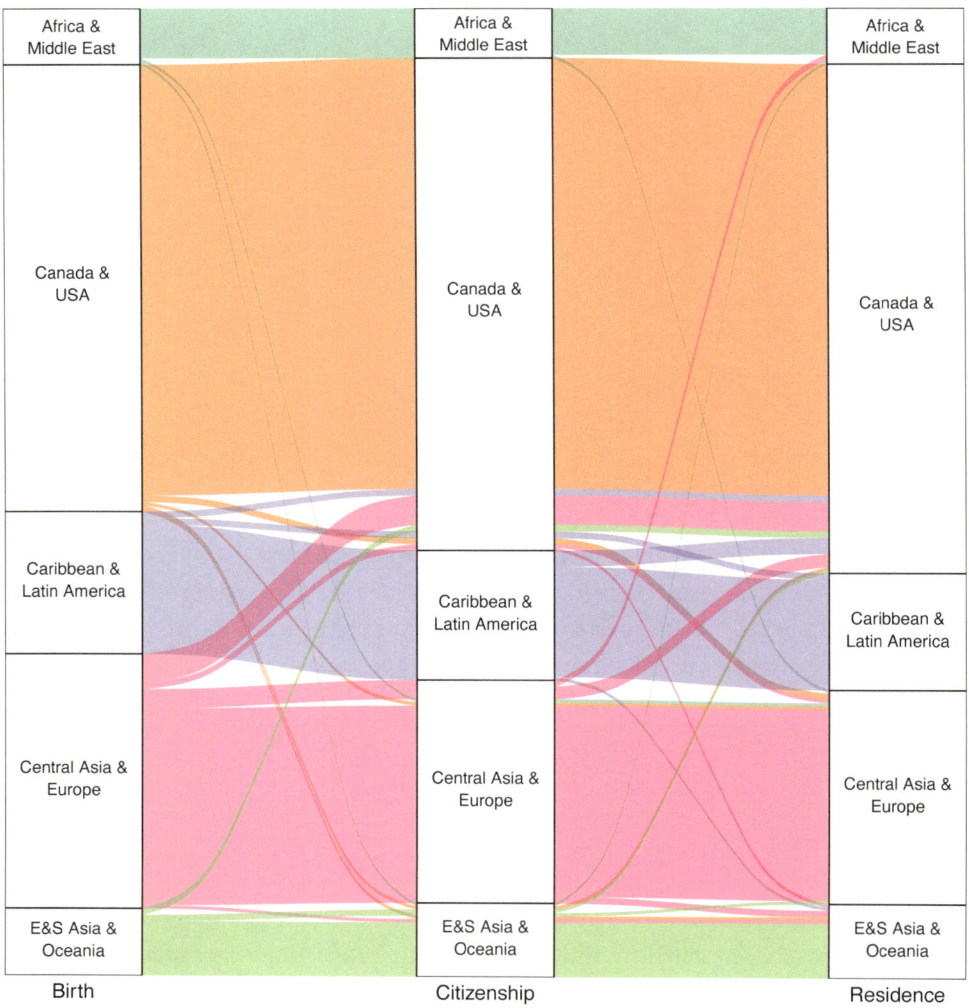

FIGURE 15.4. Connection of respondents between regions of birth, citizenship, and residence.

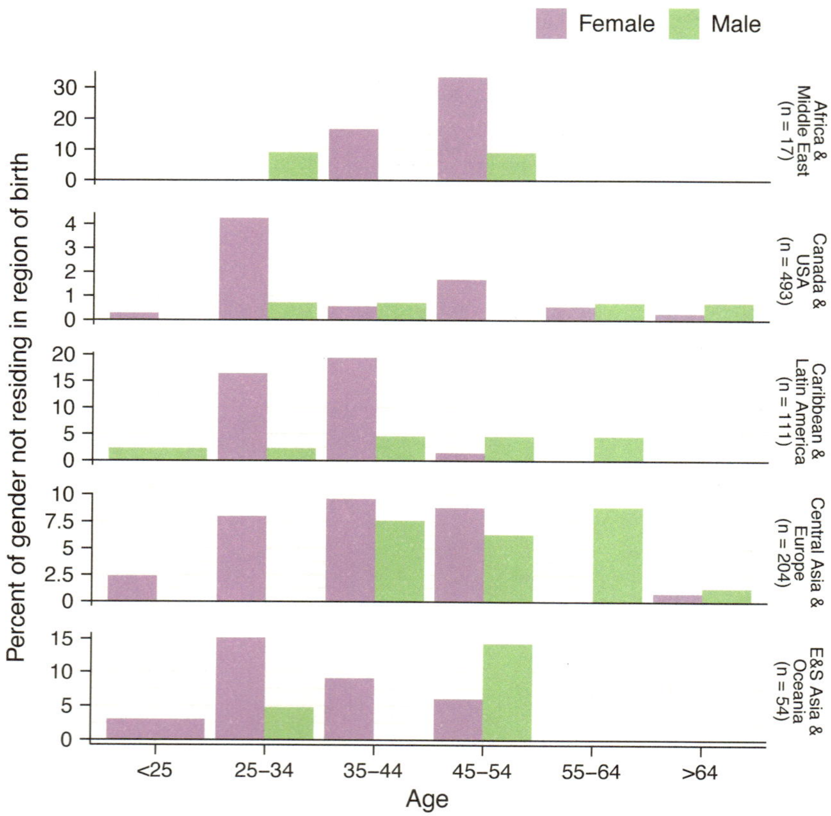

FIGURE 15.5. Percentage of each gender not residing in each region of birth (y-axis) by age (x-axis) and region (row).

through execution, analysis, publication, and dissemination in tight collaboration with local researchers (de Vos and Schwartz 2022; de Vos et al. 2023). While parachute science has and continues to be a concern in the field of marine mammalogy, this survey did not attempt to quantify its occurrence. Regardless, respondents expressed a need for continuing support for underprivileged communities. These suggestions were mainly in the form of outreach and mentorship to developing countries and areas with little scientific participation. One respondent specifically suggested to "start being better about adhering to the Nagoya protocol and consulting with Indigenous communities before coming in and sampling / doing research in their areas to avoid helicopter science."

Across the world, ethnic minorities make up proportionately lower numbers of the scientific community (Allen-Ramdial and Campbell 2014). Although people identifying as Hispanic or Latino make up approximately 22%

of the 18–34 aged population in the United States, they only make up 9% of those with doctoral degrees in STEM (NCSES 2023). Similarly, African Americans make up 7% of STEM doctoral degrees, but 14% of the US 18–34 aged population (NCSES 2023). Given the dominance of respondents from the United States, it is not surprising that a majority (78%) identified as Caucasian. A total of 90% of the US respondents identified as Caucasian. Approximately 84% identified only as Caucasian, with the remainder as Caucasian and some other ethnicity. The second most frequent (16%) identification was "Latinx / Hispanic / Spanish origin." In addition, 8% of the respondents identified with multiple ethnicities. Many were very specific in their responses, breaking them down into regions such as Eastern European or South Asian. There were also many respondents who identified as African/Asian/European American. About 1% of respondents preferred not to list their ethnicity.

In the survey, a total of 45 languages were reported being spoken as the respondent's primary language. The over-representation of the United States contributed to 68% of the respondents speaking English as their primary language. Spanish was the next most commonly spoken language at 10%. However, there were at least 32 languages spoken by three or fewer respondents, one of which was American Sign Language.

Language can represent a sizable barrier to full participation in the field by many scientists around the world (Carroll and Sironi 2022). This ranges from dissemination of findings in journals to being able to reap the full range of benefits of attending conferences and workshops. In fact, one of the common recommendations in the survey was to include providing translation services for conferences and the journal. The results of this survey indicate starting with Spanish may be a test-case for such translation services and then possibly expanding. In fact, based on this recommendation, *Marine Mammal Science* now offers translation services for authors who request them or for those articles that in the review process are noted as needing English translation services. Additionally, Carroll and Sironi (2022) have suggested moving toward publishing abstracts in other languages and facilitating conference abstract review and poster and spoken sessions in other languages.

Summary and Conclusions

As of 2018, the average marine mammalogist who responded to this survey was a heterosexual, Caucasian female, between the ages of 25 and 44, with a master's degree living and working in the United States. The results strongly suggest that there is a clear need for improvement by the field of marine mammalogy and the SMM specifically in diversity and inclusion. Even

Box 15.1 Recommendations (Based on Comments and Literature)

CONFERENCES

- Implement a code of conduct with a clear reporting structure, buy-in from all attendees, and clear consequences for violations.
- Ensure diversity of representation in plenary speakers and on panels.
- Develop and implement virtual components, including mechanisms to encourage interactions of virtual attendees.
- Consult the UN or Human Rights Watch for whether a country is safe for LGBTQ members before committing to a biennial conference there.
- Thoughtfully consider schedules of all activities, including scientific sessions and social events (see Burnett et al. 2020 for a thoughtful review of trade-offs of scheduling diversity-focused events).
- Continue to fund participants' travel, especially students and researchers from underdeveloped countries.
- When funds are available, provide additional funding for caregivers (e.g., on-site care, funds for caregiver travel, etc. See Calisi 2018 for a more comprehensive list).
- Consider innovative ways for international and underfunded researchers to network outside of the conference structure.

FIELD OF MARINE MAMMALOGY

- Monitor demographic trends to ensure activities, goals, and outcomes align.
- Adhere to the six constructs of anti-racism and perform the 20 actions within those constructs to have a more diverse and inclusive field (Ali et al. 2021).
- Encourage all scientists at all levels to attend inclusive mentoring training.
- Recognize and eliminate barriers to access, including underrepresented groups' "silent tax" by way of navigating a confusing system and having to do extra work.
- Implement code of conduct at each institution, including a clear reporting structure.
- Minimize low/unpaid jobs and internships.
- Eliminate colonialism and parachute science (Stefanoudis et al. 2021; de Vos and Schwartz 2022; de Vos et al. 2023).
- Evaluate and design spaces to improve accessibility (see Langley-Turnbaugh et al. 2009; North Carolina State University Center for Universal Design 2024).
- Document the history (see Lee 2020 and Tang-Martinez 2020).

though 78 countries were represented as a country of birth, citizenship, or residence, a majority of the countries were represented by no more than three respondents. Spoken languages are also skewed toward the Western world, with English and Spanish being the most common. However, a portion of this skew may be reflective of the fact that the survey was only offered in English. There were also comparatively low percentages of sexual/gender minorities and scientists with disabilities.

In both urban and rural areas away from water, there are few opportunities for youth to be exposed to marine mammals to spark initial interest. Minorities and members of low socioeconomic communities are additionally unlikely to encounter marine mammal researchers who look like them. This lack of role models makes it difficult to see studying marine mammals as a viable career path. Few undergraduate and graduate programs exist with a focus on marine mammals that can serve as conduits to later career opportunities, and those that do exist are very competitive. The competitive nature of these programs also means that potential students frequently have to move. This can be difficult to impossible if the students do not have sufficient resources or financial support or have caregiving responsibilities, which disproportionately falls to women.

Many universities in the United States have adopted intervention strategies to increase diversity in STEM fields (Chubin et al. 2005). The most common way to implement interventions is to focus on the educational transition point to upper-level institutions such as universities (Chubin et al. 2005). Some examples of this include summer bridge programs, mentoring and tutoring programs, research experiences for undergraduates (REUs), and career counseling and academic advising (Tsui 2007). However, all of these programs assume that the participant has had an equitable upbringing and access to resources prior to that transition point.

Organizations and societies have also started hosting diversity workshops that increase the knowledge, motivation, and skills of participants to address the issue of low diversity (Hill et al. 2010). Some societies, such as the Society for Integrative and Comparative Biology (SICB), have created committees dedicated to increasing diversity (Wilga et al. 2017). The main goal of the SICB Broadening Participation Committee (BPC) was to increase the number and support of underrepresented group members (Wilga et al. 2017). Through workshops, diversity and mentor socials, and a moderate budget, the SICB BPC was able to increase underrepresented group members to 16.2%, increase ethnic diversity by 25%, and increase gender diversity by 18% over five years (Wilga et al. 2017).

There is both a moral and scientific obligation to increase the diversity of people within the sciences by dismantling the barriers preventing them from pursuing STEM fields. Individual societies, organizations, and universities must work together to improve diversity by first improving within. One of the goals of this chapter has been to use the 2018 survey to highlight some of the issues experienced by members of the marine mammalogy community. In that survey, several respondents expressed concern about the inclusion of what may have been perceived as invasive questions in the survey such as those on ethnicity or sexual orientation. As an example, two responses to the question on ethnicity were

- "I don't find skin color to be relevant to my feeling of attachment to marine mammal science."
- "Who cares!"

These sentiments point to a need to improve communication about the importance of recognizing underrepresented members in the field. In conservation biology, it is widely recognized that it is not possible to conserve that which cannot be defined. Likewise, the field cannot begin to tackle disparities in diversity without regularly monitoring the demographics of the field. Only then can we evaluate whether efforts to improve diversity and inclusion are having a positive effect and if not, how to change course or where more focus and resources are needed.

Overall, marine mammalogy is a low diversity field, but it may be unique among STEM fields because it is female dominated (at least at early career stages). Minority groups are underrepresented in many categories, and many survey respondents shared issues they have seen or faced in this field. Addressing these problems will be difficult because improving diversity is a multifaceted issue. Any combination of factors may contribute to preventing underrepresented groups from participating in the field. Diversity and inclusion will only be improved through continuous effort. It will take time for change to be seen, but it is necessary so that researchers are more representative of a wider array of communities, bring their novel thoughts, ideas, and experiences with them to help solve complex problems, and future generations of scientists can bring their complete selves to their careers and work together in a more diverse and inclusive community.

Acknowledgments

The authors wish to thank the Society for Marine Mammalogy for their support in fielding the 2018 demographic survey, and Drs. Alexandra Curtis and Dionne Hoskins-Brown for helpful reviews and comments.

Appendix 15A: Survey Questions

Censored, raw data for the survey has been archived at Frederick Archer and Tara M. Cox, "2018 SMM Marine Mammalogist Demographic Survey," figshare, 2024, https://doi.org/10.6084/m9.figshare.26418760.v3.

SOCIETY FOR MARINE MAMMALOGY SURVEY OF MARINE MAMMALOGISTS

The Society for Marine Mammalogy (SMM) has recently created an ad hoc committee on diversity and inclusion. The first step of the committee is to better understand the demographic composition of the field of marine mammalogy and to solicit ideas for ways that SMM can help make a more diverse and inclusive community. This survey is completely anonymous and no personal information is collected or retained. The results will be used to inform the SMM as it moves forward in creating a more inclusive and diverse society. It should take less than three minutes to complete the 18 questions. Even if you are not currently a member of SMM, please feel free to fill out the survey. Our intent is to obtain as comprehensive a view of the field as possible.

1. What is your age?
2. What is your gender?
3. Do you consider yourself to be transgender?
4. Do you consider yourself to be . . . ? (options: Heterosexual or straight; homosexual, bisexual, asexual, other)
5. Do you consider yourself to have learning difficulties or disabilities?
6. Do you have a long-lasting condition resulting in severe vision, hearing, or speech impairment (not including the wearing of glasses, contact lenses, or hearing aids that correct the condition to the point that normal activities are not severely impacted)?
7. Do you have a long-lasting condition that substantially limits one or more basic physical activities such as walking, climbing stairs, reaching, lifting, or carrying?
8. What is your country of birth?
9. What do you consider to be your primary country of citizenship?
10. What is your current country of residence?
11. What is your first language?
    If you chose "Other . . ." for a language, please list it here.
12. People's ethnicity describes their feeling of belonging and attachment to a distinct group of a larger population that shares their ancestry, color, language, or religion. With which ethnic group do you identify? The list below is not intended to be globally comprehensive, so please select all that apply, or describe the ethnic group that you identify with the most in the choice labeled "Other." (options: Caucasian; Latinx / Hispanic / Spanish

origin; Middle Eastern; African; African descent [e.g., African American]; Caribbean; South Asian; East Asian; Pacific Islander)

13. What is the highest degree or level of school you have completed? (options: some high school; high school degree or equivalent; some college, no degree; associate degree; bachelor's degree; master's degree; professional degree; doctorate; other)
14. Do you consider yourself to be a . . . ? (please select all that apply; options: natural scientist, social scientist, interdisciplinary scientist, educator, whale-watching naturalist; administration/manager; science communicator; technology developer/maker)
15. What is your institutional affiliation? (please select all that apply)
16. Please select all societies of which you are currently a member. Societies or organizations not on this list can be entered in "Other."
17. Please select all membership levels that you have in the above societies. (options: student, full, associate, emeritus, life, honorary, other)
18. Please use the space below to provide suggestions for what an international marine mammal society can do to enhance diversity and inclusion.

Thank you for completing our survey.

Appendix 15B: Summary of Respondents to Survey

Average marine mammalogist is a heterosexual, Caucasian woman, aged 25–44, with a master's degree, living and working in the USA. $N = 879$.

GENDER

- More women in younger age classes (<45) with more men than women only in oldest age class (>64).
- Higher proportion of women than men have master's degrees.
- Higher proportion of men than women have doctorates.

SEXUAL ORIENTATION AND IDENTITY

- Larger proportion of younger age class reporting as LGBTQ.
- Regional differences.

DISABILITY

- 9% of respondents reported physical, mental, or sensory disability.

ETHNICITY AND GLOBAL REPRESENTATION

- 79 countries represented.
- Most represented regions: North America, Europe, and central Asia.

- Least represented region: Middle East and Africa (grouped together due to small sample size).
- Researchers in some regions were less likely to stay in their region of birth to study and work: Caribbean and Latin America; Middle East and Africa.
- 45 languages spoken.

References

Ali, Hendratta N., Sarah L. Sheffield, Jennifer E. Bauer, et al. 2021. "An Actionable Anti-Racism Plan for Geoscience Organizations." *Nature Communications* 12 (1): 3794.

Allen-Ramdial, Stacy-Ann A. and Andrew G. Campbell. 2014. "Reimagining the Pipeline: Advancing STEM Diversity, Persistence, and Success." *Bioscience* 64: 612–18. https://doi.org/10.1093/biosci/biu076.

Badgett, M. V. Lee, Christopher S. Carpenter, and Dario Sansone. 2021. "LGBTQ Economics." *Journal of Economic Perspectives* 35 (2): 141–70. https://doi.org/10.1257/jep.35.2.141.

Brown, Eryn. 2016. "Disability Awareness: The Fight for Accessibility." *Nature* 532: 137–39. https://doi.org/10.1038/nj7597-137a.

Brubaker, Rogers. 2009. "Ethnicity, Race, and Nationalism." *Annual Review of Sociology* 35: 21–42. https://doi.org/10.1146/annurev-soc-070308-115916.

Burgstahler, Sheryl. 2021. "Universal Design: Process, Principles, and Applications." DO-IT Program, University of Washington. www.washington.edu/doit.

Burnett, Nicholas P., Emily E. King, Mary K. Salcedo, Richelle L. Tanner, and Kathryn Wilsterman. 2020. "Conference Scheduling Undermines Diversity Efforts." *Nature Ecology and Evolution* 4: 1283–84. https://doi.org/10.1038/s41559-020-1276-5.

Calisi, Rebecca M., and a Working Group of Mothers in Science. 2018. "How to Tackle the Childcare-Conference Conundrum." *Proceedings of the National Academy of Sciences* 115 (12): 2845–49. https://doi.org/10.1073/pnas.1803153115.

Carroll, Emma, and Mariano Sironi. 2022. "Enhancing the Impact and Inclusivity of Research by Embracing Multilingual Science Communication." *Marine Mammal Science* 39: 351–53. https://doi.org/10.1111/mms.12991.

Cech, E. A. and T. J. Waidzunas. 2021. "Systemic Inequalities for LGBTQ Professionals in STEM." *Science Advances* 7: eabe0933. https://doi.org/10.1126/sciadv.abe0933.

Chandra, Kanchan. 2006. "What Is Ethnic Identity and Does It Matter?" *Annual Review of Political Science* 9 (1): 397–424. https://doi.org/10.1146/annurev.polisci.9.062404.170715.

Chubin, Daryl E., Gary S. May, and Eleanor L. Babco. 2005. "Diversifying the Engineering Workforce." *Journal of Engineering Education* 94 (1): 73–86. https://doi.org/10.1002/j.2168-9830.2005.tb00830.x.

De Vos, Asha, Sergio Cambronero-Solano, Sangeeta Mangubhai, Leila Nefdt, Lucy C. Woodall, and Paris V. Stefanoudis. 2023. "Towards Equity and Justice in Ocean Science." *Ocean Sustainability* 2: 25. https://doi.org/10.1038/s44183-023-00028-4.

De Vos, Asha, and Mark W. Schwartz. 2022. "Confronting Parachute Science in Conservation." *Conservation Science and Practice* 4: e12681. https://doi.org/10.1111/csp2.12681.

Ekschmitt, Klemens, Annette Klein, Barbara Pieper, and Volkmar Wolters. 2001. "Biodiversity and Functioning of Ecological Communities—Why Is Diversity Important in Some Cases and Not Others?" *Journal of Plant Nutrition and Soil Science* 164: 239–46. https://doi.org/10.1002/1522-2624(200106)164:3%3C239::AID-JPLN239%3E3.0.CO;2-0.

Flores, Andrew R. 2021. "Social Acceptance of LGBTI People in 175 Countries and Locations, 1981 to 2020." UCLA Williams Institute. https://williamsinstitute.law.ucla.edu/publications/global-acceptance-index-lgbt/.

Flores, Andrew R., and Kerith J. Conron. 2023. Adult LGBT Population in the United States. UCLA Williams Institute. Accessed July 31, 2024. https://williamsinstitute.law.ucla.edu/publications/adult-lgbt-pop-us/.

Fry, Richard, Brian Kennedy, and Cary Funk. 2021. "STEM Jobs See Uneven Progress in Increasing Gender, Racial and Ethnic Diversity." Pew Research Center. https://www.pewresearch.org/social-trends/2021/04/01/stem-jobs-see-uneven-progress-in-increasing-gender-racial-and-ethnic-diversity/.

Guterl, Fred. 2014. "The Inclusion Equation." *Scientific American* 311 (4): 38–40. https://doi.org/10.1038/scientificamerican1014-38.

Hall, Cathy, Jeremy Dickerson, David Batts, Paul Kauffmann, and Michael Bosse. 2011. "Are We Missing Opportunities to Encourage Interest in STEM Fields?" *Journal of Technology Education* 23 (1): 32–46. https://jte-journal.org/articles/10.21061/jte.v23i1.a.4.

Hanson, Sandra L., Stefan Fuchs, Silke Aisenbrey, and Natalyia Kravets. 2004. "Attitudes Toward Gender, Work, and Family Among Female and Male Scientists in Germany and the United States." *Journal of Women and Minorities in Science and Engineering* 10: 99–129. https://doi.org/10.1615/JWomenMinorScienEng.v10.i2.10.

Hartmann, Aaron C. 2019. "Disability Inclusion Enhances Science." *Science* 366 (6466): 698. https://doi.org/10.1126/science.aaz0271.

Herre, Bastian, and Pablo Arriagada. 2024. "LGBT+ Rights Have Become More Protected in Dozens of Countries, but Are Not Recognized Across Most of the World." Our World in Data. https://ourworldindata.org/progress-lgbt-rights.

Hill, Paul L., Rose A. Shaw, Jan R. Taylor, and Brittan L. Hallar. 2010. "Advancing Diversity in STEM." *Innovative Higher Education* 36: 19–27. https://doi.org/10.1007/s10755-010-9154-8.

Hooker, Sascha L., Samantha E. Simmons, Alison K. Stimpert, and Birgitte I. McDonald. 2017. "Equity and Career-Life Balance in Marine Mammal Science?" *Marine Mammal Science* 33 (3): 955–65. https://doi.org/10.1111/mms.12407.

Johnson, Dawn R. 2011. "Women of Color in Science, Technology, Engineering, and Mathematics (STEM)." *New Directions for Institutional Research* 152: 75–85. https://doi.org/10.1002/ir.410.

Koch, Amanda J., Paul R. Sackett, Nathan R. Kuncel, Jeffrey A. Dahlke, and Adam S. Beatty. 2022. "Why Women STEM Majors Are Less Likely Than Men to Persist in Completing a STEM Degree: More Than the Individual." *Personality and Individual Differences* 190: 111532. https://doi.org/10.1016/j.paid.2022.111532.

Langley-Turnbaugh, Samantha J., Glenn Wilson, and Lynn Lovewell. 2009. "Increasing the Accessibility of Science for All Students." *Journal of Science Education for Students with Disabilities* 13 (1): 2. https://doi.org/10.14448/jsesd.02.0001.

Lee, Danielle. 2020. "Diversity and Inclusion Activisms in Animal Behaviour and the ABS: a Historical View from the U.S.A." *Animal Behaviour* 164: 273–80. https://doi.org/10.1016/j.anbehav.2020.03.019.

Matsuda, Shayle B. 2023. "Centering Transgender and Gender Non-Conforming Experience, Access, and Safety in Ecological Fieldwork." *Integrative and Comparative Biology* 63 (1): 86–97. https://doi.org/10.1093/icb/icad017.

Nash, M. 2021. "National Antarctic Program Responses to Fieldwork Sexual Harassment." *Antarctic Science* 33 (5): 560–71. https://doi.org/10.1017/S0954102021000432.

NCSES (National Center for Science and Engineering Statistics, National Science Foundation). 2023. Women, Minorities, and Persons with Disabilities in Science and Engineering. Accessed September 7, 2023. https://ncses.nsf.gov/pubs/nsf23300/data-tables.

North Carolina State University Center for Universal Design. 2024. Accessed July 30, 2024. https://design.ncsu.edu/research/center-for-universal-design/.

Ortiz-Martínez, Gabriela, Patricia Vázquez-Villegas, María I. Ruiz-Cantisani, et al. 2023. "Analysis of the Retention of Women in Higher Education STEM Programs." *Humanities and Social Sciences Communications* 10(1): 1–14. https://doi.org/10.1057/s41599-023-01588-z.

Peterson, Raven J. 2021. "We Need to Address Ableism in Science." *Molecular Biology of the Cell* 32: 507–10. https://doi.org/10.1091/mbc.E20-09-0616.

Poushter, Jacob, and Nicholas Kent. 2020. "The Global Divide on Homosexuality Persists." Pew Research Center. https://www.pewresearch.org/global/2020/06/25/global-divide-on-homosexuality-persists/.

Rudzki, Elizabeth N., and Kevin D. Kohl. 2023. "Deficits in Accessibility Across Field Research Stations for Scientists with Disabilities and/or Chronic Illness, and Proposed Solutions." *Integrative and Comparative Biology* 63 (1): 114–27. https://doi.org/10.1093/icb/icad019.

Sarju, Julia P. 2021. "Nothing About Us Without Us: Towards Genuine Inclusion of Disabled Scientists and Science Students Post Pandemic." *Chemistry—A European Journal* 27: 10489–94. https://doi.org/10.1002/chem.202100268.

Sellami, Nadia, Brit Toven-Lindsey, Marc Levis-Fitzgerald, Paul H. Barber, and Tama Hasson. 2021. "A Unique and Scalable Model for Increasing Research Engagement, STEM Persistence, and Entry into Doctoral Programs." *CBE—Life Sciences Education* 20 (11): 1–11. https://doi.org/10.1187/cbe.20-09-0224.

Snyder, Thomas D., de Brey, Cristobal, and Dillow, Sally A. 2019. *Digest of Education Statistics, 2017* (NCES 2018-070). National Center for Education Statistics.

Stefanoudis, Paris V., Wilfredo Y. Licuanan, Tiffany H. Morrison, Sheena Talma, Joeli Veitayaki, and Lucy C. Woodall. 2021. "Turning the Tide of Parachute Science." *Current Biology* 31 (4): 184–85. https://doi.org/10.1016/j.cub.2021.01.029.

Tang-Martinez, Zuleyma. 2020. "The History and Impact of Women in Animal Behaviour and the ABS: A North American Perspective." *Animal Behaviour* 164: 251–60. https://doi.org/10.1016/j.anbehav.2019.12.011.

Tilman, David, Forest Isbell, and Jane M. Cowles. 2014. "Biodiversity and Ecosystem Functioning." *Annual Review of Ecology, Evolution, and Systematics* 45: 471–93. https://doi.org/10.1146/annurev-ecolsys-120213-091917.

Trewin, Shari, Vicki Hanson, Jennifer Rode, et al. 2019. "Accessible Conference Guide. Association for Computing Machinery's Special Interest Group on Accessible Computing 2.4." http://www.sigaccess.org/welcome-to-sigaccess/resources/accessible-conference-guide/.

Tsui, Lisa. 2007. "Effective Strategies to Increase Diversity in STEM Fields: A Review of the Research Literature." *Journal of Negro Education* 76 (4): 555–81. https://www.jstor.com/stable/40037228.

UN Women. 2022. *LGBTIQ+ Equality and Rights: Internal Resource Guide*. UN Women.

United Nations. 2016. *Living Free and Equal: What States Are Doing to Tackle Violence and Discrimination Against Lesbian, Gay, Bisexual, Transgender and Intersex People*. United Nations HR/PUB/16/3.

Wilga, Cheryl A. D., Michele K. Nishiguchi, and Brian Tsukimura. 2017. "Broadening participation in the Society for Integrative and Comparative Biology." *Integrative and Comparative Biology* 57 (1): 7–17. https://doi.org/10.1093/icb/icx004.

Wyatt-Nichol, Heather, and Lorenda A. Naylor. 2013. "The Policy Landscape of Sexual Orientation." *Journal of Public Management and Social Policy* 19 (1): 5–18. https://api.semanticscholar.org/CorpusID:150118154.

Acknowledgments

We extend our deepest gratitude to everyone who contributed to this book. First and foremost, our heartfelt thanks go to the authors who shared their expertise and insights, helping to shape the book's content. We also wish to acknowledge the many graduate students and postdoctoral researchers whose creativity, passion, and excellence have driven many of the research areas explored in this volume. We respectfully acknowledge the Indigenous peoples and communities on whose lands much of the work in this volume was conducted and whose waters are home to many of the cetacean societies we study.

Over the past two decades, many of the advances in our understanding of cetacean societies have emerged from long-term individual-based field studies. Many people have supported these projects over the years, and we are grateful to all who have contributed. We especially thank those whose dedication and vision helped establish and sustain these long-term studies. In particular, we acknowledge Mike Bigg, Ken Balcomb, Graeme Ellis, and John Ford for their pioneering work on killer whales in the Northeast Pacific Ocean. We also acknowledge Richard Connor, Rachel Smolker, Janet Mann, Bill Sherwin, Michael Krützen, and Simon Allen for their groundbreaking work on the Shark Bay dolphins. We thank Roger and Katy Payne, Doug Cato, Louis Herman, Phillip Clapham, Scott Baker, and many others for their foundational studies on humpback whales.

Many individuals have supported us in our academic journeys, which have ultimately led to this book. In particular, Andrew Foote would like to thank Paul Spong and Helena Symonds of OrcaLab for giving him the opportunity to spend three months with them, an experience that set him on the path to study killer whales. And he thanks all those who have mentored and supported him and his research and shared experiences, including Tom Gilbert; Filipa Samarra; the communities in the Highlands and Islands of Scotland and San Juan Island; and valued colleagues at NOAA, including Phil Morin, Kim Parsons, Bob Pitman, and John Durban. Darren Croft acknowledges the many mentors and colleagues who have guided and supported him. He is deeply grateful to Emma Foster for introducing him to the southern resident killer whales and to Ken Balcomb, whose collaboration and friendship he will always cherish. He also extends his gratitude to the dedicated staff and volunteers (past and present) at the Center for Whale Research for their tireless work and commitment. Above all, he is profoundly thankful to his family—Safi, Malila, and Malik—who have endured and supported countless field seasons studying animal societies. Ellen Garland would like to thank all members of the South Pacific Whale Research Consortium for collaborating and sharing data to allow ocean-basin scale work to be possible. She is deeply grateful to Mike Noad for taking a chance on her as a grad student and letting her gallivant around the South Pacific visiting collaborators at each field site. She is indebted to him for his mentorship and support, along with Phil Clapham, Scott Baker, Rochelle Constantine, and Luke Rendell. Stephanie King would like to thank Richard Connor, Michael Krützen, and Simon Allen for welcoming her into the Shark Bay Dolphin Research team. The greatest privilege of her career is working with the remarkable dolphins and the incredible team of researchers in Shark Bay. She is also indebted to Peter Tyack for all his support and mentorship during her career.

We acknowledge the institutions and funding agencies that made this project possible, including the Northwest Fisheries Science Center (NOAA Fisheries), the Sea Mammal Research Unit, the International Fund for Animal Welfare, and the Royal Society (funding awarded to Ellen Garland) for contributing funds for the color illustrations in this book.

We are very thankful to the University of Chicago Press for their support, especially Joseph Calamia and Matt Lang for their guidance through the publishing process and Laura Tsitlidze for her thorough copyediting.

When *Cetacean Societies* was published in 2000, we were just beginning our scientific journey. The volume was a profound source of inspiration that played a significant role in shaping our careers and influencing the questions we asked and how we approached them. We are deeply grateful to the editors and authors of the original volume, which has been a guiding light for our studies on cetacean societies. We hope that this new volume will serve a similar function for the next generation of scientists, who will build upon the knowledge presented here and continue to advance the field.

Finally, we are deeply grateful to those working tirelessly to conserve and protect cetacean societies threatened with extinction and to those working to make our society more inclusive.

Glossary

Definitions provided reflect the editors' understanding of the selected terms in the context in which they are applied within this volume. The editors acknowledge that some terms may have additional meanings not captured in these definitions.

absolute brain size. The gross mass of the brain independent of body size.

actively coordinated collaboration. When individuals attempt to coordinate in time and space to reach a common goal, by monitoring their partner's behavior and responding accordingly.

adult lifespan. The time an organism lives after reaching sexual maturity.

affiliation. Positive social relationships between individuals within a group or species.

affiliative behavior. Gentle contact behavior between the appendages or body of one individual and any part of another individual, and in the Shark Bay bottlenose dolphins (and possibly other species) affiliation may include some noncontact synchronous movements. Other forms of affiliative behavior include prey sharing.

aggregations. Any form of gathering and coming together of individuals. Groups of individuals may be of the same or unrelated species. Interactions between members of the aggregation may be minimal, unstructured, or structured.

agonism. Includes threats and overt aggressive behavior but is a broader category of behavior than aggression because it includes acts of submission, as when a subordinate individual retreats upon the approach of a dominant individual.

alliance. Traditionally defined as a relationship between two or more individuals based on repeated coalition formation, defined as acts of cooperation against conspecifics over resources or rank. Here we expand the definition to include competition against members of other odontocete species.

allonursing. Offspring suckling from a presumably lactating female who is not the mother.

alloparental care. The care provided to offspring by individuals other than the biological parents. Also referred to as allomaternal care when a female is the caregiver.

archaeocete. A paraphyletic grouping of early fossil whales including Pakicetus, Protocetus, and Basilosaurus, among others.

Artiodactyla. The mammalian order that includes not only cetaceans but hoofed mammals such as cattle, goats, sheep, antelopes, giraffes, deer, pigs, camels, and hippos.

association index. A ratio used to quantify how often individuals are observed together given how often they are observed.

assortative mating. A mating pattern in which individuals with similar phenotypes or genotypes are more likely to mate with one another.

astragalus. One of the tarsal bones of the foot that forms the lower part of the ankle. In artiodactyls it has a distinct double pulley-like shape.

autosomes. The biparentally inherited chromosomes of the genome, inherited from both parents, and excluding the sex chromosomes.

baleen. A thick plate of keratin used during filter feeding in mysticete whales.

behavioral plasticity. The extent to which individuals adjust their behavior as a function of changing conditions.

binomial mixture model. A statistical approach that models data as arising from multiple distinct groups or populations, each following a binomial distribution. It is particularly useful in estimating the number and types of social relationships within a population by analyzing the rate of associations between individuals.

bubble-net feeding. A feeding technique where humpback whales create a bubble net or cloud using their blowholes that confuses and encircles prey for easy capture.

burst-pulse vocalization. A discrete, isolated sequence of high repetition rate clicks, thought to express emotional content such as aggression, excitement, or agitation.

bycatch. The unintentional capture of species in fisheries that are not the target species.

by-product collaboration. When individuals perform the same behavior at the same time in response to the same environmental stimuli (e.g., prey), with no intentional coordination.

capital breeding. A reproductive strategy that involves the use of stored reserves to fuel growth and reproduction.

Cetancodonta. The monophyletic group that contains hippos and cetaceans.

clade. A monophyletic grouping of organisms.

clan (killer whales). A set of pods with similar vocal dialects, sharing at least one call type.

clan (sperm whales). A set of social units that share a coda dialect; whales from different clans rarely (if ever) interact.

coalescent theory. A model of genetic changes through time in a neutrally evolving, randomly mating population of constant size.

coalition. Traditionally defined as two or more individuals engaged in acts of cooperation against conspecifics over resources or rank; here we expand the definition to include competition against members of other odontocete species.

codas. Stereotyped, Morse code-like patterns of clicks produced by sperm whales in social situations.

cognition. The process of acquiring and processing knowledge.

consortships. An association between two or three adult males and an adult female for the purposes of mating, the association can last minutes to weeks and may be established and maintained by aggression.

contiguity. In genetic studies, this refers to the extent of continuous or overlapping sequence data covering a chromosome.

cooperative breeding. A social system where individuals assist in raising offspring that are not their own.

cross-modal. Involving interactions between two or more sensory modalities.

crypsis. Behaviors that reduce detection by predators such as the lowering of acoustic signal level to limit eavesdropping or cue detection of the young.

cultural drift. Cumulative random changes in the frequency of behavioral traits caused by imprecise copying during social learning.

cultural hitchhiking. A form of gene-culture coevolution in which diversity at a neutral genetic locus is reduced due to selection on culturally inherited traits that are being transmitted in parallel.

culture. Information or behaviors shared within a group and acquired from conspecifics through some form of social learning (cultural transmission).

diploid. Carrying two copies of a chromosome (in contrast to haploid: one copy of a chromosome, e.g., mammalian Y chromosome).

dispersal. The movement of individuals away from their natal group or region.

dominance (in a genetic context). The relationship between alleles at the same site of the genome on different chromosomal strands. Recessive alleles are expressed in homozygous genotypes but masked in heterozygous genotypes.

dominance hierarchies. Structure of social rankings within a group that determines access to resources, mates, or social status.

dominance rank. Determines priority of access to resources among individuals based upon participation in or observation of previous contests and is often influenced by social relationships with others. Dominance may be more often communicated with submissive behaviors by the subordinate rather than overt aggression by the dominant individual.

driver-barrier feeding. A type of role-specialized foraging behavior performed by bottlenose dolphins, in which one dolphin (the driver) herds fish toward other dolphins positioned in a line (the barriers), resulting in the trapped fish leaping into the air between them from where the dolphins can catch them.

eavesdropping. The use of information in signals by individuals other than the intended receiver.

echelon (baleen whales). A feeding strategy that involves two or more cetaceans swimming in a V formation with following whales behind and to the left or right of a lead whale. It is thought this arrangement may help funnel missed prey into the whale following behind.

echolocation. The active emission of an acoustic signal into the surrounding environment and the subsequent auditory processing of returning echoes to obtain information for navigation or foraging.

ecological knowledge. Information and understanding gained by individuals regarding their environment.

ecotype. Ecologically distinct population within a species, often differing in diet, behavior, genetic and social structure, and physical characteristics.

effective population size. The size of an idealized population (*see* coalescent theory) that experiences drift at the same rate as the study population.

encephalization. Any measure of brain size relative to body size.

encephalization quotient (EQ). One way of quantifying relative brain size, which is the ratio of the observed to predicted brain size based on its allometric relationship to body mass.

entanglements. Unintentional encounters between whales or dolphins and fishing gear that result in their accidental capture.

ethnographic method. The process of ruling out genetic or environmental causation before culture can be confirmed.

evolutionarily significant unit. Evolutionary units that show genetic or heritable phenotypic distinctiveness, and that demonstrate isolation, such that there is a restricted flow of information that determines genotype or phenotypes, from other such units.

extrinsic mortality risk. The probability of death from external factors, such as predation or environmental hazards.

first-order alliance. An association between two or three adult males, almost always from the same second-order alliance, based on the formation of consortships with individual females. The stability of first-order alliance relationships is highly variable.

fission-fusion. Refers to a social grouping pattern where individuals associate in groups that often change in composition as individuals and other groups join and leave. Within a fission-fusion grouping pattern there may be stable associations as is the case for some strongly bonded allied males in Shark Bay, or stable matrilineal groups in other species.

founder effects. The loss of genetic variation when a subset of an ancestral population establishes a new population.

F_{ST}. A measure of genetic differentiation between populations based on differences in allele frequencies.

genealogy. The genetic relationship between individuals based on the sharing of some of their genome sequence identical-by-descent from a common ancestor.

gene-culture coevolution. The correlation or interaction between traits inherited through cultural and genetic transmission.

genetic drift. Random changes in the frequency of alleles evolving under neutrality.

grandmother hypothesis. This hypothesis posits that females who have ceased reproduction can enhance their inclusive fitness by promoting the survival and reproductive success of their grand offspring. This intergenerational help by post-reproductive females is thought to play a significant role in the evolution of menopause.

heterozygous. Carrying two different copies (alleles or haplotypes) at the same gene.

high-throughput sequencing. The parallel sequencing of many DNA molecules, enabling the concurrent reading of whole genomes from many individuals.

homophily. The tendency of individuals to associate and form social bonds with others who are similar to themselves in traits such as age, sex, kinship, or behavior.

homozygous. Carrying two copies of the same alleles or haplotype at the same gene.

horizontal transmission. Cultural transmission of behavior between members of the same generation.

identical-by-descent. The sharing of a segment of the genome between two or more individuals, inherited from a common ancestor without being broken up by recombination.

identity cues. Variation in signals that may inform a receiver of identity but that was not selected for this function.

identity signals. Signals selected to inform a receiver of individual, group, population, or species identity.

inclusive fitness. The total genetic contribution an individual makes to future generations, which includes not just their own offspring but also the offspring of their relatives.

income breeding. A reproductive strategy where animals rely on energy and resources acquired during the breeding period to fuel reproduction, rather than relying exclusively on stored energy reserves.

infant position. A stereotypical swimming position of a calf in relation to its mother (echelon position), underneath and slightly behind her, which provides hydrodynamic benefits and nursing access for the calf.

infanticide. The intentional killing of young dependent offspring, often by males seeking to increase their own reproductive opportunities.

intergenerational harm. A negative fitness effect of one generation on another—for example, due to competition for limited resources.

intergenerational help. A positive fitness effect of one generation on another—for example, prey sharing across generations.

intergroup alliances. Alliances between distinct social groups. *See* third-order alliance.

intrasexual conflict. Competition between individuals of the same sex, often for access to mates.

inversion. A region of a chromosome with reverse orientation on one strand.

keratinous. Made of keratin, a waxy protein secreted from epithelial cells and found in skin, hair, horns, feathers, and baleen.

kin selection. The evolutionary process where certain traits are selected because they have positive effects on the survival and reproductive success of an individual's relatives.

kinship. Typically refers to the degree of genetic relatedness among individuals within a species.

kinship dynamics. Age-linked changes in the patterns of local genetic relatedness to other individuals in the social group.

kinship structure. The organization of social relationships or social associations relative to genetic relatedness.

leverage point. A point within a complex system where a small intervention can produce a disproportionately large change.

life history. The series of changes an organism undergoes from birth to death, including growth, reproduction, and aging.

linkage (in a genetics context). Refers to the proximity of sites in the genome and how likely alleles at two sites are to be disassociated by recombination, or inherited on the same chromosome strand.

lobtailing. A feeding strategy that involves slapping the surface of the water with tail flukes before blowing bubbles in the same place, then lunging through the water to capture prey.

Lombard response. Increasing the loudness of a vocal signal in response to increased background noise.

mandibular symphysis. The space where the left and right mandibles (bones of the lower jaw) come together. In humans, this occurs at the chin.

mate guarding. The physical defense of a mate that attempts to prevent them from reproducing with other individuals.

maternal investment. The resources a mother allocates toward the growth and survival of her offspring, including gestation, lactation, and protection. This investment enhances the offspring's fitness at a cost to the mother's allocation to other offspring.

matrilineal. Descending from a single female; social structures where maternal kin play a central role in social organization.

maxillae. Plural of maxilla; the major bones of the upper jaw in cetaceans.

meiosis. The process of cell division in sexually reproducing organisms.

menopause. The physiological termination of reproduction before the end of life in females resulting in a prolonged post-reproductive lifespan. This term is used in literature on toothed whales even in the absence of menstruation.

metagenomic. Refers to a DNA sample from a community of organisms, as opposed to a single individual or species.

mitochondria. Organelles within cells that contain their own circular DNA. The DNA of mitochondria has been used in phylogenetics and population genetics due to its high copy number within the cell and rapidly evolving regions.

mitochondrial DNA. Maternally inherited circular chromosome, associated with ATP production and metabolism.

modularity. The degree to which a network can be divided into distinct, tightly knit subgroups or communities, where connections within each subgroup are denser than those between subgroups.

monophyletic. A group of organisms descending from a common ancestor that includes all of its descendants.

mortality risk. The probability of an individual dying at a given time or age.

mother hypothesis. A hypothesis that older females cease reproduction to focus energy and resources on the survival and reproductive success of existing offspring.

multilevel societies. A social organization where individuals are grouped into multiple, hierarchically arranged levels, such as smaller family units nested within larger groups or communities.

nasofacial asymmetry. A condition in some odontocetes where certain bones or soft tissues on the left and right of the face and nose differ in size or shape. This is related to the production of sound involved in echolocation.

nomadism. A movement pattern where individuals or groups do not establish a permanent home range but instead roam over large areas, often following unpredictable or seasonal resources without returning to a fixed location.

oblique transmission. Cultural transmission of behavior from older, nonparent models to a younger generation.

organizational complexity. Complexity related to the division of labor, social roles, and distinct functions within a group.

pachyosteosclerotic. Thick, dense bones.

palate. The roof of the mouth.

paraphyletic. A group of organisms descending from a common ancestor that does not include all of its descendants.

pedigree. A record of genetic ancestors.

philopatry. Tendency of individuals to remain within their natal social group, area, or home range into adulthood.

phonic lips. Muscular folds in the nasal cavities of odontocetes that vibrate to create sounds.

Pleistocene. Geological epoch spanning 2.58 million to 11,700 years BP, and a period of repeated glacial cycles.

pod (killer whales). A set of closely related matrilines sharing the same vocal dialect.

polygyny. A mating system that involves the association of one male with multiple females.

pops. Narrowband, low-frequency (peak energy <2 kHz) pulsed vocalizations typically produced in repetitive trains, averaging 10 pops per train and 6–12 pops/sec, by male bottlenose dolphins in Shark Bay during consortships with individual females that function as a threat to keep the female close.

post-reproductive lifespan. The period of an individual's life after they have stopped reproducing.

precautionary principle. Public decision-making that requires decision makers not to use "lack of full scientific certainty" as a reason not to take measures to mitigate harm from threats to the environment or health.

prosocial behavior. Actions intended to benefit others, including helping behaviors such as food sharing, help in contests against others and affiliative behaviors that, for example, reduce recipients' stress or parasite burden.

proximate mechanism. The immediate cause (physical, cognitive, or biological) that underlies a particular behavior.

recessive. A trait expressed only when the underpinning genotype is homozygous for the trait-associated allele.

reciprocal altruism. Evolutionary theory arguing that a behavior that reduces an individual's fitness while increasing that of another may be supported by the expectation that the beneficiary will reciprocate.

recombination. The process by which parental DNA is broken down and recombined to form a new combination of the parental alleles in the offspring's genome.

relatedness. The degree of genetic similarity between individuals.

relational complexity. The diversity and variability in the types of social relationships individuals maintain within a society. It reflects how individuals interact with different types of associates in varying contexts.

relationship type. A category of social interaction between individuals, defined by its strength, duration, and the specific behaviors involved.

relative brain size. The mass of the brain relative to the total mass of the body. This is often expressed as a ratio. *See* encephalization quotient.

reproductive conflict. Competition between individuals over reproductive opportunities.

reproductive conflict hypothesis. The hypothesis that competition between different generations for reproductive opportunities drives the selection for older females to shift their effort from producing their own offspring to aiding the survival and reproductive success of their relatives. This change in investment in helping versus harming with age is thought to play a significant role in the evolution of menopause.

reproductive value. The potential future reproductive output of an individual.

resilience. The ability to maintain function in the face of exogenous disturbance.

resource holding potential (RHP). The ability of individuals or alliances to win contests over resources that may be influenced by factors such as size, number, and age as well as physical and social skill.

role specialization. A division of labor in which specific animals repeatedly take on particular roles during a group task (e.g., chaser versus blocker in cooperative hunting).

runs of homozygosity (ROH). An identical-by-descent tract of homozygous DNA, arising from both parents sharing a recent common ancestor.

Sanger sequencing. Low-throughput sequencing of a single target strand of DNA from a single individual using polymerase chain reaction (PCR) to amplify many copies of the target DNA.

second-order alliance. The core unit of male social organization in Shark Bay. Second-order alliances comprise between 4 and 14 mostly unrelated males. They are formed among peers in their early to midteens and are established and maintained by affiliative behavior. Second-order alliances may last decades and are based on cooperation against other males over estrus females. First-order alliances are formed from within the second-order alliance.

sexual coercion. The use of force, intimidation, or persistent harassment by one individual to increase mating opportunities with another.

sexual size dimorphism (SSD). Differences in body size between males and females in a population or species. Measured as ratio of male to female size, so in a population with greater SSD males are relatively larger than females compared to a population with lower SSD.

shared intentionality. When individuals attempt to coordinate in time and space with others with whom they share knowledge of their intentions and goals and also share motivation to pursue that goal together. Sometimes called "we-intentionality."

signature whistle. Individually distinctive whistles that are learned by individual dolphins of several species.

site fidelity. The tendency of individuals to return to the same location repeatedly over time, often for activities like feeding and breeding.

social bond. A positive relationship or "friendship" between two individuals or alliances established and maintained with affiliative behaviors.

social brain hypothesis. A hypothesis suggesting that the cognitive demands of managing complex social relationships drive the evolution of large brains.

social cohesion. A measure of the strength of social bonds between multiple individuals, reflecting the degree of social stability within a group.

social complexity. The degree of information needed to capture the variation of social interactions and structures within a society, including factors such as hierarchical relationships, role specialization, and group dynamics. High social complexity involves intricate networks of relationships and behaviors that support coordination, cooperation, and social organization.

social differentiation. Variation in the types or strengths of relationships within a society; often quantitatively defined as coefficient of variation among rates of association or interactions.

social learning. Learning that is influenced by observation of, or interaction with, another animal or its products.

social network. A representation of social structure, treating individuals as nodes or vertices in a graph with the relationships between them represented as edges connecting the nodes.

social repertoire. The range of behaviors exhibited by individuals within a population during social interactions. Typically characterized by the number of distinct behavioral categories, such as cooperative caregiving, alliance formation, social defense, complex vocalizations etc.

social unit (sperm whale). A stable, matrilineal group of female sperm whales and their offspring that stay together and move together.

sociality. A measure of the gregariousness of individuals within an animal population, i.e., how social they are.

socially influenced collaboration. When an individual pursues a specific goal or performs a specific behavior due to the presence or behavior of others, with no intentional coordination.

socioecological model. A model of how social structures and behaviors interact with ecological factors, such as resource availability and environmental conditions, to influence the organization and dynamics of animal groups.

sociogenetic. Interactions between social environments and genetic factors that influence behavior and organization.

sociosexual behavior. Sexual behavior in social, non-conceptive contexts.

sperm competition. A form of postcopulatory sexual selection where the ejaculates of two or more males simultaneously physically compete to fertilize eggs.

standing genetic variation. Alternative genetic variants found in natural populations.

stem-cetaceans. Fossil taxa that show some but not all characteristics of extant cetaceans ("crown cetaceans").

structural variants. >1 Kb regions of the genome that differ within and between individuals, caused by deletion, insertion, copy number variation, translocation, and inversion of part of the genetic code.

super alliance. Name given to the first 14-member second-order alliance discovered in Shark Bay in the 1990s.

symbolic markers. Arbitrary traits that function as reliable indicators of cultural group membership to conspecifics.

synchrony. Two or more individuals engaged in simultaneous movements and vocalizations.

synonymous (in a genetics context). A genetic change that results in the same amino acid sequence. As opposed to nonsynonymous, a DNA change that codes a change in amino acid.

Synrhina. An odontocete clade consisting of the common ancestor of Platanista and Delphinus and all of its descendants. This excludes all sperm whales.

telescoping. A process during cetacean evolution by which bones moved over one another in relation to the posterior movement of the nares (nostrils).

third-order alliance. Relationships established and maintained by affiliative behavior between members of different second-order alliances in Shark Bay that have the same function as second-order alliances, i.e., cooperation against other males over access to females. Third-order alliances are equivalent to intergroup alliances in this population.

ultimate explanations. Explanations that deal with the evolutionary history and adaptive value of a behavior.

vertical transmission. Cultural transmission of behavior from parent to offspring.

vocal dialects. Learned stereotyped variants of calls produced by a subset of individuals within a population.

vocal matching exchange. An exchange of vocal signals between two or more animals in which the vocal signal emitted by one animal is repeated or "matched" by the other animal.

vocal production learning (or more loosely, vocal learning). The production of modified or novel vocalizations, as a result of learning from the experience of the acoustic signals of others. Central to this definition is that auditory input from a model sound leads to the formation of a memory of the sound (a template), to which the vocal output is compared. The auditory model may be a range of sounds including vocalizations produced by another animal (recorded or live), synthesized playbacks or even nonvocal mechanical sounds.

whistle contour. The modulation pattern of the fundamental frequency of a whistle.

Contributors

ALANA ALEXANDER, Department of Anatomy, University of Otago

SIMON J. ALLEN, Department of Evolutionary Anthropology, University of Zurich; and School of Biological Sciences, University of Bristol

FREDERICK IVAN ARCHER, Marine Mammal and Turtle Division, Southwest Fisheries Science Center, National Marine Fisheries Service

SHANAN ATKINS, School of Animal, Plant and Environmental Sciences, University of the Witwatersrand

PHILIPPA BRAKES, Centre for Ecology and Conservation, Biosciences, Faculty of Environment, Science and Economy, University of Exeter; and Cetacean Ecology Research Group, Massey University

MAURICIO CANTOR, Marine Mammal Institute, Department of Fisheries, Wildlife, and Conservation Sciences, Oregon State University

EMMA L. CARROLL, School of Biological Sciences, University of Auckland–Waipapa Taumata Rau

YACINE BEN CHEHIDA, Department of Biology, University of York

CHRISTINE M. KONRAD CLARKE, Department of Biology, Dalhousie University

RICHARD C. CONNOR, Institute of Environment and Department of Biological Sciences, Florida International University

TARA M. COX, Department of Marine and Environmental Sciences, Savannah State University

DARREN PAUL CROFT, Center for Whale Research; and Centre for Research in Animal Behaviour, Department of Psychology, University of Exeter

VOLKER B. DEECKE, Institute of Science and Environment, University of Cumbria

REBECCA DUNLOP, School of the Environment, University of Queensland

SAMUEL ELLIS, Centre for Research in Animal Behaviour, Department of Psychology, University of Exeter

OLGA A. FILATOVA, Department of Biology, University of Southern Denmark

MICHAEL C. FONTAINE, UMR MIVEGEC, Université de Montpellier, CNRS, IRD

ANDREW D. FOOTE, Centre for Ecological and Evolutionary Synthesis, Department of Biosciences, University of Oslo

VIVIENNE FOROUGHIRAD, Department of Marine Biology, Texas A&M University at Galveston

DANTE FREEMAN, Department of Marine Sciences, Savannah State University

ELLEN C. GARLAND, Sea Mammal Research Unit, Scottish Oceans Institute, School of Biology, University of St. Andrews

LIVIA GERBER, Australian National Wildlife Collection, Commonwealth Scientific and Industrial Research Organisation

INGERID HARANG HELGESTAD, Centre for Research in Animal Behaviour, Department of Psychology, University of Exeter

TAYLOR A. HERSH, Marine Mammal Institute, Department of Fisheries, Wildlife, and Conservation Sciences, Oregon State University

KELLY JAAKKOLA, Research Department, Dolphin Research Center

VINCENT JANIK, Scottish Oceans Institute, School of Biology, University of St. Andrews

EVE JOURDAIN, Department of Biosciences, Norwegian Orca Survey

STEPHANIE L. KING, School of Biological Sciences, University of Bristol

MICHAEL KRÜTZEN, Department of Evolutionary Anthropology, University of Zurich

MARIE LOUIS, Department of Birds and Mammals, Pinngortitaleriffik, Greenland Institute of Natural Resources

JANET MANN, Department of Biology, Georgetown University

KAREN K. MARTIEN, Marine Mammal and Turtle Division, Southwest Fisheries Science Center, National Marine Fisheries Service

MADALINA MATEI, SMRU Consulting, University of St. Andrews

MORGAN L. MCCARTHY, Section for Molecular Ecology and Evolution, Globe Institute, University of Copenhagen

MOLLY H. F. MCENTEE, Department of Ecology and Evolutionary Biology, University of California, Santa Cruz

MIA LYBKÆR KRONBORG NIELSEN, Department of Ecoscience, Aarhus University

JULIE N. OSWALD, Scottish Oceans Institute, School of Biology, University of St. Andrews

SUSAN E. PARKS, Department of Biology, Syracuse University

KIM M. PARSONS, Conservation Biology Division, Northwest Fisheries Science Center, National Marine Fisheries Service

LUKE RENDELL, Sea Mammal Research Unit, School of Biology, University of St. Andrews

PERNILLE MEYER SØRENSEN, School of Biological Sciences, University of Bristol

DIPANI SUTARIA, School of Marine Biology and Aquaculture, James Cook University

PETER LLOYD TYACK, Scottish Oceans Institute, School of Biology, University of St. Andrews

MICHAEL WEISS, Center for Whale Research

HAL WHITEHEAD, Department of Biology, Dalhousie University

SONJA WILD, Department of Environmental Science and Policy, University of California, Davis

Index

Page numbers in italics refer to figures and tables.